Mathematics Education as a Research Domain: A Search for Identity

An ICMI Study

T0137381

Book 1

Edited by:

Anna Sierpinska

Concordia University, Montreal, Quebec, Canada

and

Jeremy Kilpatrick

University of Georgia, Athens, Georgia, USA

Kluwer Academic Publishers

Dordrecht / Boston / London

Library of Congress Cataloging-in-Publication Data

Mathematics education as a research domain : a search for identity :
    an ICMI study / edited by Jeremy Kilpatrick and Anna Sierpinska.
        p.   cm. — (New ICMI studies series ; v. 4)
    Includes index.
    ISBN 0–7923–4599–1 (hardcover : alk. paper)
    1. Mathematics—Study and teaching—Research. I. Kilpatrick,
Jeremy. II. Sierpinska, Anna. III. Series.
QA11.M3757   1997
510'.71—dc21   97–20240

ISBN (set) 0-7923-4599-1 (hardback)   ISBN (Book 1) 0-7923-4945-8 (hardback)
ISBN (set) 0-7923-4600-9 (paperback)   ISBN (Book 1) 0-7923-4947-4 (paperback)

Published by Kluwer Academic Publishers,
P.O. Box 17, 3300 AA Dordrecht, The Netherlands.

Sold and distributed in the U.S.A. and Canada
by Kluwer Academic Publishers,
101 Philip Drive, Norwell, MA 02061, U.S.A.

In all other countries, sold and distributed
by Kluwer Academic Publishers,
P.O. Box 322, 3300 AH Dordrecht, The Netherlands.

Printed on acid-free paper

TABLE OF CONTENTS

BOOK 1

Part VI: Mathematics Education and Mathematics

FOREWORD

No one disputes how important it is, in today's world, to prepare students to understand mathematics as well as to use and communicate mathematics in their future lives. That task is very difficult, however. Refocusing curricula on fundamental concepts, producing new teaching materials, and designing teaching units based on 'mathematicians' common sense' (or on logic) have not resulted in a better understanding of mathematics by more students. The failure of such efforts has raised questions suggesting that what was missing at the outset of these proposals, designs, and productions was a more profound knowledge of the phenomena of learning and teaching mathematics in socially established and culturally, politically, and economically justified institutions – namely, schools.

Such knowledge cannot be built by mere juxtaposition of theories in disciplines such as psychology, sociology, and mathematics. Psychological theories focus on the individual learner. Theories of sociology of education look at the general laws of curriculum development, the specifics of pedagogic discourse as opposed to scientific discourse in general, the different possible pedagogic relations between the teacher and the taught, and other general problems in the interface between education and society. Mathematics, aside from its theoretical contents, can be looked at from historical and epistemological points of view, clarifying the genetic development of its concepts, methods, and theories. This view can shed some light on the meaning of mathematical concepts and on the difficulties students have in teaching approaches that disregard the genetic development of these concepts.

All these theories are interesting and important for a mathematics educator, but the knowledge needed to understand (and act upon) the phenomena of teaching and learning mathematics in the school institution must deal simultaneously with cognition, social institutions, and mathematics in their complex mutual relationships. Theories cannot just be imported ready-made from other domains: Original theories have to be elaborated if we are to understand better the problems of mathematics education.

Efforts in this direction have given rise to a growing body of research that is called research in mathematics education or research in the didactics of mathematics. People who do this research, whether working in university mathematics departments or not, are always interested in maintaining good relations with other communities, including the community of mathematicians. But good relations can only be built on mutual understanding. The work reported herein has been aimed at contributing to the growth of such understanding.

The present book is one of a series of publications resulting from ICMI Studies. The first such study was launched by the International Commission on Mathematical Instruction (ICMI) in 1984. Its theme was The Influence of Computers and Informatics on Mathematics and Its Teaching. Subsequent studies were concerned with such themes as: School Mathematics in the 1990s,

Mathematics as a Service Subject, Mathematics and Cognition, Assessment in Mathematics Education, and Gender and Mathematics. Reports of these studies have been published by Cambridge University Press, Kluwer Academic Publishers, and the University of Lund Press. A study on the problems of teaching geometry has been conducted, and studies are planned on the relations between the history and the pedagogy of mathematics and on the specific problems of tertiary mathematics education.

The theme of the ICMI Study reported in this book was formulated as a question: 'What is Research in Mathematics Education and What are Its Results?' No single agreed-upon and definite answer to the question, however, is to be found in these pages. What the reader will find instead is a multitude of answers, various analyses of the actual directions of research in mathematics education in different countries, and a number of visions for the future of that research. Given this diversity, the reader may ask why a study on this theme was held.

In his letter of invitation to the International Program Committee for the Study Mogens Niss (the Secretary of ICMI and an ex-officio member of the committee) wrote:

Mathematics education has been in existence as a field of academic research for roughly a century. Its growth within the last three decades has been enormous not only in the number of research studies undertaken but also in the number of researchers, places in which scholarly work is being done, and of academic fields represented in that work. It is time to review the state of the field and to begin a dialogue with other scientific communities, in particular the mathematical research community. The International Commission on Mathematical Instruction is undertaking a study on research in mathematics education that will describe the work being done, analyze developments and trends within the field, and assess what research in mathematics education has produced. Major outcomes of the study will be presented at the International Congress of Mathematicians in Zürich in 1994.

Have the goals as defined above been met? As it turned out, the participants at the Study Conference were more interested in looking at the evolution of ideas and at the problems yet to be solved than in taking stock of the field's achievements or results. The questions of assessing what research in mathematics education has produced and of developing criteria for the quality of research were particularly difficult, and the divergence of opinion on these questions was particularly wide. Moreover, concerning the question of a dialogue with other scientific communities, the program committee found that in some countries, the mathematical research community is not the main interlocutor of mathematics education researchers. For example, in the USA, much mathematics education research is conducted within colleges, schools, or departments of education.

Preliminary results from the study were presented at the ICM 1994 in Zürich. Although it was then too early to synthesize the major outcomes of the study, the presentation clarified for many of the mathematicians in the audience some of the differences between research in mathematics education and such domains of activity as the popularization of mathematics, historical studies of mathematics, inventing challenging problems for more gifted students, writing textbooks, or teaching in an innovative way. A subsequent presentation and discussion at the Eighth International Congress on Mathematics Education in Seville in 1996 was valuable in putting the final chapters into context.

The study officially started with a meeting of the program committee (for which we served as co-Chairs) on 19 August 1992, during the Seventh International Congress on Mathematical Education in Québec. Members of the program committee then collaborated to produce a 'Discussion Document' (reprinted in Part I), describing the reasons for the study and its aims, and generally laying a framework for discussion in the form of five questions: 'What is the specific object of study in mathematics education?'; 'What are the aims of research in mathematics education?'; 'What are the specific research questions or *problématiques* of research in mathematics education?'; 'What are the results of research in mathematics education?'; and 'What criteria should be used to evaluate the results of research in mathematics education?'. The Discussion Document appeared in the December 1992 *Bulletin of the International Commission on Mathematical Instruction* (No. 33). In early 1993, it was also published in a number of international mathematics education journals, including *L'Enseignement Mathématique*, *Educational Studies in Mathematics*, *Recherches en Didactique des Mathématiques*, and the *Zentralblatt für Didaktik der Mathematik*. A shortened version appeared in the *Journal for Research in Mathematics Education*.

The Discussion Document contained a call for papers to be submitted by 1 September 1993. The papers and other expressions of interest were used to develop a program and invitation list for a Study Conference organized in Washington, DC, from 8 to 11 May 1994. The conference was sponsored by the ICMI together with the Mathematical Sciences Education Board of the U.S. National Academy of Sciences and the University of Maryland. The participants included more than 80 people from around the world (see the participant list in Part I). Most of the sessions were held at the Adult Education Center of the University of Maryland in College Park, but part of the conference was a half-day symposium for invited U.S. mathematicians and mathematics educators at the National Academy of Sciences building in Washington, DC.

The conference comprised plenary sessions, working groups that addressed the five questions posed in the Discussion Document, and paper sessions in which specific examples of research were discussed in the light of the questions of the study. The plenary sessions were devoted to general themes: definitions of the domain, problems of balancing theory and practice in research, the place of teaching design in research in mathematics education, the training of researchers, and views of mathematics education research held by mathematicians. The conference did not end with a final resolution of the type: *This is what research in mathematics education should be.* Rather, the outcome was more like: *Yes, in spite of all the differences that divide mathematics education researchers (in terms of theoretical approaches, views on relations between theory and practice, philosophies of mathematics, etc.), they still constitute a community, and it is necessary to search for what constitutes its identity.*

The present book is not a set of proceedings from the conference. Most of the authors of the chapters in Parts II to VI, however, participated in the conference, and their chapters reflect the discussions there. It was not possible to organize the book along the five questions posed in the Discussion Document because the authors often addressed all five. In fact, there are two categories of chapters: those that respond, in a way, to the whole Discussion Document, and those that focus on some specific question.

The book is organized as follows: After a section containing materials related to the ICMI Study Conference (Part I), several chapters offer a broad perspective on the definition of mathematics education as a research discipline (Part II). Within this broad view, Part III focuses on the goals, orientations, and envisioned or actual outcomes of mathematics education research. The chapters in Part IV then provide examples or counterexamples for the general theses offered in the previous two parts. They contain descriptions and analyses of the evolution of different research paradigms in mathematics education in different countries or cultures.

Parts V and VI are shorter. Part V is devoted to the question of criteria for judging the quality of research in mathematics education, and Part VI is concerned with the interactions, particularly along social and epistemological dimensions, of mathematics education with mathematics and mathematicians. Both of these treat difficult topics. In fact, one important outcome of the study was a greater recognition of the reasons for the difficulty of the questions that the study posed, leading possibly to another set of questions better suited to the actual concerns and research practices of mathematics education researchers.

ACKNOWLEDGMENTS

We would like to express our appreciation to the other members of the International Program Committee for their help in organizing the Study Conference and preparing the present volume:

Nicolas Balacheff, IMAG & Université Joseph Fourier, Grenoble, France
Willibald Dörfler, Universität Klagenfurt, Austria
Geoffrey Howson, University of Southampton, UK
Mogens Niss, Roskilde University, Denmark (ex-officio)
Fidel Oteiza, Universidad de Santiago, Chile
Toshio Sawada, National Institute for Educational Research, Japan
Anna Sfard, University of Haifa, Israel
Heinz Steinbring, Universität Bielefeld, Germany

Support for the Study Conference was provided by the U.S. National Science Foundation, the University of Maryland, and the National Academy of Sciences. We are particularly grateful to James T. Fey, University of Maryland, for undertaking the local arrangements for the conference; to Richard Herman, University of Maryland, for providing resources for logistical support; to Midge Cozzens,

National Science Foundation, for assisting with a grant from the Foundation to cover many of the travel expenses; and to Linda Rosen, Mathematical Sciences Education Board, and her staff for handling the conference registration. Special support with organization and arrangements was provided by Patricio Herbst, University of Georgia; Monica Neagoy, University of Maryland; and Ramona Irvin, Mathematical Sciences Education Board. Virginia Warfield, University of Washington, provided valuable assistance in translation.

Finally, thanks are also due to Kluwer Academic Publishers, in particular to Peter de Liefde and Irene van den Reydt, for their understanding and forbearance as the book has made its belated way to publication.

Anna Sierpinska and Jeremy Kilpatrick

THE ICMI STUDY CONFERENCE

THE ICMI STUDY CONFERENCE

In this part of the volume, materials are presented that pertain to the Study Conference 'What is Research in Mathematics Education, and What Are Its Results?' held in Washington, DC, in May 1994. The chapters in subsequent parts were initially prepared in response to a Discussion Document, which is reproduced in the next section. The document also served as a point of departure for the conference, and most of the chapters were revised, at times extensively, in light of discussions there. The second section below lists the conference participants. The remainder of Part I contains the reports of the five working groups, which met for several hours over three days of the conference. Each group was asked to discuss and respond to one of the five principal questions in the Discussion Document. The working group reports are followed by a chapter by Alan Bishop that is based on a paper he and Dudley Blane prepared for the conference and on the summary remarks he made at the final session.

DISCUSSION DOCUMENT

The following people have contributed to the present document: N. Balacheff, A. G. Howson, A. Sfard, H. Steinbring, J. Kilpatrick, and A. Sierpinska.

As mathematics education has become better established as a domain of scientific research (if not as a scientific discipline), exactly what this research is and what its results are have become less clear. The history of the past three International Congresses on Mathematical Education demonstrates the need for greater clarity. At the Budapest congress in 1988, in particular, there was a general feeling that mathematics educators from different parts of the world, countries, or even areas of the same country often talk past one another. There seems to be a lack of consensus on what it means to be a mathematics educator. Mathematics education no longer means the same as *didactique des mathématiques* (if it ever did). French *didacticiens* refuse to translate their *didactique des mathématiques* into 'mathematics education': a special English edition of the journal *Recherches en Didactique des Mathématiques* bears the title 'Research in *Didactique* of Mathematics.' *Die Methodik* (or the Polish *metodyka*, the Slovak *metodika*, and the like) have become obsolete. Does *research* mean the same as *recherche* or *investigación?* How do these words translate into other languages? Standards of scientific quality and the criteria for accepting a paper vary considerably among the more that 250 journals on mathematics education published throughout the world.

3

Sierpinska, A. and Kilpatrick, J. Mathematics Education as a Research Domain: A Search for Identity, 1–32.
© *1998 Kluwer Academic Publishers. Printed in Great Britain.*

Despite this lack of consensus, publications appear that endeavor to depict the 'state of the art' in mathematics education research. Individuals try to construct didactical theories. But reviewers never have trouble demonstrating the one-sidedness or incompleteness of such publications. Attempts to describe research in mathematics education or *didactique des mathématiques* or whatever other name is used may resemble the accounts of the legendary blind men exploring the legs of a huge elephant.

The ICMI Study *What is research in mathematics education, and what are its results?* does not seek to describe the state of the art. Nor does it intend to tell anyone what research in mathematics education is or is not, or what is or is not a result. Instead, the organizers of the study propose to clarify the different meanings these ideas have for mathematics educators – to pinpoint the different perspectives, goals, research problems, and ways of approaching problems. The study will bring together representatives of the different groups of researchers, allow them to confront one another's views and approaches, and seek a better mutual understanding of what we might be talking about when we speak of research in mathematics education.

SOME QUESTIONS ABOUT RESEARCH

Such a wide-ranging discussion is badly needed in a community increasingly divided into specialized groups and cliques that are not always tolerant of each other. Besides mutual understanding within the community, however, there is also a need to explain the domain to representatives of other scientific communities, among which the community of mathematicians seems to be the most important. Nicolas Balacheff has observed:

Most of us want to develop this research field within the academic community of mathematicians; this implies both the explanation of our purpose on a social ground (is there any need to develop such research?) and its relevance within the narrow academic world. For this reason, although it is not my sole concern, I have in mind the question of scientific standards, theses, publications, congresses, the employment of young academics in the field, and the connection between our research and research done in other fields.

Thus we need an 'inner' identification of the research domain of mathematics education, as well as an outer vision from the perspective of other domains.

One external domain, for example, is sociology. How is mathematics education organized and institutionalized? Where is research on mathematics education conducted? Where are theses on mathematics education defended? If a mathematics educator employed by a mathematics department has acquired his or her habilitation degree in, say, a department of pedagogy or philosophy (such a degree being unavailable at the employing institution), is he or she accepted as a full member of the community of mathematicians that awards doctoral or master's degrees in mathematics? Are mathematics educators viewed as a part of the mathematics community? Similar questions arise when research in mathematics education is surveyed from other domains, including history, philosophy, anthropology, and psychology.

An approach from both within and outside the field of research in mathematics education raises the following questions, among others, to be discussed:

(1)   What is the specific object of study in mathematics education?
The object of study (*der Gegenstand*) in mathematics education might be, for example, the teaching of mathematics; the learning of mathematics; teaching–learning situations; didactical situations; the relations between teaching, learning, and mathematical knowledge; the reality of mathematics classes; societal views of mathematics and its teaching; or the system of education itself.

If a mathematics educator studies mathematics, is it the same object for him or her as it is for a mathematician who studies mathematics? What is mathematics as a subject matter? What is 'elementary mathematics'? Analogous questions could be asked concerning the learner of mathematics as an object of study. Is it the same object for a mathematics educator as it is for a psychologist or a pedagogue? Is the mathematics class or the process of learning in the school viewed in the same way by a mathematics educator and a sociologist, anthropologist, or ethnographer? Are questions of knowledge acquisition viewed the same way by a mathematics educator and an epistemologist?

The variety of activities offered at the ICMEs certainly distinguishes these congresses from, say, the international congresses of mathematicians. ICME 7 was compared by some to a supermarket. Is there a unity in this variety? What gives unity to different kinds of study in mathematics education? Is this the object of research? Or is the object of research perhaps not even something held in common? Might the commonality lie in the pragmatic aims of research in mathematics education?

(2)   What are the aims of research in mathematics education?
One might think of two kinds of aims: pragmatic aims and more fundamental scientific aims. Among the more pragmatic aims would be the improvement of teaching practice, as well as of students' understanding and performance. The chief scientific aim might be to develop mathematics education as a recognized academic field of research.

What might the structure of such a field be? Would it make sense to structure it along the lines of mathematical subject matter (e.g., the didactics of algebra or the didactics of geometry), of various theories or approaches to the teaching and learning of mathematics, or of specific topics or *problématiques* (research on classroom interaction and communication, research on students' understanding of a concept, etc.)?

Both kinds of aims seem to assume that it is possible to develop some kind of professional knowledge, whether that of a mathematics teacher, a mathematics educator, or a researcher in mathematics education. The question arises, however, whether such professional knowledge can exist at all. Is it possible to provide a teacher, say, with a body of knowledge that would, so to say inevitably, ensure the success of his or her teaching? In other words, is teaching an art or a profession (*un métier*)? Or is it perhaps a personal conquest? As Luigi Campedelli used to say, '*La didattica è, e rimane, una conquista personale*'.

What does successful teaching depend on? Are there methods of teaching so sure, so objective, that they would work no matter who the teacher and students were? Are there methods of teaching that are teacher-proof and methods of learning that are student-proof? If not, is there anything like objective fundamental knowledge for a researcher in mathematics education – something that any researcher could build upon, something accepted and agreed upon by all? Or will the mathematics educational community inevitably be divided by what is considered as belonging to this fundamental knowledge, by philosophies and ideologies of learning, by what is considered worth studying?

Many mature domains of scientific knowledge have become highly specialized into narrow subdomains. Is this the fate of mathematics education as well? Or rather, in view of the interdisciplinary nature of mathematics education, must every researcher necessarily be a 'humanist,' knowing something of all domains and problems in mathematics education?

Although we aim at clarifying the notion of research in mathematics education as an academic activity, we should be careful not to fall into needlessly 'academic' debates. After all, the ultimate goal of our research may be for a specific teacher in a specific classroom to be better equipped to guide his or her students as they seek to understand the world with the help of mathematics.

(3)   What are the specific research questions or problématiques *of research in mathematics education?*
Mathematics education lies at the crossroads of many well-established scientific domains such as mathematics, psychology, pedagogy, sociology, epistemology, cognitive science, semiotics, and economics, and it may be concerned with problems imported from these domains. But mathematics education certainly has its own specific *problématiques* that cannot be viewed as particular cases or applications of those from other domains. One question the ICMI study might address is that of identifying and relating to each other the various *problématiques* specific to mathematics education.

There are certainly two distinct types of questions in mathematics education: those that stem directly or almost directly from the practice of teaching and those generated more by research. For example, the question of how to motivate students to learn a piece of mathematics (inventing interesting problems or didactic situations that generate a meaningful mathematical activity), or how to explain a piece of mathematics belong to the first kind. The question of identifying students' difficulties in learning a specific piece of mathematics is also directly linked to practice. But questions of classifying difficulties, seeing how widespread a difficulty is, locating its sources, or constructing a theoretical framework to analyze it already belong among the research-generated questions. The problem is, however, that a difficulty may remain unnoticed or poorly understood without an effort to answer questions of the latter type; that is, without more fundamental research on students' understanding of a topic. Is it, therefore, possible to separate so-called practical problems from so-called research-generated problems?

Is it possible to admit the existence of two separate types of knowledge: the theoretical knowledge for the scientific community of researchers and the practical knowledge useful in applications for teachers and students? It might be helpful to reflect on the nature of these two types of knowledge, on relations between them, and on whether it would be possible to have a unified body of knowledge encompassing them both.

(4)   What are the results of research in mathematics education?
Any result is relative to a *problématique,* to the theoretical framework on which it is directly or indirectly based, and to the methodology through which it was obtained. This relativity of results, though commonplace in science, is often forgotten. One often interprets findings from biology, sociology, or mathematics education as if they were a kind of absolute truth. The reason may be that in these domains we really want to know the truth and not simply whether, if one proposition is true, some other proposition is also true. Questions of biology, sociology, or mathematics education can be of vital importance and fundamental to our survival and well-being.

Two types of 'findings' can be distinguished in mathematics education: those based on long-term observation and experience and those founded on specially mounted studies. Are the former less 'scientific' than the latter? Geoffrey Howson offers an example:

In the seventeenth century, Spinoza set out three levels of understanding of the rule of three (which, incidentally, can be viewed as an elaboration of the instrumental-relational model of Skemp and Mellin-Olsen expounded over three centuries later). This, like the well-known levels of the van Hieles, was based on observation and experience. On the other hand, for example, CSMS [Concepts in Secondary Mathematics and Science] used specially mounted classroom studies to develop and investigate similar hierarchies of understanding. Do we rule out the work of Spinoza as research in mathematics education? If we do, then we lose much valuable knowledge, especially that resulting from curriculum development. If we do not, then it becomes difficult to find a workable definition [of research in mathematics education].

Balacheff points out that it may be difficult to contrast, in this way, the hierarchies obtained by the van Hieles and the CSMS group. Besides the different ways in which these hierarchies were obtained, the van Hieles and the CSMS group may not have been asking the same kind of question. 'What are these questions?' asks Balacheff. 'What is the validity of the answers they provide? How is it possible to relate them?'

Can a new formulation of an old problem be a research result? Can a problem be a result? Or a questioning of the theory related to a problem, a methodology, or a whole *problématique?* Can a concept be a result? It might be useful to have a definite categorization of the things we do in mathematics education, and of the things we thereby 'produce'.

Most people would probably agree that *making empirical investigations* is research. But is the *doing of practical things* research? Is *thinking* research? Can these activities be separated? Can a result be obtained without thinking and the doing of practical things? Should mathematics education be considered a science? Perhaps it is a vast domain of thought, research, and practice. What qualifies a domain of activity as scientific is the kind of validation and justification methods it uses. Proofs and experiments are considered scientific. But there are thoughts not validated in either of these ways that are valuable because they are filled with meaning.

What examples are there of what we consider results in mathematics education to be? What do we know today that we did not know before? What have we learned about the processes of learning and teaching? What do we know about mathematics that mathematicians were not aware of before?

Can we identify some categories of results? One category might be *economizers of thought.* Any facts, laws, methods, procedures, or theories that are general enough to direct our experience and predict its results will give us increased power over our teaching and learning. Another category might be *demolishers of illusions.* Results that undermine our beliefs and assumptions are always valuable contributions to the field. A third category might be *energizers of practice.* Teachers welcome research that helps them understand what they teach and provides them with ideas for teaching. The development of teaching materials, activities, and challenging problems belongs to this category. Other categories of results might emerge from epistemological, methodological, historical, and philosophical studies.

(5) What criteria should be used to evaluate the results of research in mathematics education?
How do we assess the validity of research findings? How do we assess their worth? Should we use the criterion of relevance? What about objectivity? Or originality? Should we consider the influence research has had on the practice of teaching? What other criteria should we use?

The first problem is to clarify the meaning of terms such as *truth, validity,* and *relevance* in the context of mathematics education. A related issue is the question of what is knowledge as such. This is an even more fundamental question than that of validation. If we knew what kind of knowledge mathematics education aims at, we would be better equipped for answering the question of methods of validation.

It is also useful to understand the ways in which research results are used. How have the results of research in mathematics education been applied? How do teachers use the research? How do policy-makers use it? By clarifying the uses to which research is put, can we develop better criteria for assessing its validity?

PARTICIPANTS

The following people took part in the conference:

Josette Adda (France)
Gilbert Arsac (France)
Michèle Artigue (France)
Ferdinando Arzarello (Italy)
Nicolas Balacheff (France)
Mariolina Bartolini-Bussi (Italy)
Jerry Becker (USA)
Alan Bishop (Australia)
Ole Björkqvist (Finland)
Dudley Blane (Australia)
Morten Blomhøj (Denmark)
Lenore Blum (USA)
Paolo Boero (Italy)
Guy Brousseau (France)
Margaret Brown (UK)
Ronnie Brown (UK)
Jere Confrey (USA)
Beatriz D'Ambrosio (USA)
Willibald Dörfler (Austria)
Tommy Dreyfus (Israel)
Nerida Ellerton (Australia)
Paul Ernest (UK)
James Fey (USA)
Claude Gaulin (Canada)
Gunnar Gjone (Norway)
Juan Godino (Spain)
Pedro Gómez (Colombia)

Koeno Gravemeijer (Netherlands)
Miguel de Guzmán (Spain)
Gila Hanna (Canada)
Kath Hart (UK)
Yoshihiko Hashimoto (Japan)
Gillian Hatch (UK)
Milan Hejny (Czech Republic)
Patricio Herbst (USA)
James Hiebert (USA)
Bernard Hodgson (Canada)
Geoffrey Howson (UK)
Brian Hudson (UK)
Bengt Johansson (Sweden)
David Johnson (UK)
Christine Keitel (Germany)
Carolyn Kieran (Canada)
Jeremy Kilpatrick (USA)
Anna Kristjánsdóttir (Iceland)
Colette Laborde (France)
Ewa Lakoma (Poland)
Gilah Leder (Australia)
Frank Lester, Jr. (USA)
Claire Margolinas (France)
John Mason (UK)
Eric Muller (Canada)
Roberta Mura (Canada)
Monica Neagoy (USA)

Mogens Niss (Denmark)
Iman Osta (Lebanon)
Fidel Oteiza (Chile)
Michael Otte (Germany)
Erkki Pehkonen (Finland)
David Pimm (UK)
Susan Pirie (UK)
Norma Presmeg (USA)
Thomas Romberg (USA)
Linda Rosen (USA)
Toshio Sawada (Japan)
Yasuhiro Sekiguchi (Japan)
Anna Sfard (Israel)
Shizumi Shimizu (Japan)
Christine Shiu (UK)
Anna Sierpinska (Canada)
Edward Silver (USA)
Claudie Solar (Canada)
Judith Sowder (USA)
Lynn Steen (USA)
Heinz Steinbring (Germany)
Hans-Georg Steiner (Germany)
Juliana Szendrei (Hungary)
Gérard Vergnaud (France)
Virginia Warfield (USA)
David Wells (UK)
Erich Wittmann (Germany)

In addition, invited officials from the National Academy of Sciences, the National Science Foundation, the University of Maryland, and various organizations in the Washington, DC, area attended plenary sessions.

WHAT IS THE SPECIFIC OBJECT OF STUDY IN MATHEMATICS EDUCATION?

REPORT OF WORKING GROUP 1

Leader: Jere Confrey
Reporters: Dudley Blane and Anna Kristjánsdóttir[1]
The work of the group started with two presentations, one by Anna Sfard and the other by Christine Keitel. Sfard spoke on whether mathematics is the same

object for the mathematics educator as it is for the mathematician.[2] For a mathematician, she said, mathematics is a *disciplina mentis,* a way of understanding and knowing. The views of mathematics educators on what mathematics is have been changing over the years: Platonism, constructivism, social constructivism, situated cognition, and others. The question is: What are the educational implications of Platonism or of any other philosophy of mathematics? 'Misconception' makes sense within a Platonistic frame of mind; 'learning in groups' is a consequence of a belief in social constructivism, and so on. The much-debated situated-cognition approach to the learning and teaching of mathematics views learning as 'legitimate peripheral participation in a social practice'. It postulates that mathematics will be learned by more students if taught in integration with other subjects and in real-world contexts and if practical learning apprenticeships are developed. Each of these philosophies poses its own problems and dilemmas. In particular, situated cognition leads to the question: Doesn't contextualization contradict abstraction, and isn't mathematical thinking based mainly on abstraction? Sfard concluded her talk by saying that, because philosophical positions towards mathematics have implications for the decisions we make in research on teaching, learning, and creating mathematics, we have to be aware of these positions. We must make them explicit and known to ourselves and others.

Christine Keitel spoke on common sense and mathematics. One of the questions she posed was: How are mathematics and common sense related? It is known that mathematics shapes common sense (e.g., by the use of numbers in speaking about values, so that 80% on a test means *good;* 20% means *bad*). Does learning mathematics mean overcoming common sense? Is it true that scientific thinking is based on undoing commonsense thinking? On the other hand, maybe it makes sense to speak of a *mathematical common sense* that would be worthwhile developing in students. She proposed a change from seeing the task of mathematics education as modifying, preparing, or transforming ('elementarizing') scholarly mathematics for teachers and students to seeing it as an investigation starting from an analysis of the social system, including the educational system, in which the relationship between mathematics and society is embedded. This analysis is then a basis on which a new understanding, organization, and systematic conceptualization of the 'objects' of research in mathematics education could be obtained. In particular, this new approach would lead us to studying 'the literature' of mathematics instead of its 'grammar'.

Mathematics as an Object of Research in Mathematics Education

The theme of the group was designed broadly, but, perhaps as a result of the questions raised by the two presentations in the group, the discussion focused on whether mathematics can be regarded as an object of research in mathematics education, and if so, in what sense?

The participants generally agreed with Sfard's point that if mathematics is an object of research in mathematics education, it is not the same object for mathematics educators as it is for mathematicians, philosophers, and logicians. The discussion brought up several aspects and consequences of this difference.

Borrowing Charles Morris's distinction between syntactics, semantics, and pragmatics, one could say that while philosophers, logicians, and mathematicians may make abstractions from the learning and uses of mathematics and occupy themselves only with its semantic and syntactic aspects, mathematics educators have to practice the 'pragmatics' of mathematics: they have to study the relations between mathematics and its learners and users. That implies the necessity, in mathematics education, to take into account the social, cultural, and institutional contexts of learning and teaching mathematics. It also requires us to broaden the notion of mathematics to encompass mathematical practices that are relevant for the society at large but are other than those of university mathematicians. There is, in certain cultures, a kind of mathematical common sense (e.g., the notion of zero forms a part of it) shared by the society. Is this common sense worth developing, or is it so incompatible with the mathematics that we want to teach children that it needs to be regarded as an enemy?

'What is the nature of mathematics?' was a much debated question in this group. Some participants voiced their doubts as to whether this is a relevant question for mathematics education, which challenged the opponents to formulate stronger arguments.

It was not obvious to everyone that the notion of mathematics should be broadened to the extent of encompassing, for example, the mathematics of carpenters (who, according to Wendy Millroy's study, reported in a 1992 monograph of the *Journal for Research in Mathematics Education,* do not themselves consider what they are doing as mathematics). But, it was said, if we see mathematics as 'the science of pattern and order', then there is a reason to accept certain professional practices and tacit competences as mathematics. The debate on the nature of mathematics continued as the topic of situated-cognition learning theory was discussed.

Situated Cognition

The situated-cognition theory mentioned by Sfard in her presentation raised a debate in the group. The question was: Is situated cognition a learning theory that can be reasonably adopted by mathematics education? What might our reservations be about such a theory?

Situated-cognition theory has become known in North America through the works of Jean Lave. It has remained relatively unknown for example, in France. The reactions of the French participants to Sfard's presentation of its basic tenets were the following: C. Margolinas said that similar ideas had been proposed 20 years before by G. Brousseau in France (she mentioned particularly the role attributed to the milieu in Brousseau's theory of didactic situations). G. Vergnaud stated that the idea of situated cognition is trivial: All knowledge is, of course, situated, contextual. But, he said, it is not always situated in real-life situations: There is an important part of mathematics that has to be learned not in authentic but in artificial situations.

Vergnaud's statement provoked opposing arguments from some participants. For example, J. Confrey said that there is strong evidence from the history of mathematical discoveries that all knowledge comes into being in contexts that are often real, concrete, or experimental. Confrey wrote this summary of her remarks:

I took Vergnaud's statement as denying situated learning as offering any serious challenge to the character of mathematics, but only faddish, poor or weak mathematics. I felt he said real mathematics is abstract and decontextualized.... I tried to say that history itself is retold through secondary sources and as a result it supports the view of decontextualization too much. Newton used Hooke's physical demonstration to create mathematics and formalize them but later destroyed the designs and equipment. Probability evolved as a discipline to describe what scientists do. We must take a critical stance towards the view of mathematics reinforced and put forth by mathematicians. In the US we are constitutionally democratic even though we often fail to accomplish this goal. The fundamental inclusion of a broad set of situations into our teaching of mathematics may be necessary to enfranchise the majority of students into mathematics education, and its resulting challenges to the existing perceptions of mathematics must be taken seriously.

What Is 'Common Sense'?

The discussion revealed that people with different mother tongues have different understandings of the phrase *common sense*. Participants then tried to find what is common to all these understandings: views or notions that are shared in a given culture, taken for granted, obvious, unsayable; implicit knowledge; everyday rationality? They also asked about theoretical constructs that, in different research paradigms, have meanings close to that of *common sense*. The following were mentioned: 'the system of personal practice' (by J. D. Godino); '*connaissance*' as opposed to '*savoir*' (Margolinas); '*illusion de la transparence*' (Vergnaud). P. Ernest mentioned that 'ethnomathematics' can be seen as a tacit mathematical knowledge related to common sense. Sfard proposed that common sense is a sociological concept analogous to the psychological concept of intuition.

M. Otte stated that the fact that, for mathematics educators, common sense is a social phenomenon that should be studied, and not just taken as an a priori whose existence is to be confirmed or as an obstacle to be overcome, indicates that mathematics education is as much a science as it is a socially oriented philosophy.

What Is the Specific Object of Research in Mathematics Education?

The question of the group was generally found as difficult, if not impossible, to answer. It was remarked that, in this respect, mathematics education shares the lot of all human sciences. Michel Foucault's words from *The Order of Things* were evoked, in which he identifies the source of the difficulty of human sciences in the 'complexity of the epistemological configuration' in which they are placed.

One can think of the mathematical and physical sciences as forming one vector of the space of scientific knowledge; linguistics, biology, and economics as another vector; and philosophy as a third. The human sciences then find themselves on none of these vectors but instead occupy the whole space spanned by all three.

What explains the difficulty of human sciences is not their precariousness, their uncertainty as sciences, their dangerous familiarity with philosophy, their ill-defined reliance upon other domains of knowledge, their perpetually secondary and derived character, and also their claim to universality is not, as is often stated, the extreme density of their object, it is not the metaphysical status or the inerasable transcendence of this man they speak of, but rather the complexity of the epistemological configuration in which they find themselves placed, their constant relation to the three dimensions that give them space. (Foucault, M.: 1973, *The Order of Things: An Archeology of the Human Sciences*, Vintage Books, New York, p. 348)

Several participants expressed in writing their views on the specific object of research in mathematics education. For example, J. Godino wrote that 'one of the specific objects of study for mathematics education is the transformation or adaptation of mathematical knowledge to the different teaching institutions, from the descriptive and prescriptive points of view'. According to H. Steinbring,

mathematics education as a scientific discipline has to start from the interrelations between mathematical knowledge and social demands (contexts, requirements, etc.). It has to ask itself what are the theoretical perspectives on this 'socially contextualized' knowledge that take this dialectic relationship systematically into account. The problem is not to reduce the social side to the side of mathematical knowledge, nor vice versa. And then the question is: How could such a perspective be made operative in didactic research?

D. Pimm wrote that 'in trying to get one all-embracing object, we risk losing sight of particular objects'. One particular object he wanted to speak about was related to the issue of common sense: 'How are we involved in enabling mathematics to become a common sense (as language is) while becoming increasingly aware of the ways in which it is implicated in creating an apparently natural (and, therefore, invisible) overly shaping common sense?' He added:

With regard to an object for mathematics education, one is the relation between a human individual and the mathematical knowledge (seen on a par with linguistic knowledge) that resides within our community. There is also a more obscure object of desire, that of perfect 'understanding' as well as of 'fluency' (think of Leibniz's dream of a rational calculus). My object recently is the roots of mathematics in the human unconscious and its influences that are visible.

W. Dörfler attempted to pinpoint the main difference between mathematics and mathematics education, and thus to obtain a definition of the object of research in mathematics education:

Mathematics, among other roles, provides means for making sense of the experiential phenomena. That is, one can look at the world through the lenses of mathematical concepts. In mathematics education the direction has to be reversed: One has to look at mathematics through an empathy with the learners and the teachers. Thereby a multitude of views on mathematics will result, even a multitude of 'personal mathematics'. In other words, mathematics education may study the ways in which people use mathematical ways of thinking for organizing their personal experience and how they organize and develop those ways of thinking and reflecting.

Sfard used a comparison, and her conclusion was that it is necessary for mathematics education to clarify its position with respect to mathematical knowledge:

Our ultimate objective is the enhancement of the learning of mathematics. However, as researchers, we are producing knowledge (about how people create mathematics for themselves), and as educationists, we are inducing certain knowledge in others. Therefore, we are faced with the crucial question: What is knowledge, and, in particular, what is mathematical knowledge for us? Here, we find ourselves caught between two incompatible paradigms: the paradigm of human sciences (to which we belong as mathematics education researchers) and the paradigm of mathematics. These two are completely different: Whereas mathematics is a bastion of objectivity, of clear distinction between TRUE and FALSE (for practicing mathematicians at least), there is nothing like that for us. For us, mathematics is social, intersubjectively constructed knowledge.... But we feel somewhat schizophrenic between these two paradigms because our commitment to teach MATHEMATICS makes us, to some extent, dependent on [the philosophies of mathematics held by mathematicians]. Therefore, we must make the problem explicit and cure the illness by making clear where we stand with respect to the issue of mathematical knowledge.

R. Mura gave a very personal view of the topic of the group and the discussions:

I came to this group because I had written a paper for this conference on mathematics educators' definitions of *mathematics education*. This experience, in a sense, makes it more difficult for me now to give my own definition. In general, I would say that the object of mathematics education is less problematic than the object of mathematics. Maybe what we want to concentrate on is the border cases. Some of us have had our own work challenged as not being research in mathematics education. The first issue (work not really being *research*) is common to all the social sciences and humanities. The second issue (not research in *mathematics education*) is for us to decide. Some of us have been criticized by people saying that our work is in linguistics, in women's studies, in philosophy, and so forth, rather than in mathematics education. Could we behave in a way similar to our colleagues in mathematics and say that mathematics education is what mathematics educators do?

WHAT ARE THE AIMS OF RESEARCH IN MATHEMATICS EDUCATION?

REPORT OF WORKING GROUP 2

Leader: Ole Bjorkqvist
Reporters: Pedro Gómez and Thomas Romberg
This group was asked to consider at length possible answers to the question of aims in order to clarify the notion of research in mathematics education as an academic activity. In particular, the group was asked to examine 'two kinds of aims: pragmatic aims and more fundamental scientific aims'.

The issue was addressed by first considering two papers; then each was discussed. The two papers served to focus the groups' thoughts on the question. Gilah Leder addressed the diversity of research aims in the field of mathematics education.[3] She argued that the purposes for doing research have changed during the past half century; that scholars who conduct research have diverse, often pragmatic, and occasionally scientific perspectives about the aims of their research; that the perspectives have been shaped throughout history by the Eurocentric male-dominated majority culture; and that three pragmatic considerations (esteem for research within academic circles, social or cultural priorities, and allocation or availability of resources) often shape the kinds of research carried out.

Julianna Szendrei presented a classification of four different kinds of 'results' produced by research in mathematics education. The paper had been jointly prepared with Paolo Boero.[4] Furthermore, she related each type of result to pragmatic or fundamental scientific aims and to three intended outcomes: energizers of practice, economizers of thought, and demolishers of illusions. The first type of a result, which she labeled *innovative patterns,* would include teaching materials, reports about projects, and so forth. Obviously, such results have practical consequences and are designed to 'energize practice'. The second type of result is *quantitative information* about the choices concerning the teaching of a peculiar mathematical content, general or specific learning difficulties, possible relationships with factors influencing learning, and so forth. Such results have both practical and scientific aims and are designed to both 'energize practice' and 'demolish illusions' about current practices or beliefs. The third type of result is *qualitative information* about the consequences of some methodological or content innovation, and so on. These results are related primarily to scientific aims and designed to 'demolish illusions'. The final type of result is *theoretical perspectives* regarding reports that reflect on descriptions and classifications or interpretations of phenomena, models, historical or epistemological analyses of content, and so on. Obviously, such results have scientific aims designed to 'economize thought' and perhaps to 'demolish illusions'.

Following these presentations, the members of the working group entered on three occasions into a spirited discussion of the ideas presented in the conference position paper and these papers. The discussion was also fueled by the plenary talks on the balance between theory and practice, the social and cultural conditions under which each of the members of the working group operate, and the other sessions and discussions each member participated in during the conference. The contents of the discussions ranged over several issues related to all five working groups. Working group members submitted written comments to summarize thoughts; a first-draft synthesis was written, points discussed, and format agreed upon, and after two chances to revise the report, this final version was completed.

Throughout the sessions some issues related to aims emerged again and again. These can be summarized under three headings:

Research as a Human Process

The term *research* refers to a process – something people do, not objects one can touch or see. Furthermore, research cannot be viewed as a set of mechanical procedures to be followed. Rather it is a craft practiced by scholarly groups whose members have agreed in a broad sense on what procedures are to be followed and on the criteria for acceptable work. These facts led us to the following assertions:

• An important aim of all research should be 'to satisfy the curiosity of the researcher about some situation'. [Note, first, that *situation* is used here to refer to all the objects of study being specified by Working Group 1, and second, that the researcher's 'interest' is often influenced by policymakers, school boards, and so forth.]
• That curiosity should lead to an understanding of situations. Many situations involve the teaching and learning of mathematics in classrooms with the expectation that understanding such situations could lead to improved practice. Other situations may be outside schools and may lead to improvements in the workplace. In this regard, we recognize that there are several levels of understanding such as describing or explaining.
• The actual situations a researcher might investigate are embedded in the institutional, social, political, and cultural conditions in which the researcher operates. The personal aims of different researchers will differ because of different beliefs and their membership in particular scholarly groups with differing notions about disciplined inquiry. [Note also that these groups may have differing aims.] And there may be a difference in the aims for a particular study and a set of studies or a research program.

One member of the group proposed that Figure 1 be used to illustrate the variety of things a scholar may be influenced by when deciding on the aim of a particular study.

Diversity of Aims

The teaching and learning of mathematics in schools at any level in any country is complex. When one also considers mathematics outside schools and in adult education, the complexity is compounded. These facts, when added to individual curiosity, make it clear that there has been and will be a diversity of aims. Individual studies and even research programs conducted by different persons or groups will inevitably have different aims. The concern of the group was that such diversity might make impossible any coherent compilation of findings.

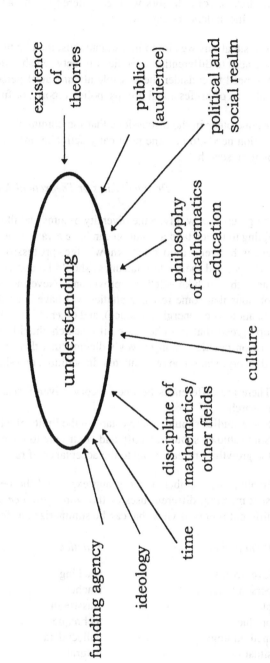

Figure 1 *Things to consider about the aims of a study*

Nevertheless, some factors were considered by the group as helpful in making the specific understandings useful:

- The situations we aim to investigate must include mathematics.
- We need to differentiate specific aims between short-term and long-term aims.
- We need to consider the possible alignment of personal aims with the external aims of professional groups, policy reports, or funding agencies.

It became clear in the discussions that the community of mathematical sciences education needs to become politically active in order to shape external expectations for research.

Practical Aims or Theoretical Aims

In the group's view, given the diversity of aims and the fact that the results of attempting to understand a situation can have a variety of implications, both the differences between theoretical knowledge, professional knowledge, practical knowledge, and their interrelatedness should be appreciated. The group also understood that such knowledge is provisional. Nevertheless, pragmatically it should be obvious that some research studies will have been designed to have practical implications (i.e., energize practice), and others to contribute to theory. In fact, the group agreed that in either case all research should eventually have a positive impact on practice. John Dewey's dictum 'that there is nothing as practical as a good theory' should be remembered. In addition, one should recognize that:

- There are differences between theory-driven research and theory-generating research.
- Some studies should aim to establish the limits of a theory.
- Some studies should identify and contribute to the elimination of obstacles to the growth of research and to the acceptance of research results.

Finally, one member of the group expressed the belief that there are, in the present meeting, different uses of the words *practice* and *theory* corresponding to different points of view that can be summarized as follows:

Point of view	Practice	Theory
level (role)	teaching	researching
person (status)	teacher	researcher
place	classroom	laboratory
product	technique	knowledge
methodology	collect data	analysis
situation	natural	experiment
time	short-term	long-term
generality	projects	fundamental knowledge
research vs. development	development	research

To paraphrase D. Lacombe in the article *'Didactique des disciplines'* in the *Encyclopaedia Universalis*: In the end, if the researcher wants 'recognition' for improving practice, it is enough to become a salesperson for a brand of instruments, or even better, several brands of instruments put together. On the other hand, if researchers want to leave a 'trace' of their intellectual contribution, they must be prepared to see their work the object of criticism or even derision and to undermine the comfort of the establishment.

Sharing the Findings of Research

One aspect of conducting a research study is that a study is not complete until a report is written explaining the results and anticipating actions of others. Thus, one important aim must be with respect to sharing the results with others. This aim must involve:

- deciding on an audience (or audiences), and
- considering the potential consequences of the results for that audience.

In conclusion, the following two questions need to be considered by all when talking about the aims of research in mathematics education:

1. To whom are we (mathematics education researchers) talking when we list the aims of our research?
2. Are we trying to determine what the aims of mathematics education research have been and are, or are we trying to make proposals about what the aims could be in the future?

The first question is relevant since there are 'outside factors' (funding agencies, government bodies) that have in the past shaped at least partially what the aims have been. On the other hand, the research community could and should influence the way these external agencies have an impact on those aims.

Researchers can have a programmatic or a descriptive approach concerning the aims of research in mathematics education. The former could amount to the proposal of some kind of research agenda, whereas a descriptive approach could require an analysis and description of how the aims of research in mathematics education have evolved and how we can use this evolution as an indication of their future status.

The group saw research in mathematics education as the activities and results of a community of scholars. Therefore, from a descriptive point of view, the aims of research in mathematics education are the aims of this community as far as its activities and results. In this respect, a programmatic approach is not very helpful if one sees the research community as a body that evolves according to multiple interests and perspectives.

This multiplicity of research perspectives tends to characterize the way researchers see the aims of their work. One can see the aims as to explain, predict, or control (empirical-analytic); or to understand (interpretive); or to improve or revise practice (critical). However, these perspectives are not specific to mathematics education research. What makes the aims of mathematics education research specific are the phenomena and the practices we are trying to explain, predict, control, understand, and improve.

WHAT ARE THE SPECIFIC RESEARCH QUESTIONS OR *PROBLÉMATIQUES* OF RESEARCH IN MATHEMATICS EDUCATION?

REPORT OF WORKING GROUP 3

Leader: Mariolina Bartolini-Bussi
Reporters: Bernard Hodgson and Iman Osta
A general problem of the group was the ambiguity of the task: Is it possible or even meaningful to discuss 'specific research questions' without considering at the same time the problems of objects of research in mathematics education, aims of research in mathematics education, results of research in mathematics education, and criteria for research in mathematics education? However, the group tried to accomplish the task.

Two presentations introduced the discussion: Nicolas Balacheff spoke on the case of research on mathematical proof, and Ed Silver spoke on research questions in the international community of researchers. In his presentation, Balacheff elicited three different components of a didactical situation that establish relationships with different research fields: the content (defined by a pragmatic epistemology of proof, where researchers in mathematics education can question mathematicians' practices); the learning process (where social interaction can act as a catalyst for developing proofs or counterexamples, and hence where psychology and sociology can provide elements of frameworks); and the classroom situation as an object of study (entity) (where questions specific to didactics occur, e.g. related to the didactical contract and to the *milieu;* the latter concept may lead to the idea of mathematical phenomenology, which in turn leads one to question mathematicians).

In his presentation, Silver discussed two sources of research questions: (a) theory and prior research; and (b) educational practice and problems. The relationship between theory and practice can make international communication difficult (the example of the QUASAR project was presented as context-bound research, difficult to discuss and evaluate in an international forum when compared with the author's research on problem posing). In mathematics education research, in addition to disciplinary issues, it is necessary to consider political, financial, and other societal issues.

After the two presentations the group started the discussion. The main points of the discussion were the following (different streams went on with continuous intersection; what follows is not a chronicle but a kind of reconstruction of the outcome of the discussion).

Complexity of Research Questions

This point was raised by several participants: From the two examples presented by the speakers (QUASAR and proof), it was clear that each research question addresses a complex relationship between different components (e.g., disciplinary, political, social, psychological, and didactical issues). This complexity requires making reference to and drawing upon different research fields (namely, mathematics, epistemology, psychology, sociology, didactics, and so on).

Some research questions deal with only one of the components (e.g., epistemology in the case of proof), but most deal with relationships among several components. It is necessary to avoid 'unjustified reduction of complexity'. What about 'justified reduction of complexity'?

One of the causes of complexity is, for instance, the relationships between mathematics education and mathematics. We cannot reduce research in mathematics education to part of research in mathematics, as researchers in mathematics education are responsible for their own methods, theories, and *problématiques*. But mathematicians and researchers in mathematics education share responsibility for the relevance of the research done with respect to mathematics.

Considering the practice of mathematics education in the fullest sense (planning and reflection, curriculum design and development, teacher preservice and inservice education and not only classroom teaching–learning situations) provides examples of the complexity of relationships between mathematics and mathematics education. (See also the problem of theory and practice below.)

A Dichotomy

A major (false?) dichotomy (THEORY versus PRACTICE) was included in the Discussion Document as a source of different sets of research questions. This dichotomy was a critical point in the group discussion. Some participants supported the separation between theory and practice (identifying them, respectively, as RESEARCH and INNOVATION) in order to allow to understand their specific constraints and so as to allow links to be built between them in a non-naïve way. Other participants expressed a need to consider theory and practice (without identifying them with research and innovation) as being in a true dialectical relationship that also included research problems generated by practical needs.

Examples were offered of relevant research questions raised by the links between theory and practice (e.g., the role of the teacher, considered as central in the teaching–learning process; the design and analysis of long-term processes; the analysis of the constraints on teachers that prevent them from being innovative, and ways to cope with these constraints). This contrast is related to either implicit or explicit assumptions about *paradigms*. Provocative questions were posed: Are the relationships between research and innovation research questions? If so, who is in charge of coping with them? Teachers? Researchers? Both in cooperation? Other people? Is this a general question? Is it dependent on the given paradigm?

Criteria for Judging Questions

Another point concerned *criteria*. Do we need criteria to judge the validity of specific research questions? Two contrasting positions emerged in the discussion: We need criteria, and it is better not to have criteria.

Communication and Relevance Internationally

Another point concerned the problems of communication and relevance with respect to the international community. This point had been raised by Silver as one among several in his presentation, but it was transformed by the group into a major issue to be discussed. That was a partial shift from the assigned task, as the problem of internationalization was not a specific topic for the group, but is a general problem of research in mathematics education. Not only research questions, but also objects, aims, results and criteria could and should be submitted to discussion by the international community.

A first problem concerns communication (of specific questions) because of the difficulty of giving all the relevant information concerning contextual features of a given research, and so on. This problem can be solved pragmatically by publishing papers with a suitable length to allow these assumptions to be shared with the readers. A second major problem concerns relevance: The way the international community influences the choice of local research questions, the relevance of local research for the international community, and the relevance of theoretical international research for local communities. Two complementary positions emerged: On the one hand, from the experience of the past the international community may be supposed to be willing to conduct only certain kinds of discussions (related to content or to theoretical issues and avoiding questions that are too context-bound). On the other hand, several participants underlined the richness of having discussion from different perspectives in order to identify potential controversies or points of consensus. The interest of the international community in such discussion is obvious, as recognizing differences can lead to a better understanding. Consider the following metaphor from zoology: The existence of kangaroos in Australia enlarges the European knowledge of mammals, even if it makes no sense to plan the importation of kangaroos to Europe. The importance of very specific (hence context-bound) questions was also underlined by the analogy with some 'good ideas' that have been developed in science (according to history) in very specific contexts, before becoming universals. So, whether or not we have a problem of communication, we need to make explicit and accept differences in contexts, assumptions, and paradigms.

A Proposal

Because so many problems need to be addressed by the international community of researchers in mathematics education, a final suggestion of this working group was the proposal of establishing an acknowledged community of researchers in this field, similar to the International Group for the Psychology of Mathematics Education or the International Study Group on the Relations Between History and Pedagogy of Mathematics. Such a group could become a Special Interest Group of the International Commission on Mathematical Instruction.[5]

WHAT ARE THE RESULTS OF RESEARCH IN MATHEMATICS EDUCATION?

REPORT OF WORKING GROUP 4

Leader: Susan Pirie
Reporters: Tommy Dreyfus and Jerry Becker

The working group was initiated by two introductory papers. Carolyn Kieran made reference to accumulated research results and divided them into immediate versus long-term results.[6] Long-term results include the development of new theoretical frameworks. Kieran traced the history of the development of such frameworks during the past quarter century. Ferdinando Arzarello presented a general frame from which to view (Italian) research in mathematics education.[7] He described the first period of research in mathematics education in Italy as considering static (snapshot) phenomena (including concept-based didactics and innovation in the classroom) and made the transition to a dynamic (movie) view that stresses relationships between the components of the static description.

The discussions of the group revolved essentially around the following question: Under what conditions should a fact be considered a research result? The role of questions and their position within a theoretical framework was considered, as was the role of interpretation; the specificity to mathematics education of these theoretical frameworks, questions and interpretations; the role of teachers in the research process; and the domain of validity of the results.

An attempt was made to categorize various results – for example, in terms of the categories suggested in the Discussion Paper. This attempt, however, was not seen as successful by the working group participants and was therefore discontinued.

Theory and the Nature of Mathematics Education

There was widespread agreement within the group that without a *question*, there can only be facts but no results. It was stated repeatedly that effects (e.g., statistical differences in achievements between different groups) alone are not results. In didactics, we want to *explain* the differences, not just show them. We need to identify the variables of the didactic situation in order to combine the different facts into a coherent network of reasons.

There was less agreement on how research questions may arise and how their appropriateness should be judged. Some participants argued that questions must arise and be tied together within a theoretical framework. Others felt that as long as one defines what one is looking for, the framework does not matter. It was noted that anthropologists do not necessarily start from a theory but generate a theory. If we borrow their research methods, can we remain with theory as pre-requisite to result?

The content of the questions was seen as important: just about anything that is research based can count as a result under the condition that it informs the circular process of understanding the learning of mathematics and thus improving its teaching. ('Research based' means generated by disciplined scientific inquiry.)

Although some participants saw theory as necessarily underlying research, it was agreed that there are not only experimental, but also theoretical results. It was suggested that a theory is a result of different results. Also, a theory on which a piece of research is based may be modified as a result of the research. In order for that modification to be admissible, it was seen as essential that the researchers know and state in advance what they are going to do with the results. Otherwise, they are doing exploratory research at best. In other words, theory should not be modified on the basis of a particular research study, unless the possible modification of theory was stated at the outset as a potential outcome.

Context was also seen as relevant: the character of mathematics education research is specified by the community and by the questions asked, not by the results. It is not the result itself, but the conditions under which it was obtained, that make it significant. For example, 'understanding' cannot be considered as something defined a priori. It has a place in the didactic process, and must be defined by the role it takes in the teaching–learning process.

The group also discussed whether mathematics education was a *design science,* like the engineering sciences. In a design science, the extraordinary idea that leads to new development and the ensuing development (or at least its design) are considered results. Biotechnology was cited as a field in a similar situation: one central idea might produce a technological solution of a certain problem; the rest may be routine. If we do not take seriously the solution of teachers' problems by effective design, we will fail as a field. Others objected and claimed that while teaching may be an art, research is not: its results must fall within a framework. Design has creative, artistic aspects; research does not.

The question thus arose of whether, and under what conditions, a piece of software – or a textbook – could be considered as a result. Again, there was a wide range of opinions. Most participants agreed that if design principles are based on previous research results and are systematically implemented, the product of the design should be considered a result. In other words, not a single piece, but a type or class of software (e.g., microworlds), may count as a result. However, design or development in our field cannot possibly be based on confirmed results if only because of the extremely limited number of confirmed results in the field. Thus, some participants concluded that an ingenious single design was enough for a creation to count as a result of mathematics education, whereas others expressed doubts as to whether any designer is ever able to explain how a piece of software was designed.

This discussion led to the emergence of yet another criterion for results, namely that they have some predictive power. If someone is able to describe the pertinent variables and the effect of a piece of software on the learner under various conditions (and only then), the software should be considered a result.

Interpretation of Results and Specificity of Mathematics Education

Can one have a result without interpretation? Several *levels of interpretation* were identified. At the lowest level, facts may exist as facts. The next level is the researcher analyzing and interpreting research data. The next is putting them into a framework – for example, when writing a paper. A further level may occur when someone else reads the paper (some disagreed that reinterpretation was possible on the basis of reading only). At every level there is a result and an interpretation.

A complementary opinion was that only the change from one theoretical frame to another allows a real interpretation of results. This view led to the question of how researchers from different fields look at the same results. Can we borrow results from other fields? Can we look at a psychologist's or a sociologist's results and make them useful? Can we interpret them in terms of our background as mathematicians?

For instance, when a psychologist and a mathematics educator look at a contingency table, they probably see two different sides of the same coin; the questions they ask are completely different. The mathematics educator's question could be 'What mathematical means is there to explain a connection between two phenomena?' A model of implication is needed to answer this question, whereas the psychologist may need only correlation.

Thus, we can borrow results from another field when we already have a question to which this result gives a (partial) answer. In the process, we are also reinterpreting, which is typical for interdisciplinary fields.

This observation naturally led to the question of the specificity of mathematics education as a field: Is it a *scientific discipline* or only a combination of mathematics and psychology? Here again, there was no agreement. Some participants pointed to a far-ranging parallelism to, say, physics education. They saw the fields as similar, with the same types of questions and results; they saw us as borrowing and adapting methodologies from other disciplines; and they saw the differences only in the specific questions (e.g., naïve conceptions in physics education do not seem to have a counterpart in mathematics education) and in the corresponding results. Others felt that there is a tension between interdisciplinarity and mathematics; that we have specificities; that we cannot transfer directly from, say, language education; and that the types of questions we ask are specific to our field. It became clear that in order to attempt to discuss this dichotomy seriously, one needs to get down to the lower level and talk specifics.

Any discussion of research in mathematics education must occur in a context involving mathematics. This relevance of the mathematics may express itself in the types of questions asked. We have also put more mathematics into our models of understanding than three decades ago. The extraordinary role of the notion of representation was seen as typical for mathematics education research. We are working not only on concepts (as others do), but also on symbols (and other representations). In physics, much less depends on complex abstract representations, at least at the school level. Similarly typical is the specific epistemological status of mathematical objects and their recursive structure: What are mathematical objects? How are they being conceived by students? In mathematics, there is hardly any distinction between objects and relationships: relationships can become objects at the next higher level: number, function, operator, functor. We (mentally) manipulate manipulations on lower level objects. Such a recursive multilayered structure is not encountered in other topics.

We also use some theoretical frameworks that are specifically mathematical, such as the process–object distinction. Because our frameworks are specific, even if we use general terms, they describe results specific to mathematics education. Thus, mathematics and its specificities are inherent in the research questions from the outset. One is looking at mathematics learning, and one cannot ask these questions outside of mathematics.

Communication and the Role of Teachers

Communication was mentioned as a possible necessary condition: a research or design/development outcome should not be considered as a result unless it has been effectively communicated to relevant audiences: fellow researchers (locally, nationally, or internationally), teachers, school and policy boards, and so on. An essential part of research is that results must be communicable and communicated. In this light, attention was drawn to the fact that researchers, teachers, curriculum developers, parents, and policy boards all have different questions. Where the questions come from is essential to what the results are. And since the results depend on the questions, the results are different for different audiences.

The question of how far results must be communicated to be considered results led to the realization that the communication requirement was somewhat naïve; for example, does one need to demonstrate that 10 or 120 teachers applied one's research results? Does one need to show that teachers will further influence other teachers to use one's results? Isn't it possible to do research, say, on tertiary mathematics teaching and not have any university teachers apply it? Is this a requirement for something to be a result? Moreover, just by writing a paper one does not ensure communication. People may not read the paper. And if they do, they may give the results a very different interpretation from the intended one. The opinion was even expressed that what researchers find out cannot be transmitted in writing. It must be transmitted face-to-face. Thus, we need long exchanges, not brief one-phrase discussions. Only such direct interactions can allow us to find out whether others ask the same questions we do.

Communication to teachers does not usually take the form of papers, but may appear in the form of collections of tasks, exercises, and so forth. Such collections may, for example, show future teachers what discussions about mathematics are, and enable them to implement similar tasks in their classrooms. We were thus led to the question of whether professional development is research and, more generally, what the teachers' role in producing research results could be.

Although it was generally felt that professional development should be considered research, if the theoretical background is explicit, opinions and experiences on *teacher research* ranged widely. Much research is carried out nowadays with teacher participants. For example, in the Italian framework, the collaboration between researchers and teachers was a critical component influencing the transition from snapshots to movies. Classroom observations by teachers are often very revealing, but teacher participation in a study may significantly change the relevance and meaning of its results – which has been found to be the case by the Japanese, too.

One argument was that results are tied to the institution that required the result. A teacher's 'result' is attached to his or her school and is therefore not research. On the other hand, the questions teachers can and should ask can be treated by researchers. But that must not be a restricting factor. The tension can be resolved by a complete separation between a person's role as teacher and as researcher. Then teachers can produce research results.

To realize this separation, it is important to emphasize the connection and distinction between the responsibility of the teacher and that of the researcher. Even if it is the same person, there must be a complete change of behavior and responsibility: in research one has to overcome the inhibition to question the question. Research provides questions and answers that allow one to construct teaching situations. A situation can be proposed by a researcher to a teacher, but it will be the teacher's responsibility to implement it in the classroom. Teachers find themselves in a societal framework. Often they are forced to give answers beyond the area we already know from research investigations. They are forced to give 'engineering answers' without having a theory from which to draw the answers. This phenomenon points back to the design science aspect of mathematics education and, perhaps, provides it with more credence.

Field of Validity

Finally, validity was mentioned as a criterion for a fact to be considered as a result, and that led to the question of the field of validity of a result in space and time. The notion of field of validity of a result was seen as essential. It was also noted that we are not usually sensitive to determining fields of validity: Even if we try to fully separate teaching from researching, the tendency to generalize results beyond their field of validity is widespread.

To what extent can research results from one environment or culture (e.g., Japan) be linked to those from another (e.g., the USA), and to what extent are results culture-specific? It was pointed out that whether a result is valid depends on the theory within which it was obtained, not on the place where it was obtained. The value of a result thus depends on the interpretation of the theoretical (or philosophical) climate through which it is viewed. This climate changes with time and location.

Results are results within a certain theory. A mathematical statement is not necessarily a result of mathematics. The statement should also be unexpected for the person to whom it is addressed (or at least it should be sufficiently complex to draw the person's attention): it needs to be within the sphere of attention of the one to whom it is addressed.

Results cannot become nonresults; they are permanent. But while results themselves remain, their relevance may well be ephemeral. They can be interpreted differently in view of different theories. As a consequence, they may lose their interest when other theoretical frameworks become predominant. It was suggested that facts are perhaps more ephemeral in mathematics education than in other sciences. But even in mathematics, results are only results in terms of a theory. When you enlarge a theory, theorems may become nontheorems. Quaternions were mentioned as a possible example. Something that was an established and interesting result can disappear as a result. The fact remains. The theory is not there any more and the result thus disappears. In any case, results are neither universal nor eternal, and we should not look for fixed answers to what are *the* results of mathematics education research.

WHAT CRITERIA SHOULD BE USED TO EVALUATE THE RESULTS OF RESEARCH IN MATHEMATICS EDUCATION?

REPORT OF WORKING GROUP 5

Leader: Bengt Johansson
Reporters: Margaret Brown and Morten Blomhøj

Significance of the Topic

Why discuss this issue now? Mathematics education is developing toward a position of attaining a degree of status as a scientific discipline. This discussion provides the necessary background for coherent development of the discipline of mathematics education. Part of the reason is the need to attain some agreement as to what constitutes quality within research in mathematics education. If we see mathematics education as a discipline defined sociologically by those who claim to study it, is such agreement a necessary requirement (compare with the sociology of science, e.g., Thomas Kuhn)?

More essential than the product is the continuing process of discussing criteria for quality within the research community. Nevertheless, the generation of specific criteria in a variety of decision situations is important for making necessary judgments that critically affect the careers of those within the community. The ethics of such judgments demands that the criteria are open and that there is some mechanism for renegotiation. Such situations include acceptance of journal articles (discussed in Gila Hanna's and Frank Lester's presentations),[8] but also judgments about funding of research proposals, evaluation during the process of research, final reports, conference presentations, and so forth. Discussion of specific criteria appropriate to each situation also allows professional development of researchers and reviewers, who may otherwise have different interpretations of guidelines.

Formulating a Framework

We have chosen to reinterpret the meaning of our brief more broadly in terms of the considerations that inform judgments about the quality of research in mathematics education. Judgment has within it the notion of a tacit dimension building from experience and personal values as well as more explicit criteria.[9] This dimension goes beyond the evaluation of research results, and involves a discussion about the nature and development of our field. It does not presuppose that a set of agreed criteria are desirable, necessary, or even possible.

In this effort, we found it fruitful to take a point of departure in the description of the domain of mathematics education as a scientific discipline given by Hans-Georg Steiner in his presentation to the group. Mathematics education is seen as a field covering the practice of mathematics teaching and learning at all levels in (and outside) the educational system in which it is embedded (Figure 2).

Mathematics education is a field of study in its own right. It lies within the field of education and has interdisciplinary relations to a set of foundation

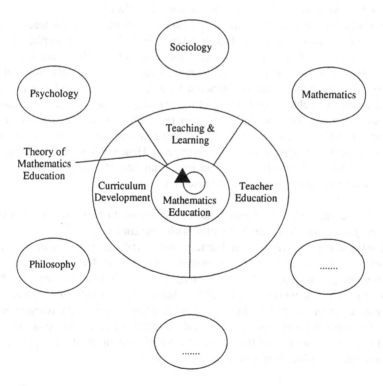

Figure 2

disciplines such as psychology, anthropology, history, and philosophy. But it also has a special relationship with the discipline of mathematics itself. This relationship distinguishes it from neighboring fields such as science education. Guy Brousseau discussed this relation in referring to the irreducible mathematical part of research in mathematics education.[10]

Within this domain of mathematics education, Steiner pointed to complexity, interdisciplinarity, complementarity and critique. The study of mathematics education can encompass paradigms and paradigm shifts, as in the development of science. The matter of different contexts for mathematics education and different cultural traditions must also be considered in relation to the framework.

Sources for Criteria for Quality in Research in Mathematics Education

The framework within which mathematics education is located provides a structure for examining criteria for the quality of research in mathematics education. The remainder of the report is organized around questions that were discussed by the group, referring back to this framework.

(1)   Could the relation of mathematics education to classroom practice generate criteria for quality in research in mathematics education?
The whole domain of mathematics education is concerned with the teaching and learning of mathematics in the broadest sense, including, for example, learning outside school, system-wide implementation, and so forth. Therefore it is likely that research within the field of mathematics education will have implications for the development of mathematics teaching. On the other hand, the practical implication for a change of practice in the teaching of mathematics is a dangerous criterion for assessing the quality of research; it is neither necessary nor sufficient. Although the research may not have immediate and obvious implications, benefits may appear only in the long term. Hence although it should not be a strict criterion, it is likely to be an important consideration for judging the worthwhileness and relevance of the research.

(2)   Could the relation of mathematics education to MATHEMATICS generate criteria for quality in research in mathematics education?
MATHEMATICS is written in capital letters here to indicate that it is to be understood as mathematics taken in the widest sense – to include, for instance, its history, its philosophy, and its various applications. Whatever views are held about the nature of MATHEMATICS will be a basis for judgments about the validity of research in mathematics education. This condition requires the research to be specific about the views of mathematical knowledge and activity that underpin it. It requires awareness of the general nature of mathematical concepts, and of the special difficulties that arise in acquiring these.

*(3)   Could the relation of mathematics education to foundation disciplines gen-
erate criteria for quality in research in mathematics education?*
Although research in mathematics education gains its autonomy in the act of
problem posing, in considering problems in mathematics education we are able
to select and define both theories and methodologies from other foundation disciplines (background sciences such as psychology, sociology, and philosophy).

Imported criteria for these disciplines must form part of the criteria for the
quality of research in mathematics education. For example, research in mathematics education could form the subject of articles publishable in journals related
to these disciplines. Such 'imported' criteria could deal with the nature of evidence
(a point made by Paolo Boero), or with the judgment of validity. The debate on
qualitative/quantitative methods relates to their appropriate application to relevant
research questions. This point highlights the interdisciplinarity of mathematics education, although the process is not only one-way. The use of either the theory or
the methodology may produce new developments in the parent discipline.

*(4)   Could the concept of progress in mathematics education generate criteria
for quality in research in mathematics education?*
The aim of research in mathematics education must be to extend our understanding and knowledge of some aspect of mathematics education. This requirement
means that we need criteria to enable us to identify progression of theory.
Theory is here to be understood in a relatively informal sense as a network of
concepts and relationships. One aspect of progression of theory can be interpreted as the forming of new concepts and relations within an existing paradigm.
It is important that researchers are aware of the current state of the development
of theory at least in the paradigm within which they are working. This awareness
forms a criterion for progress and therefore for quality within that paradigm.

Much profit can arise also, however, from comparison of the interpretative
power of different paradigms in respect to a set of situations, leading perhaps to
a more general comparison or even a synthesis. Clearly such comparisons
require knowledge, understanding, and application of different paradigms. The
generation of totally new paradigms ('home grown theories') within mathematics education remains an overall aim. Again, the criterion of interpretative power
is important in the evaluation of new paradigms.

*(5)   How do we take cultural differences within mathematics education into
account?*
Criteria can be generated from the previous considerations. However, it is necessary to allow for the fact that there are substantial differences in cultural traditions relating to mathematics education, arising from different communities and
sub-communities. In time, we would hope that there is a growth in shared
knowledge within the research community. This growth is assisted by bi-lateral
and multi-lateral conferences in which the main agenda is to open up and

discuss the differences between traditions. In the meantime, we can only use criteria on which there is a broad consensus.

Criteria for Quality in Research in Mathematics Education

In this report we have emphasized the framework for the process of setting up criteria rather than describing the product in the form of a set of operational criteria themselves. But because it is necessary to make judgments relating to research in mathematics education, the challenge of trying to specify such sets of criteria does have to be faced in specific fields.

In facing this challenge, we should consider the advantages and disadvantages of the effects of using criteria for different purposes. For example, there is a risk that they might lead to stereotyping with researchers playing safe. For this reason, it is important that criteria if stated are illustrated by exemplars, and that the exemplars are sufficiently rich to indicate the range of possible diversity within, and not the narrowness of the criteria.

The fragility of the process of using criteria to make judgments within mathematics education requires that such criteria have a wider validity than just in this field. Competition with other disciplines – for example, for research funding – calls for the use of criteria for quality that apply to research in general. Criteria are needed for example in judging products such as:

- research proposals
- interim or final reports of research
- journal or conference papers
- doctoral theses
- applications for jobs or promotions.

Related criteria are also needed to guide the ongoing research process. Some criteria relate to the quality of the research itself, and are hence common to all these items. Other criteria relate specifically to one item.

NOTES

1. Documentation for the report was gathered during the conference by A. Kristjánsdóttir, N. Presmeg, J. Confrey, and A. Sierpinska. A. Sierpinska wrote the final text.
2. A revised version of Sfard's paper is in Part VI.
3. A revised version of Leder's paper is in Part III.
4. A revised version of the Boero and Szendrei paper is in Part III.
5. This proposal was subsequently sent to the ICMI Executive Committee, and an ad hoc committee, chaired by Nicolas Balacheff, was appointed to address it.
6. A revised version of Kieran's paper is in Part III.
7. A revised version of the Arzarello paper, written with Mariolina Bartolini-Bussi, is in Part IV.
8. See the chapters by Hanna and by Lester and Lambdin in Part V, which are based on papers presented in the working group.
9. See Hanna's guidelines, which include personal concept of quality, in Part V.
10. Brousseau gave an address in the first plenary session of the conference.

ALAN J. BISHOP

RESEARCH, EFFECTIVENESS, AND THE PRACTITIONERS' WORLD

1. INTRODUCTION

The ICMI Study Conference on Research was a watershed event with a great deal of significant interaction between the participants. It was an energizing and involving experience, but as one of my tasks was to 'summarize' at the end of the conference, I tried to take a more objective stance during my involvement. I reported in my summary that I could detect certain emphases in the discourses together with some important silences. Here are some of them:

Emphases	*Silences*
analyses	syntheses
critiques	consensus-building
talking to ourselves	awareness of other audiences
political arguing (to persuade)	researched arguing (to convince)
individual cases	over-arching structures
local theory	global theory
well-articulated differences	well-articulated similarities
disagreements	agreements

It seems as if the researcher's training encourages one to analyze, to look for holes in arguments, to offer alternative viewpoints, to challenge and so forth. Or the pattern could reflect the fact that the idea of the conference itself and of this ICMI Study was seen as a challenge to the participants' authority. Certainly the 'politics of knowledge' was alive and well in all its manifestations.

So what is the concern of this chapter? Am I seeking just a nice, warm, collaborative engagement, and feeling that it is a pity that we seem to disagree so much? I have to admit that with so much conflict in the world today I do wish that there was more obvious peaceful collaboration. I also believe that with the whole idea of research into education under attack from certain ignorant politicians and bureaucrats, those who engage in research should at least collaborate more, and should spend less time 'attacking' each other.

My real concern, however, is with what I see as researchers' difficulties of relating ideas from research with the practice of teaching and learning mathematics. In the discussion of this ICMI Study at the Eighth International Congress on Mathematical Education in Seville, many people spoke about the dangers of

33

Sierpinska, A. and Kilpatrick, J. Mathematics Education as a Research Domain: A Search for Identity, 33–45.
© *1998 Kluwer Academic Publishers. Printed in Great Britain.*

researchers just talking to each other, and thereby ignoring the practical concerns of teachers. Moreover, with the general tendency towards greater accountability in education, we find increasing pressures for more *effective* modes of mathematics education. This pressure needs to be responded to by the research community, and my concern is that, in general, it is not.

This chapter, then, is concerned with the researchers' relationship with the practitioners' world, and it starts by considering the issues of 'effectiveness'.

2. EFFECTIVENESS: THE PRESSURE AND THE RESPONSES

Mathematics has increasingly become a significant part of every young person's school curriculum, and as mathematics has been growing in importance, so has the public pressure to make mathematics teaching as effective as possible. 'Mathematics for all' (see, e.g., Damerow et al. 1986) has become a catch phrase which has driven educators in many countries to constantly review their mathematics curricula and their teaching procedures. The challenge of trying to teach *mathematics* to all students, rather than, say, arithmetic, has been one element, the other being that of teaching *all* students regardless of possible disadvantage, handicap or obstacle.

The pressure for greater effectiveness in mathematics teaching now comes both from the application of business-oriented approaches in education, supported by the theory of economic rationalism, and also from the increasingly politicized nature of educational decision making, driven by the economic challenges faced by all industrialized countries since the last world war. Education is now seen as an expensive consumer of national funds, and concerns over the perceived quality of mathematical competence in the post-school population have fueled the pressure.

Research has also contributed (perhaps inadvertently) to this pressure, firstly by supplying a range of evidence on competence, and secondly by raising expectations about the potential for achievements throughout the whole school population. It has also contributed to the concern for effectiveness by helping to generate alternative approaches in mathematics teaching. There are now many possible permutations of teaching styles, instructional aids, grouping arrangements, print media and so forth, all available within the modern classroom and school.

Add to these the possible curricular contents and emphases, curricular sequences, and examinations and assessment modes, which themselves have been stimulated by research, and the variety of *potential* mathematics educational experiences becomes bewildering, not just to teachers. Moreover, these only refer to formal mathematics education – one could continue to contemplate the further large number of possibilities if informal and non-formal educational experiences were included, along with the World Wide Web.

In the face of such a plethora of educational alternatives, the demand for knowledge about effectiveness is easily understandable, and needs to be taken seriously. However, there is little evidence that researchers are addressing the issue seriously enough.

The editor of the recent *Handbook of Research on Mathematics Teaching and Learning* (Grouws 1992) advised:

The primary audience for the handbook consists of mathematics education researchers and others doing scholarly work in mathematics education. This group includes college and university faculty, graduate students, investigators in research and development centers, and staff members at federal, state, and local agencies that conduct and use research within the discipline of mathematics.... Chapter authors were not directed to write specifically for curriculum developers, staff development coordinators, and teachers. The book should, however, be useful to all three groups as they set policy and made decisions about curriculum and instruction for mathematics education in schools. (p. ix)

Indeed, nowhere in that volume are the issues of effectiveness specifically and systematically addressed, although the chapters by Bishop (1992, ch. 28) and Davis (1992, ch. 29) do intersect with that domain of concern. Bishop reflects on the critical relationship between teacher and researcher, and between researcher and the educational system. Davis argues also for greater interaction between the education system and researchers.

The lack of relationship between mathematics education research and practice is documented by many references in the literature (e.g., Brophy 1986; Crosswhite 1987; Freudenthal 1983; Kilpatrick 1981). There are, however, some signs that research and researchers are relating more closely to the ideas of reform in mathematics teaching (Grouws, Cooney & Jones 1989; Research Advisory Committee 1990, 1993).

The pressure for effectiveness in mathematics teaching supports and encourages the 'reform' goal of research. In this sense, effectiveness is achieved by changing mathematics teaching since the present practice is assumed to be relatively ineffective. Thus the often quoted dichotomy between research and practice is becoming refocused onto issues concerning the role of research in changing and reforming practice. It is tempting to see 'reform' as being merely 'change' in line with certain criteria, but that analysis loses the essential dynamic associated with a reform process.

3. RESEARCHERS AND PRACTITIONERS

There are several key issues to be faced here. Can research reform practice? Should that be its role? Clearly there are other sources for knowledge with which to inform change. What is the relationship of those to the knowledge generated by researchers?

If we consider the people involved, it is clear that it is only practitioners who have it in their power to change, and therefore to reform, practice. Sadly the researcher/practitioner relationship has frequently not been a close one, nor has the knowledge each generates. As Kilpatrick (1992) says: 'The actions of a practitioner who interprets classroom events within their contexts could not be further removed from the inferences made by a researcher caught up in controlling variation, quantifying effects, and using statistical models' (p. 31).

Can these two sets of activities be reconciled? More important perhaps, is whether practitioners and researchers are dealing with such different kinds of knowledge that communication becomes impossible.

Furthermore, the Research–Development–Dissemination model still lives in many assumptions about the relationship between research and practice, and reflects an assumed power structure which accords the researcher's agenda and actions greater authority than the practitioner's. The increasing moves to involve teachers in research teams are to be applauded, but currently only serve to reinforce the existing power structure. We hear little about researchers being invited to join teaching or curriculum planning teams. If the two knowledge domains are at present so mutually exclusive, then what hope is there for research to be influential in reforming practice?

Researchers clearly need to take far more seriously than they have done the fact that reforming practice lies in the practitioners' domain of knowledge. One consequence is that researchers need to engage more with practitioners' knowledge, perspectives, work and activity situation, with actual materials and actual constraints, and within actual social and institutional contexts. We will look at a good example of this later in the chapter.

There are some encouraging signs that researchers are engaging more with actual classroom events within actual classrooms. We are learning more about the teachers' knowledge, perspectives and so forth. The research agendas though are still dominated by the researchers' questions and orientations, not the practitioners'. Researchers tend to be interested in communication patterns, constructivist issues, group processes and so forth, any of which *may* produce ideas which could have reform implications. But we know precious little about *teachers'* perspectives on reform-driven issues which researchers could seriously address.

Dan Lortie's (1975) conclusions still ring true: 'Teachers have an in-built resistance to change because they believe that their work environment has never permitted them to show what they can really do' (p. 235). Lortie's view is that as a result, teachers often see the proposals for change made by others as 'frivolous' when they do not actually affect their working constraints.

Schools are more than just a collection of one-teacher classrooms. Consideration of the social dimension of mathematics education forces us to realize the institutional constraints which shape students' learning. Arfwedson (1976) analyzed the 'goals' and the 'rules' of a school, from the teacher's perspective, and showed that the goals of a school are not seen to be accompanied by sanctions since they are related to the pedagogical methods and attitudes adopted, whereas rules *do* carry sanctions (e.g., a teacher cannot disregard keeping to a timetable nor recording pupils' attendance with impunity). He also points out that the power of the teacher is somewhat superficial since there would appear to be inevitable conflict between rules and goals, and although the teacher has apparent pedagogical freedom, the rules impinge strongly. 'On the one hand the teacher is a part of the hierarchical power-structure of the school organization, on the other hand it is his duty to realize goals that are mainly democratic and anti-authoritarian' (pp. 141–142). Thus any desire on the part of the teacher to bring about change can become inhibited or destroyed.

Much clearly depends on the other practitioners who shape mathematics education. At the government, state, and local levels, there are curriculum and assessment designers. Textbook writers who receive governmental contracts are equally influential. School administrators, who may be part-time teachers, shape whole-school curricula and structure timetables and schedules which constrain what classroom teachers are able to do. Add parents, local employers, and politicians into the mix with their different agendas, and Arfwedson's and Lortie's comments assume an even greater significance.

Where is the research on the influence of different timetable and scheduling patterns on mathematical learning? Where are the researchers who are prepared to engage in the real practitioners' world of time constraints, local politics, and petty bureaucracies? There seems to be a certain amount of coyness on the part of many researchers, perhaps rationalized in terms of 'academic freedom', to join the practitioners' world. There is also a large amount of academic snobbery together with plenty of wishful thinking.

If researchers are to stand any chance of helping to reform practice, then they surely must enter the practitioners' world, derive more of their agendas from the problems of that world, conform more to practitioners' criteria and norms for solving those problems, and communicate more within practitioners' communication mechanisms, such as teachers' journals and newspapers.

4. RESEARCH APPROACHES AND PRACTICE

Are certain kinds of research activity better suited to improving practice than others? In Bishop (1992) I elaborated my ideas about the three components of the research process:

- Enquiry, which concerns the reason for the research activity. It represents the systematic quest for knowledge, the search for understanding, and gives the dynamism to the activity. Research must be *intentional* enquiry.
- Evidence, which is necessary in order to keep the research related to the reality of the mathematical education situation under study, be it classrooms, syllabuses, textbooks, or historical documents. Evidence samples the reality on which the theorizing is focused.
- Theory, which recognizes the existence of values, assumptions, and generalized relationships. It is the way in which we represent the knowledge and understanding that comes from any particular research study. Theory is the essential product of the research activity, and theorizing is, therefore, its essential goal. (p. 711)

Of the three main traditions influencing research (see Bishop 1992), the pedagogue and the scholastic philosopher seem to differ most markedly in respect of the relationship with practice (see Table 1):

Theory	Goal of enquiry	Role of evidence	Role of theory
Pedagogue tradition	Direct improvement of teaching	Providing selective and exemplary children's behavior	Accumulated and shareable wisdom of expert teachers
Empirical scientist tradition	Explanation of educational reality	Objective data, offering facts to be explained	Explanatory, tested against the data
Scholastic philosopher tradition	Establishment of rigorously argued theoretical position	Assumed to be known. Otherwise remains to be developed	Idealized situation to which educational reality should aim

Table 1
Source: Bishop 1992, p. 713.

The pedagogue tradition is overtly concerned with improving practice, while the scholastic philosopher is not. The pedagogue tends to involve teachers *in* the research process, whereas the scholastic philosopher tends to treat teachers as agents of practice who are largely irrelevant to the research process. The pedagogue tradition puts a large premium on 'knowing' the educational reality – the scholastic philosopher tradition takes that reality as a generalized assumption on which to base theorized possibilities.

Perhaps the pressure for increased effectiveness is reflected in the increasing dominance of pedagogue-influenced research approaches. As one example, we now find much less reliance on surveys and questionnaires and much more emphasis on case studies and anthropologically stimulated research.

Perhaps also the pressure is reflected in a tendency to choose research methods appropriate for a particular problem rather than to stay wedded to a particular research method. Begle's (1969) rallying call, to which many responded, is summed up in this statement:

I see little hope for any further substantial improvements in mathematics education until we turn mathematics education into an experimental science, until we abandon our reliance on philosophical discussion based on dubious assumptions and instead follow a carefully correlated pattern of observation and speculation, the pattern so successfully employed by the physical and natural scientists. (p. 242)

As various writers in the *Handbook of Research on Mathematics Teaching and Learning* indicate, that view no longer predominates, largely because of the failure of the experimental method to produce the benefits sought by the providers of research funds, and by the educational system at large.

The question to reflect on is whether *any* one research method can meet the demands of improving the effectiveness of mathematics teaching. Recently, action research has been promoted as the only research approach which will have lasting effects in schools (Carr & Kemmis 1986). Action research, as described by critical-theorist proponents such as Carr, Kemmis and McTaggart, is concerned with research by people on their own work, with the explicit aim of improvement, and following an essentially critical approach to schooling: 'Philosophers have only interpreted the world in various ways ... the point is to change it' (Carr & Kemmis 1986, p. 156).

More than any other approach, action research takes 'change' as its focus and encourages practitioners of different kinds to research collaboratively their shared problems. As Kemmis and McTaggart (1988) point out, 'the approach is only action research when it is collaborative' (p. 5). Action research thus emphasizes group collaboration by the participants rather than the specific adoption of any one particular method. There is also a strong ideological component to action research, and it is thus more appropriate to refer to it as a 'methodology' rather than as one specific method (such as case study).

The action research methodology goes a stage further towards combining research and practice than other approaches where teachers join research teams. The latter practice, although encouraged by many, still tends to perpetuate the center–periphery model of educational change, defined by Popkewitz (1988) as consisting of the following stages:

1. initial research identifies, conceptualises and tests ideas without any direct concern for practice;
2. development moves the research findings into problems of engineering and 'packaging' of a program that would be suitable for school use;
3. this is followed by dissemination (diffusion) to tell, show and train people about the uses and possibilities of the program;
4. the final state is adoption/installation. The change becomes an integral and accepted part of the school system. (p. 131)

This is a model which mathematics education research has often adopted, either deliberately or accidentally. Indeed, some projects have stayed at the first stage, while others have moved to Stages 2 and 3, often in the form of textbooks or computer programs. Few, however, have reached Stage 4.

Partly the reasons are that the research has tended to focus mainly on the *learning* of the particular topic in question, or on the *curriculum* questions. If this happens, then there is no particular reason why any *teaching* approaches derived from or implied by the research will be at all successful.

Choosing to focus research on school-based curriculum topics does not necessarily produce improvements in teaching either. The content-oriented chapters in Grouws (1992) bear testimony to that:

1. After an exhaustive analysis of both the semantics of, and children's understanding of, rational number concepts, Behr et al. (1992) state that 'little is known about instructional situations that might facilitate children's ability to partition' (p. 316).

2. Concluding the chapter on algebra, Kieran (1992) says:

> As we have seen, the amount of research that has been carried out with algebra teachers is minimal.... Teachers who would like to consider changing their structural teaching approaches and not to deliver the material as it is currently developed and sequenced in most textbooks are obliged to look elsewhere for guidance. (p. 413)

3. Concluding the chapter on geometry, Clements and Battista (1992) say: 'We know a substantial amount about students' learning of geometric concepts. We need teaching/learning research that leads students to construct robust concepts' (p. 457).
4. In the concluding section of the chapter on probability, Shaughnessy (1992) says: 'It is crucial that researchers involve teachers in future research projects, because teachers are the ultimate key to statistical literacy in our students' (p. 489).
5. Concluding the chapter on problem-solving, Schoenfeld (1992) says: 'There is a host of unsolved and largely unaddressed questions dealing with instruction and assessment' (p. 365). As Brophy (1986) says: 'Mathematics educators need to think more about *instruction,* not just curriculum and learning' (p. 325).

As was stated above, theory is the way in which we report the knowledge and understanding that comes from any particular research study. However, in relation to the theme of this chapter, the central issue about theory is to what extent should it shape the research itself? This issue highlights the role of theory in determining the research questions, in shaping the research process, and in determining the research method.

The thrust of the arguments so far in this chapter is that it should be the practitioners' problems and questions which should shape the research, not theory. So should research be any more than just practical problem-solving? And to what extent should the knowledge which is research's outcome be any more valid as a form of knowledge-for-change as any other form of knowledge?

Action research has certainly been interpreted by some as just practical problem-solving. However, as Ellerton et al. (1989) point out: 'It involves more than problem solving in that it is as much concerned with problem posing as it is with problem solving' (p. 285). They go on to show that action researchers do *not*

> identify with what has come to be known as 'the problem-solving approach' to achieving educational change. By this latter approach, a school staff identified problems which need to be addressed in their immediate setting: they design solutions to these problems, and in so doing, become trained in procedures for solving future problems. (p. 286)

For Carr and Kemmis (1986, p. 159), the term *research* implies that the perspectives of the participants in the research are changed. Thus, even in action research, the research process should be a significant learning experience for the participants. In that sense, the research problems and questions need to be couched in terms within the participants' schemes of knowledge. Theory enters through the participants' knowledge schemes, and insofar as these schemes involve connections with published theory, so will that theory play a part in shaping the research. It is this theory which offers the dimensions of generality which make the difference between research and problem-solving.

This pattern is well illustrated by a powerful study of mathematics teaching in first schools (Desforges & Cockburn 1987), which followed the analysis of Doyle (1986). In that sense, it built on a theoretical notion coming from earlier work, but the research questions are centered in the practitioners' world, particularly: 'What factors do teachers take into consideration in adopting management and teaching techniques and what factors force her to amend her goals or behaviors in the day-to-day world of the classroom?' (p. 23). The study describes and interprets the teachers' behaviors in the context of the teachers' reality and looks generally at the activities relating to stimulating higher-order thinking in mathematics, e.g. in problem-solving.

After documenting the teachers' practices in detail, the study concludes that it was clearly dealing with teachers of quality, who yet failed to deliver many of the aspirations which they and other mathematics educators fully endorsed. The authors ask:

Why was there so much pencil and paper work and so little meaningful investigation? Why was there so little teacher–pupil and pupil–pupil discussion? Why was there so little diagnostic work? Why was the curriculum so dominated by formal mathematics schemes and so little influenced by children's spontaneous interests? Why did teachers with such an elaborate view of children's thinking cast their pupils into passive-receptive roles as learners or permit them to adopt such roles? (p. 125)

After further analysis the researchers point to the classroom and institutional realities which shape the practices, and comment that those realities are not designed for the conscious development of higher-order thinking. Indeed rather than criticizing the teachers for failing, they point out just how daunting it is to establish and sustain higher-order skills in a mathematics curriculum. The teachers' achievements are thereby that much more impressive. They say: 'We conclude that classrooms as presently conceived and resourced are simply not good places in which to expect the development of the sorts of higher-order skills currently desired from a mathematics curriculum' (p. 139).

I had reached the same conclusions in Bishop (1980) when I said:

The problem is that classrooms appear not to be particularly appropriate environments in which to learn mathematics. Classroom learning can be characterized by the following constraints:

(a) It must take place in a limited time
(b) It is often incomplete learning
(c) There are multiple objectives
(d) The conditions are 'noisy'
(e) The atmosphere is one of mutual evaluation

(f) Presentation sequences are a compromise
(g) Teaching is a stressful occupation

Research is developing rapidly and our knowledge of learning is becoming more and more sophisticated. Meanwhile classrooms are becoming more of a challenge for teachers and many feel that the quality of teaching is declining. (pp. 339–340)

Desforges and Cockburn's (1987) theoretical analysis led them to conclude that, for practice,

our prescriptions for change are directed more at those who provide – materially and especially conceptually – for practice and only tangentially at those who execute it. That is to say that in so far as contemporary mathematics teaching practices in the infant school may be seen to fall short of expectations, the burden of responsibility, in the terms of our analysis, lies with the educational managers who – whether deliberately or by default – provide the crucial psychological parameters of the teaching environment to which teachers and children alike must adapt. It is on these same groups that the onus for change must lie. (p. 143)

The Desforges and Cockburn study demonstrates vividly what researchers can contribute to the development of practice not only by contextualizing the research in the classroom realities, but also by couching the whole study in terms of practitioners' knowledge schemes. It also demonstrates that theory development is a goal, in that the study is both an analysis of practice and a search for explanations.

Perhaps one of the most important consequences for theory development is that researchers should pay more attention to synthesizing results and theories from different studies. As was said at the start of this chapter, the researchers at the ICMI Study Conference illustrated the tendency of researchers everywhere to analyze, critique and seek alternatives to each others' ideas, rather than trying to synthesize, build consensus, or recognize agreements. It is no good expecting teachers, or any other practitioners, to do the synthesizing, as they are frequently not the ones with access to the different ideas, results or approaches. An implication of a study like the one above is that it is the practitioners' epistemologies which should provide the construct base of the synthesized theory.

5. RESEARCHERS' ROLES AND THE PRACTITIONERS' WORLD

Research is big business in some countries, while in others it is another arm of government. In some situations, researchers can do whatever they like, while in others their practices are heavily proscribed either by external agencies or by ethical codes. Most would probably still yearn longingly for the academic ideal of the disinterested researcher, defining their own research in their relentless pursuit of knowledge. No researchers worth their salt would have any difficulty generating a research agenda for well into the 21st century. 'Knowledge for knowledge's sake' may sound old-fashioned but would still find appeal with many today.

However, the climate is changing. The days of the disinterested researcher are both ideologically and realistically numbered. For the big-ticket researcher, large research budgets are probably a thing of the past. In some countries, researchers have been deliberately marginalized from mainstream educational debate, while in others they have deliberately excluded themselves from it, and have found their research aspirations severely blunted. Increasingly, educational research and researchers are having to justify their continued existence in a world that is financially competitive, often politically antagonistic to institutionalized critique, and increasingly impatient with 'time-wasting' reflection and questioning.

If research is to have the kind of impact on practice and on the practitioners' world that many want it to have, then there is a clear need for much more disciplined enquiry into the practitioners' situation and an urgency to grapple with the issues of the theory/practice relationship. As a conclusion to this chapter and as a contribution to the debate, I offer the following ideas:

- *Researchers need to focus more attention on practitioners' everyday situations and perspectives.* The research site should be the practitioners' work situation, and the language, epistemologies, and theories of practitioners should help to shape the research questions, goals and approaches.
- *Team research by researchers/practitioners should be emphasized.* The work and time balance of the research activity will need to be negotiated, and the roles of the members clarified. The team should also include practitioners from other parts of the institution other than those whose activities are the focus. They are often the people who set the constraints on the development of teaching, as the study by Desforges and Cockburn (1987) showed. It seems to be of little value to involve them only at a dissemination stage, since their activities might well have contributed in an indirect way to the outcomes of the research.
- *The institutional context and constraints should be given greater prominence in research.* This is the 'practice' counterpart to the 'practitioner' point above. Institutions develop their own rules, history, dynamics and politics, and these need to be recognized and taken account of in the research.
- *Exceptional situations should be recognized as such, and not treated as 'normal' or generalizable.* Indeed, it is better to assume that every situation is exceptional, rather than assume it is typical. Typicality needs to be established before its outcomes can be generalized.
- *The process of educational change needs to be a greater focus in research in mathematics teaching.* It is rather surprising that, although many researchers assume a goal of change in their research, there has been relatively little research focus on the process of change itself.
- *Social and anthropological approaches to research should increase in prominence.* These approaches seem likely to offer the best way forward if researchers hope to make significant advances in how practitioners change

their ideas and activities. Again it is no accident that they are already coming into greater prominence.

● *Conclusions and outcomes should be published in forms which are accessible to the maximum number of practitioners.* Researchers should resist the pressure to publish only in research journals, as these are rarely read by practitioners. If a team approach is adopted more frequently, then the practitioner members of the team can, and should, help with appropriate publication and dissemination.

It seems appropriate to finish this chapter by quoting a few more sentences from Desforges and Cockburn's (1987) study because they address the need for researchers to enter the practitioners' world, to admit their ignorance and to struggle to develop new theoretical interpretations:

Rather than creating the aura that the only factor preventing the attainment of our aspirations in early years' mathematics teaching is the conservative practice of teachers, experts in the field should admit that they have yet to equip themselves – let alone the profession – with the conceptual tools adequate to the job.... Such an admission might draw more first school teachers into the kind of research work necessary. We have shown that teachers have a vast knowledge of children's responses to tasks. They are also very self-critical. Because they care about children it is very easy to make them feel guilty and feeling guilty they withdraw in the face of self-confessed experts. In this way researchers throw away their best resource, leave teachers open to cheap political jibes and make teaching more difficult. (p. 154)

REFERENCES

Arfwedson, G.: 1976, 'Ideals and Reality of Schooling', in M. Otte, R. Bromme, D. Kallòs, U. Lundgren, T. Mies & D. Walker (eds.), *Relating Theory to Practice in Educational Research* (Materialien und Studien, Band 6), Institut für Didaktik der Mathematik, Universität Bielefeld, Bielefeld, Germany, 139–146.

Begle, E. G.: 1969, 'The Role of Research in the Improvement of Mathematics Education', *Educational Studies in Mathematics* 2 (2/3), 232–244.

Behr, M. J., Harel, G., Post, T. & Lesh, R: 1992, 'Rational Number, Ratio, and Proportion', in D. A. Grouws (ed.), *Handbook of Research on Mathematics Teaching and Learning*, Macmillan, New York, 296–333.

Bishop A. J.: 1980, 'Classroom Conditions for Learning Mathematics', in R. Karplus (ed.), *Proceedings of the Fourth International Conference for the Psychology of Mathematics Education*, University of California, Berkeley, CA, 338–344.

Bishop, A. J.: 1992, 'International Perspectives on Research in Mathematics Education', in D. A. Grouws (ed.), *Handbook of Research on Mathematics Teaching and Learning*, Macmillan, New York, 710–723.

Brophy, J.: 1986, 'Teaching and Learning Mathematics: Where Research Should Be Going', *Journal for Research in Mathematics Education* 17 (5), 323–336.

Carr, W. & Kemmis S.: 1986, *Becoming Critical: Education, Knowledge and Action Research*, Deakin University Press, Geelong, Victoria.

Clements, D. H. & Battista M. T.: 1992, 'Geometry and Spatial Reasoning' in D. A. Grouws (ed.), *Handbook of Research on Mathematics Teaching and Learning*, Macmillan, New York, 420–464.

Crosswhite, F. J.: 1987, 'Cognitive Science and Mathematics Education: A Mathematics Educator's Perspective', in A. H. Schoenfeld (ed.), *Cognitive Science and Mathematics Education*, Erlbaum, Hillsdale, NJ, 165–277.

Damerow, P., Dunkley, M. E., Nebres, B. F. & Werry, B.: 1986, *Mathematics for All*, Unesco, Paris.

Davis, R. B.: 1992, 'Reflections on Where Mathematics Education Now Stands and on Where It May Be Going', in D. A. Grouws (ed.), *Handbook of Research on Mathematics Teaching and Learning*, Macmillan, New York, 724–734.

Desforges, C. & Cockburn, A.: 1987, *Understanding the Mathematics Teacher: A Study of Practice in First Schools*, Falmer, London.

Doyle, W.: 1986, 'Academic Work', *Review of Educational Research* **53**, 159–200.

Ellerton, N. F.: 1986, 'Children's Made-up Mathematics Problems – A New Perspective on Talented Mathematicians', *Educational Studies in Mathematics* **17** (3), 261–271.

Ellerton, N. F., Clements, M. A. & Skehan, S.: 1989, 'Action Research and the Ownership of Change: A Case Study', in N. F. Ellerton & M. A. Clements (eds.), *School Mathematics: The Challenge to Change*, Deakin University Press, Geelong, Victoria, 184–302.

Freudenthal, H.: 1983, 'Major Problems of Mathematics Education', in M. Zweng, T. Green, J. Kilpatrick, H. Pollak & M. Suydam (eds.), *Proceedings of the Fourth International Congress on Mathematical Education*, Birkhäuser, Boston, 1–7.

Grouws, D. A. (ed.): 1992, *Handbook of Research on Mathematics Teaching and Learning*, Macmillan, New York.

Grouws, D. A., Cooney T. J. & Jones, D.: 1989, *Perspectives on Research on Effective Mathematics Teaching*, Erlbaum, Hillsdale, NJ.

Kemmis, S. & McTaggart, T. (eds.): 1988, *The Action Research Planner*, Deakin University Press, Geelong, Victoria.

Kieran, C.: 1992, 'The Learning and Teaching of School Algebra', in D. A. Grouws (ed.), *Handbook of Research on Mathematics Teaching and Learning*, Macmillan, New York, 390–419.

Kilpatrick, J.: 1981. 'The Reasonable Ineffectiveness of Research in Mathematics Education', *For the Learning of Mathematics* **2** (2), 22–29.

Kilpatrick, J.: 1992, 'A History of Research in Mathematics Education', in D. A. Grouws (ed.), *Handbook of Research on Mathematics Teaching and Learning*, Macmillan, New York, 3–38.

Lortie, D. C.: 1975, *Schoolteacher: A Sociological Study*, University of Chicago Press, Chicago.

Popkewitz, T. S.: 1988, 'Institutional Issues in the Study of School Mathematics', *Educational Studies in Mathematics* **19** (2), 221–249.

Research Advisory Committee: 1990, 'Mathematics Education Reform and Mathematics Education Research: Opportunities, Obstacles and Obligations', *Journal for Research in Mathematics Education* **21** (4), 287–292.

Research Advisory Committee: 1993, 'Partnership', *Journal for Research in Mathematics Education* **24** (4), 324–328.

Schoenfeld, A. H.: 1992, 'Learning to Think Mathematically: Problem Solving, Metacognition, and Sense Making in Mathematics', in D. A. Grouws (ed.), *Handbook of Research on Mathematics Teaching and Learning*, Macmillan, New York, 334–370.

Shaughnessy, J. M.: 1992, 'Research in Probability and Statistics: Reflections and Directions', in D. A. Grouws (ed.), *Handbook of Research on Mathematics Teaching and Learning*, Macmillan, New York, 334–370.

Alan J. Bishop
Faculty of Education,
Monash University,
Clayton,
Victoria 3168,
Australia

PART II

MATHEMATICS EDUCATION AS A RESEARCH
DISCIPLINE

JOSETTE ADDA

A GLANCE OVER THE EVOLUTION OF RESEARCH IN MATHEMATICS EDUCATION

As the aim of this ICMI Study is to examine the situation of the current research in mathematics education, I think that it could be useful to look at it from the perspective of an 'old hand' and to balance the things that have changed and those that remained the same. Of course, this will be a very personal view; other people may see things very differently.

1. THE PROBLEMS

As with any scientific domain, mathematics education is characterized by the research issues it addresses. At the Fourth International Congress on Mathematical Education in Berkeley (1980) Hans Freudenthal was challenged to present the 'major problems of mathematics education' of the moment, as Hilbert was at the Paris International Congress of Mathematicians in 1900 for the major mathematical problems of his time. Freudenthal began by explaining the difference in the nature of these two classes of problems. Some people back then perhaps imagined that one could solve the problems in the same way. Today, I think that what Hans Freudenthal said in his talk is clear for everybody:

[P]roblems, problem solving, problem solvers, mean different things in mathematics education from what they mean in mathematics.... Moreover, in mathematics, you can choose one major problem, say from Hilbert's catalogue, solve it, and disregard the remainder. In education, all major problems, and, in particular, those I am going to speak about, are strongly interdependent' (Freudenthal 1983, p. 1).

Looking at Freudenthal's list of thirteen problems, the thing that strikes one the most is that not only are none of them solved yet today, but they are still of major interest and, in addition, that they have produced important new problems.

The first problem was: *Why can Jennifer not do arithmetic?*, where 'rather than an abstraction like John and Mary, Jennifer is a living child'. This is still, of course, the first problem not only for researchers, but for any teacher. A little more is known today about the errors done by students than what Freudenthal quoted in his talk, but many more difficulties are taken into account today.

We can note also that when he evoked John and Mary, he asked, 'Does it sound sexist?'. This seemed like a strange remark at that time, but we are now very concerned about the sexist aspect. Much research is being done in this area

Sierpinska, A. and Kilpatrick, J. Mathematics Education as a Research Domain: A Search for Identity, 47–56.
© *1998 Kluwer Academic Publishers. Printed in Great Britain.*

and even, in 1993, an ICMI Study was devoted to 'Gender and Mathematics Education'.

The second problem was: *How do people learn?* Not only is this always a question for mathematics educators, but this problem has now spread to many other researchers in diverse domains of the cognitive sciences. H. Freudenthal insisted on the fact that to work on this problem the greatest difficulty was 'learning to observe learning processes'. Presently more and more researchers are concerned with this question, as evidenced, for example, by the existence of the international research group called 'Classroom Research'.

The third problem was: *How to use progressive schematization and formalization in teaching any given mathematical subject?* This is precisely the question underlying most research done today under the heading of 'didactic engineering'.

The fourth problem referred to 'insight': *How to keep open the sources of insight during the training process; how to stimulate the retention of insight, in particular, in the process of schematizing?* We are all aware of the many recent books and papers on insight outlining the numerous problems entailed.

The fifth problem was: *How to stimulate reflection on one's own physical, mental and mathematical activities?* It looked new at that time but today many researchers refer to metalevels, and the theme of *metacognition* has produced very important works – though the problem has not yet been answered.

The sixth problem was: *How to develop a mathematical attitude in students?* The existence of the working group, 'Improving students' attitudes and motivation', coordinated by Gilah Leder at the Seventh International Congress on Mathematical Education in Québec in 1992, revealed the vitality of the theme. Through most research papers published in the last few years we can see that the interest of mathematics education is now more on problem solving than on mathematical knowledge and it may even be noted that 'problem posing' is now a category of the subject index in the mathematics education journal of reviews and abstracts, *Zentralblatt für Didaktik der Mathematik*.

The seventh problem: *How is mathematical learning structured according to levels and can this structure be used in attempts at differentiation?*, was quite a political one, leading, in many countries, to diverse kinds of experiments related to differentiation, grouping or streaming (homogeneous or heterogeneous, etc.). The results were sometimes contradictory, and no unique solution has yet been found.

The eighth problem was formulated as: *How to create suitable contexts in order to teach mathematizing?* We all know how vast is the research on that problem, in particular in The Netherlands, and this problem is probably the one in which the Freudenthal Institute is mostly engaged (see Gravemeijer, this volume). In general, the interest to produce problem-situations is great but didactic errors are frequent and the so-called 'concrete situations' are often artificially built *a-posteriori* as 'dressings' of mathematical themes, instead of being *a-priori* real-life situations which can be mathematized. This leads to experiments linking the classroom and the environment of students' lives. In addition, we see now the role of contexts (and specially the classroom context) in the understanding of mathematics and many of us are engaged in analyzing the distortions of mathematics in this classroom context.

From today's perspective, I think that the ninth problem: *Can one teach geometry by having the learner reflect on his spatial intuitions?*, can be considered as a special case of the eighth problem above.

Hans Freudenthal was a little reticent about his tenth problem: 'I am obliged to say something about calculators and computers. You would protest if I did not.' Now, many groups of researchers around the world are involved in the question he asked: *How can calculators and computers be used to arouse and increase mathematical understanding?* (see the ICMI Study on this theme, Churchman et al. 1986), and the domain has had such a rapid development that sometimes it looks as if the objectives of the mathematics educators are inverted and the target of study is more the computing tool than the human mathematical understanding.

The eleventh major problem was: *How to design educational development as a strategy for change?* Hans Freudenthal said: '*Curriculum* development viewed as a strategy for change is a wrong perspective. My own view, now shared by many people, is *educational* development.' I am afraid that in many countries this 'wrong perspective' is still the standard one, and it is clear to everybody that there is little change in this domain.

The twelfth problem focused on *textbooks and teacher training*. These two aspects are still present in research, especially in 'teacher training'. There have been institutional changes and experiments in many countries and some failures have been analyzed but no entirely satisfactory solution has yet been found.

The thirteenth problem, the last of the list, was presented as: *Educational research itself as a major problem of mathematics education*. As regards this problem, all of us are so convinced of its present importance that we are meeting in the current ICMI Study to work on it.

2. THE METHODS

The problems have developed and given birth to new ones, but have the approaches to them really changed? For unsolved mathematical problems, the research on how to solve them has at least led to changes in methods. The same has occurred in mathematics education but here it is even more evident because, in fact, at the beginning of mathematics education there was very little reflection on its methods. They were simply transplanted from other disciplines – mainly from psychology.

To bring the changes to light, I find it useful to refer again to Hans Freudenthal, in particular to his book published in 1978: *Weeding and Sowing –*

Preface to a Science of Mathematical Education. This book contains a very sharp criticism of research in mathematics education at that time. Can we say that *weeding* is now over? Of course not. Are the errors of the youth gone? Most of them have resisted to change and, for example, the so-called 'arm chair pedagogy', which Freudenthal denounced, still exists. But it is less practiced today. So many times in the past, for instance, we could see the deplorable results of the creation of curricula in the offices of ministries or of universities as they were implemented in real classrooms. We have learnt from this experience (cf. 'New Math' reforms!).

One of the most frequent methods, adopted from the psychologists, was to ask the students questions (e.g. multiple choice questions). H. Freudenthal insisted on showing the naïveté of the conclusions drawn from the artifacts produced in the behaviors of students subjected to such tests. We all remember the bias he brought to light in some of Piaget's experiments or in the IEA Study. Again for many years we could find, for instance, in *Educational Studies in Mathematics*, reports of statistical research based only on counting the answers to specific questions received from some specific populations in some specific situations. Of course, it must be said that some evolution at least has taken place: we cannot find, as we did earlier, publications about answers without the text of the questions being asked; the representativity of the population tested is now less often contestable and the best improvement is that the limits of the study are now often made explicit. In fact, the ways in which to ask questions, the role of the context, the linguistic aspects..., all these specific difficulties of the method have given rise to new fields of research. We know now that the *interpretation* of answers to questions is not evident at all. On the contrary, there lies all its complexity. Thus there are now papers comparing ways of asking questions, ways of evaluating answers, comparisons of populations, gender, age, etc.

As the taking into account of the role of the context of questioning increased, researchers were led to examine the object 'mathematics classroom', first as external observers (with an aspiration to objectivity), but, later, more and more as ethnographers. This evolution came about through the increase of papers describing new methods now often influenced by sociologists more than, as earlier, by psychologists. The importance is given to *qualitative* aspects more than to quantitative ones: What precisely is said and in what conditions it is said instead of how many times it is said. Thus, for instance, in the journal *Educational Studies of Mathematics* one finds less tables of numbers and more transcripts of verbal exchange in classrooms and copies of rough drafts of the students' work. It can be noted that, even back then, H. Freudenthal wished to see more classroom chronicles (or daily journals). New methods, however, had to fight for their existence against the tradition of other disciplines previously called 'sciences': The belief that a research has to be quantitative in order for it to be taken seriously was very strong.

3. SCIENTIFICITY AND UNIVERSALITY

The question *Is mathematics education a science?* has not yet been completely answered, and the present ICMI Study on what is research in mathematics education could perhaps consider, as a title, 'Preface to a Science of Mathematical Education', as in Freudenthal's book of 1978. Actually the boundaries between science and scientism are not obvious in all human sciences and I think that, for still a long time, mathematics education will have to be careful to avoid the *temptations of scientism.*

At least we have some criteria for what are 'scientific domains'. The most easy to use is the criterion of *universality.* Many local actions or innovations may be interesting, politically important and so on, but have no universal value; they are providing a momentum for the local educational system without contributing to the growth of universal knowledge. It is an error to call them 'science'. But to be local, selective or limited in scope, is not always contrary to universality. For instance, if we consider the experiments of P. U. Treisman (1985) with minority students at Berkeley, it is clear that, although local, they produced meaning for every situation of the same type around the world and are of universal interest.

Some years ago we used to hear the expressions, 'English research', 'French research', 'German research', 'Italian research', etc. I think these expressions are beginning to be less fashionable. It is clear that if some applied research can perhaps be national (practically 'applied' to a national school system, for example), scientific research can have no frontiers and this is why, as in any science, most of the scientific journals are international.

Of course, some historical reasons may produce a specific orientation of research in one country during one period (as can happen in mathematics research) but, to construct a science, the ways of thinking of researchers have to be understood by any member of the international comunity. This is actually the reason for the importance of international commissions or groups such as ICMI (International Commission on Mathematical Instruction), CIEAEM (Commission Internationale pour l'Étude et l'Amélioration de l'Enseignement Mathématique), PME (The International Group for the Psychology of Mathematics Education), etc., and international journals, to prevent encystment due to isolation. But we cannot ignore that we are then confronted with the serious linguistic problem of translation. This problem exists for any scientific domain: for instance, a French chemist has to read many papers in English (and perhaps in German), but he is prepared to do it when he first decides to study chemistry. On the other hand, in France for example, people are often first engaged in learning and then teaching mathematics in French only. Later, they decide to engage in mathematics education research, with the implicit condition that they will stay inside the frontiers of French territory and the French language. We still find this situation existing among doctoral students today but I think that it will necessarily have to change in the near future.

4. THE SPECIFICITY

At the time when *L'Enseignement mathématique* was the only journal on mathematics education, its papers were quite *mathematical*; not generally offering new
results but rather new presentations, a synthesis of different notions or analyses
of proofs, etc. Mathematicians engaged in mathematics education were still
doing mathematics. Later on, psychology, its interests, its problems and its
methods, influenced mathematics education. We began to understand that if it is
a science it is only as a 'human science'. But, as mathematicians, it was difficult
for us to understand that a human science cannot be, as is mathematics, a 'hard
science'. At its best, it can only be considered a 'soft science' (see the section on
'Methods' above).

But now, some people (coming often from social sciences) are considering
mathematics education as 'soft' as any other 'science of education'. But this is
also an error because mathematics education has a specificity. It refers to human
thinking related to 'hard' objects: the mathematical objects.

This explains why there are so many misunderstandings and why neither researchers in mathematics nor researchers in general education can clearly understand what is considered as research in mathematics education.

5. THE RELATION WITH OTHER DISCIPLINES

In his definition of 'didactique des disciplines' for the *Encyclopaedia
Universalis*, in 1970, D. Lacombe described 'didactics of a discipline' as using,
on the one hand, one 'object-discipline' (*discipline-objet*), which, for 'didactics
of mathematics' is the discipline of mathematics and, on the other hand, many
'tool-disciplines' (*disciplines-outils*). The general situation is, of course, the
same as it was in 1970 but I think it is interesting to note some of the differences. Firstly, each of the tool-disciplines used at that time made progress, and,
secondly, we are also using new tool-disciplines. In 1970, the tools quoted were
some parts of mathematics itself (such as statistics and logic), computer science
(for statistics and also for the Computer Aided Instruction), psychology (cognitive and relational), sociology (namely, at that time, the place of school and students in society), and linguistics (for instance, the study of mathematical and of
school discourses). Today all these disciplines have evolved, especially in inter-
faces that are of particular usefulness to mathematics educators, like psycho-
sociology and socio-linguistics. Linguistics has developed the pragmatics and
semiotics which help us, better than did earlier semantics, to understand the
problems of communication in the classroom and especially the role of context
in interpretation. It is now well-known that 'The meaning is not the message',
and therefore we are now led to use other aspects of sociology, namely micro-
sociology and even ethnography, to analyze the conditions of mathematical
learning. For computer sciences, not only has CAI developed but Artificial

Intelligence is a new tool-discipline for our research. In the papers from the 70s, we often find references to the 'blackbox': What was done by the human mind (especially in the learning of mathematics) had to be approached by relations between *inputs* and *outputs* (performances in answers to questions, for instance). Today we are just beginning to find the possibilities of learning from a new discipline, neurobiology, which, hopefully, will enable us to 'look inside the brain' of the learners and the problem-solvers and to know, perhaps, how neurons receive and interpret mathematical messages, how solutions of problems are found, etc.

Another new aspect can be noted concerning the relations between mathematics education and other disciplines. Mathematics educators not only use these disciplines in their research, but followers of these disciplines use the results of mathematics educators in theirs: for instance, psychologists look at the concepts of problem-solving, cognition and metacognition, and sociologists study the role of the teaching of mathematics in societal hierarchies.

6. CONCLUSION

As this evolution of mathematics education is presented here from a subjective and personal point of view, it will be clearer if, as a conclusion, I briefly relate the story of my personal itinerary in this field.

The question I met early in life, even as a student in high school, was: 'Why are there people who say that they understand nothing in mathematics?' Later on, I decided to teach mathematics and to do research in order to try to answer the questions, 'How do we understand mathematics?' and 'What is mathematics understanding?' I began to approach these questions with mathematical logic, then with mathematics education related to logico-linguistic aspects. Then I had to study the influence of the school classroom context on understanding in relation to the context of the whole society in which it is embedded. The understanding of mathematics – but, above all, the misunderstanding which was easier to study at first – was taking place in 'blackboxes' receiving and producing messages in a specific environment. This could, at best, lead us to know how to improve mathematics understanding, but not at all to know how it works in the human brain, how strategies of problem-solving are created, etc. I am not so naïve as to imagine that this fundamental question will be answered during my lifetime, but I am happy to see a great evolution and progress between what I heard, read or wrote in the 60s and what is presented now, especially in interdisciplinary studies. I am convinced that the development of 'cognitive sciences', with exchanges between linguists, psychologists, computer scientists, neurobiologists, logicians and other mathematicians, will be very fruitful in the future and if that cooperation succeeds in producing a specific science called, with the singular, 'Cognitive Science', then I am sure that the best future to wish for mathematics education is that it becomes a part of this Cognitive Science.

REFERENCES

Churchman, R. F. et al. (eds.): 1986, *The Influence of Computers and Informatics on Mathematics and its Teaching*, ICMI Study series, Cambridge University Press, Cambridge.
Freudenthal, H.: 1978, *Weeding and Sowing – Preface to a Science of Mathematical Education*, Reidel, Dordrecht, Holland.
Freudenthal, H.: 1983, 'Major Problems of Mathematics Education', in *Proceedings of the Fourth ICME-Berkeley*, Birkhäuser Inc., Boston, MA, 1–7.
Lacombe, D.: 1970, 'Didactique des Disciplines', in *Encyclopaedia Universalis*, Paris, 113–116.
Treisman, P. U.: 1985, A Study of the Mathematical Performance of Black Students at the University of California at Berkeley. Doctoral dissertation, University of California, Berkeley.

Josette Adda
Tour Béryl
40 Avenue d'Italie,
75013 Paris,
France

NORMA C. PRESMEG

BALANCING COMPLEX HUMAN WORLDS: MATHEMATICS EDUCATION AS AN EMERGENT DISCIPLINE IN ITS OWN RIGHT

1. INTRODUCTION

I taught a course on informal geometry – a content course – to students at The Florida State University who are prospective high school mathematics teachers. In the first week I asked them to bring or wear to the next class, something which had geometry in it, and to come to class prepared to tell why they had chosen that particular item and to talk about its geometry.

In an interview, one of the students, Dena (who hopes to teach algebra rather than geometry), told me about her reactions to this task, as follows.

Dena: I noticed when you said, for us to bring something to class or wear something that had geometry in it, for a little while I was having a difficult time, because, everything I picked up had geometry in it. And, I said, maybe there's something I misunderstood about the directions. Y'know.
Interviewer: In fact, even just the shape of a piece of clothing, any clothing.
Dena: Yeah. Anything, has geometry in it. So, for a little while I was confused. I didn't know what to bring to class, until, until I realized that, everything is going to have.... I said to myself, everything, of course everything is going to have geometry to it because, y'know, anytime.... You're going to make a desk. I mean, you draw, y'know. Your plans, for making the desk, involves geometry. And everything, that is, just everywhere. I think that geometry is taught as something abstract, sketching things with proofs and rules and, not as very, everyday.

Dena's recollections of her high school geometry experiences were negative ones. 'I didn't like it at all!', she concluded.

Implicit in this episode are several points which are relevant to the emergence of mathematics education as a discipline in its own right, separate from but not unrelated to other disciplines such as mathematics, psychology, sociology, philosophy, linguistics, history, and (relatively recently) anthropology.

Firstly, the disciplines of mathematics and mathematics education differ substantially because their subject matters are different. Mathematics education is concerned with the complex 'inner' and 'outer' worlds of human beings (Bruner 1986) as they engage in activities associated with learning of mathematics. Dena's agonizing over the nature and boundaries of geometry is fruitful and

57

Sierpinska, A. and Kilpatrick, J. Mathematics Education as a Research Domain: A Search for Identity, 57–70.
© *1998 Kluwer Academic Publishers. Printed in Great Britain.*

provocative subject matter to a mathematics education researcher interested in the teaching and learning of geometry. The avenues along which this research may lead depend not only on the data, but also on the interests and interpretations of the researcher. The tendency of such hermeneutic research to use progressive focusing rather than pre-ordinate design (Hartnett 1982) makes this kind of research as interesting as a mystery story, even if the mystery is to some extent self-created. In this respect, mathematics education research may have elements in common with mathematics research.

A second point is that the inner and outer worlds of a student – while in this context specific to learning geometry – relate to the disciplines of psychology and sociology respectively, and to the interactions between their elements, as they concern an individual such as Dena. A balance between elements of these two disciplines is required in mathematics education, as witnessed in recent debates on the necessity of steering a course between Piaget and Vygotsky, representing individual and social aspects of learning respectively, in constructing theory in mathematics education (Confrey 1991, 1994, 1995a, 1995b; Ontiveros 1991). It is significant that Confrey believes that neither Piaget's nor Vygotsky's theory alone is adequate to model the complex processes of human learning. She elaborated as follows:

What I argue is that proposing an interaction between the two strands will constitute a significant change in both theories, and will require a theory which is neither Piagetian nor Vygotskian, but draws heavily on both. I argue this due to Vygotsky's rejection of the possibility of the development of many of the basic processes of Piaget, such as the development and awareness of schemes, of operations and of reflective abstraction, until social interaction is established. An alternative theory will have to propose a much stronger and more detailed description of how the 'natural' and the 'socio-cultural' activities of the child are linked, allowing for the complexity of each and probably requiring a renaming of the natural strand to reflect a more constructivist view (Confrey 1992, pp. 5–6).

In this paragraph, Confrey gave expression to the need which is strong in the emerging discipline of mathematics education, for its own theories and models which take into account, but integrate and extend theories from, other disciplines such as sociology and psychology.

A third point implicit in Dena's pondering is that philosophy is ubiquitous in questions which are of concern to mathematics education researchers. The nature of geometry is an ontological issue, while how it was taught in Dena's school experience relates to issues of epistemology. Both components are essential in mathematics education theory-building, since one's beliefs about the nature of mathematics and mathematical knowledge are the 'spectacles' through which one looks at its teaching and learning.

Tension between the view that 'Everything is mathematics' (as Dena expressed it, 'Everything is going to have geometry to it'), and the rigorous mathematical position that 'Only formal mathematics is valid', was well expressed by Millroy (1992) in her monograph on the mathematical ideas of a group of carpenters, as follows:

... it became clear to me that in order to proceed with the exploration of the mathematics of an unfamiliar culture, I would have to navigate a passage between two dangerous areas. The foundering point on the left represents the overwhelming notion that 'everything is mathematics' (like being swept away by a tidal wave!) while the foundering point on the right represents the constricting notion that 'formal academic mathematics is the only valid representation of people's mathematical ideas' (like being stranded on a desert island!). Part of the way in which to ensure a safe passage seemed to be to openly acknowledge that when I examined the mathematizing engaged in by the carpenters there would be examples of mathematical ideas and practices that I would recognize and that I would be able to describe in terms of the vocabulary of conventional Western mathematics. However, it was likely that there would also be mathematics that I could not recognize and for which I would have no familiar descriptive words (Millroy 1992, pp. 11–13).

On the basis of her research results, Millroy argued strongly for the broadening of traditional ideas of what constitutes mathematics. She wrote, 'We need to bring nonconventional mathematics into classrooms, to value and to build on the mathematical ideas that students already have through their experiences in their homes and in their communities' (ibid. p. 192). Steen's (1990) and the National Research Council's (1989) view of mathematics as the science of pattern and order opens the door to this lifting of the limiting boundaries of mathematics. Millroy's recommendation is consonant with those in the National Council of Teachers of Mathematics Curriculum and Evaluation Standards for School Mathematics (1989). A related point is that a 'mathematical cast of mind' may be a characteristic of students who are gifted in mathematics (Krutetskii 1976). This mathematical cast of mind enables these students to identify and reason about mathematical elements in all their experiences; they construct their worlds with mathematical eyes, as it were. But unless teachers are aware of the necessity of encouraging students to recognize mathematics in diverse areas of their experience, only a few students will develop this mathematical cast of mind on their own. Many more will continue to regard mathematics as 'a bunch of formulas' to be committed to short term memory for a specific purpose such as an examination, and thereafter forgotten (Presmeg 1993).

The foregoing sets the scene for a fourth point which emerges from these considerations, namely, the links which mathematics education research is building with various branches of anthropology, particularly with regard to methodology and construction of theory. Millroy's (1992) study was ethnographic. Entering to some extent into the worlds of Cape Town carpenters in order to experience their 'mathematizing' required that Millroy become an apprentice carpenter for what she called an extended period, although the four-and-a-quarter months of this experience might still seem scant to an anthropologist (Eisenhart 1988). But the point is that the ethnographic methodology of anthropological research is peculiarly facilitative of the kinds of interpreted knowledge which are valuable to mathematics education researchers and practitioners. After all, each mathematics classroom may be considered to have its own culture (Nickson 1992). In order to understand the learning, or, sadly, the prevention of learning which may take place there, the ethnographic mathematics education researcher needs to be part of this world, interpreting its events for an extended period, as in the research described by Schifter and Fosnot (1993). Hermeneutic methodology and construction of theory are both revisited in what follows.

2. AIMS OF MATHEMATICS EDUCATION RESEARCH

In addition to the scientific goal of theory-building which is an important aspect of research in any discipline, mathematics education research has had and will continue to have a pragmatic goal, namely, the improvement of teaching and learning of mathematics at all levels. This pragmatic goal is implicit in the meaning of the word 'hermeneutic'. There is a creative tension between these two aims, but like the complex human worlds which are the subject of mathematics education research, they are indivisible except as a device for analysis. Each aim is sterile without the other. For example, as in the high school geometry experience which Dena started to describe in the opening transcript, it is not uncommon for high school mathematics teachers in the USA and elsewhere to teach as though geometry means the axiomatic deductive system of Euclid. Typically, students plunge into such a course without the well-meaning teacher ascertaining that students have had experiences with 'everyday' informal geometry which gradually prepare their thinking for the deductive level of geometric thought needed for Euclidean geometry and later for the rigorous abstract level which the mathematician takes for granted (van Hiele 1986). What is tragic about this state of affairs is that a teacher may thereby prevent students from growing to higher levels in their geometrical understanding. Once a student starts memorizing definitions and theorems for tests without striving to construct personal meaning for the mathematics, the impetus for mathematical growth is taken away.

The idea of five levels of geometrical thought (which may be characterized as recognition, analysis, ordering, deduction and rigor) grew out of the research in the 1950s of two high school mathematics teachers in The Netherlands, Pierre van Hiele and his late wife Dina van Hiele Geldof. Their twin doctoral dissertations reflected the two complementary aims of constructing theory and of improving practice respectively. The research problem, namely the difficulty they perceived their students to be experiencing in learning high school Euclidean geometry, was identified in their own classrooms.

In mathematics education research as in that of other disciplines, the foundational ideas of a constructed theory do not remain where they were, but they are modified and explicated – reinterpreted – by other researchers. In the case of the theory of the van Hieles, once the importance of their work was perceived by English-speaking mathematics education researchers in the 1970s, following descriptions by Freudenthal (1973) and Wirszup (1976), groups of mathematics educators in several countries took up research in this area. Following seminal work by several researchers including Burger and Shaughnessy (1986) in Oregon, and de Villiers (1987) in South Africa, interpretations of how the levels of geometric thought interweave in the complexities of learning geometry, are ongoing. An example is the 'degrees of acquisition' theory proposed by Jaime and Gutierrez (1990) as a result of their research in Spain, which they contrasted further with interpretations of US researchers (Gutierrez, Jaime, Shaughnessy & Burger 1991; Gutierrez, Jaime & Fortuny 1991). Further reinterpretations and connections via metaphors with other theories were taken up in Portugal (Matos 1991). Although the theories of geometric levels of thinking have received substantial attention in mathematics education research, not much attention has yet been paid to the phases between the levels, which the van Hieles suggested are necessary for students to grow from each level to the next one. For their practical classroom implications, the phases may turn out to be just as important as the levels (Presmeg 1990). In this way, the scientific and pragmatic aims of mathematics education research are illustrated to be obverse faces of the same coin.

There is much mathematics education research, however, which is conducted in classrooms but with individual students or groups of students, which also has this pragmatic goal of improving teaching practice for the purpose of fostering students' increased understanding and enjoyment of mathematics, with consequent 'better performance'. The issue of more meaningful assessment of student performance (de Lange 1993; Ginsburg et al. 1993) is another story which will not be pursued here.

The work of the powerful team of Cobb, Yackel & Wood (1991, 1992) represents the unity which exists not only between the twin aims of theory-building and of improvement of mathematical pedagogy, but also between research conducted in classrooms and that conducted in interviews with individual students or pairs of students. They are engaged in a research and development project 'that addresses the problem of developing a coherent framework within which to talk about both teaching and learning' (1991, p. 83). Basic philosophical issues are addressed as they construct a framework 'that makes it possible for us to cope with the complexity of classroom life' (ibid.). Their work is discussed further in a later section of this chapter. Mathematics education research which does not aim to take into account the complexity of the worlds of teachers and learners of mathematics is of limited value in those worlds. It follows that all disciplines which relate to human beings and their activities will have relevance in this endeavor.

3. METHODOLOGY

Mathematical understanding may be characterized as 'interconnected knowledge' (Fennema et al. 1991, p. 5). As mathematics education matures as a discipline, an important component of its research, too, is the pursuit of connections. The present qualitative research paradigm is more facilitative of the connections helpful to teachers, between the scientific and pragmatic aims, than was the extensive quantitative paradigm which preceded it (ibid.). The construction of theory is grounded in mathematics education research with teachers and students, and the complexities of classroom life are usually not ignored in this theory-building (Bauersfeld 1991; Eisenhart et al. 1993). In the USA and elsewhere, there are groups of mathematics education researchers who conduct their research programs with these complexities in mind and draw creative energy from the interconnections between disciplines. Maher, Davis and Alston (1993) are another team who acknowledge the complexity of elements involved in the learning of mathematics. They wrote, 'In all of this, it has literally amazed us to see how much more complexity is involved in "simple" mathematical thinking than we had previously imagined' (p. 210). They characterized their research as studies of '*situated* concept development' (Davis & Maher 1993, pp. 54–55; their italics), thereby emphasizing the interconnectedness of inner and outer worlds of students in such learning. They expressed the interconnectedness of elements of their research as follows:

When one goes into a school to videotape students doing mathematics, and then analyzes these tapes in careful detail, one is simultaneously engaged in research, data collection, data analysis, theory building, curriculum design (since you will surely try to find ways to avoid or remediate any observed weaknesses and to take advantage of any observed possibilities), pedagogical innovation (for the same reason), and teacher education. No one of these can be (or should be) separated off from the others. The basic questions you are addressing are: How do students think about mathematics and how can we make this process work better? All the components listed are necessarily a part of this process. Indeed, there are really no definable boundaries between the various parts (Davis & Maher 1993, p. 29).

Again, the balance of unity and complexity of mathematics education research is expressed, in parallel with the complex human worlds which are its subject matter.

Although qualitative researchers in mathematics education are aware of their own subjectivity in this interpretive endeavor (Eisner & Peshkin 1990), it is by no means the case that rigor is totally compromised, that we make up fairy stories – even if we sometimes use the narrative mode, which is valuable on occasion in reporting this research (Bruner 1990). In my own research on the roles of visual imagery in the teaching and learning of high school mathematics (Presmeg 1985, 1992b), I have tried to maintain rigor in several ways. During the period of data collection, triangulation of viewpoints of teachers, students, and researcher, at several stages of the research, ensured the tempering of my interpretations. In the reporting of the research I used teachers' and students' own words wherever possible, to enable readers to make their own interpretations and compare these with mine. I agree that 'It is necessary for researchers to confront their own subjectivity and to be explicit regarding the role of subjectivity in their interpretations' (Research Advisory Committee 1992). But having said this I have to admit that I experience annoyance when a researcher is so explicit about his or her own subjective interpretations that the reporting does not enable me as a reader to get close to the people in the research, be they students, teachers, or grocery shoppers (Lave 1988). Again, a balance is required, in this case between reporting of interpretations of the phenomena investigated, and of evidence for these interpretations, for instance in actual transcript data.

I have been inspired by the viewpoint of Heshusius (1994) who sees qualitative research as, in a sense, moving beyond issues of subjectivity. The researcher, in an empathic encounter, becomes the child or adult whose sense-making is the focus of the research. In the reporting of sensitive interpretations of such experience, the person interviewed or observed is not lost but found through the 'participatory mode of consciousness' of the researcher. This phenomenon is reminiscent of T. S. Eliot's description of

... music heard so deeply
That it is not heard at all, but you are the music
While the music lasts
(T. S. Eliot 1971, p. 44).

This, to me, is the pinnacle of interpretive research. Research in this vein is an art rather than a science, and its artistry (to which I aspire without coming close) is not attained overnight. Even on a more mundane level, according to Ginsburg et al. (1993, p. 28), 'Piaget claimed that it required a year's daily practice to acquire skill in clinical interviewing'.

4. CONSTRUCTING THEORY

The dilemma of constructing theory is that any theory, by its nature, is a simplified model which facilitates understanding of some elements of a phenomenon but excludes others. Thus no theory can be complete or final, and any theory which is put forward as final and infallible approaches what Habermas (1978) called ideology. In the process of constructing theory, it is easy to lose the complexity of a human phenomenon. This dilemma is particularly acute in mathematics education research by virtue of the nature of the endeavor, which, as I have argued in this chapter, is an attempt to understand the balancing of the complex human worlds involved in mathematics education, whether or not the students believe this process to be one aimed at constructing mathematical meaning.

Therefore I regard as particularly valuable the development of theory which takes into account the dilemmas, tensions, and contradictions of mathematical classrooms – in a phrase, 'the paradox of teaching' (Cobb, Yackel & Wood 1991, pp. 84–85). This paradox arises from the 'tension between encouraging students to build on their current understandings on the one hand and initiating students into mathematical culture of the wider community on the other' (p. 84). Cobb et al. called this initiation process 'acculturation into mathematical culture'. But since this initiation of students is not into the mathematics of a foreign society but in most cases into the mathematics of their own, I believe with Bishop (1988) that this process should rather be called 'enculturation'. In fact, in later papers Cobb (1994, 1995) does refer to this process as enculturation. Wendy

Millroy's (1992) apprenticeship into the mathematics of Cape Town carpenters was in some ways an acculturation. In order to enter into the taken-as-shared beliefs which characterized the mathematical culture of the carpenters, she had to suspend, to some extent, her enculturated beliefs about the nature of mathematics. The carpenters did not believe they were doing 'mathematics' at all!

Mathematics education research traditions are still evolving (Research Advisory Committee 1992). In line with the need to broaden the ontological views of what mathematics is and of what it means to 'do mathematics', mathematics education research is concerned with both the enculturation and the acculturation processes. With respect to the former, that is, enculturation, some radical constructivists have argued that the teacher's roles as mathematical enculturator and as facilitator of students' construction of individual mathematical meaning, are essentially contradictory. But some leaders of the radical constructivist movement have moved away from this position (if they ever espoused it) towards the view that a contradiction is not necessarily the case here. Von Glasersfeld (1993) admitted the possibility of two essentially different kinds of mathematical knowledge, one of which – knowledge of conventions – can be constructed only in processes such as enculturation.

This movement towards an epistemology which takes into account both enculturation and the personal construction of meaning, is less radical and more acceptable to me. It is significant that some mathematics education scholars have in recent years developed perspectives which integrate and move beyond these two 'opposing' views; Confrey (1995a, 1995b) in her dialectic process, and Cobb and Yackel (1995) in their 'emergent' perspective, have given examples of such syntheses. I see the value of radical constructivism as lying in its role as an antidote to the previously dominant paradigm – unfortunately still prevalent in most American high schools – in which all mathematical knowledge was believed to be transmitted by enculturation, largely ignoring the personal construction of meaning which is at the heart of mathematical activity. This construction of meaning is often idiosyncratic, often involves imagery of various kinds (Presmeg 1985, 1992a, 1992b); but these processes, complex as they are, are only part of the story. I see no essential contradiction between the roles of enculturator and facilitator of students' individual constructions. As several researchers have documented, students can come to see the necessity for adopting generally accepted mathematical conventions, without compromising the intellectual excitement generated as they construct and experiment with their own, in the consensual domain of their own groups or classrooms (Lampert 1991; Cobb, Yackel & Wood 1992; Schifter & Fosnot 1993; Cobb 1994, 1995; Cobb & Yackel 1995).

The foregoing illustrates how mathematics education theories, by their nature as simplified models, can generate apparent contradictions and incompatibilities where none necessarily exist in the complex collective 'webs of significance' (Geertz 1973, p. 5) of mathematical cultures. This is not to say that theory-building is not needed; theory provides the lens through which we examine these issues.

The point is that we cannot afford to lose sight of the complexity of the human worlds involved in mathematics education, in our construction of theory. The theory-building of Guy Brousseau in France, often used in analysis of classroom complexities by researchers such as Arsac, Balacheff & Mante (1991), demonstrates a fine-tuned awareness of how complex classroom processes of mathematical teaching and learning are. In a similar vein, Gregg (1995) wrote about the tensions and contradictions of the school mathematics tradition, as evidenced in the practices of a novice teacher whose coping strategies exacerbated the situations she was attempting to cope with. When Brousseau and Otte (1991) wrote about 'the fragility of knowledge', they epitomized the paradoxes, contradictions, and complexities of teaching and learning mathematics in school situations.

With regard to the second process, acculturation in mathematics education, there is a relatively neglected area in mathematics education research, especially in its theory-building aspect. The ethnomathematics movement (D'Ambrosio 1985, 1987) is groping its way in this direction. The work of mathematics educators such as Gerdes (1986, 1988b), and of anthropologists such as Saxe (1991), is of value in sensitizing the mathematics education research community to the evidence of non-institutional mathematical thinking in the activities of people who, if asked, might not attach mathematical connotations to their activities at all – like Millroy's (1992) carpenters. But unlike Millroy's and Saxe's studies, which characterize acculturation into the mathematical worlds of Cape Town carpenters and Brazilian candy sellers respectively, in some ethnomathematical projects there is a tendency to impose the mathematics of the researcher on the practices of those studied – and this is not acculturation at all. In such a project, Gerdes (1988a) studied the sand drawings of the Tchokwe in Angola not for the purpose of entering into their mathematical thinking, but to find illustrations of his own. In any case, a lacuna still exists in the construction of theory about how insights from these important anthropological elements of mathematics education might enhance classroom practice, for instance in the mathematics education of minority students (Secada 1990, 1991). Both the scientific (theory-building) and pragmatic (classroom practice) aspects are required.

In the cases of Millroy's and Saxe's anthropological mathematics education research, the acculturation aspect was present because these researchers attempted to 'enter into' the unfamiliar mathematical cultures of carpenters and candy sellers respectively. But acculturation on the part of the researcher is also taking place when the adult tries to understand in detail the mathematical meaning-making of a child, even if both belong to the same macroculture. When Davis (1993) reports detailed mathematical thinking of Brian (Grade 7) and Tony (Grade 5), he is describing acculturation, too – his own acculturation into the worlds of these children. It is in this sense that acculturation may take place in a teacher or researcher at the same time that a student experiences enculturation into the mathematics of his or her home culture. Piagetian research has made us aware that to an adult the ways of thinking of children at various stages of development may appear as exotic as a foreign culture.

In concluding this section on the construction of theory, I want to mention in passing, four further theory-building developments which I see as exciting and full of promise for further rich research ideas. A description of the first of these can be found in papers by Voigt (1995) and Yackel (1992, 1995). In these deep papers, the balancing of the complex sociological, psychological, anthropological and philosophical elements of mathematics education is centered on discussions of social interactionism, mathematical meaning, consensual domains, indexicality and reflexivity, and the nature of mathematical knowledge.

Still concerned with the nature of mathematical knowledge, but with more emphasis on psychological aspects of learning mathematics, is Sfard's (1991) paper which links the history of certain mathematical concepts to the way in which mathematical processes become reified as mental objects to be operated on, in turn, in mathematical processes at new levels of mathematical abstraction. Sfard's work in Israel resonates with the independent theory-building in Sweden of Bergsten (1990, 1992) who suggested a three-dimensional crystal model, incorporating processes and structures, for the learning of mathematics.

A third exciting theoretical development which bridges, *inter alia*, the disciplines of linguistics and philosophy, is the work on prototypes, metaphors, metonymies, integrated cognitive models, and imaginative rationality, by Johnson (1987) and Lakoff (1987). Although not developed for this purpose by its authors, this work has tremendous significance for understanding mathematical thinking (Presmeg 1992a, 1992b), and therefore for research and practice in mathematics education. It is not that the processes they describe are useful adjuncts in mathematical cognition: significance for mathematics education lies in their claim that these processes and models underlie human reasoning itself, which is a *sine qua non* of mathematics (Krutetskii 1976).

Finally, linking these last two developments (mathematical reification, and processes such as metaphor and metonymy), are recent papers which use semiotic models as frameworks for theory-building and analysis of research data (Whitson 1992; Presmeg 1997). Semiosis also casts light on another vital element of mathematical cognition, namely, the use of symbolism (Cobb et al. 1997).

5. CONCLUSION

The entire thrust of this chapter has been the complexity of the interrelationships among the elements which are grist to the mill of qualitative mathematics education research. This is an exciting time to be part of this endeavor, as theories which are specific to this discipline with its many facets are constructed, argued about, reconstructed and negotiated by mathematics education researchers such as Vergnaud (1990) in France, Dörfler (1991) in Austria, Dreyfus (1991) in Israel, Goldin (1992) in the USA, and others from diverse geographical and theoretical backgrounds. If the diversity calls for tolerance, it is also a potential source of creative energy which mirrors the complexity of the diverse concerns of our discipline.

REFERENCES

Arsac, G., Balacheff, N. & Mante, M.: 1991, 'Teacher's Role and Reproducibility of Didactical Situations', *Educational Studies in Mathematics* 23 (1), 5–29.

Bauersfeld, H.: 1991, 'Integrating Theories for Mathematics Education', plenary paper, *Proceedings of the 13th Annual Meeting of the North American Chapter of the International Group for the Psychology of Mathematics Education, Blacksburg, Virginia, October 16–29, 1991*, Vol. 1, 269–290.

Bergsten, C. R.: 1990, Matematisk Operativitet. Unpublished Ph.D. dissertation, University of Linköping.

Bergsten, C. R.: 1992, 'Schematic Structures of Mathematical Form', poster presented at the 16th Annual Meeting of the International Group for the Psychology of Mathematics Education, Durham, New Hampshire, August 6–11.

Bishop, A. J.: 1988, *Mathematical Enculturation: A Cultural Perspective on Mathematics Education*, Kluwer Academic Publishers, Dordrecht.

Brousseau, G. & Otte, M.: 1991, 'The Fragility of Knowledge', in A. J. Bishop, S. Mellin-Olsen, and J. van Dormolen (eds.), *Mathematical Knowledge: Its Growth through Teaching*, Kluwer Academic Publishers, Dordrecht, 13–36.

Bruner, J.: 1986, *Actual Minds, Possible Worlds*, Harvard University Press, Cambridge, MA.

Bruner, J.: 1990, *Acts of Meaning*, Harvard University Press, Cambridge, MA.

Burger, W. F. & Shaughnessy, J. M.: 1986, 'Characterizing the van Hiele Levels of Development in Geometry', *Journal for Research in Mathematics Education* 17 (1), 31–48.

Cobb, P.: 1994, 'Where is the Mind? Constructivist and Sociocultural Perspectives on Mathematical Development', *Educational Researcher* 23 (7), 13–20.

Cobb, P.: 1995, 'Continuing the Conversation: A Response to Smith', *Educational Researcher* 24 (7), 25–27.

Cobb, P., Gravemeijer, K., Yackel, E., McClain, K. & Whitenack, J.: 1997, 'Mathematizing and Symbolizing: The Emergence of Chains of Signification in One First-Grade Classroom', in D. Kirshner & A. J. Whitson (eds.), *Situated Cognition Theory: Social, Semiotic and Neurological Perspectives*, Lawrence Erlbaum Associates, Hillsdale, NJ.

Cobb, P., Yackel, E. & Wood, T.: 1991, 'Curriculum and Teacher Development: Psychological and Anthropological Perspectives', in E. Fennema, T. P. Carpenter & S. J. Lamon (eds.), *Integrating Research on Teaching and Learning Mathematics*, State University of New York Press, Albany, NY, 83–119.

Cobb, P., Yackel, E. & Wood, T.: 1992, 'A Constructivist Alternative to the Representational View of Mind in Mathematics Education', *Journal for Research in Mathematics Education* 23 (1), 2–33.

Cobb, P. & Yackel, E.: 1995, 'Constructivist, Emergent, and Sociocultural Perspectives in the Context of Developmental Research', plenary paper, *Proceedings of the 17th Annual Meeting of the North American Chapter of the International Group for the Psychology of Mathematics Education, October 21–24, 1995, Columbus, Ohio*, Vol. 1, 3–29.

Confrey, J.: 1991, 'Steering a Course between Piaget and Vygotsky', *Educational Researcher* 20 (8), 28–32.

Confrey, J.: 1992, 'How Compatible are Radical Constructivism, Social-Cultural Approaches and Social Constructivism?', in L. P. Steffe and J. Gale (eds.), *Constructivism in Education*, Lawrence Erlbaum Associates, Hillsdale, NJ, 185–226.

Confrey, J.: 1994, 'A Theory of Intellectual Development: Part 1', *For the Learning of Mathematics* 14 (3), 2–8.

Confrey, J.: 1995a, 'A Theory of Intellectual Development: Part 2', *For the Learning of Mathematics* 15 (1), 38–48.

Confrey, J.: 1995b, 'A Theory of Intellectual Development: Part 3', *For the Learning of Mathematics* 15 (2), 36–45.

D'Ambrosio, U.: 1985, 'Ethnomathematics and its Place in the History and Pedagogy of Mathematics', *For the Learning of Mathematics* 5 (1), 44–48.

D'Ambrosio, U.:1987, 'Ethnomathematics, What it Might Be', *International Study Group on Ethnomathematics Newsletter* 3 (1).

Davis, R. B.: 1993, 'Student Discovery and Teacher Intervention', paper presented at the Research Presession of the 71st Annual Meeting of the National Council of Teachers of Mathematics, Seattle, Washington, March 30.

Davis, R. B. & Maher, C. A. (eds.): 1993, *Schools, Mathematics, and the World of Reality*, Allyn and Bacon, Boston, MA.

De Lange, J.: 1993, 'Real Tasks and Real Assessment', in R. B. Davis & C. A. Maher (eds.), *Schools, Mathematics, and the World of Reality*, Allyn and Bacon, Boston, MA.

De Villiers, M.: 1987, 'Research Evidence on Hierarchical Thinking, Teaching Strategies and Van Hiele Theory; Some Critical Comments', Research Unit for Mathematics Education at the University of Stellenbosch, Stellenbosch, *Research Report No. 10*, November 1987.

Dörfler, W.: 1991, 'Meaning, Image Schemata and Protocols', plenary paper, *Proceedings of the 15th Annual Meeting of the International Group for the Psychology of Mathematics Education, Assisi, June 29–July 4, 1991*, Vol. 1, 17–32.

Dreyfus, T.: 1991, 'On the Status of Visual Reasoning in Mathematics and Mathematics Education', plenary paper, *Proceedings of the 15th Annual Meeting of the International Group for the Psychology of Mathematics Education, Assisi, June 29–July 4, 1991*, Vol. 1, 33–48.

Eisenhart, M. A.: 1988, 'The Ethnographic Research Tradition and Mathematics Education Research', *Journal for Research in Mathematics Education* 19 (2), 99–114.

Eisenhart, M., Borko, H., Underhill, R., Brown, C. & Jones, D.: 1993, 'Conceptual Knowledge Falls through the Cracks', *Journal for Research in Mathematics Education* 24 (1), 8–40.

Eisner, E. W. & Peshkin, A. (eds.): 1990, *Qualitative Enquiry in Education: The Continuing Debate*, Teachers College Press, Columbia University, New York.

Eliot, T. S.: 1971: *Four Quartets*, Harcourt Brace Jovanovich, New York.

Fennema, E., Carpenter, T. P. & Lamon S. J. (eds.): 1991, *Integrating Research on Teaching and Learning Mathematics*, State University of New York Press, Albany, NY.

Freudenthal, H.: 1973, *Mathematics as an Educational Task*, D. Reidel Publishing Co., Dordrecht.

Geertz, C.: 1973, *The Interpretation of Cultures*, Basic Books, New York.

Gerdes, P.: 1986, 'How to Recognize Hidden Geometrical Thinking: A Contribution to the Development of Anthropological Mathematics', *For the Learning of Mathematics* 6 (2), 10–17.

Gerdes, P.: 1988a, 'On Possible Uses of Traditional Angolan Sand Drawings in the Mathematics Classroom', *Educational Studies in Mathematics* 19 (1), 3–22.

Gerdes, P.: 1988b, 'On Culture, Geometrical Thinking and Mathematics Education', *Educational Studies in Mathematics* 19 (2), 137–162.

Ginsburg, H. P., Jacobs, S. F. & Lopez, L. S.: 1993, 'Assessing Mathematical Thinking and Learning Potential', in R. B. Davis & C. A. Maher (eds.), *Schools, Mathematics, and the World of Reality*, Allyn and Bacon, Boston, MA, 237–262.

Goldin, G. A.: 1992, 'On Developing a Unified Model for the Psychology of Mathematical Learning and Problem Solving', plenary paper, *Proceedings of the 16th Annual Meeting of the International Group for the Psychology of Mathematics Education, Durham, New Hampshire, August 6–11, 1992*, Vol. 3, 235–251.

Gregg, J.: 1995, 'The Tensions and Contradictions of the School Mathematics Tradition', *Journal for Research in Mathematics Education* 26 (5), 442–466.

Gutierrez, A., Jaime, A. & Fortuny, J. M.: 1991, 'An Alternative Paradigm to Evaluate the Acquisition of the van Hiele Levels', *Journal for Research in Mathematics Education* 22 (3), 231–251.

Gutierrez, A., Jaime, A., Shaughnessy, J. M. & Burger, W. F.: 1991, 'A Comparative Analysis of Two Ways of Assessing the van Hiele Levels of Thinking', *Proceedings of the 15th Annual Meeting of the International Group for the Psychology of Mathematics Education, Assisi, June 29–July 4, 1991*, Vol. 2, 109–116.

Habermas, J.: 1978, *Knowledge and Human Interests*, Heinemann, London.

Hartnett, A.: 1982, *The Social Sciences in Educational Studies: A Selective Guide to the Literature*, Heinemann, London.

Heshusius, L.: 1994, 'Freeing Ourselves from Objectivity: Managing Subjectivity of Turning towards a Participatory Mode of Consciousness?', *Educational Researcher* **23** (3), 15–22.

Jaime, A. & Gutierrez, A.: 1990, 'A Study of the Degree of Acquisition of the van Hiele levels in Secondary School Students', *Proceedings of the 14th Annual Meeting of the International Group for the Psychology of Mathematics Education, Oaxtepec, Mexico, July 15–20, 1990*, Vol. 2, 251–258.

Johnson, M.: 1987, *The Body in the Mind: The Bodily Basis of Meaning, Imagination and Reason*, University of Chicago Press, Chicago.

Krutetskii, V. A.: 1976, *The Psychology of Mathematical Abilities in Schoolchildren*, University of Chicago Press, Chicago.

Lakoff, G.: 1987, *Women, Fire, and Dangerous Things: What Categories Reveal about the Mind*, University of Chicago Press, Chicago.

Lampert, M.: 1991, 'Connecting Mathematical Teaching and Learning', in E. Fennema, T. P. Carpenter and S. J. Lamon (eds.), *Integrating Research on Teaching and Learning Mathematics*, State University of New York Press, Albany, NY, 121–152.

Lave, J.: 1988, *Cognition in Practice: Mind, Mathematics and Culture in Everyday Life*, Cambridge University Press, Cambridge.

Maher, C. A., Davis, R. B. & Alston, A.: 1993, 'Brian's Representation and Development of Mathematical Knowledge: The First Two Years', in R. B. Davis & C. A. Maher (eds.), *Schools, Mathematics, and the World of Reality*, Allyn and Bacon, Boston, 173–211.

Matos, J. M.: 1991, Changes in the van Hiele Theory to Accommodate Prototype Effects Originating from Image-Schematic, Metaphoric and Metonymic Models. Paper presented at the 5th International Conference on Theory of Mathematics Education, Paderno Del Grappa, Italy, June 20–27, 1991.

Millroy, W. L.: 1992, *An Ethnographic Study of the Mathematics of a Group of Carpenters*, Monograph 5, National Council of Teachers of Mathematics, Reston, VA.

National Council of Teachers of Mathematics: 1989, *Curriculum and Evaluation Standards for School Mathematics*, National Council of Teachers of Mathematics, Reston, VA.

National Research Council: 1989, *Everybody Counts: A Report to the Nation on the Future of Mathematics Education*, National Academy Press, Washington, DC.

Nickson, M.: 1992, 'The Culture of the Mathematics Classroom: An Unknown Quantity?', in D. A. Grouws (ed.), *Handbook of Research on Mathematics Teaching and Learning*, Macmillan, New York, 101–114.

Ontiveros, J. Q.: 1991, Piaget and Vygotsky: Two Interactionist Perspectives in the Construction of Knowledge. Paper presented at the 5th International Conference on Theory of Mathematics Education, Paderno Del Grappa, June 20–27, 1991.

Presmeg, N. C.: 1985, The Role of Visually Mediated Processes in High School Mathematics: A Classroom Investigation. Unpublished Ph.D. dissertation, University of Cambridge.

Presmeg, N. C.: 1990, 'Applying van Hiele's Theory in Senior Primary Geometry: Uses of Phases between the Levels', *Proceedings of the 10th National Congress on Mathematics Education, Durban, South Africa, July 4–6, 1990*, 1–8.

Presmeg, N. C.: 1992a, 'Review of Mark Johnson's Book, "The Body in the Mind: The Bodily Basis of Meaning, Imagination and Reason"', *Educational Studies in Mathematics* **23** (3), 307–314.

Presmeg, N. C.: 1992b, 'Prototypes, Metaphors, Metonymies, and Imaginative Rationality in High School Mathematics', *Educational Studies in Mathematics* **23** (6), 595–610.

Presmeg, N. C.: 1993, 'Mathematics – "A Bunch of Formulas"? Interplay of Beliefs and Problem Solving Styles', *Proceedings of the 17th Annual Meeting of the International Group for the Psychology of Mathematics Education, Tsukuba, Japan, July 18–23, 1993*, Vol. 3, 57–64.

Presmeg, N. C.: 1997, 'Reasoning with Metaphors and Metonymies in Mathematics Learning', in L. D. English (ed.), *Mathematical Reasoning: Analogies, Metaphors, and Images*, Lawrence Erlbaum Associates, Hillsdale, NJ.

Research Advisory Committee: 1992, 'Issues Concerning Evidence', *Journal for Research in Mathematics Education* **23** (4), 341–344.

Saxe, G. B.: 1991, *Culture and Cognitive Development: Studies in Mathematical Understanding*, Lawrence Erlbaum Associates, Hillsdale, NJ.

Schifter, D. & Fosnot, C. T.: 1993, *Reconstructing Mathematics Education: Stories of Teachers Meeting the Challenge of Reform*, Teachers College Press, Columbia University, New York.

Secada, W. C.: 1990, 'Needed: An Agenda for Equity in Mathematics Education', *Journal for Research in Mathematics Education* **21** (5), 354–355.

Secada, W. C: 1991, 'Diversity, Equity, and Cognitivist Research', in E. Fennema, T. P. Carpenter, & S. J. Lamon (eds.), *Integrating Research on Teaching and Learning Mathematics*, State University of New York Press, Albany, NY, 17–53.

Sfard, A: 1991, 'On the Dual Nature of Mathematical Conceptions: Reflections on Processes and Objects as Different Sides of the Same Coin', *Educational Studies in Mathematics* **22** (1), 1–36.

Steen, L. A.: 1990, *On the Shoulders of Giants: New Approaches to Numeracy*, National Academy Press, Washington, DC.

Van Hiele, P. M.: 1986, *Structure and Insight: A Theory of Mathematics Education*, Academic Press, London.

Vergnaud, G.: 1990, 'Epistemology and Psychology of Mathematics Education', in P. Nesher & J. Kilpatrick (eds.), *Mathematics and Cognition: A Research Synthesis by the International Group for the Psychology of Mathematics Education*, Cambridge University Press, Cambridge, 14–30.

Voigt, J.: 1995, 'Thematic Patterns of Interaction and Sociomathematical Norms', in P. Cobb and H. Bauersfeld (eds.), *The Emergence of Mathematical Meaning, Interaction in Classroom Processes*, Lawrence Erlbaum Associates, Hillsdale, NJ, 163–202.

Von Glasersfeld, E.: 1993, Piaget's Approach to the Psychology of Learning. Paper presented in the Science and Mathematics Education Seminar Series, The Florida State University, February 2.

Whitson, J. A.: 1992. Cognition as a Semiosis Process: Grounding, Mediation, and Critical Reflective Transcendance. Paper presented at a symposium on Situated Cognition: Social, Semiotic and Neurological Perspectives, at the Annual meeting of the American Educational Research Association, San Francisco, April.

Wirszup, I.: 1976, 'Breakthroughs in the Psychology of Learning and Teaching Geometry', in J. L. Martin (ed.) *Space and Geometry: Papers from a Research Workshop*, Educational Resources Information Center, Columbus, Ohio.

Yackel, E.: 1992, Reaction Paper to the Plenary Presentations by J. Voigt and G. Saxe in the Subgroup on Sociological and Anthropological Perspectives, Working Group 4, Seventh International Congress on Mathematical Education, Quebec City, August 16–23.

Yackel, E.: 1995, 'Children's Talk in Inquiry Mathematics Classrooms', in P. Cobb and H. Bauersfeld (eds.), *The Emergence of Mathematical Meaning, Interaction in Classroom Processes*, Lawrence Erlbaum Associates, Hillsdale, NJ, 131–162.

Norma C. Presmeg
Mathematics Education Program, Box 3032,
The Florida State University,
Tallahassee, Florida,
U.S.A. 32306-3032

PAUL ERNEST

A POSTMODERN PERSPECTIVE ON RESEARCH IN MATHEMATICS EDUCATION

Over the past quarter century mathematics education has become established world-wide as a major independent area of knowledge and research. The field now has numerous dedicated journals, book series and conferences serving national and international communities of scholars. Many countries offer specialist master's and doctoral programs of study in mathematics education, and new entrants often receive their postgraduate education within the field itself. Given this coming of age, as we approach the 21st century it is appropriate to engage in a period of critical reflection and self-scrutiny. Thus the present volume embodying the ICMI-sponsored inquiry into the nature of research in mathematics education and its products provides a welcome opportunity for our field to take stock of itself, and its outcomes and effects.

This chapter reflects on the nature of research in mathematics education and offers a particular perspective with postmodern elements. An argument is offered for the need for, and the legitimacy of, a multiplicity and variety of viewpoints, theories, frameworks, methodologies, and interests within mathematics education. This is a plea for a decentered view of the subject. My claim is that since it is an interdisciplinary field, drawing on the sciences, social sciences, humanities and other fields of professional knowledge, such multiplicity is inevitable. But for all the strengths of this multiplicity, it leads inevitably to problems concerning some of the questions that need to be addressed. In particular, it means that no single unique set of criteria for evaluating the results of research in mathematics education can be found and that no overall consensus about the field can be achieved.

As a background to this discussion, there are a number of central problems and issues related to the nature of research in mathematics education that need clarification, and the chapter addresses some of them. These include the problem of the ambiguity of the term 'mathematics education'. There is also that of determining the specific object of study in mathematics education, and the relationship between the object of study and the means by which it is studied, who does the studying, and the social contexts of study. This raises the issue of the aims of research in mathematics education, and the extent to which the field can be mapped out as a set of concerns or problems in any coherent way.

Sierpinska, A. and Kilpatrick, J. Mathematics Education as a Research Domain: A Search for Identity, 71–85.
© *1998 Kluwer Academic Publishers. Printed in Great Britain.*

1. THE SPECIFIC OBJECT OF STUDY IN MATHEMATICS EDUCATION

Is there a specific object of study in mathematics education? This question is even more difficult to answer than it first appears, because the term 'mathematics education' is ambiguous. It signifies both a practice (or rather a set of practices) and a field of knowledge. Mathematics education describes both the teaching of mathematics, to elementary school children for example, and an academic specialism with its own Ph.D.s, conferences and journals. Even the term 'mathematics educator' commonly used to signify someone involved in mathematics education research (such as in Balacheff et al. 1992) is used across this ambiguity. For the dictionary definition of 'educator' is someone who teaches, trains or provides schooling. Such ambiguity is by no means unique to mathematics education. 'Statistics', for example, describes both the data studied and the particular mathematical field of knowledge involved.

Thus there are real differences underpinning the ambiguity. To wholly identify the object of study in mathematics education with classroom teaching and learning practices is already problematic for a number of reasons. First of all, it is not possible to completely separate the object of study from the means by which it is studied, from who does the studying and from the social contexts of study. Thus even a narrow focus on the teaching and learning of mathematics brings in the nature, methodology and social context of the researchers involved. Secondly, an attempt to pre-specify the scope of a field of study is usually a narrowing move, an attempt to exclude certain kinds of theorizing about the field which do not seem to have immediate payoffs in terms of practical knowledge or application. But 'relevance' is a notoriously short sighted criterion and it is often 'blue sky' research that turns out to have the greatest utility. At the turn of the century Poincaré remarked that at last in non-Euclidean geometry we had a theory that was truly inapplicable. A few years later Einstein published his theory of relativity.

Although relevance is often used as if it were a simple adjective, it is better viewed as a ternary relation in which X is described as relevant by Y with reference to goal Z (Ernest 1991). The two commonly suppressed arguments are, who is making the judgment and what goal they are presupposing in their criteria.

Despite the above cautions, I do think it is worth trying to list (although clearly not exhaustively) some of the immediate practices involved in mathematics education, or in which mathematics education is involved, as a means of indicating its nature and multiplicity. In my view these include the following:

- The teaching and learning of mathematics at all levels in school and college
- Out of school learning (and teaching) of mathematics
- The design, writing and construction of texts and mathematics learning materials
- The study of mathematics education in pre-service teacher education

- The graduate study of mathematics education texts and results
- Research in mathematics education at all levels.

What is immediately apparent is that this preliminary listing of the practices of mathematics education, as the objects of study of mathematics education, is recursive. For these practices are defined in terms of the field of mathematics education itself. However this recursive loop is not a vicious circle. What it means is that research in mathematics education studies itself; it is reflexive.

On the issue of the reflexivity of research in mathematics education it is interesting to note that this is a property that largely distinguishes the humanities and arts from the sciences. Traditionally the physical sciences have not included the cognitions of humanity amongst their objects of study. More recently psychology and cognitive science treat aspects of human knowing directly; but they are not usually seen to be reflexive by accounting for themselves as areas of knowing within their overall field of interest. In contrast, sociology, philosophy, anthropology, education, and social studies of science all try to account for themselves within their field of study. These are all *reflexive* fields, and indeed Bloor's (1976) 'strong program' in the sociology of knowledge explicitly includes a reflexivity criterion.

What does this mean for mathematics education? As a reflexive field of study it suggests that it is more akin to the humanities and social sciences than the physical or hard sciences. This may seem a trite conclusion, but it is important to bear in mind when calls are heard to identify mathematics education as a branch of mathematics and to regard mathematicians as our natural peers and audience. They are distinct fields of study, and their relations are more complex.

My discussion of the reflexivity of research in mathematics education may be disturbing to some. By claiming that mathematics education is its own object of study and research, and not just the teaching and learning of mathematics, I may have strayed beyond the bounds that some of my fellow mathematics educators prefer to remain within. Or more strongly, it may be that these broader concerns go beyond what some would consider to be the legitimate objects of study of mathematics education. This is not my view, but it is necessary to note areas of dissent, and to acknowledge their legitimacy, if mathematics education is to be a genuine conversation which acknowledges the validity of multiple viewpoints.

Perhaps there is a useful distinction that can be drawn from a consideration of this possible area of dissent. Perhaps I should tentatively distinguish the primary objects of study of mathematics education including phenomena directly concerned with the teaching and learning of mathematics, and secondary objects of study involving the field of mathematics education itself. This division respects the metaphor of reflection implicit in the concept of 'reflexivity'. For the so-called primary objects of study are in some sense direct, whereas the secondary ones involve a further (second) level of reflection upon the primary objects. Of course, this distinction is only of limited use, because it is not so clear how much use it is or if studies or practices could be assigned unambiguously to one or other level. What the distinction might do, however, is to reassure those in the mathematics education community who are concerned that research in mathematics education is or might be losing touch with its central concerns, namely the teaching and learning of mathematics. By labeling such concerns as primary, their priority is established; something with which I doubt anyone would disagree.

Different views as to what are the proper foci or objects of study of research in mathematics education might traditionally be attributed to different ideologies, epistemologies, beliefs, etc. But perhaps it is more fruitful to take a postmodern perspective and consider instead what are the social practices, roles and organizations that support mathematics educators institutionally, and to whom they are accountable or allied. Thus, for example, full-time researchers in mathematics education usually have to justify their projects in terms of improvements to the teaching and learning of mathematics. Researchers who are primarily teacher educators can instead justify their research in terms of increased knowledge of or possible improvements to teacher education, or even as 'energizers of thought' for teachers or themselves. The 'best' justification for research is thus a function of context, and by no means a transcendent truth.

One immediate consequence of the above discussion is that according to my perspective there is no unique object of study for research in mathematics education. Rather there are multiple foci, some primary and some secondary. What is clear is that the ICMI enquiry into the nature of research in mathematics education, represented by this volume, is unambiguously directed at the second of the two levels. By adopting a reflective meta-level consideration of the field as its goal, it asserts the value and importance of the secondary object of study of mathematics education, namely mathematics education itself, and in particular, research in mathematics education.

In accounting for the objects of research in mathematics education in terms of the practices it refers to I am making a move to situate knowledge in concrete practices. Such a postmodern shift is currently taking place in many fields of thought. Cognitive science and the philosophy of mind have suggested that mind is modular, with local knowledges, skills and agencies taking the place of a single controlling intelligence. Psychology, both cognitive and social, is looking at forms of situated learning which emphasizes the priority of context. Poststructuralism elaborates on this perspective and argues that the self and knowledge are distributed through different discursive practices in which the discourse has a key role. Postmodernism has challenged epistemological metanarratives in philosophy and suggests that local practices are the sole determinants of knowledge. Wittgenstein's late philosophy looks at meaning and knowledge as situated in 'language games' embedded in pre-existing 'forms of life'. The philosophy of science is looking at local practices of scientists instead of overarching theories of scientific method which fail to describe actual laboratory life. The philosophy of mathematics is likewise beginning to consider the practices of mathematicians instead of theories of mathematical knowledge and truth (Ernest 1994a). Thus, an attempt to define mathematics education in terms of its various multicentered practices, instead of some analytic or rationally derived prespecified mission, has strong support from many currents of contemporary thought. More than that, such postmodern perspectives on knowledge suggest that anything but a multicentered perspective on the field would be a misrepresentation, forcing the multiplistic and polycentric practices of our field onto a single Procrustean bed.

To return to the concrete, it is perhaps worth trying to list, in a necessarily tentative and incomplete way, some of the objects of research in mathematics education, from my perspective. These include as primary objects:

- The nature of mathematics and school mathematical knowledge
- The learning of mathematics
- The aims and goals of mathematics teaching and schooling
- The teaching of mathematics, including the methods and approaches involved
- The full range of texts, materials, aids and electronic resources employed
- The human and social contexts of mathematics learning/teaching in all their complexity
- The interaction and relationships between all of the above factors.

The secondary objects might be taken to include:

- The nature of mathematics education knowledge: its concepts, theories, results, literature, aims and function
- The nature of mathematics education research: its epistemology, theoretical bases, criteria, methodology, methods, outcomes and goals
- Mathematics education teaching and learning in teacher education, including practice, technique, theory and research
- The social institutions of mathematics education: the persons, locations, institutions (universities, colleges, research centers), conferences, organizations, networks, journals, etc. and their relationships with its overall social or societal contexts.

My claim is that the proper object of study of research in mathematics education includes at least all of these components, interpreted broadly. However, as has always been the case, far more attention will be and should be given to the primary components, but the secondary ones are important too, and should not be neglected.

2. THE AIMS OF RESEARCH IN MATHEMATICS EDUCATION

Beyond the proper *objects of study* of research in mathematics education, there are also the *means* by which such research is carried out. This is a central and

essential concern when considering the process of research, its intended outcomes, and the presuppositions on which it rests. But first, it is necessary to pay
a little attention to the concept of research itself.

There are many views about the nature of research and of educational research in particular. Ultimately, according to postmodernism, it is the practice of
knowledge-making as it takes place in different contexts alone that specifies
what it is. However, a brief, rationalistic account of the 'nature' of research, in
mathematics education or in any other field, can serve as a useful, if tentative,
starting point. From this perspective it is systematic and critical enquiry carried
out with the aim of producing knowledge. The key words are *knowledge*,
systematic, *critical*, and *enquiry* (Ernest 1994b). What I shall argue below is that
the particular way that these are understood is not a simple matter, but depends
on the reader or researcher's underlying assumptions, and, in particular, on their
underlying research paradigm and the community practices of which they are a
part.

The nature and basis of knowledge is one of the central concerns of philosophy – epistemology, in particular – but it also lies at the heart of every discipline
or field of study. Each has its own forms of knowledge, and accepted means of
justifying knowledge claims. Knowledge, in any particular area of study, is what
is known and deliberately accepted by practitioners in that field. What is known
and accepted is that which is appropriately supported by evidence: by proof in
mathematics, by empirical test in science, by argumentation and/or tests in
social sciences such as mathematics education. Each field also has its body of
theoretical knowledge, the theories, hypotheses, questions, and so on, known by
its practitioners. The practitioners in any field also have tacit knowledge
(Polanyi 1958) which includes knowledge of the specialist language of the field,
and meta-knowledge of the field. Kuhn (1970) discusses this for science, Kitcher
(1984) and Ernest (1992b, 1997) for mathematics.

'Enquiry' concerns the processes of knowledge-getting or making. Systematic
enquiry should:

- link with and build on existing knowledge in the relevant educational research literature, thus adding to the body of knowledge, fitting into the
  'system of knowledge';
- use organized processes of enquiry, systematic methods of research, linked to
  existing methodology, providing a justification for knowledge claims;
- result in a systematically organized text, document or other public communicative form, so that others can access the results of the educational
  research;
- possibly engage in theory-building resulting in the construction of some
  systematically organized body of reflective knowledge.

Of course the structure which such 'systems' attach to are all the while organ-
ically evolving and growing as the associated practices and community cultures
change.

Such enquiry needs to be critical and self-critical, which results from the systematic and warranted character of knowledge claims, according to the community criteria of the day, and the need for the products of research to be subjected to critical review prior to acceptance by the community. This social process of critical review in any research community, such as our own, also results in, or is coupled with internal, self-critical faculties. During the research process these are employed, to a greater or lesser extent, to look for weaknesses (lack of conformity to explicit or more usually tacit community standards and expectations) in the conduct and reports of research in progress, and to attempt to mold and express the research in ways satisfying the criteria for rigor and justification of the field.

One further aspect of educational research can be stressed. The term 'educational research' applies to both the process of research and its product. To research is to engage in the process of searching for or trying to build knowledge. But 'research' is also the outcome of this process, knowledge itself.

Educational research rests on a number of fundamental presuppositions made by the researcher, and these need to be considered before the aims of research in mathematics education can be treated with any thoroughness.

Research is often understood to take place within a recognized or unconsciously assumed overall theoretical research perspective or *paradigm* (Kuhn 1970; Bassey 1990–91). There are multiple research paradigms, each with its own assumptions about knowledge and learning (epistemology), about the world and existence (ontology), and about how knowledge is obtained (methodology). Following the work of Habermas (1972), educational researchers are in the habit of distinguishing three main educational research paradigms: the scientific, interpretative and critical-theoretic research paradigm (Schubert 1986; Galbraith 1991; Ernest 1994b). These vary in a number of significant ways, including the research methodology they employ, but a brief and simplified account is as follows.

The scientific research paradigm originates with rationalism and the scientific method as employed in the physical sciences, experimental psychology, etc. It is concerned with objectivity, prediction, replicability, and the discovery of scientific generalizations or laws describing the phenomena in question. The forms of enquiry used include survey, comparative experimental, quasi-experimental methods, and so on. There is often an emphasis on quantitative data, but qualitative data can, of course, also be used, as and when appropriate. What is central to the scientific research paradigm is the search for general laws predicting future educational outcomes. Thus 'process–product' research in mathematics teaching is typical work in the scientific paradigm. It examines classroom and learner variables and seeks to correlate them with mathematics learning outcomes.

The second paradigm is that of interpretative (or naturalistic) research, which developed from the methods used in sociological and social science research, including anthropology and ethnography. It is primarily concerned with human understanding, interpretation, intersubjectivity, lived truth (i.e. truth in human terms), and so on. It uses ethnographic, case study, largely qualitative forms of enquiry, and attempts to overcome the weaknesses of subjectivity through triangulating multiple viewpoints. Of course, quantitative methods are also used, as and when appropriate. In mathematics education research this paradigm is especially evident in the experimental work of constructivist researchers. A seminal early use of the method is that of Erlwanger (1973) in his celebrated case study of a single child's learning ('Benny').

Third, there is the critical-theoretic paradigm. The central feature of this position is the desire not just to find out, but to engage in social critique and social and institutional change to improve or reform aspects of social life. In education, this often concerns working on social justice issues, such as redressing gender or racial inequalities. To this end, it often involves participant engagement and validation. One of the best known discussions of this approach applied to educational research is that of Carr and Kemmis (1986).

The critical-theoretic paradigm is closely associated with Action Research, which is popular among the 'teacher-as-researcher' movement, with teachers working to change their teaching or school situations to improve classroom learning. In my view, Action Research, however, too often baulks at addressing oppression in society, to fit comfortably under the critical-theoretic paradigm. Such developments as Paolo Freire's (1972) project for emancipating Brazilian peasants through literacy, although not explicitly critical-theoretic, perhaps serve as its best examples. Likewise, in mathematics education, the paradigm is reflected in the work of Gerdes (1985) in Mozambique and researchers such as Mellin-Olsen (1987), and Skovsmose (1985, 1994).

Habermas (1972) argues that there is no such thing as pure knowledge, research or quest for 'truth'. He makes the radical claim that underpinning every knowledge-seeking enterprise there is a particular type of interest or desire at work, even in the case of science where many maintain that knowledge is 'interest- and value-free'. He distinguishes three types of interest that underlie the quest for knowledge. These are: to predict and control the phenomena under study (technical interest), to understand and make sense of them (practical interest), and to achieve social justice (emancipatory interest). These correspond directly to the interests underlying the three educational research paradigms: the desire to predict and control educational processes (scientific paradigm), the desire to understand educational phenomena (interpretative paradigm), and the desire to change education and through it society for the better (critical-theoretic paradigm).

Coupled with these interests are the intended outcome, the type of knowledge or other outcome (social change) intended to follow from the enquiry. These include first, objective knowledge and scientific generalizations and truths; second, subjective understanding, personal truths, and illuminating studies of unique individuals and situations; and third, social changes and improved social institutions and conditions (respectively).

Thus a discussion of the aims of research in mathematics education should take into account the issue of the underlying educational research paradigm involved, and the interests and intended outcomes of that paradigm. Certain different specific knowledge-types or social changes may be the planned outcomes of mathematics education research projects.

Needless to say, this account would be seen by some to be controversial. For rather than acknowledging that there are multiple valid paradigms and sets of assumptions underpinning research, each with different strengths and aims, some researchers have preferred to fight for their own paradigm as the sole one that is valid. Scientific research paradigm supporters have argued that they own the sole route to objectivity and truth. Supporters of the interpretative research paradigm argue that there is no objectivity and truth, and that only they can offer valid understanding. Critical-theoretic research paradigm supporters have argued that the others are victims of 'false consciousness' and that only they can reveal the ideologically-induced distortions in education and society. This account may be a caricature, but Gage (1989) has written of the 'paradigm wars' waged between supporters of the three paradigms in the educational research community in the USA.

The view of Gage, which I endorse, is that educational research paradigms are tools that should serve our practical ends in education, and that acknowledging their multiplicity but judging them by their fruits is the best policy. Beyond this, the postmodern view is that they represent the legitimate (but criticizable) outlooks of communities with different practices and theoretical discourses. To argue for one and against the others is to adopt the perspective of one community including its antagonisms. Instead we need to acknowledge the perpetual presence of multiple perspectives.

In a parallel way, research in mathematics education has multiple aims. These include the development of a body of professional knowledge about the results of research, including students' thinking and activity, mathematics teaching approaches and resources. Some of this can be systematized and organized into theoretical structures. Some make up a rich body of case-knowledge which provide exemplars and insights. In addition, there are a whole range of theoretical concepts and frameworks that help us understand how children learn mathematics, how mathematics texts subliminally incorporate sexist and racist messages, how the hidden curriculum of the mathematics classroom results in certain learner conceptions of mathematics and mathematical activity. Such insights play a vital, if unquantifiable, role in teacher education, sensitizing new teachers to go beyond their unexamined assumptions about mathematics, and its learning and teaching.

There is also the importance of mathematics education research, reflection and theory for teacher educators. The excitement of being at the leading edge of new thinking, of making new connections, of grasping exciting new intellectual vistas, and contributing to them, perhaps in a modest way, serves as a vital 'energizer of thought' for teacher educators, such as myself. I am sure that the enthusiasm such involvement generates in me is reflected in better and more inspirational teaching to my pre-service student teachers. On occasion they explicitly remark on the excitement of working with ideas and theories in mathematics education from the vanguard of thought.

Overall, I am claiming that there is no unique aim that research in mathematics education should serve. To better the teaching and learning of mathematics is of course a shared, if vague, goal of us all, but there are many ways of doing this through research in mathematics education, some well known and some as yet undreamed of.

3. ARE THERE SPECIFIC QUESTIONS OR PROBLÉMATIQUES OF RESEARCH IN MATHEMATICS EDUCATION?

Is there a unique and essential feature of research in *mathematics education*? Is it that part of *education and educational research* which deals primarily with the teaching and learning of mathematics? Or is it that part of *mathematics*, or at least of the thought and activity of mathematicians, which primarily concerns education in mathematics, that is its teaching and learning? One way to try to answer this is to enquire into the essential character of mathematics education. But I consider this to be largely fruitless. It seems to me that mathematics education is defined in terms of a variety of realized practices, and not in terms of essential characteristics. In terms of practice, what can be said is that researchers in mathematics education are located institutionally both in faculties of education and in faculties of mathematics.

Thus both of the above answers hold true in terms of institutional location, and in terms of disciplinary culture. What both share is a concern with the specific concepts, methods, modes of thought and problems of mathematics, and the unique ways in which mathematical knowledge is structured, organized, and validated, and how these features impact upon learners when they are confronted with mathematical activities and representations of mathematical knowledge.

What I have listed are perhaps the very well-known general issues of interest and problem locations in mathematics education research. Beyond these, there are a number of specific issues that arise for me from the unique characteristics of mathematics and its special social role, which give rise to further problem-areas and might enlarge the problématique of mathematics education. Of course, the following is a personal view.

Mathematics is a discipline perhaps uniquely based solely in text and concepts, which does not describe or refer to the experiential world except indirectly. What problems arise from learning this uniquely imaginary and abstracted field? Many are concerned with signification since meaning is defined internally, without reference to experienced objects in the world (except at the level where mathematics is part of everyday language and practices). Further problems arise from the strong objectivity of mathematical language which excludes all reference to persons, places and events.

Mathematics shares with the sciences an extreme precision with which its concepts, methods and ideas are defined. Consequently it is required that mathematics learning results in very precise and narrowly defined behaviors and capabilities, manifested in specific contexts, and learners are judged in terms of the precision of their performances against very high standards. Such expectations are very demanding and doubtless contribute to learning problems in mathematics.

Perhaps because of the former characteristics, there is a great disparity in the range of conceptions of mathematics. These vary from seeing it as the mechanical and dreary rules of book-keeping, to a language and way of thinking of tremendous power, which opens access to a rich universe of beautiful structures, abstract concepts and patterns; a field with an overwhelming aesthetic and a source of fascination and wonder.

Such ranges of conceptions are held by learners and teachers of mathematics, and by citizens at large in society. A consequence of this great variation is the lack of communication, misunderstandings and affective problems concerning mathematics. Mathematicians and mathematics educators often share the latter, aesthetically-oriented view of the discipline, but address many persons with other perceptions of mathematics. This disparity of perceptions is a major source of problems.

Mathematics has a unique social role as a prestige subject indicating rationality and intellectual ability. It also has an arbitrarily high social status attached to it, resulting in part from this role and its relationship with science and technology, but also as a result of historical contingency. This leads to the real problems concerning mathematics and equity. To a significant extent, social rewards (including wealth, status and power) and life-chances are distributed to members of society according to success in the learning of mathematics and its certification. The function of mathematics as a 'critical filter' in society brings with it real social and moral problems, especially with regard to gender, race, social class and 'ability'.

Mathematics has a unique relationship with computing and informatics. It gave birth to digital computing and provides the language for its operation and comprehension. Mathematics has also been dramatically transformed by the impact of computers in a number of well-known ways. Consider, for example, the impact of computers on number theory, numerical calculation, statistics, operational research, optimization theory, linear algebra (each transformed by the power of computation), mathematical modeling (e.g. finite element analysis), proof (e.g. the 4-color theorem), symbolic manipulation (e.g. using *Derive* or *Mathematica*), and the birth of Chaology. This relationship and the increasingly central role of computers in the modern world give rise to a variety of problems and a problématique for mathematics education. Of particular note is the problem of how the power of the computer can or should be employed in mathematics education to best effect. A second problem is how mathematics and school mathematics should be reformed in the light of the computer revolution. There is also a moral or political problem about the relationship between computers and research in mathematics education: namely, that because of the biases of funding bodies there may be a distortion of the field which favors research in mathematics education with computers as a central part. This distortion might be seen as reflecting commercial, industrial and military interests, possibly at the expense of the improvement of the teaching and learning of mathematics.

These represent some possible question-sets or problématiques for research in mathematics education. They represent a brief and tentative attempt to identify some of the problems that arise from the specific characteristics of mathematics as a discipline, and its role in society. But there is also the issue of the problems facing mathematics education which are shared by education in other domains. One criticism I would make at some research in mathematics education is that it takes too little notice of what is done outside of its own narrow ambit, and it consequently is perpetually 'reinventing the wheel'. A very proximate field of study sharing many of the concerns of mathematics education is that of science education. These two fields of study for too long have had a parallel but separate existence. Thus, for example, recently I remarked on ten interdisciplinary themes which are the shared concern of both science and mathematics education (Ernest 1992a), as follows:

1. Constructivism as learning theory and approach to epistemology, and alternative conceptions in mathematics and science.
2. Applications of exciting new humanistic theories from hermeneutics, post-structuralism, postmodernism, anthropology, ethnomethodology and relativistic philosophies of science and mathematics.
3. The impact of recent fallibilistic developments in the philosophy of mathematics and science in challenging the 'certainty culture'.
4. A recognition of the intertwined cultural, historical and philosophical development of science, mathematics and technology.
5. Critical citizenship, epistemological empowerment and enhanced democracy through science and mathematics for all.
6. A concern with the gender and race imbalance in the participation rates and assessment outcomes of school and college science and mathematics.
7. The move away from traditional teaching styles and the adoption of group-work, discussion, and enquiry approaches facilitated by teachers' new roles and by the new technology, especially calculators and microcomputers.
8. Problem-solving, investigation, the processes of enquiry and the nature of mathematics and science.
9. The unique contribution of mathematics and science to rationality, via the rational appraisal and rigorous testing in the warranting of knowledge.
10. The critical appraisal of theories of education, science, mathematics and technology and of their conceptual presuppositions.

This list illustrates that a significant part of what we think is the unique problématique of mathematics education very closely resembles, if it is not identical with, the problématique of an adjacent field.

4. THE RESULTS OF RESEARCH IN MATHEMATICS EDUCATION

In general, the results of research in mathematics education will be significantly affected by the underlying educational research paradigm (see section 2 above). However there is one very important general issue concerning these results which needs discussion and clarification. This is the relationship between the results of research, the knowledge produced, and organized fields of study and disciplines.

I argued above that research in mathematics education should result in knowledge that links with and builds on existing knowledge, fitting into the 'system of knowledge', resulting in the construction of some systematically organized body of reflective knowledge (possibly including new theory).

This raises a number of questions. What is the relationship between mathematics education as a field of study and other areas of knowledge such as mathematics, education, philosophy, psychology, sociology, social studies of science, history, anthropology, cognitive science, linguistics, semiotics, and so on? Should the results of research in mathematics education be 'home-grown' theories, or should mathematics education theories relate closely to existing theories in other fields? It is undeniably the case that local or small scale theories are home-grown in the field of mathematics education itself. For example, whatever their current status, there is no doubt that the hierarchical theories of the van Hieles and of the CSMS project are home-grown. More such theories could be cited, and more should be encouraged to develop.

In addition, I am of the opinion that we need to pay serious attention to developments outside of the mathematics education community, as a disciplined source and resource for our reflections and theorizing. Mathematics education as an emergent field has an especial need and responsibility to look outside itself for new ideas, and for sources of approaches and methodology, and for means of giving our knowledge increased validity. Most concepts and methods employed in research in mathematics education originate in other fields of study. We need to acknowledge this more explicitly. Currently, the boundaries between existing disciplines and areas of knowledge are being questioned, and so the very issue of what is a part of our field is problematic.

Overall, what I am proposing is that research in mathematics education needs to look more systematically at concepts and developments outside the field as a means of making progress in theory. In my own work I have found that developments in the philosophy of mathematics (and in the sociology of knowledge, philosophy of science, rhetoric of the sciences, semiotics, logic, etc.) have been a great source of insight for mathematics education (e.g. Ernest 1991, 1997). Others, too numerous to mention, have added significantly to knowledge in mathematics education through a consideration of and application of the knowledge, theory and methods of the history of education, psychoanalysis, semiotics and linguistics, sociology, ethnomethodology, mathematical structure, psychology, post-structuralism, computer science, cognitive science, and so on. In my view, we need to always keep aware of developments across the whole field of human thought, and take and apply whatever ideas are useful for our researches.

5. CONCLUSION

An analysis of the nature of research in mathematics education raises many issues, but again and again the issue of the complexity and multifaceted nature of the field keeps emerging. There are various objects of enquiry, research approaches and paradigms, traditions, institutional locations, practices, evaluation criteria, adjacent fields of knowledge and a growing body of theories, results, and publications that deserve a full and thorough consideration. To acknowledge this requires recognition of and tolerance for this multiplicity. Without consensus it is too early to try to delimit or pre-specify a canonical central conception of the field of knowledge. Research in mathematics education has multiple aims and audiences, and remains healthy whilst it continues to surprise us. Ten years ago the idea of a whole journal issue devoted to psychoanalysis and mathematics education was unthinkable. Yet a recent issue of *For the Learning of Mathematics* (vol. 13, no. 1, February 1993) was devoted to just that.

However, the postmodern multicentered conception of research in mathematics education that I have argued for must not be seen as a license so that 'anything goes'. We must acknowledge the existence of multiple communities of interest and expertise whose judgments, whilst never infallible, reflect shared concerns and judgments and admit the possibilities of innovation and growth while striving to maintain (multiple sets of) high standards in our field.

REFERENCES

Balacheff, N., Howson, A. G., Sfard, A., Steinbring, H., Kilpatrick, J. & Sierpinska, A.: 1992, 'Discussion Document for an ICMI Study. What is Research in Mathematics Education and What are its Results?', *Educational Studies in Mathematics* 23 (6), 625–630.

Bassey, M.: 1990–91, 'On the Nature of Research in Education' (Parts 1–3), *Research Intelligence* 36, 35–38; 37, 39–44; 38, 16–18.

Bloor, D.: 1976, *Knowledge and Social Imagery*, Routledge and Kegan Paul, London.

Carr, W. & Kemmis, S.: 1986, *Becoming Critical*, Falmer Press, London.

Erlwanger, S. H.: 1973, 'Benny's Conception of Rules and Answers in IPI Mathematics', *Journal of Children's Mathematical Behaviour* 1 (2), 7–26.

Ernest, P.: 1991, *The Philosophy of Mathematics Education*, Falmer Press, London:.

Ernest, P.: 1992a, 'Letter to Editor', *Science and Education* 1 (4), 396–397.

Ernest, P.: 1992b, 'The Relationship between the Objective and Subjective Knowledge of Mathematics', in F. Seeger and H. Steinbring (eds.), *The Dialogue between Theory and Practice in Mathematics Education: Overcoming the Broadcast Metaphor*, Institut für Didaktik der Mathematik der Universität Bielefeld, Materialen und Studien Band 38, 33–48.

Ernest, P. (ed.): 1994a, *Mathematics, Education and Philosophy: An International Perspective*, Falmer Press, London.

Ernest, P.: 1994b, *Educational Research, Its Philosophy and Purpose: An Introduction to Research Methodology and Paradigms*, University of Exeter School of Education, Exeter.

Ernest, P.: 1997, *Social Constructivism as a Philosophy of Mathematics*, SUNY Press, Albany, New York.

Freire, P.: 1972, *Pedagogy of the Oppressed*, Penguin Books, Harmondsworth.

Gage, N. L.: 1989, 'The Paradigm Wars and Their Aftermath: A "Historical" Sketch of Research on Teaching Since 1989', *Teachers College Record* 91 (2), 135–150.

Galbraith, P.: 1991, 'Paradigms, Problems and Assessment: Some Ideological Implications'. Paper presented at MERGA Conference, Perth, Australia, July.

Gerdes, P.: 1985, 'Conditions and Strategies for Emancipatory Mathematics Education in Underdeveloped Countries', *For the Learning of Mathematics* 5 (1), 15–20.

Habermas, J.: 1972, *Knowledge and Human Interests*, Heinemann, London.

Kitcher, P.: 1984, *The Nature of Mathematical Knowledge*, Oxford University Press, New York.

Kuhn, T. S.: 1970, *The Structure of Scientific Revolutions* (2nd ed.), Chicago University Press, Chicago.

Mellin-Olsen, S.: 1987, *The Politics of Mathematics Education*, Reidel, Dordrecht.

Polanyi, M.: 1958, *Personal Knowledge*, Routledge & Kegan Paul, London.

Schubert, W. H.: 1986, *Curriculum: Perspective, Paradigm, and Possibility*, Macmillan, New York.

Skovsmose, O.: 1985, 'Mathematical Education versus Critical Education', *Educational Studies in Mathematics* 16, 337–354.

Skovsmose, O.: 1994, *Towards a Philosophy of Critical Mathematics Education*, Kluwer, Dordrecht.

Paul Ernest
University of Exeter,
School of Education,
Heavitree Road,
Exeter EX 1 2LV,
UK

ERICH CH. WITTMANN

MATHEMATICS EDUCATION AS A 'DESIGN SCIENCE'[1]

In a paper presented to the Twenty-second Annual Meeting of German Mathematics Educators in 1988 Heinrich Bauersfeld presented some views on the perspectives and prospects of mathematics education. It was his intention to stimulate a critical reflection 'among the members of the community' on what they do and on what they could and should do in the future (Bauersfeld 1988). The early 1970s had witnessed a vivid programmatic discussion on the role and nature of mathematics education in the German speaking part of Europe (see the papers by Bigalke, Griesel, Wittmann, Freudenthal, Otte, Dress and Tietz in the special issue 74/3 of the *Zentralblatt für Didaktik der Mathematik* as well as Krygowska 1972). Since then the status of mathematics education has not been considered on a larger scale despite the contributions by Burscheid (1983), Bigalke (1985) and Winter (1986). So the time is overdue for redefining the basic orientation for research; therefore, Bauersfeld's talk could hardly have been more appropriate.

In recent years the interest in a better understanding of the nature and role of mathematics education has also grown considerably at the international level as indicated, for example, by the ICMI Study on 'What is research in mathematics education and what are its results?' launched in 1992 (cf. Balacheff et al. 1992).

The following considerations are intended both as a critical analysis of the present situation and an attempt to capture the specificity of mathematics education. Like Bauersfeld, the author presents them 'in full subjectivity and in a concise way' as a kind of 'thinking aloud about our profession'.[2]

1. THE 'CORE' AND THE 'RELATED AREAS' OF MATHEMATICS EDUCATION

The sciences should influence the outside world only by an enlightened practice; basically they all are esoteric and can become exoteric only by improving some practice. Any other participation leads to nowhere.

(J. W. v. Goethe, *Maximen und Reflexionen*)

Generally speaking, the task of mathematics education is to investigate and to develop the teaching of mathematics at all levels including its premises, goals and societal environment. Like the didactics of other subjects, mathematics education requires the crossing of boundaries between disciplines and depends on results and methods of considerably diverse fields, including mathematics,

Sierpinska, A. and Kilpatrick, J. Mathematics Education as a Research Domain: A Search for Identity, 87–103.
© *1998 Kluwer Academic Publishers. Printed in Great Britain.*

general didactics, pedagogy, sociology, psychology, history of science and others. Scientific knowledge about the teaching of mathematics, however, cannot be gained by simply combining results from these fields; rather it presupposes a *specific* didactic approach that *integrates* different aspects into a coherent and comprehensive picture of mathematics teaching and learning and then transposing it to practical use in a constructive way.

The specificity of this task necessitates, on the one hand, sound relationships to the disciplines related to mathematics education, and on the other hand, a balance between practical proximity and theoretical distance with respect to schools. Bauersfeld (1988, p. 15) refers here to the 'two cultures' of mathematics education. How we can integrate the variety of aspects, and, at the same time, set weights and deal with the tensions that exist between theory and practice is not at all clear *a priori*. This is why it is so difficult to arrive at a generally shared conception of mathematics education.

In my view, the specific tasks of mathematics education can only be actualized if research and development have specific linkages with practice at their *core* and if the improvement of practice is merged with the progress of the field as a whole.

This *core* consists of a variety of components, including, in particular:

- analysis of mathematical activity and of mathematical ways of thinking,
- development of local theories (for example, on mathematizing, problem solving, proof and practising skills),
- exploration of possible contents that focus on making them accessible to learners,
- critical examination and justification of contents in view of the general goals of mathematics teaching,
- research into the pre-requisites of learning and into the teaching/learning processes,
- development and evaluation of substantial teaching units, classes of teaching units and curricula,
- development of methods for planning, teaching, observing and analysing lessons, and
- inclusion of the history of mathematics education.

Work in the core necessitates the researcher's interest and proximity to practical problems. A caveat is in order, however. The orientation of the core towards practice may easily lead to a narrow pragmatism that focuses on immediate applicability and may therefore become counterproductive. This hazard can only be avoided by connecting the core to a variety of related areas that bring about an exchange of ideas with related disciplines and that allow for investigating the different roots of the core in a systematic way (see Figure 1). Of course, the core and the related areas overlap, and the ill-defined borders between them change over time. Thus, a strict separation is not possible.

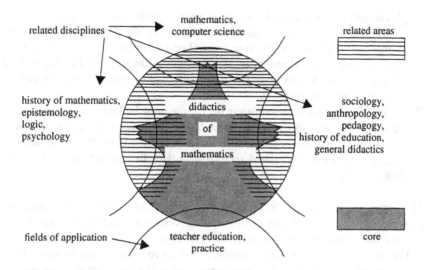

Figure 1 *The core and the areas related to mathematics education, their links to the related disciplines and the fields of application*

Although the related areas are indispensable for the whole entity to function in an optimal way, the specificity of mathematics education rests on the core, and therefore the core must be the central component. Actually, progress in the core is the crucial element by which to measure the improvement of the whole field. This situation is comparable to music, engineering and medicine. For example, the composition and performance of music must take precedence over the history, critique and theory of music; in mechanical engineering the construction and development of machines is paramount to mechanics, thermodynamics and research of new materials; and in medicine the cure of patients is of central importance when compared to medical sociology, history of medicine or cellular research.

However, the division between the core and the related areas does not imply that the core is restricted to practical applications since the related areas have to develop the necessary theory. In fact, building theories or theoretical frameworks related to the design and empirical investigation of teaching is an essential component of work in the core (cf. Freudenthal 1987).

As in engineering, medicine and art, the different status of the core and the related areas is also clearly indicated in mathematics education by the following facts:

1. The core is aimed at an *interdisciplinary, integrative view of different aspects and at constructive developments* whereby the ingenuity of mathematics educators is of crucial importance. The related areas are derived much more from the corresponding disciplines. Therefore research and

development in didactics in general get their *specific* orientation from the requirements of the core. Theoretical studies in the related areas become significant only insofar as they are linked to the core and thus receive a specific meaning. In particular, the research problems listed in Bauersfeld (1988, pp. 16–18) can be tackled in a sufficiently concrete and productive way only from the core.

2. Teacher education oriented towards practice must be based on the core. The related areas are indispensable for a deeper understanding of practical proposals and for their application in an appropriate way. However, also in teacher education, the related areas realize their full impact only if they are linked to the core.

The central position of the core is mainly an expression of the *applied status* of mathematics education. Emphasizing the core does not diminish the importance of the related areas, nor does it separate them from the core. As clearly indicated in Figure 1, it is the core, the related areas and a lively interaction between them that represent the full picture of mathematics education and that also necessitate the common responsibility of mathematics educators independent of their special fields of interest.

Work in the core must start from mathematical activity as an original and natural element of human cognition. Further, it must conceive of 'mathematics' as a broad societal phenomenon whose diversity of uses and modes of expression is only in part reflected by specialized mathematics as typically found in university departments of mathematics. I suggest a use of capital letters to describe MATHEMATICS as mathematical work in the broadest sense; this includes mathematics developed and used in science, engineering, economics, computer science, statistics, industry, commerce, craft, art, daily life, and so forth according to the customs and requirements specific to these contexts. Specialized mathematics is certainly an essential element of MATHEMATICS, and the broader interpretation cannot prosper without the work done by these specialists. However, the converse is equally true: specialized mathematics owes a great deal of its ideas and dynamics to broader scientific and societal sources. By no means can it claim a monopoly for 'mathematics'.

It should go without saying that MATHEMATICS, not specialized mathematics, forms the appropriate field of reference for mathematics education. In particular, the design of teaching units, coherent sets of teaching units and curricula has to be rooted in MATHEMATICS.

As a consequence, mathematics educators need a lively interaction with MATHEMATICS and they must devote an essential part of their professional lives to stimulating, observing and analyzing genuine MATHEMATICAL activities of children, students and student teachers. Organizing and observing the fascinating encounter of human beings with MATHEMATICS is the very heart of didactic expertise and forms a natural context for professional exchange with teachers.

As a part of MATHEMATICS, specialized mathematics must be taken seriously by mathematics educators as one point of view that, however, has to be balanced with other points of view. The history of mathematics education clearly demonstrates the risks of following specialized mathematics too closely: On the one hand, subject matter and elements of mathematical language can be selected that do not make much sense outside specialized mathematics – perhaps a lasting example of this mistake is the New Maths movement. On the other hand, the educationally important fields of MATHEMATICS that are no longer alive in specialized research and teaching may lose the proper attention – perhaps the best example for this second mistake is elementary geometry.

Mathematics educators must be aware that school mathematics cannot be derived from specialized mathematics by a 'transposition didactique du savoir savant au savoir enseigné' (cf. Freudenthal 1986). Instead, they must see school mathematics as an extension of pre-mathematical human capabilities which develop within the broader societal context provided by MATHEMATICS (cf. Schweiger 1994, p. 299 and Dörfler 1994, as well as the concept of 'ethnomathematics' in D'Ambrosio 1986). It is only from this perspective that the unity of mathematics teaching from the primary through the upper secondary level can be established and that reasonable mathematical courses in teacher training can be developed which deserve to be called a scientific background of teaching.[3]

2. A BASIC PROBLEM IN THE PRESENT DEVELOPMENT OF MATHEMATICS EDUCATION: THE NEGLECT OF THE CORE

The 'hard sciences' are successful, as they deal with 'soft problems'. The 'soft sciences' are badly off, as they are confronted with 'hard problems'.

(Heinz v. Foerster)

An approach to the study of problems of learning and teaching in mathematics education requires a scientific framework that includes both research methods and standards. As a young discipline, mathematics education is under considerable pressure from different directions. How to establish standards is as controversial as the status of mathematics education itself and can likewise be addressed in different ways.

One tempting approach is to adapt methods and standards from the hard sciences and the humanities. I dare say that all around the world quite a number of mathematics educators are taking this approach wherein the scientific background and their personal interests might be as influential as the wish to be recognized and supported by scientists in the related disciplines. However, approaches, methods and standards adopted from related disciplines are more easily applied to problems in the neighborhood of these disciplines than to problems in the core. Consequently, a great deal of didactic research adheres to mathematics, psychology, pedagogy, sociology, history of mathematics and so forth. Thus the holistic origin of didactic thinking, namely mathematical activity in social contexts, is dissolved into single strands, and the specific tasks of the core are neglected. In my view this is a big problem that presently inhibits major progress in mathematics education. The problem is by no means restricted to mathematics education, however. For example, Clifford and Guthrie (1988, p. 3) have identified it as a universal problem in education:

Our thesis is that schools of education, particularly those located on the campuses of prestigious re-search universities have become ensnared improvidently in the academic and political cultures of their institutions and have neglected their own worlds. They have seldom succeeded in satisfying the scholarly norms of their campus letters and science colleagues, and they are simultaneously estranged from their professional peers. The more they have rowed toward the shores of scholarly research the more distant they have become from the public schools they are bound to serve (Clifford and Guthrie 1988, p. 3).

The movement away from the core and towards the related areas may also be problematic because very often the adoption of frameworks and standards from related disciplines is linked to the dogmatic claim that these frameworks and standards were the only ones possible for didactics. From this position follows a blindness towards the central tasks of mathematics education and a systematic underestimation of the *constructive achievements* brought about in the core. Sometimes the core is even denied a scientific status. Mathematics educators who retreat into a 'mathematical garden' (H. Meschkowski, oral communication) tend of course to trivialize the educational aspects of mathematics education; similarly, those working in the areas related to psychology and pedagogy neglect the mathematical aspects. These tendencies are reinforced by voices from the related disciplines that argue against the scientific status of didactics more or less publicly. As a result we have an unreasonable set back into reductionist positions analyzed as unfounded many years ago (cf. Bigalke 1985; Winter 1985). It is ironic that mathematics education set out in the late sixties to overcome exactly these polarized positions. What is urgently needed therefore is a methodological framework that does justice to the core of mathematics education.

3. MATHEMATICS EDUCATION AS A SYSTEMIC-EVOLUTIONARY 'DESIGN SCIENCE'[4]

It is the yardstick that creates the phenomena.... A religious phenomenon can only be revealed as such if it is captured in its own modality, i.e., if it is considered by means of a religious yardstick. To locate such a phenomenon by means of physiology, psychology, sociology, economics, linguistics, art, etc. means to deny it. It means to miss exactly its uniqueness and its irreducibility.

(Mircea Eliade, *The Religions and the Sacred*)

Establishing scientific standards in mathematics education by adopting standards from related disciplines is, as mentioned, unwise because problems and tasks of mathematics education tend to be tackled only insofar and to the extent that they are accessible to the methods of the related disciplines. As a consequence, the core is not sufficiently recognized as a scientific field in its own right.

Fortunately there is a silver lining in this dilemma if one abandons the fixation on the traditional structures of the scientific disciplines and instead looks at the specific character of the core, namely the constructive development of and research into mathematics teaching. Here mathematics education is assigned to the larger class of 'design sciences' (cf. Wittmann 1975) whose scientific status was clearly delineated from the scientific status of natural sciences by the Nobel Prize Winner Herb Simon. The following quotation from Simon (1970, pp. 55–58) explains also the resistance offered to the design sciences in academia. In this way the present situation of mathematics education is embedded into a wider context and becomes accessible to a rational evaluation.

Historically and traditionally, it has been the task of the science disciplines to teach about natural things: how they are and how they work. It has been the task of engineering schools to teach about artificial things: how to make artifacts that have desired properties and how to design...

Design, so construed, is the core of all professional training; it is the principal mark that distinguishes the professions from the sciences. Schools of engineering, as well as schools of architecture, business, education, law and medicine, are all centrally concerned with the process of design.

In view of the key role of design in professional activity, it is ironic that in this century the natural sciences have almost driven the sciences of the artificial from professional school curricula. Engineering schools have become schools of biological science; business schools have become schools of finite mathematics...

The movement toward natural science and away from the sciences of the artificial has proceeded further and faster in engineering, business and medicine than in the other professional fields I have mentioned, though it has by no means been absent from schools of law, journalism and library science...

Such a universal phenomenon must have a basic cause. It does have a very obvious one. As professional schools ... are more and more absorbed into the general culture of the university, they hanker after academic respectability. In terms of the prevailing norms, academic respectability calls for subject matter that is intellectually tough, analytic, formalizable and teachable. In the past, much, if not most, of what we knew about design and about the artificial sciences was intellectually soft, intuitive, informal and cookbooky. Why would anyone in a university stoop to teach or learn about designing machines or planning market strategies when he could concern himself with solid-state physics? The answer has been clear: he usually wouldn't ...

The older kind of professional school did not know how to educate for professional design at an intellectual level appropriate to a university; the newer kind of school has nearly abdicated responsibility for training in the core professional skills ...

The professional schools will reassume their professional responsibilities just to the degree that they can discover a science of design, a body of intellectually tough, analytic, partly formalizable, partly empirical, teachable doctrine about the design process.

It is the thesis of this chapter that such a science of design not only is possible but is actually emerging at the present time (Simon 1970, pp. 55–58).[5]

In my opinion the framework of a design science opens up to mathematics education a promising perspective for fulfilling its tasks and also for developing an unbroken self-concept of mathematics educators. This framework supports the position described in part 2, for the core of mathematics education concentrates on constructing '*artificial objects*', namely teaching units, sets of coherent teaching units and curricula as well as the investigation of their possible effects in different educational 'ecologies'. *Indeed the quality of these constructions depends on the theory-based constructive fantasy, the 'ingenium', of the designers, and on systematic evaluation, both typical for design sciences.* How well this conception of mathematics education as a design science reflects the professional tasks of teachers is shown, for example, by Clark and Yinger (1987, pp. 97–99) who have identified teaching as a 'design profession'.

The clear structural delineation of mathematics education as a *design* science from the related sciences underlines its specific character and its relative independence. Mathematics education is not an appendix to mathematics, nor to psychology, nor to pedagogy for the same reason that any other design science is not an appendix to any of its related disciplines. Attempts to organize mathematics education by using related disciplines as models miss the point *because they overlook the overriding importance of creative design for conceptual and practical innovations.*

As far as research frameworks and standards are concerned, mathematics educators working in the core should primarily start from the achievements in the core already available. There is no doubt that during the past 25 years a significant progress, including the creation of theoretical frameworks, has been made within the core and that standards have been set which are well-suited as an orientation for the future. 'Developmental research' as suggested by Freudenthal and elaborated by Dutch mathematics educators is a typical example (cf. Freudenthal 1991, pp. 160–161; Gravemeijer 1994). Of course, it is reasonable also to adopt methods and standards from the related disciplines to the extent that they are appropriate to the problems of the core.

It is no surprise that the objections to the view of mathematics education as a 'design science' emerge, for the simple reason that the design sciences have traditionally followed – and are still widely following – a mechanistic paradigm whose harmful side effects are becoming more and more visible. This approach would certainly be detrimental to education. However, we are presently witness to the rise of a new paradigm for the design sciences that is based on the 'systemic-evolutionary' development of living systems and takes the complexity and self-organization of these systems into account (cf. Malik 1986). Even if researchers in the design sciences in general hesitate to adopt this new paradigm, there is no reason why mathematics educators should not follow it, even more so since this paradigm corresponds to recent developments in the field. The systemic-evolutionary view on the teacher–student and the theorist–practitioner relationships differs greatly from the traditional view. Knowledge is no longer seen as the result of a transmission from the teacher to a passive student, but is conceived of as the productive achievement of the student who learns in social interaction with other students and the teacher. Therefore the materials developed by mathematics educators must be construed so as to acknowledge and allow for this interactive approach. In particular, they must provide teachers and students the freedom to make choices of their own. In order to facilitate and stimulate a flexible use of the materials designed in this way, teachers have to be trained and regarded as partners in research and development and not as mere recipients of results (cf. Schupp 1979; Schwab 1983; Fischer & Malle 1983; and the papers by Brown & Cooney, von Harten & Steinbring, Voigt, and others in *Zentralblatt für Didaktik der Mathematik* 4 (91) and 5 (91)). As a consequence, teacher training receives a new quality. An important orientation for innovations along these lines is the approach developed by Schön (1987) for the training of engineers that is based upon the idea of the 'reflective practitioner'.

As a systemic-evolutionary design science mathematics education can follow different paths. It is certainly not reasonable to develop it into a 'monoparadigmatic' form as postulated, for example, for the natural sciences. In a design science the simultaneous appearance of different approaches is a sign of progress and not of retardation as stated by Thommen (1983, p. 227) for management theory:

Because of a continuously changing economic world it is possible to (re-)construct an economic context within different formal frameworks or models. These need not be mutually exclusive, on the contrary, they can even be complementary, for no model can take all problems and aspects into account as well as consider and weigh them equally. The more models exist, the more problems and aspects are studied, the greater is the chance for mutual correction. Therefore we consider the variety of models in management theory as an indicator for an advanced development of this field moving on in an evolutionary, not a revolutionary process in which new models emerge and old ones disappear.

4. THE DESIGN OF TEACHING UNITS AND EMPIRICAL RESEARCH

That, in concrete operation, education is an art, either a mechanical art or a fine art, is unquestionable. If there were an opposition between science and art, I should be compelled to side with those who assert that education is an art. But there is no opposition, although there is a distinction.

(John Dewey, *On the sources of a science of education*)

For developing mathematics education as a design science it is crucial to find ways how design on the one hand and empirical research on the other can be related to one another. In the following I propose a specific approach to empirical research, namely empirical research centered around teaching units.

It cannot be denied that teaching units, and on a wider scale curricula, have found attention in mathematics education in the past. In fact, curriculum development held a prominent place in the late sixties and early seventies. Nevertheless, I contend that the design of teaching units has never been a focus of research. At best teaching units have been used as more or less incidental examples in investigating and presenting theoretical ideas. Many of the best units were published in teachers' journals, not in research journals, and were hardly noticed by the research community. For this phenomenon the following explanation is offered: in contrast to 'research', the design of teaching has been considered as a mediocre task normally done by teachers and textbook authors. To rephrase Herb Simon: why should anyone anxious for academic respectability stoop to designing teaching and put him- or herself on one level with teachers? The answer has been clear: he or she usually wouldn't.

In order to overcome this fundamentally incorrect view we have to recognize that in all fields of design there is – by the very nature of design – a wide spectrum of competence and experience ranging from the amateur, to the novice, the less or more skilled worker, the experienced master, up to the creative inventor. Typically, the bulk of design on a larger scale is done in special centers for research and development. As a design science mathematics education can be no exception from this rule. That teachers take part in design can be no excuse for mathematics educators to refrain from this task. On the contrary: the design of substantial teaching units, and particularly of substantial curricula, is a most difficult task that must be carried out by the experts in the field. By no means can it be left to teachers, though teachers can certainly make important contributions within the framework of design provided by experts, particularly when they are members of or in close connection with a research team. Also, the adaptation of teaching units to the conditions of a special classroom requires design on a minor scale. Nevertheless, a teacher can be compared more to a conductor than to a composer or perhaps better to a director ('metteur en scène') than to a writer of a play. For this reason there should exist strong reservations about 'teachers' centers' wherein teachers meet to make their own curriculum.

We should be anxious to delineate teaching units of the highest quality from the mass of units developed at various levels for various purposes. These 'substantial' teaching units can be characterized by the following properties:

1. They represent central objectives, contents and principles of mathematics teaching.
2. They provide rich sources for mathematical activities.
3. They are flexible and can easily be adapted to the conditions of a special classroom.
4. They involve mathematical, psychological and pedagogical aspects of teaching and learning in a holistic way, and therefore they offer a wide potential for empirical research.

Typically, a substantial teaching unit always carries a name. As examples I mention 'Arithmogons' by Alistair McIntosh and Douglas Quadling, 'Mirror cards' by Marion Walter, 'Giant Egbert' and other units developed in the Dutch Wiskobas project, and Gerd Walther's unit 'Number of hours in a year'. Other examples and a systematic discussion of the role of substantial teaching units in mathematics education are given by Wittmann (1984).

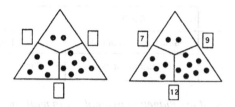

Figure 2

For the sake of clarity, one example of a substantial teaching unit is sketched below. In our primary school project 'Maths 2000' the following setting of arithmogons is used in grade 1:

A triangle is divided in three fields by connecting its midpoint to the midpoints of its sides. We put counters or write numbers in the fields. The simple rule is as follows: Add the numbers in two adjacent fields and write the sum in the box of the corresponding side (see Figure 2). Various problems arise: when starting from the numbers inside, the numbers outside can be obtained by addition. When one or two numbers inside and respectively two or one number outside are given, the missing numbers can be calculated by addition and subtraction. When the three numbers outside are given, we have a problem that does not allow for direct calculation but requires some thinking. It turns out that there is always exactly one solution. However, it may be necessary to use fractions or negative numbers.

The mathematics behind arithmogons is quite advanced: the three numbers inside form a vector as well as the three numbers outside. The rule of adding numbers in adjacent fields defines a linear mapping from the three-dimensional vector space over the reals into itself. The corresponding matrix is non-singular. One can generalize the structure to n-gons as shown in McIntosh and Quadling (1975).

The teaching unit based upon arithmogons consists of a sequence of tasks and problems that arise naturally from the mathematical context. The script for the teacher may be structured as follows:

1. Introduce the rule by means of examples and make sure that the rule is clearly understood.
2. Present some examples in which the numbers inside are given.
3. Present some examples in which some numbers inside and some numbers outside are given.
4. Present a problem in which the numbers outside are given.
5. Present other problems of this kind.

As can be seen, a substantial teaching unit is essentially open. Only the key problems are fixed. During each episode the teacher has to follow the students' ideas in trying to solve the problems. This role of the teacher is completely

	Tools	Method
Piagetian Psychology	Structured sets of tasks	Clinical interviews
Mathematics Education	Teaching units	Clinical teaching experiments

Table 1   *Adaptation of the Piagetian psychology to mathematics education*

different from traditional views of teaching. Teaching a substantial unit is basically analogous to conducting a clinical interview during which only the key questions are defined and the interviewer's task is to follow the child's thinking.

The structural similarity between substantial teaching units on the one hand, and clinical interviews on the other, suggests an adaptation of Piaget's method for studying children's cognitive development to empirical research on teaching units (see Table 1). As a result we arrive at 'clinical teaching experiments' in which teaching units can be used not only as research tools, but also as objects of study.

The data collected in these experiments have multiple uses: they tell us something about the teaching/learning processes, individual and social outcomes of learning, children's productive thinking, and children's difficulties. They also help us to evaluate the unit and to revise it in order to make teaching and learning more efficient.

The Piagetian experiments were repeated many times by other researchers. Many became a focus of extended psychological research. Some even established special lines of study; for example, the 'conservation' experiments. It is no exaggeration to say that Piaget's experiments and the patterns he observed in children's thinking survived much longer than his theories, in many cases until the present. In the same way, clinical teaching experiments can be repeated and thereby varied. By comparing the data we can identify basic patterns of teaching and learning and derive well-founded specific knowledge on teaching certain units. Much can be learned here from Japanese research in mathematics education (cf. Becker & Miwa 1989).

In conducting such studies, existing methods of qualitative research can be effectively used, particularly those developed by French mathematics educators in connection with 'didactic situations' and with 'didactic engineering' (cf. Brousseau 1986; Artigue & Perrin-Glorian 1991; Arsac et al. 1992). Concerning the reproducibility of results it is very instructive to look at the social sciences. Friedrich von Hayek, another Nobel Prize winner in economics, has convincingly pointed out that empirical research on highly complex social phenomena yields reproducible results if directed towards revealing general patterns beyond special data (von Hayek 1956). To admit that the results of teaching and learning depend on the students and on the teacher does not preclude the existence of patterns related to the mathematical content of a specific teaching unit (cf. also Kilpatrick 1993, pp. 27–29; Sierpinska 1993, pp. 69–71). Of course, we must not expect all these patterns to arise on any occasion, nor under all circumstances. It is quite natural that patterns will occur, varying with the educational ecologies. One should be reminded here of the well-known fact that Piagetian interviews also reveal recurring content-specific patterns which, however, do not occur with every individual child.

Research centered around teaching units is useful for several reasons. First, it is related to the subject matter of teaching (cf. the postulate of 'relatedness' in Kilpatrick 1993, p. 30). Second, knowledge obtained from clinical teaching experiments is 'local'. Here we need to be more careful in generalizing over contents than we have been in the past. In the future we can certainly expect to derive theories covering a wide range of teaching and learning. But these theories cannot emerge before a variety of individual teaching units has been investigated in detail. For studying the mathematical theory of groups the English mathematician Graham Higman stated in the fifties 'that progress in group theory depends primarily on an intimate knowledge of a large number of special groups' (Higman, informal paper). The striking results achieved in the eighties in the classification of finite simple groups showed that he was right. In a similar way, the detailed empirical study of a large number of substantial teaching units could prove equally helpful for mathematics education.

Third, theory related to teaching experiments is meaningful and applicable. However, we should be aware that, due to the inherent complexity of teaching and learning, the data and theories that research might provide may never provide complete information for teaching a certain unit. Only the teacher is in a position to determine the special conditions in his or her classroom. Therefore there should be no sharp separation between the researcher and the teacher as stated earlier. As a consequence, teachers have to be equipped with some basic competence in doing research on a small scale. My experience in teacher training indicates that introducing student teachers into the method of clinical interviews is an excellent way towards that end (Wittmann 1985).

In my opinion, the most important results of research in mathematics education are sets of carefully designed and empirically studied teaching units that are based on fundamental theoretical principles. It follows that these units should form a major part of the professional training of teachers. Teachers who leave the university should have in their baggage a set of substantial teaching units that represent the standards of teaching. From the experiences with our primary school project 'Maths 2000' it is clear that such units are the most efficient carriers of innovation and are well-suited to bridge the gap between theory and practice.

In concluding this section, it is important to again emphasize that the design of teaching units and empirical research centered around them can only be successfully carried out within a system of mathematics education that consists of the core, related areas, related disciplines and lively interactions among these components. In particular, strong links to mathematics in the broader sense (i.e. MATHEMATICS) are necessary.

5. AND THE FUTURE OF MATHEMATICS EDUCATION?

The frogs tend to forget that once they were tadpoles, too.

(Korean proverb)

Generally speaking, it may be taken for granted that dealing in an intelligent way with complex systems on a scientific basis will become inevitable in all parts of human life. Very often the methods offered by the specialized disciplines are not sufficient. Riedel (1988) recently pleaded for a more context-related, more practical and less-formal 'second philosophy', in contrast to the traditional 'first philosophy' that aims at complete descriptions and deductions and that is bound to fail when applied to complex systems, because of its 'ideology of self-restriction' (Fischer 1980). This seems to be a signal for a critical reflection in all sciences from which mathematics education as a systemic-evolutionary design science can take profit in the long range, since society will have to accept the fact that the development of human resources is at least as important for economic prosperity as the development of new technologies and new marketing strategies.

In the short run the status of didactics in the universities will remain arduous. The resistance from the specialized sections within the related disciplines to establishing didactics in teacher training programs at all levels and to funding research in didactics is likely to continue. The history of the universities shows many instances in which scholars of established disciplines displayed their ignorance and acted in an unfair way towards newly evolving disciplines. The resistance of the old universities towards the technical schools, and the resistance of pure mathematicians towards applied ones at the turn of the century and the vote of the German Philosophical Society against the establishment of chairs of pedagogy at the universities in the fifties are only a few examples. Obviously it is difficult, if not impossible, for specialists to understand and to appreciate new developments on the very borderline of their discipline.

In order to strengthen their position at the universities and to acquire funds from research foundations, mathematics educators need support from society. In this respect the relationships of mathematics education to the schools play a fundamental role. The use and the indispensability of didactic research for improving practice have to be convincingly demonstrated to teachers, supervisors, administrators, parents and the public. This can only be achieved from the core, that is, by concentrating on central tasks and by organizing design, empirical research and teacher education accordingly.

At the same time, there is potential in establishing a network of 'Public–School–School Administration–Teachers' Unions–Teacher Training–Design, Research, Development' people in which the core of mathematics education will naturally find its proper place. In other words, organizing a systemic effort involving all the constituent groups.

This is consistent with the advice given by Clifford and Guthrie to schools of education in general (cf. Clifford & Guthrie 1988, pp. 349–350):

The major mission of schools of education should be the enhancement of education through the preparation of educators, the study of the educative process, and the study of schooling as a social institution. As John Best has observed, the challenge before schools of education is quite different from that confronting the specialist in politics in a department of political science; concerned with building the discipline, he or she is under no obligation to train county clerks, city managers, and state legislators, and to improve their performance by conducting research directed toward that end. In order to accomplish their charter, however, schools of education must take the profession of education, not academia, as their main point of reference. It is not sufficient to say that the greatest strength of schools of education is that they are the only places available to look at fundamental issues from a variety of disciplinary perspectives. They have been doing so for more than half a century without appreciable effect on professional practice. It is time for many institutions to shift their gears (Clifford & Guthrie 1988, pp. 349–350).

NOTES

1. This chapter is a revised and extended version of the paper 'Mathematikdidaktik als "design science"' , published in *Journal für Mathematikdidaktik* 13 (1992), 55–70. I am indebted to Jerry P. Becker, P. Bender, H. Besuden, W. Blum, E. Cohors-Fresenborg, Th. J. Cooney, L. Führer, H. N. Jahnke, A. Kirsch, G. N. Müller, H.-Chr. Reichel, H. Schupp, Ch. Selter, H.-J. Vollrath, J. Voigt, G. Walther and H. Winter for critical remarks to earlier drafts. The paper was published in 1995 in *Educational Studies in Mathematics* **29**, 355–374, and is reprinted in this volume with the permission of the publishers.

2. The present paper concentrates on the didactics of mathematics although the line of argument pertains equally to the didactics of other subjects and also to education in general (cf. Clifford & Guthrie 1988, a detailed study on the identity crisis of the Schools of Education at the leading American universities).

3. I do not intend to give mathematical specialists an advice they have not asked for. However, in my opinion, it would also be beneficial for them to perceive themselves as partners in a larger mathematical system described by MATHEMATICS. Without some change of awareness on their part, all attempts to change the public image of 'mathematics' are nothing but cosmetic and bound to fail.

4. The term 'design' and related terms used subsequently in this chapter might cause irritation, for in traditional understanding these terms are linked to mechanistic procedures of making tools and controlling systems (cf. Jackson 1968, 163 ff.). In the third part of this paper we will show, however, that, in striking contrast to the 'mechanistic' paradigm of design and management, there is a new 'systemic-evolutionary' paradigm based on the appreciation of the complexity and self-organization of living systems. It is in the context of this new paradigm that the term 'design' and similar ones are being used.

5. The underestimation of the 'skills of designing and making' is deeply rooted in our culture (cf. Smith 1980, p. 22):

> Throughout the whole of our society we show little respect for the skills of designing and making. Indeed in many of our schools these very skills are looked down upon and are referred to as the noddy subjects, fit only for the less able in our community.
> I remember, during my years as chairman of the Schools Council, visiting a school where, after I had been shown the fairly conventional range of school work, I was taken into the workshops and there on the bench was a most beautiful and competent piece of metal work. It was a joy to look at, but it was described to me as a piece of work 'by one of our less able pupils'. It was an extraordinary description, which spoke volumes about our distorted scale of values. There was a

piece of work which expressed ability, as fine in its way as the best essay written by the highest flyer in English, but never seen by academic people as such. To write things with pen on paper is an up-marked, respectable activity; to conceive pattern in your mind and to make them with your hands is a down-marked activity, less worthy of respect.

REFERENCES

Arsac, G. et al.: 1992, 'Teacher's Role and Reproducibility of Didactical Situations', *Educational Studies in Mathematics* **23**, 5–29.

Artigue, M. & Perrin-Glorian, M.-J.: 1991, 'Didactic Engineering, Research and Development Tool: Some Theoretical Problems Linked to this Duality', *For the Learning of Mathematics* **11**, 13–18.

Balacheff, N. et al.: 1992, 'What is Research in Mathematics Education and What Are Its Results? Discussion Document for an ICMI Study', *Bulletin of the International Commission on Mathematical Instruction* **33**, December, 17–23.

Bauersfeld, H.: 1988, 'Quo Vadis? Zu den Perspektiven der Fachdidaktik', *Mathematica Didactica* **11**, 3–24.

Becker, J. P. & Miwa, T.: 1989, *Proceedings of the U.S.–Japan Seminar on Mathematical Problem Solving*, Columbus, Ohio: ERIC Clearinghouse for Science and Mathematics (ED 304 3/5).

Bigalke, H.-G.: 1974, Sinn und Bedeutung der Mathematikdidaktik, *Zentralblatt für Didaktik der Mathematik* 3, 109–115.

Bigalke, H.-G.: 1985, 'Beiträge zur wissenschaftstheoretischen Diskussion der Mathematikdidaktik', in M. Bönsch, and L. Schäffner, *Theorie und Praxis. Schriftenreihe aus dem FB Erziehungswissenschaften I der Universität Hannover.*

Brousseau, G.: 1986, Fondéments et Méthodes de la Didactique des Mathématiques. Thèse d'État, Université de Bordeaux I.

Brown, S. I. & Cooney, Th. J.: 1991, Stalking the dualism between theory and practice, *Zentralblatt für Didaktik der Mathematik* 20, 112–117.

Burscheid, H.-J.: 1983, 'Formen der wissenschaftlichen Organisation in der Mathematikdidaktik', *Journal für Mathematikdidaktik* 4, 219–240.

Clark, Ch. M. & Yinger, R. J.: 1987, 'Teacher Planning', in J. Calderhead (ed.), *Exploring Teachers' Thinking*, The Falmer Press, London.

Clifford, G. J. & Guthrie, J. W.:1988, *Ed School, A Brief for Professional Education*, University of Chicago Press, Chicago.

D'Ambrosio, U.: 1986, 'Socio-Cultural Bases for Mathematical Education', *Proceedings of the Fifth International Congress on Mathematical Education*, Birkhäuser, Boston, MA, 1–6.

Dörfler, W.: 1994, 'The Gulf between Mathematics and Mathematics Education', in *What is Research in Mathematics Education and What Are its Results?* (Background papers for an ICMI Study Conference), International Commission on Mathematical Instruction, University of Maryland, College Park, MD.

Dress, A.: 1974, Spekulationen über Aufgaben und Möglichkeiten einer Didaktik der Mathematik, *Zentralblatt für Didaktik der Mathematik* 3, 129–131.

Fischer, R.: 1980, 'Zur Ideologie der Selbstbeschränkung im Mathematikstudium', in *Mathematikunterricht an Universitäten. Zweiter Teil. Zeitschrift für Hochschuldidaktik*, Vienna, Sonderheft S3, 32–72

Fischer, R. & Malle, G.: 1983, *Mensch und Mathematik*, Bibliographisches Institut, Mannheim.

Freudenthal, H.: 1974, Sinn und Bedeutung der Didaktik der Mathematik, *Zentralblatt für Didaktik der Mathematik* 3, 122–124.

Freudenthal, H.: 1986, Review of Yves Chevallard, 'La Transposition Didactique du Savoir Savant au Savoir Enseigné', *Educational Studies in Mathematics* **17**, 323–327.

Freudenthal, H.: 1987, 'Theoriebildung zum Mathematikunterricht', *Zentralblatt für Didaktik der Mathematik* 3, 96–103.

Freudenthal, H.: 1991, *Revisiting Mathematics Education. China Lectures*, Kluwer Academic Publishers, Dordrecht.

Gravemeijer, K.: 1994, 'Educational Development and Development Research in Mathematics Education', *Journal for Research in Mathematics Education* **25** (5), 443–525.

Griesel, H.: 1974, Überlegungen zur Didaktik der Mathematik als Wissenschaft, *Zentralblatt für Didaktik der Mathematik* 3, 115–119.

Jackson, Ph.A.: 1968, *Life in Classrooms*, Holt, Rinehart and Winston, New York.

Kilpatrick, J.: 1993, 'Beyond Face Value: Assessing Research in Mathematics Education', in G. Nissen, and M. Blomhøj (eds.), *Criteria for Scientific Quality and Relevance in the Didactics of Mathematics*, Roskilde University, Denmark, 15–34.

Krygowska, A. Z.: 1972, 'Mathematik-didaktische Forschung an der Pädagogischen Hochschule Krakau', *Beiträge zum Mathematikunterricht 1971*, Schroedel Hannover, 117–125.

McIntosh, A. and Quadling D.: 1975, 'Arithmogons', *Mathematics Teaching* **70**, 18–23.

Malik, F.: 1986, *Strategie des Managements komplexer Systeme*, Haupt, Bern.

Otte, M.: 1974, Didaktik der Mathematik als Wissenschaft, *Zentralblatt für Didaktik der Mathematik* 3, 125–128.

Riedel, M.: 1988, *Für eine zweite Philosophie*, Suhrkamp, Frankfurt a.M.

Schön, D.: 1987, *Educating the Reflective Practitioner*, Jossey-Bass, San Francisco, CA.

Schupp, H.: 1979, 'Evaluation eines Curriculums', *Der Mathematikunterricht* **25**, 22–42.

Schwab, J.: 1983, 'The Practical 4: Something for Curriculum Professors to Do', *Curriculum Inquiry* **13**, 239–265.

Schweiger, F.: 1994, 'Mathematics is a Language', in D. F. Robitaille, et al. (eds.), *Selected Lectures from the 7th International Congress on Mathematical Education Québec 1992*, Les Presses de l'Université Laval, Sainte Foy, 197–309.

Sierpinska, A.: 1993, 'Criteria for Scientific Quality and Relevance in the Didactics of Mathematics', in G. Nissen, and M. Blomhøj (eds.), *Criteria for Scientific Quality and Relevance in the Didactics of Mathematics*, Roskilde University Denmark, 35–74.

Simon, H. A.: 1970, *The Sciences of the Artificial*, MIT-Press, Cambridge, MA.

Smith, A.: 1980, *A Coherent Set of Decisions*, the Stanley Lecture, Manchester Polytechnic.

Thommen, J.-P.: 1983, *Die Lehre von der Unternehmensführung*, Haupt, Bern and Stuttgart.

Tietz, H.: 1974, Zur Didaktik der Mathematik, *Zentralblatt für Didaktik der Mathematik* 3, 131–132.

Voigt, J.: 1991, Interaktionsanalysen in der Lehrerbildung, *Zentralblatt für Didaktik der Mathematik* 20, 161–168.

von Harten, G. & Steinbring, H.: 1991, Lesson transcripts and their role in the in-service training of mathematics teachers, *Zentralblatt für Didaktik der Mathematik* 20, 169–177.

Von Hayek, F. A.: 1956, 'The Theory of Complex Phenomena', in F. A. Von Hayek (ed.), *Studies in Philosophy, Politics, Economics*, Routledge & Kegan Paul, London 1967, 22–42.

Winter, H.: 1985, 'Reduktionistische Ansätze in der Mathematikdidaktik', *Der Mathematikunterricht 31*, 75–88.

Winter, H.: 1986, 'Was heißt und zu welchem Ende studiert man Mathematikdidaktik?', in H. Schanze (ed.), *Lehrerbildung in Aachen – Geschichte, Entwicklungen, Perspektiven*, Pädagogische Hochschule, Aachen, 174–194.

Wittmann, E. Ch.: 1975, 'Didaktik der Mathematik als Ingenieurwissenschaft', *Zentralblatt für Didaktik der Mathematik* 3, 119–121.

Wittmann, E . Ch.: 1984, 'Teaching Units as the Integrating Core of Mathematics Education', *Educational Studies in Mathematics* **15**, 25–36.

Wittmann, E. Ch.: 1985, 'Clinical Interviews embedded in the "philosophy of teaching units" – a means of developing teachers' attitudes and skills', in B. Christiansen (ed.), *Systematic Cooperation between Theory and Practice in Mathematics Education*, Mini-Conference at ICME 5, Adelaide 1984, Copenhagen: Royal Danish School of Education, Dept. of Mathematics 1985, 18–31.

Erich Ch. Wittmann
University of Dortmund,
Department of Mathematics,
D-44221 Dortmund,
Germany

ROBERTA MURA

WHAT IS MATHEMATICS EDUCATION?
A SURVEY OF MATHEMATICS EDUCATORS IN CANADA

This chapter is about some results from a survey of mathematics educators in Canadian universities that was conducted in 1993. The aim of the survey was to collect information about their social backgrounds, education, careers and views about mathematics and mathematics education, in order to gain a sense of the identity of this professional community. The results presented here are mainly those concerning the views about mathematics education. The remaining results are discussed in Mura (1995a, 1995b).

After giving a few methodological details about the survey and making some comments on the translation of the terms 'mathematics education' and 'didactique des mathématiques', I shall briefly describe the kind of university degrees and employment held by mathematics educators. I will then turn to the main subject of this chapter, namely how mathematics educators define mathematics education and which books they consider to have had the most influence on the field.

1. THE SURVEY

My intention was to reach all mathematics educators who were faculty members of a Canadian university. In order to do so, I sent questionnaires to all those whose name appeared in the mailing list of the Canadian Mathematics Education Study Group or in its directory of current research (Kieran & Dawson 1992). I also asked each of these people to name all mathematics educators in their own universities, and I then sent questionnaires to the additional individuals identified in this way.

The cover page of the questionnaire contained the following two questions: 'Do you hold a tenured or tenure-track position at a Canadian university?' and 'Is mathematics education your primary field of research and teaching?'. Those who did not answer positively both questions were not part of the target population and were invited to return the questionnaire without completing it any further.

Altogether 158 questionnaires were sent off by mail. After two reminders, 106 (67%) were returned. Of these, 63 were completed by respondents belonging to the target population and were retained for the present study. The sample consisted of 44 men (70%) and 19 women (30%). The median age of the group was 50 years, with a range from 30 to 64. Forty-one respondents (65%) spoke English at work and 22 (35%) spoke French.

Sierpinska, A. and Kilpatrick, J. Mathematics Education as a Research Domain: A Search for Identity, 105–116.

The instrument designed to collect the data was a questionnaire comprising 54 questions, 7 of which were open-ended.[1] An English or a French version of the questionnaire was used as appropriate.

2. MATHEMATICS EDUCATION VS. DIDACTIQUE DES MATHÉMATIQUES

The French word 'didactique' does not have the same pejorative connotation that the English word 'didactics' seems to have. In the questionnaire, I used the term 'mathematics education' in English and 'didactique des mathématiques' in French, that is I considered the two terms to be translations of each other. This is common usage in Canada: for instance, the French name of the Canadian Mathematics Education Study Group is 'Groupe Canadien d'Étude en Didactique des Mathématiques'. However, the term 'didactique des mathématiques' is more specific than 'mathematics education', the latter being liable to carry all the various meanings of 'education': system of teaching, process by which one gains knowledge, knowledge gained, general area of work concerned with teaching, or field of study concerned with all of these. Such a variety of meanings can cause misunderstanding. For example, the term 'educator' applies to all teachers, hence, all mathematics teachers are, in a sense, mathematics educators.

Perhaps for this reason, some authors opt for the use of the words 'didactics' and 'didactician' in English, in spite of their present formality and negative connotation. Indeed, the expression 'didactics of mathematics' occurs in the title of a recent publication in Kluwer's series 'Mathematics Education Library' (Biehler et al. 1994). This choice might have been influenced by the fact that, although the volume is written in English, its four co-editors are German. Not all of the contributing authors follow the editors' example in this respect (most of the native English speakers do not), but those who do, having two terms at their disposal, are able to make a linguistic distinction between the discipline, 'didactics of mathematics', and its object of study, 'mathematics education' – that is, education in mathematics.[2]

3. WHO ARE MATHEMATICS EDUCATORS?

Since the vast majority of universities do not have mathematics education departments, 'mathematics educator' is a label that individual members of various departments may or may not choose to apply to themselves. A few of those who received my questionnaires hesitated before deciding whether mathematics education was their primary field of research and teaching. In most cases it was an interest of theirs, but was it a strong enough interest to be identified as the primary one? A member of a mathematics department wrote: 'I almost returned your questionnaire with a NO for mathematics education being my primary field, but I hesitated and now I am lost: I feel like I'm coming out of the closet! Confessing to mathematics education as my primary interest'.

Of the 63 who did acknowledge mathematics education as their primary field, and who were thus retained for the present study, 47 (75%) worked in education departments, 13 (21%) in mathematics departments and 3 had joint appointments. Eleven of the 13 who were employed by mathematics departments were concentrated in two Quebec institutions that assign to their mathematics departments the task of teaching mathematics education courses.

Fifty-five (87%) of the respondents had worked, some time in the past, at a school or a two-year college: 25 (40%) at elementary schools, 45 (71%) at secondary schools and 23 (36%) at two-year colleges (some had worked at more than one level).

Concerning their education, 56 (89%) of the respondents held doctoral degrees: 46 in education (including mathematics education), 8 in mathematics and 2 in psychology. Thirty-five (56%) held university degrees both in education and in mathematics, 16 (25%) did not have any degree in education and 12 (19%) did not have any degree in mathematics. Without exception, those who specified that their doctoral degrees were in mathematics education (32) classified them as degrees in education.

In summary, as a group, mathematics educators in Canada are closer to the field of education than to that of mathematics: they are more likely to have earned their doctoral degrees in education programs and to be employed by education departments. Thus, mathematics education tends to be institutionalized in a way that is not conducive to its becoming 'an integral part of mathematics' as advocated by Brousseau (1994). Furthermore, I am not sure that the mathematics community would be ready to encourage such a development. For instance, a participant in the present survey who had turned from mathematics to mathematics education mentioned 'the barbs from colleagues about how [he] was wasting [his] time in mathematics education'. Another mathematician, who participated in a survey that I conducted previously, in response to a request for a definition of mathematics, wrote: 'Mathematics is what people who call themselves mathematicians do. I would draw the line however at so-called Math Education. "Math Methods" is not mathematics' (underscoring in the original).

4. WHAT IS MATHEMATICS EDUCATION?

The question concerning views about mathematics education was open-ended and it read: 'How do you define mathematics education?', in English, and 'Comment définissez-vous la didactique des mathématiques?', in French. A space of eight blank lines was provided for the answer. Of the 63 respondents, 13 (21%) skipped this particular question. The 50 responses that were offered (33 in English and 17 in French) ranged in length from one single question mark to 72 words.

I analyzed these data by listing all the ideas that I could detect in each response. I then defined an initial set of eleven general themes emerging from this list and used them to produce a first classification of the responses. The themes were not meant to be exclusive categories, as each response could contain references to several of them. I then submitted the data and the set of themes to two independent judges, asking them to classify the data according to the given themes and to comment on the themes themselves. One of the judges was a history educator and the other one a mathematics educator who had not participated in the study.

The judges' comments and reactions led me to revise my original classification and set of themes, merging similar and overlapping themes, clarifying the definitions and dropping the themes that occurred very infrequently. This produced a final set of five themes that I shall present below, together with my revised classification. Taking into account the merging and dropping of themes, agreement of the two judges' original (and only) classifications with my revised one (that is, percentage of the occurrences of the themes identified by the judges that I too recognized) was respectively 90% and 92%. Conversely, of the occurrences of the themes identified by myself, all were recognized by at least one judge and 74% were recognized by both judges. The numbers in parentheses after each theme described below are respectively the number of respondents who made reference to it (N), and the numbers of these respondents who were English-speaking (NE) or French-speaking (NF). To interpret the data, one must keep in mind that the French responses constitute 34% of the sample.

Theme 1: *Mathematics education concerns the teaching of mathematics* (N = 36, NE = 25, NF = 11)
    Examples:

- The study of teaching contexts in the instruction of mathematics. (Classified also under Theme 3.)
- The process by which students are guided in their learning of mathematics concepts [...] (Classified also under Theme 4.)
- La didactique de la mathématique est l'étude des approches et des moyens à prendre pour aider les jeunes à construire leurs concepts mathématiques [...] (Classified also under Theme 4.)

Theme 2: *Mathematics education concerns the learning of mathematics* (N = 23, NE = 15, NF = 8)
    Examples:

- Systematic study (with students) of how persons come to know and understand mathematics [...] (Classified also under Theme 3.)
- An understanding of the way people learn mathematics. (Classified also under Theme 3.)
- [...] La didactique de la mathématique est l'étude de la compréhension de chaque notion mathématique, de chaque concept mathématique par l'apprenant [...] (Classified also under Theme 3.)

Theme 3: *Mathematics education is a theoretical pursuit. It is a (pure) science. It is the study of the teaching, learning and creation of mathematics; the analy-*

sis of the construction of mathematical concepts by the learners (N = 22, NE = 15, NF = 7)
Examples:

- The study of the philosophical, psychological and pedagogical issues raised by the teaching, learning and creation of mathematics. (Classified also under Themes 1, 2 and 5.)
- [...] Domaine qui s'intéresse à l'analyse des phénomènes d'enseignement et d'apprentissage des mathématiques avec ses propres concepts et théories fournissant un cadre explicatif possible à ces phénomènes. (Classified also under Themes 1 and 2.)
- [...] Domaine de l'activité scientifique qui s'intéresse particulièrement à la compréhension des phénomènes associés à l'apprentissage des mathématiques [...] (Classified also under Theme 2.)

Theme 4: *Mathematics education is a practical pursuit. It is an art, an applied science, an applied discipline. It is the search for ways to improve the teaching and learning of mathematics. It is a way of teaching mathematics. Mathematics education is mathematics teaching* (N = 21, NE = 14, NF = 7)
Examples:

- The study of ways to improve the teaching of mathematics. (Classified also under Theme 1.)
- Mathematics education should not seek to implant facts, rules, formulae, theorems, etc., but rather to awaken in the learner a personal drive for insights into the pattern, beauty and personal relevance of mathematics. (Classified also under Theme 1.)
- L'art de faire comprendre les mathématiques. (Classified also under Theme 1.)

Theme 5: *Mathematics education borrows from various disciplines; it is the interface between various disciplines* (N = 14, NE = 6, NF = 8). *The disciplines mentioned are: psychology, developmental psychology, psychology of the learner (12 times); education, pedagogy, theory of learning, curriculum theory (9 times); mathematics (9 times); epistemology (6 times); philosophy (3 times); sociology (twice); history (once) and the sciences and engineering (once)*
Examples:

- Mathematics education is not a discipline *per se*. It is an integrated application of mathematics, psychology (mainly epistemology), sociology and philosophy.
- C'est cette science [...] qui permet d'intégrer les diverses approches tant mathématiques, épistémologiques que psychologiques pour en faire une discipline en soi avec ses propres balises. (Classified also under Theme 3.)
- Point de rencontre de la discipline mathématique avec l'épistémologie, la pédagogie et la psychologie.

Themes 1 and 2 concern the object of mathematics education. Thirty-six respondents mentioned the teaching of mathematics and 23 its learning. The two categories, of course, are by no means mutually exclusive; indeed, 17 respondents mentioned both teaching and learning and one referred specifically to 'the relationship between the two'. However, in the present context of an open-ended question, teaching has come to the respondents' minds more frequently than learning.

Several responses classified under Themes 1 and 2 do not contain the actual words 'teaching' or 'learning' ('enseignement' or 'apprentissage'). Particularly in the case of 'teaching', the authors chose expressions that probably better convey something of their philosophies, like 'guiding', 'leading', 'awakening', 'supporting', 'helping', 'providing experiences' or 'construction of conditions'. One was explicit in her rejection of the word 'teaching': 'La didactique des mathématiques est une façon de faciliter la pensée mathématique. Nous ne pouvons pas "enseigner" la pensée mathématique, mais nous pouvons offrir des situations et des stratégies pour guider vers une pensée mathématique.' Similarly, learning was sometimes expressed as 'coming to know', 'understanding', 'construction' or 'appropriation des concepts'.

Two main tendencies of approximately equal weight appear in mathematics educators' descriptions of the goal of their field. The first one (Theme 3) is more theoretical and was put forward by 22 respondents. It identifies the aim of mathematics education as analyzing, understanding and explaining the phenomena of the teaching and learning of mathematics (some included the creating of mathematics as well).

The second tendency (Theme 4) is a more practical one and can be traced in the answers of 21 respondents. It assigns to mathematics education the goal to improve the teaching of mathematics and to facilitate its learning. Such a goal, of course, implies a value judgment on what constitutes improvement. Thus, for some mathematics educators, mathematics education means promoting or putting into practice particular methods and philosophies of teaching and learning mathematics. A few responses seem to me to use the expression 'mathematics education' as a synonym for 'mathematics teaching' (see, for instance, the second example illustrating Theme 4 above). In this perspective, mathematics education is action oriented and it is thought of as an art or an applied science.

Again, the two tendencies identified here are not mutually exclusive. In fact, four respondents integrated elements of both tendencies in their definitions of mathematics education. Contrary to what one might expect, even withdrawing these four individuals, the group who expressed a theoretical orientation and the group who expressed a practical orientation do not differ substantially from each other in their involvement in research as measured by the number of publications and communications, the number of theses supervised and manuscripts reviewed, membership in editorial boards, participation in joint research projects, co-authored publications and exchange of information with colleagues in Canada and abroad.

A final important idea that emerges from the analysis of the present data is the need for mathematics education to make use of several other disciplines (Theme 5), especially psychology, mathematics, general education theories and epistemology. This theme is the only one that showed a difference between the two language groups: it occurred 8 times among the French-speaking respondents compared to 6 times among the twice-as-large English-speaking group. Some people went as far as defining mathematics education as some sort of inter-section of various combinations of other disciplines. For most, this peculiar situation did not preclude mathematics education from being a discipline or a science in its own right (or an applied discipline or science). Others chose different terms, like 'domaine de l'activité scientifique' or 'plus un champ disciplinaire qu'une discipline', but one squarely stated that 'mathematics education is not a discipline *per se*'. Only two responses (one of them classified under this theme, both French) contained some comments asserting the autonomy of mathematics education, mentioning that it has its own concepts, theories or guide marks.

To pursue the issue of whether or not mathematics education is viewed by its practitioners as an autonomous discipline, I list below the actual terms that they employed to refer to it (I indicate in parentheses the number of responses in which a term occurs):

English: 'study' (11), 'systematic study' (1), 'applied discipline' (1), 'art and science' (1), 'interface of ...' (1), 'integrated application of ...' (1), 'process' (2).

French: 'étude' (4), 'examen' (1), 'domaine' (1), 'science' (5, once in quotes), 'domaine de l'activité scientifique' (1), 'science appliquée' (1), 'science et art' (2), 'art' (1), 'discipline' (2), 'champ disciplinaire' (1), 'point de rencontre de ...' (1), 'carrefour' (1), 'processus' (1).

The phrasing of several responses avoided categorizing mathematics education in this way, mostly by defining it as an activity (e.g. teaching, thinking, etc.).

Apparently, French-speaking respondents were more inclined to label their field a science than their English-speaking colleagues. This result may reflect a difference in the ways the two groups view mathematics education, or it may simply be due to a difference in the usage of the word 'science' in the two languages.

I shall close this section by quoting three definitions of mathematics education that at first puzzled me, but that make sense when one thinks of the various meanings carried by the word 'education' mentioned earlier:

- 'Teaching and learning mathematics at all levels and in all forms.'
- 'The integration and re-integration of mathematics with the living.'
- 'See "mathematics".'

The first one might correspond to a broad interpretation of 'education'. The second one remains rather cryptic, but it might allude to the process of acquiring an education in mathematics. As to the third one, its author, by referring to the definition of mathematics that he gave elsewhere in the questionnaire, seems to imply that mathematics education is identical with mathematics – that is the knowledge gained through one's education in mathematics.

5. BOOKS THAT INFLUENCED MATHEMATICS EDUCATION

The question about books that influenced mathematics education read: 'Please identify some of the books which, in your opinion, have had the most influence on the development of mathematics education. (Maximum of ten books.)' Twenty (32%) of the 63 respondents did not answer this question at all. Two of them explained their silence: 'Le domaine est tout neuf. Je ne suis pas sûr qu'il y ait eu un ouvrage décisif encore' and 'Answering this question would take too much thought and too much time'.

Most of those who did answer chose to name authors rather than book titles. One person justified his choice: 'I do not think that individual books have an impact as much as the collected works of an individual or possibly a group.' For this reason, I decided to compile statistics on authors rather than on book titles (in practice, I had to allow a few special cases, like the Soviet Studies series or the NCTM publications). When a respondent mentioned more than one book by the same author, I counted it as a single citation of that author.

The 43 respondents who answered the question discussed here mentioned on average 5.6 authors each. Altogether, 105 different authors were named. I shall present the results both globally and separately by language group, for, whether or not 'mathematics education' and 'didactique des mathématiques' mean the same thing, there is a difference in the writings that English-speaking and French-speaking mathematics educators consider to have had the most influence on their field.

The eight authors most frequently cited by English-speaking respondents were: NCTM (16 times), Polya (14 times), Piaget (10 times), Freudenthal (7 times), Bruner and Dienes (6 times each), Skemp (5 times) and Euclid (4 times). As for the French-speaking respondents, the eight authors most frequently cited were: Dienes and Piaget (8 times each), Vergnaud (6 times), Brousseau and Freudenthal (5 times each), Ginsburg, Polya and Skemp (3 times each). The combined list of the 'top eleven' authors for the whole set of respondents reads: Piaget (18 times), Polya (17 times), NCTM (16 times), Dienes (14 times), Freudenthal (12 times), Bruner and Skemp (8 times each), Vergnaud (6 times), Brousseau (5 times), Euclid and Skinner (4 times each).

This list is consistent with the ideas expressed by mathematics educators in response to the question of defining mathematics education discussed in the previous section. In fact, both the theoretical and the practical perspectives that were prominent in the responses to that question are also well represented among the authors cited here. Moreover, consistently with the view of mathematics education as a junction of various disciplines, the list features authors coming from diverse backgrounds, especially mathematics and psychology.

The results also bear out the opinion expressed by one of the respondents that mathematics education is a very new field: with the exception of Euclid, all of the most frequently cited authors belong to the 20th century. Indeed, some of the authors are still alive today and some of the books quoted have only just been published.

6. CONCLUSION

What is mathematics education? The answer to this question evolves through negotiation within our professional community and with other groups interested in our activities. Rather than analyzing the major contributions to this debate, by means of the present study, I wanted to capture the opinions of the whole mathematics education community in Canadian universities – its 'grass roots' as well as its leaders. The method that I used to collect the data was neither suited nor meant to explore in depth the views of individual respondents, however it has allowed me to identify a few trends in the opinions of a rather large and scattered population.

In defining mathematics education, the participants in the survey have focused on its object and goal, but have paid little attention to its method, except for stating that various other disciplines must be taken into account. The object of mathematics education is rather easily described (much more easily, for instance, than the object of mathematics[3]): 17 respondents pointed to the complementary phenomena of teaching and learning mathematics, 6 mentioned learning alone and 19 teaching alone, a few included the creation of mathematics. Some respondents spelled out that this object should be interpreted broadly: 'The study of *all* aspects of learning and teaching mathematics [...]' (my emphasis). I am altogether in favor of such an attitude of openness, for our object of study remains narrowly defined compared to that of other disciplines like psychology, sociology or mathematics itself. Our field is in no danger of losing its identity because of a lack of definition of its object of study. On the contrary, the inclusion of all possible aspects of it can only help us to better grasp its complexity.

As for the goal of mathematics education, some respondents were oriented more towards theory and others more towards action. Thus, at one end of the spectrum, there are mathematics educators whose main interest is, say, a theoretical understanding of how people learn mathematics and who are operating more or less within the territory of psychology or cognitive science, and, at the other end, there are those whose overwhelming concern is classroom practice and for whom mathematics education is almost identical with mathematics teaching. This kind of double goal is not exceptional. If we look again at other fields, we find, for instance, pure and applied mathematics, experimental and clinical psychology, and so on.

None of the respondents broached the subject of method – except indirectly, by mentioning the contribution of other disciplines. This might be due, in part, to the fact that the majority of the mathematics educators surveyed have studied and work in the field of education. They publish their writings in general education journals as well as in more specialized ones devoted to mathematics education. They may therefore consider their own specialty as a branch of education and assume that they have access to the whole of the ever-growing array of methods available in this eclectic field. Only its object and goal, not its method, would characterize mathematics education within the more general field of education.[4]

French-speaking respondents were more likely than English-speaking ones to mention the role of other disciplines in defining mathematics education – maybe a legacy of Lunkenbein (1979, 1983), who already emphasized this idea several years ago. I was somewhat surprised by the low frequency with which philosophy and sociology were named in this connection as well as by the omission of anthropology. Perhaps ethnographic methods, being a relatively recent acquisition in education, do not yet come to mind as readily as better established ones, such as those borrowed from psychology.

What is the relationship between mathematics education and the contributing fields? Two views have been expressed: for some, it is simply a case of a discipline making use of another's tools (like physics using mathematics, or psychology using statistics); for others, it is a mixing of disciplines that creates a new interdisciplinary field (like biochemistry, environmental studies or women's studies).

These considerations lead to a further question: Is mathematics education a science? In the 1970s, Freudenthal (1978) gave one of his books the subheading 'Preface to a science of mathematical education' and explained that it had 'the function of accelerating the birth of a science of mathematical education, which is seriously impeded by the unfounded view that such already exists' (p. v). He insisted that his book did not contain 'even the first rudiments' of such a science (p. vi). He made his position very clear: 'There is no science of mathematical education. Not yet' (p. 170). At about the same time, Lunkenbein (1979) wrote slightly more optimistically that 'la didactique de la mathématique' was a science '*in statu nascendi*', 'une discipline professionnelle [...] en train d'émerger en tant que science'. Four years later he described it without hesitation as 'science professionnelle de l'enseignant de la mathématique' (Lunkenbein 1983, p. 28). However, progression of mathematics education towards increasingly high scientific standards cannot be taken for granted. According to Kilpatrick (1992, pp. 29–31), a mere decade after Begle called, in 1996 at the First International Congress on Mathematical Education, for mathematics education to become an experimental science, doubts were being expressed about the capability of education to become a science at all.

Kilpatrick sets the beginning of mathematics education as a field of study at the end of the 19th century. He notices that, early in its history, research in mathematics education moved away from philosophical speculation towards a more scientific approach, but that, at present, the empirical-analytic model is no longer the dominant one, as researchers become interested in interpretive and critical approaches. In view of these diverse tendencies, past and present, he chooses to define research in mathematics education as 'disciplined inquiry', specifying that it 'need not be "scientific" in the sense of being based on empirically tested hypotheses, but like any good scientific work, it ought to be scholarly, public, and open to critique and possible refutation' (ibid., p. 3). In contrast, Biehler et al. (1994) head their anthology 'Didactics of Mathematics as a Scientific Discipline' and claim that scientific work has been done in the field of teaching and learning mathematics since the beginning of the 20th century. They grant that scientific communication and the sense of belonging to a separate scientific discipline were slower to develop, and they acknowledge that the role of the didactics of mathematics among other sciences at the university is still disputed, but they consider the scientific status of this discipline to be firmly established (pp. 1–2).

Where do the participants in the present survey stand on this issue? A few applied the word 'science' to mathematics education without reservation. One of them wrote: 'La didactique de la mathématique est une science toute jeune, qui a beaucoup évolué depuis 25 ans.' Most, though, did not use the word 'science' in their definitions, either because it did not occur to them or because they did not think it appropriate.

In the end, of course, deciding whether mathematics education is a science or not depends on one's idea of what qualifies as a science, and the extent to which the answer matters depends on one's need for the status conferred by the scientific label. However, the range in goals and the variety of methods that characterize our field, as highlighted by the present study, must be taken into account in setting standards that can be applied fairly to the whole of mathematics educators' research activities.

ACKNOWLEDGMENTS

I wish to thank all my mathematics education colleagues who kindly accepted to participate in the study and took the time to answer a rather long questionnaire. I am also grateful to the executive committee of the Canadian Mathematics Education Study Group for letting me use their mailing list, to Myreille St-Onge for helping me with data processing, to Christian Laville and Harry White for acting as 'judges', discussing the study with me and suggesting relevant readings, to Jean Dionne for commenting on an earlier draft of this chapter and to Catherine Clark for editing it.

NOTES

1. The questionnaire is available from the author. Several questions were common to a previous survey of mathematicians (Mura 1991, 1993) in order to make it possible to compare the two groups.

2. Before adopting a position on this linguistic issue, it might be useful to survey the situation in sister fields. The expression 'history education', for instance, is less widespread than 'mathematics education' or 'science education'. At least one international society uses 'history didactics' in its English name. Although some of its English-speaking members are uneasy with the term because of its connotation of pedantry, they have not been able to reach a consensus on what to do about it (Fontaine 1986, 1987, 1988).

3. Several mathematicians, asked to define mathematics, responded by stating that 'mathematics is what mathematicians do' (Mura 1993). No mathematics educator in the present survey defined mathematics education as 'what mathematics educators do'.

4. The situation is probably different for those mathematics educators, described by Balacheff in Balacheff et al. (1992), who 'want to develop this research field within the academic community of mathematicians'. They may have to work harder to define themselves and to gain credibility within

that community: they would have to explain not only their object of study and their goal, but, most importantly, their methods of doing research.

REFERENCES

Balacheff, N., Howson, A. G., Sfard, A., Steinbring, H., Kilpatrick, J., & Sierpinska, A.: 1992, 'Discussion Document for an ICMI Study. What Is Research in Mathematics Education, and What Are its Results?', *Educational Studies in Mathematics* 23 (6), 625–630.

Biehler, R., Scholz, R. W., Sträßer, R. & Winkelmann, B. (eds.): 1994, *Didactics of Mathematics as a Scientific Discipline*, Mathematics Education Library, Vol. 13, Kluwer Academic Publishers, Dordrecht.

Brousseau, G.: 1994, 'Problems and Results of Didactique of Mathematics. Resumen', *ICMI Study Conference 'What is Research in Mathematics Education and What Are Its Results?'*, The University of Maryland, College Park, MD, 8–11 May 1994.

Fontaine, P. F. M.: 1986, 'What is History Didactics?', *Communications* (of The International Society for History Didactics) 7 (2), 90–102.

Fontaine, P. F. M.: 1987, 'What is History Didactics? The Second Round', *Communications* (of The International Society for History Didactics) 8 (1), 9–35.

Fontaine, P. F. M.: 1988, 'What Is in an (English) Name?', *Communications* (of The International Society for History Didactics) 9 (1), 25–26.

Freudenthal, H.: 1978, *Weeding and Sowing. Preface to a Science of Mathematical Education*, D. Reidel Publishing Company, Dordrecht.

Kieran, C. & Dawson, A. J. (eds.): 1992, *Current Research on the Teaching and Learning of Mathematics in Canada*, Canadian Mathematics Education Study Group, Montréal.

Kilpatrick, J.: 1992, 'A History of Research in Mathematics Education', in D. A. Grouws (ed.), *Handbook of Research on Mathematics Teaching and Learning*, MacMillan Publishing Company, New York, 3–38.

Lunkenbein, D.: 1979, 'Qu'est-ce que la Didactique de la Mathématique?', *Bulletin de l'Association Mathématique du Québec (AMQ)* 19 (2), 14–35.

Lunkenbein, D.: 1983, 'Didactique de la Mathématique: Science Professionnelle de l'Enseignant', *Bulletin de l'Association Mathématique du Québec (AMQ)* 23 (1), 27–32.

Mura, R.: 1991, 'Sans distinction de sexe? Les carrières universitaires en sciences mathématiques', *The Canadian Journal of Higher Education* 31 (3), 59–95.

Mura, R.: 1993, 'Images of mathematics held by university teachers of mathematical sciences', *Educational Studies in Mathematics* 25 (4), 375–385.

Mura, R.: 1995a, 'Images of Mathematics held by University Teachers of Mathematics Education', *Educational Studies in Mathematics* 28 (4), 385–399.

Mura, R.: 1995b, 'Les Didacticiens et les Didacticiennes des Mathématiques au Canada: Un Portrait de Famille', in M. Quigley (ed.), *Proceedings of the 1994 Annual Meeting of the Canadian Mathematics Education Study Group, The University of Regina, 3–7 June 1994*, The University of Calgary, 91–113.

Roberta Mura
Faculté des sciences de l'éducation,
Université Laval, Cité universitaire, Québec,
Canada G1K 7P4

GUNNAR GJONE

PROGRAMS FOR THE EDUCATION OF RESEARCHERS IN MATHEMATICS EDUCATION

1. MATHEMATICS EDUCATION

The Field

Mathematics education is a discipline concerning mathematics teaching and learning on all levels. Mathematics education is related to other fields, both theoretical and practical, in various ways.

Mathematics is one of the basic sciences for mathematics education – in one sense, mathematics provides content. Mathematics education research concentrates on topics that are mathematically important. For example, the study of fundamental concepts, reasoning and proof, and problem solving are much in focus in mathematics education. Discussion and developments within mathematics (the mathematics community) are important in mathematics education.

The research methods used in mathematics education are mainly the methods used in the social sciences – pedagogy and psychology in particular. However, in contrast to much pedagogical or psychological research, there is a stronger link to teaching practice in mathematics education research. The two roots of research in the field – mathematics and psychology – we also find expressed in Kilpatrick (1992).

We consider mathematics education to be an applied science and not a basic science; it is in some ways comparable to engineering. We have fundamental research in the field as well as application and 'construction'. A similar view of the field is expressed in Wittmann (1984). I will use the term *educational sciences* to denote the various sciences focusing on education. The discussion is summarized in Figure 1.

Mathematics education has developed into an academic discipline studied at universities. The question *What is research in mathematics education?* has for some time been the focus of the mathematics education community (Sierpinska, Kilpatrick, Balacheff, Howson, Sfard & Steinbring 1993).

Research Methods

For a long time, it has been the task of mathematics educators to try to define their field and the research methods. At the Third International Congress on

117

Sierpinska, A. and Kilpatrick, J. Mathematics Education as a Research Domain: A Search for Identity, 117–127.
© *1998 Kluwer Academic Publishers. Printed in Great Britain.*

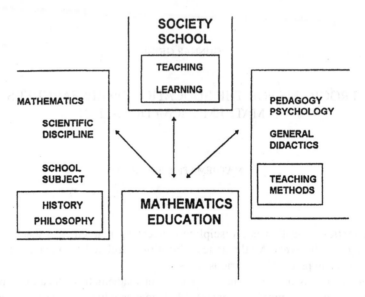

Figure 1

Mathematical Education (ICME) in Karlsruhe, Heinrich Bauersfeld (1977) of the Institut für Didaktik der Mathematik in Bielefeld pointed to the problem:

Since it is difficult to get a discipline to adapt its research strategies for use in other disciplines, mathematics educators will have to develop their own approaches.... Research and development within a discipline of mathematics education therefore will have to develop its own standards and methods. This means that an adequate precision will have to developed which on the one hand makes the outcomes accessible to the practitioner without further interpretation and on the other hand is open to scientific control. (p. 242)

Since then, we have had movements to try to establish a theoretical foundation for the field of mathematics education. Theory of Mathematics Education (TME) has been the subject of several conferences, as well as part of others; for example, the ICMEs. The education of researchers is an element much focused on: 'One of the crucial points for the development of theoretical foundation of mathematics education is, without doubt, the preparation of researchers in the field' (Batanero, Godino, Steiner & Wenzelburger 1992, p. 2).

Research methods in education are often divided into main categories. We have qualitative and quantitative research methods as two main categories, but there also exist methods that are combinations of these. Education has an extensive literature and a long tradition on research methods (see, e.g., Cohen & Manion 1980, 1985). Categories of research methods can be further subdivided into historical and developmental research, surveys, case studies, and so on, to follow the outline in Cohen and Manion (1980).

The use of research methods in education is the subject of much attention. We have several research paradigms and an ongoing debate and development.

In mathematics research, methods do not receive the same attention as in the social sciences. We might describe the goal of a major part of mathematical research as theorem proving, and the type of courses that come closest to being methods courses in mathematics would often be problem-solving seminars, where the focus is on solving mathematical problems. A reflection on method is usually not a dominant theme in published mathematical research.

Closely linked to the different research methods is the question: What is acceptable research in mathematics education – and how is research being approved? Concerning this question, I claim that the most important factor is the mathematics education community, nationally or internationally. Mechanisms provided are university degrees and publications: research reports, journal articles and books.

Mathematics Educators

The community of mathematics educators is an important factor for the development of the field of mathematics education. The relations among mathematics educators are continually changing, hence changing the field itself. The influence of the mathematics education community is crucial. I would like to rephrase Paul Ernest (1991, brackets indicate my additions):

At any one time, the nature of mathematics [education] is determined primarily by a fuzzy set of persons: mathematics [educators]. The set is partially ordered by the relations of power and status. The set and the relations on it are continually changing, and thus mathematics [education] is continuously evolving. The set of mathematics [educators] has different strengths of membership. This includes 'strong' members (institutionally powerful or active research mathematics [educators]) and 'weak' members (teachers of mathematics). (p. 98)

The case for mathematics educators is not quite the same as for mathematics. The teacher will have a different position within mathematics education – being the practitioner in the field. Also, the field itself does not have a long tradition, making it more open to changes.

2. PROGRAMS OF STUDY

Higher Education Studies in the Basic Sciences

Within higher education we find in many countries three levels of study: The *first level* is the basic training level, giving students basic skills in their subjects of choice. Students cover two (often related) subjects in this initial stage. The subjects could be mathematics and one of the natural sciences, or two languages, or some other combination. Some students would also be taking basic courses in education and mathematics. The term most often used is *undergraduate studies,* leading to a bachelor's degree. Satisfactory completion of this level is required to enter the next level.

The *second level* is the first specialization level. Students will usually concentrate within one field (discipline) that was part of their first-level syllabus. They will advance further in depth of the subject chosen. In many universities, this is also a level where research is required. It gives an initial training in research. The student is often expected to write a (research) thesis. The scope of these theses may vary from one university (country) to another – some programs do not require (research) thesis work. Only extended written presentations may be sufficient. It is usual to call this *graduate studies,* leading to a master's degree.

The *third level* is further specialization. Completion of a research thesis is required. The doctor's degree completes this level. This degree (or equivalent) is, moreover, often the requirement for a university (research) position. There are various traditions concerning study at this level. One tradition is an almost unstructured individual study. A university might provide a grant, but as research fellows in this tradition the 'students' are on their own and have to make their own contacts and investigations. A thesis is expected within a certain period of time. In the humanities and social sciences, the average age for completing a thesis of this kind is usually quite high. Another tradition is a – more or less rigid – program of study, with coursework and examinations, the thesis occupying a certain percentage of the program. The second tradition is by many considered to be the more efficient – producing more doctorates in a shorter time. Because of the present stress on efficiency some countries have introduced the second type (tradition) – when previously the first type was the only one possible.

The three levels are also linked to teacher training. The first level is usually the requirement for teaching in primary and lower secondary schools, the second level for upper secondary and higher education (colleges), and the third level for teaching at a university. In many countries, a typical upper secondary teacher in mathematics would have a master's degree in mathematics, or one of the natural sciences with mathematics courses in addition.

Mathematics Education

Programs of study for mathematics education do not always conform to programs of study in the basic disciplines. We find programs of study in mathematics education mostly in higher level studies, building upon studies in basic sciences, such as mathematics or psychology. The programs, however, might be subject to the same requirements as programs in the basic sciences – educating researchers within rigid time limits. Time constraints might be in conflict with the need for education in several basic subjects and experience in teaching, as a prerequisite.

For the Seventh International Congress on Mathematical Education in Quebec in 1992, an international survey of research programs was presented (Batanero et al. 1992). In this study, various characteristics of programs were given. What is probably the most striking feature of this documentation is the wide variation in the programs:

> The estimated time in doctoral programs necessary to finish a dissertation varies between 1 (3 cases) and 6 years (average 2.8). This time has to be added to the time for course work. 21 doctoral programs include complementary research reports as part of credits necessary. In the master the estimated time to finish a dissertation varies between 1 and 4 years, with an average of 1.5. (p. 16)

Many hold the view that teaching practice is a key factor for educating researchers in mathematics education. Some institutions require the students to have extensive school practice before entering graduate study. This requirement means the program is of a different type than study in a basic science; it will take more time to complete. Another possibility is to require teacher training only, thus making it possible to have a research education comparable, in time used, to research education in other areas. Some programs do not require teaching practice – the majority in the survey by Batanero et al. (1992, p. 11).

In mathematics education, there are probably few academic first-level courses. There may be a few second-level courses, mainly as preparations for third-level studies. These could be designed for teachers or for students having completed teacher training. There are also several categories of students studying mathematics education.

The Students

The students in mathematics education come mainly from three fields: mathematics, the educational sciences, and teaching – with a background in general teacher training. These categories of students have different backgrounds and have different strengths and weaknesses.

Students with a background in mathematics have had little or no contact with educational research. They are not used to giving written presentations as essays; however, they have a knowledge of mathematics to a certain degree. The weakness of the second group is often lack of knowledge of mathematics. On the other hand, they may have had some methods courses in education (or other social sciences) and some training in writing presentations. They may be more used to oral elements in their education. Both these groups may not have had teaching experience. Hence concerning research education, they have very different needs. Students from any one field need to concentrate in the two others; for example, students with a mathematical background need to study some pedagogy and psychology, and need to have teaching practice.

An important question is whether there should be different programs for educating researchers in mathematics education for different groups of students. The discussion above suggests that at least two different programs should be considered.

3. EDUCATING RESEARCHERS: PROBLEMS TO BE CONSIDERED

The Case of Norway

Mathematics education as a research discipline mainly developed internationally after 1945. Some countries have had programs for the education of researchers for quite a long time. In Norway and the rest of the Nordic countries, we are just now starting to introduce more structured programs.

It has, however, for a long time been a possibility to write a thesis in mathematics education. Since this was an individual 'program', one was in a sense free also to choose the thesis topic. One problem with the situation was the assessment of such a thesis. Since no one had an academic position in mathematics education, the assessment had to be made by researchers in other fields, mostly in mathematics or the educational sciences.

If we look at the situation in Norway, there are two traditions in the types of mathematics education research. One tradition is based on an extensive study of the subject; that is, a master's degree in mathematics. Another is based on the study of general pedagogy. Up to the end of the 1970s, we have had several master's theses in both these traditions. At the beginning of the 1980s, structured doctoral programs started to appear. They were modeled after American Ph.D. programs and started in Norway within civil engineering. Most of the basic sciences now have structured doctoral programs, and the traditional doctoral degree coexists with the new programs. Such new programs have been established in mathematics and natural science education. To establish such programs in mathematics education, however, has not been without problems.

Especially at the master's level, there have been problems with graduate programs in mathematics education. One problem is the length of study. It is felt that a master's degree in mathematics education should have a program of the same length as a master's degree in a basic science like mathematics. The problem then arises at the first level (in undergraduate studies) if we want to integrate mathematics, education, another school subject besides mathematics – to complete teacher training – and last, but not least, teaching practice. In addition, there is a first-level mathematics education course that gives an introduction to the field.

Aims

Before discussing models, I should briefly comment on one key question: What are the aims of educating researchers in mathematics education? To be educated as a researcher in mathematics education should be no different from being educated as a researcher in any other academic field. Researchers should have a broad knowledge in the field: they should be introduced to the mathematics education community, know available 'tools of the trade', and also have a specialized knowledge – an in-depth knowledge of some part of the field – and to have carried through, under supervision, a research project.

I will not discuss here the applicability of research in mathematics education, but concentrate on the researcher. In several countries, researchers are likely to play important roles in developing mathematics education on a national level. For models, we should look to the study of other educational sciences. There is a wide variety of educational research in pedagogy and psychology, much of which would be a model for mathematics education as well.

Structure of Programs

There is a variety of different programs in mathematics education (Batanero et al. 1992). Some are linked closely to mathematics, others to other educational sciences, and some to science education. Let us consider some of the possible structures, beginning with different types of books on research methods:

> There are three common types of books on research methods. The first leads the reader through the stages of research as if there was a mechanical sequence that, if followed, arrives invariably at reliable and valid evidence. The second type supports one approach in opposition to others. Here the reader is made aware that there are a variety of approaches and that there is a disagreement over their relative virtues. But there is still an attempt to sell one of many possible ways to collect evidence. The third type of book introduces the reader to the variety of research methods without taking sides. (Marten Shipment, Foreword to Cohen & Manion 1985)

Although this quotation concerns books on research methods, the same classification, with some adjustments, holds for programs as well. Should a program focus on leading a student through 'the stages of research' as a mechanical process? Should the program focus on only one research approach? Or should it introduce the student to a variety of research approaches?

The main difference between a book and a program is that in a program, research has to be carried out, and a method has to be chosen. Ideally, the method used should follow from an evaluation of different methods applied to a problem. With programs aiming at a high efficiency in producing doctorates, the ideal is perhaps not always possible to implement.

This discussion points to an important problem concerning the education of researchers in a field that is not a basic science itself but that builds on several basic sciences. The programs will easily become too extensive. The same conflict also arises in other elements of a program. Let us first consider the content of *coursework*.

It is easy to agree that the study of mathematics should be a fundamental part of mathematics education study, but it should perhaps be a different type of course than courses offered by the mathematics department. I claim that there is a need for a different type of mathematics course, linking the subject to history and philosophy. Is perhaps the division between the various departments in a university not so suitable for the needs of mathematics education? The same discussion is also important concerning the role of pedagogy and psychology – is there a need for special psychology courses in mathematics education? We should also pose the same question of whether teaching practice should be required.

Perhaps the most important and least controversial element of educating the researcher is the writing of a *research thesis*. The writing of a doctor's (or to some degree also the master's) thesis is the researcher's entry work into the research community.

Requirements to be met by *prospective students* are in danger of being too extensive. The following elements would be desirable qualifications:

- background in mathematics (at least an undergraduate degree)
- background in pedagogy and psychology
- teacher training and some teaching experience
- proven academic record.

It is close to impossible to find students with all these qualifications. As mentioned above, students tend to fall into three categories: those with a background (at least an undergraduate degree) in mathematics, those with a corresponding background in pedagogy and psychology, and students (teachers) with extensive teaching practice. Each group has strengths and weaknesses.

One especially important type of student is the returning teacher, with teaching experience, interest in mathematics education, a practical knowledge of pedagogy, and in some cases very little formal mathematics. Our experience shows that students with this background can carry out important research, so the question becomes: To what extent should these students have to prove themselves academically in the various fields?

To sum up this discussion, I would like to focus on some important questions: Should we work towards a more uniform model for educating researchers in mathematics education – as we will find in the basic sciences? Could we define a core of common knowledge that should be present in all programs? Or should we consider a *wide* variety of programs to be a good thing in itself, with the possibility of having very strong 'subcultures'?

4. MODELS FOR EDUCATING RESEARCHERS IN THE FIELD OF MATHEMATICS EDUCATION

Structure

When we consider educating for research in general, we can identify at least two different models. One model is the open academic study we find in the basic sciences – where the decision to specialize is postponed to the second level. Another is the structure we find in some professional studies. In many countries, studying to become a civil engineer or a psychologist is a study where the aim is set at the beginning. All courses given – also in the basic sciences – are somewhat influenced by the profile of the education. Mathematics courses might be specially designed for engineering students.

In many ways, professional studies might be more efficient. Courses can be adapted to the need of the student to a higher degree than in general academic studies. However, considering the range of different types of students – especially practicing teachers – likely to enroll in a research program in mathematics education, I claim that the open academic programs are most suitable. However, this approach poses several problems.

Courses for Mathematics Education

The study of *mathematics* is an important element and should not be left out. The question is how much mathematics is necessary? For most programs, the mathematics necessary for teacher qualifications for upper secondary should be sufficient. Another problem is that the courses offered by most mathematics departments are not so suitable for the purpose of mathematics education. Most courses are designed with the research mathematician in mind, giving the foundation for further studies in mathematics. In the initial training, little weight is placed on the history and philosophy of mathematics. Communication in mathematics is also generally neglected. These elements would be very important in educating researchers in mathematics education. This observation suggests that mathematicians and mathematics educators should cooperate in designing mathematics courses for students of mathematics education.

In the above discussion, attention was directed towards the different groups of students likely to enter a study of mathematics education. This diversity leads to the question: Could there be a program in mathematics education without mathematics? The question is almost contradictory, but allowing for a variety of research, there clearly could be a research thesis within mathematics education with no more mathematics than found within basic education in schools. A *program* in mathematics education, on the other hand, should contain about a year's study of mathematics at university level.

Much of the argument regarding mathematics also holds true for courses in pedagogy and psychology. However, the important contribution from these fields is in the area of research methods. Clearly, methods courses are important and cannot be left out. However, here the field is open to various possibilities. What is stated above for books on research methods above gives an important classification for method courses. Within present programs in mathematics education, as well as in pedagogy, two approaches are clearly visible: the survey-of-methods course or the single-project course.

An important question concerning all courses is: *Who should teach them?* Should the mathematics courses and education courses for students of mathematics education be taught by mathematics educators only, or by mathematicians and educators, respectively? There are good arguments for both cases. Whereas the mathematics educator would be more attentive to the needs of the students, the mathematicians and educators would provide contacts with two related fields. We will probably see a variety of different solutions to this problem, dictated by what is practical at different institutions.

The Research Thesis

The research thesis is the most important part of the researcher's education. Theses vary a great deal from one university to another. Even within one institution, two doctor's theses in mathematics education written in different departments may vary widely. Concerning the thesis, there is the question of whether the community of mathematics educators can agree on some criteria of quality. It is difficult, however, to specify special criteria for mathematics education that would not also apply to other academic fields such as pedagogy or psychology; for example:

- provide an explicit discussion of the method used in the investigation, arguing for the choice of method and weighing the method used against alternative methods, and
- show mastery of the tools of the trade in the structure and format of the thesis.

We can also formulate somewhat specific criteria; for example, publishable in national or international research journals. Such criteria are at present used in mathematics and science education at the University of Oslo.

5. CONCLUDING REMARKS

Earlier the question was raised about the variation of programs for educating researchers. My personal view is that the field of mathematics education is in a state where variety is desirable. The field itself has many facets, and students with very different backgrounds are attracted to it. There should be room for programs that cater for the student of mathematics as well as the student of education. Special care should be given to the needs of the returning teacher.

REFERENCES

Batanero, M. C., Godino, J. D., Steiner, H. G. & Wenzelburger, E.: 1992, 'Preparation of Researchers in Mathematics Education: An International TME-Survey', *Institut für Didaktik der Mathematik*, Occasional Paper 135.
Bauersfeld, H.: 1977, 'Research Related to the Mathematical Learning Process', in H. Athen & H. Kunle (eds.), *Proceedings of the Third International Congress on Mathematical Education*, Organizing Committee of the 3rd ICME, Karlsruhe, 231–245.
Cohen, L. & Manion, L.: 1980, *Research Methods in Education*, Routledge, London.
Cohen, L. & Manion, L.: 1985, *Research Methods in Education* (2nd ed.), Routledge, London.

Ernest, P.: 1991, *The Philosophy of Mathematics Education,* Falmer Press, London.

Kilpatrick, J.: 1992, 'A History of Research in Mathematics Education', in D. A. Grouws (ed.), *Handbook of Research on Mathematics Teaching and Learning,* Macmillan, New York, 3–38.

Sierpinska, A., Kilpatrick, J., Balacheff, N., Howson, A. G., Sfard, A. & Steinbring, H.: 1993, 'What Is Research in Mathematics Education, and What Are Its Results?', *Journal for Research in Mathematics Education* **24,** 275–278.

Wittmann, E.: 1984, 'Teaching Units as the Integrating Core of Mathematics Education', *Educational Studies in Mathematics* **15,** 25–36.

Gunnar Gjone
University of Oslo,
Department of Teacher Education and School Development,
P.O. Box 1099 Blindern,
N-0316 Oslo,
Norway

PART III

GOALS, ORIENTATIONS, AND RESULTS OF
RESEARCH IN MATHEMATICS EDUCATION

GILAH C. LEDER

THE AIMS OF RESEARCH

1. INTRODUCTION

Educational research has become a huge, multi-purpose enterprise. Annual conferences such as those of the American Educational Research Association illustrate this vividly. In 1994, for example, in addition to the 11 different divisions listed in the meeting program – each with its own particular focus – an ever-increasing number of interest groups, presentation formats, and presenters vied for a place on the program. The titles of the presentations revealed a multiplicity of intentions and purposes, at least at the level of detail. Is it possible to discern, or even reasonable to expect, a commonality of objectives at the broader level? What is educational research, and what are its aims?

2. THE NATURE OF EDUCATIONAL RESEARCH

Bulging shelves in many university libraries suggest that research, or the search for new knowledge, not only continues but that its findings are also amply documented. There is also evidence of a substantial increase in recent years in the volume, scope, and diversity of educational research in general, and research in mathematics education in particular.

Definitions of educational research generally show overlap as well as subtle differences. Romberg (1992) listed 10 activities characteristic of research and which 'almost every research method text outlines' (p. 51). These are: identifying a phenomenon of interest, building a tentative model, relating the phenomenon and model to others' ideas, asking specific questions or making reasoned conjectures, selecting a general research strategy for gathering evidence, selecting specific procedures, collecting the information, interpreting the information collected, transmitting the result to others, and anticipating the action of others. These elements are reflected in the following definition.

Educational research is the

systematic and in-depth inquiry concerning the purposes of education; the processes of teaching, learning and personal development; the work of educators; the resources and organisational arrangements to support educational work; the policies and strategies to achieve educational objectives; and the social, political, cultural and economic outcomes of education. (Review Panel, Strategic Review of Research in Education, 1992, p. 8, submission from the Australian College of Education)

Sierpinska, A. and Kilpatrick, J. and Mathematics Education as a Research Domain: A Search for Identity, 129–140.
© *1998 Kluwer Academic Publishers. Printed in Great Britain.*

Striking features of the definition are the implied diversity of the nature and scope of educational research, the focus on disciplined inquiry, and on the pragmatic aims of research.

3. DIVERSITY OF RESEARCH

In the natural sciences it has typically been assumed that a mature discipline will have only one dominant research paradigm at any one time. But, to repeat questions voiced by many others:

What role do interests, individual and collective, play in the evolution of scientific knowledge? Do all scientists seek the same kinds of explanations? Are the kinds of questions they ask the same? Do differences in methodology between different subdisciplines even permit the same kind of answers? And when significant differences do arise in questions asked, explanations sought, methodologies employed, how do they affect communication between scientists? (Keller 1983, p. xii)

These questions are particularly relevant in the social sciences and in education where 'it is far more likely that the coexistence of competing schools of thought is a natural and quite mature state' (Shulman 1986, p. 5).

That this diversity exists within mathematics education can be inferred from the approaches listed by Kilpatrick (1992). Relevant details are summarized in Table 1.

Perspective	Main aims
Empirical-analytic (the traditional aims of science)	To explain, predict, or control
Ethnographic (cf. anthropology)	To understand the meanings of the learning and teaching of mathematics for participants in these activities
Action research (cf. critical sociology)	To improve practice and involve the participants in that improvement

Table 1 *Research approaches and their aims*
Source: Adapted from Kilpatrick 1992, pp. 3–4.

Over time, domination of research within the empirical-analytic tradition has given way to an acceptance and appreciation of investigations more consistent with the methods and aims of the interpretative and critical techniques.

As implied above, the purposes for doing educational research and the methods selected for this research are not static but likely to change with time. A useful, though certainly not unique, summary of changes in the purposes and methods of educational research is shown in Table 2.

The different theoretical frameworks, methodologies, settings, assumptions and constraints encourage different conceptions of key issues. The implications of this for the aims of research are illustrated in different ways in the next sections.

Time line	1950s & 1960s	1970s	1980s	1990s
Economic trends	Expansion (1945–59)	Rapid growth (1960–75)	Stasis (1976–88)	Recession (1989–)
Socio-political discourse	Interventionist idealism	Pragmatic realism	Instrumental rationalism	Nationalism/ internationalism Global competitiveness
Research culture	Amelioration innovation	Pure and applied R & D	Applied consolidation	Strategic priority setting
Operational patterns	Investigator driven	Team-oriented, National R & D centres and labs	Focused centres, Key centres	Networks and partnerships Cooperative research centre

Table 2 *Changing patterns of educational research*
Source: Poole 1992, p. 4. :

4. THE AIMS OF RESEARCH

Individual Perspectives[1]

In March of 1981, David Wheeler invited proposals for 'a "research programme" for some particular problem within mathematics education' (Wheeler 1981, p. 27). Outlines of 'a programme which research should be concerned with' were also welcome. Several years later, in January of 1983, another challenge was issued to mathematics educators: to formulate 'a number of specific problems whose solution would be likely to advance substantially our knowledge about mathematics education' (Wheeler 1984, p. 40). Both these requests, it seems to me, are implicit invitations to reflect on the aims of research in mathematics education. The responses received and subsequently published are summarized in Table 3.

Thus the research problems selected ranged from general and broad to quite specific, were set in a variety of mathematical domains which generally reflected the interests of the proposer, and could be investigated in various ways. The aims of research they collectively represent are diverse, more often pragmatic, occasionally scientific.

Perspectives on Disciplined Inquiry

The traditional aims of research, as noted earlier, are to explain, to predict, or to control. But from whose or what perspective? It has been claimed (e.g., Belenky, Clinchy, Goldberger & Tarule 1986; Gilligan 1982; Harding 1991; Harding & O'Barr 1987) that

conceptions of knowledge and truth that are accepted and articulated today have been shaped throughout history by the male-dominated majority culture. Drawing on their own perspectives and

Author	Summary of Research Program/Problems
Pearla Nesher	To understand the processes involved in the acquisition of mathematics as a language system – using an interdisciplinary approach
Alan Bell	To achieve a discrete and permanent change in the level of students' understanding in a key conceptual task
Caleb Gattegno	To identify the mental structures and activities needed for the production of mathematical structures
Geoffrey Howson	To know more about: factors that influence student motivation, teachers' perceptions of their role, mathematical ability, a theoretical framework for studying and investigating mathematics education
Tom Kieren	What are the constructive mechanisms of mathematics and how do they function in knowledge building?
Nicolas Balacheff	To understand better how characteristics of individual students, components of the school setting, and methods of teaching affect the learning of mathematics
Jeremy Kilpatrick	How are 'meaning' and 'automaticity' to be orchestrated? Can higher-level objectives be achieved given imperfect attainment of related 'low-level' objectives? What, and how, are general intellectual abilities affected by the study of mathematics?
Dick Tahta	What are the criteria for recognizing a solution to a problem in mathematics education? What factors facilitate and hinder the learning of mathematics?
Jere Confrey	What structures influence the development of alternate conceptions and misconceptions? What influences the match/mismatch between the teachers' intentions and the mathematics learnt by students in class?
Alan Bishop	What factors influence students' learning of mathematics (e.g., student goals, the media, medium of instruction, computers, expected levels of performance, communication skills)?
Efraim Fischbein	What are the objectives of mathematics education? Can we learn from research? How should different topics be organized and presented? What factors ensure problem-solving success? Can they be taught? Can we reconcile the logical structure of mathematics with its power to solve practical problems or the role of intuition in understanding or creating mathematics?
Willem Kuijk	What physical/neurological activities occur during mathematical activities? Is the acquisition of mathematical knowledge necessarily hierarchical? Can computer simulation replace mathematical imagery?
Gerard Vergnaud	How are children's conceptions of mathematical operations formed and changed? How will computer science affect mathematics education?
Merlyn Behr, John Bernard, Diane Briars, George Bright, Judy Threadgill-Sowder, Ipke Wachsmuth	Can students' success with formal mathematics be improved through instruction in reading mathematical symbols: identifying critical barriers, optimum teacher characteristics, and effective instructional materials; focusing on affective components of mathematics learning, on links between mathematics and other subject areas? What is the impact of computer technology on mathematics teaching and learning?

Table 3   *Implied aims of research in mathematics education*[2]

Author	Summary of Research Program/Problems
John Mason	How can personal and societal images of mathematics be modified, updated, and balanced? What (personal) experiences influence mathematics learning? What does it mean to 'know' something in mathematics education? And how can that 'knowing' be passed on to others?
David Tall	Will computer technology change mathematics learning and teaching?
David Robitaille	How can we evaluate and capitalize on the mathematical strengths of each individual? How will, and should, calculators and computers affect the teaching and learning of mathematics? What factors facilitate/hinder mathematics problem solving? Are these factors unique to mathematics?
Tom Carpenter	Mathematics education research is most fruitfully used to address problems which are theory driven, and focused and specific rather than general and broad.
W. M. Brooks	What is mathematics education?

visions, men have constructed the prevailing theories, written history, and set values that have become the guiding principles for men and women alike.... Relatively little attention has been given to modes of learning, knowing, and valuing that may be specific to, or at least common in, women. (Belenky et al. 1986, pp. 5–6)

In this view, sexism has influenced the aims and methodology of research in the framing of problems, in the methods of gathering information, in the coding and analyses of data and in the interpretation of results.

While rejecting the notion of a distinctive feminist method, Harding (Harding and O'Barr 1987; Harding 1991) has nevertheless argued that feminist empiricism begins with the position that science and its global methods are basically sound, but that some practices, procedures, assumptions and therefore findings of scientists are biased against females. Because these practices are detrimental to females and to science they must, she maintains, be identified and curtailed. How this stance might affect the aims of research is illustrated in Table 4 through summaries of important features of two views of intellectual development: those proposed by Perry (1970) and Belenky et al. (1986).

Despite the inevitable simplification of the two frameworks, the overlap and differences in their value positions are apparent. Subtle differences in expected behaviors emerge if we accept the proposition that 'all our thoughts and actions are influenced by what we are, by our subjectivity, by our expectations, by our place in society and by the material and cultural context in which we live' (Mura 1991, p. 35). How these differences affect research in mathematics education per se, its aims, perspectives on learning, acceptable explanations and predictions is still keenly debated in the education and mathematics education research communities. A study reported recently in the Australian

Men's forms of intellectual and ethical development (Perry 1970)	Women's ways of knowing (Belenky et al. 1986)
Basic dualism – passive learners depend on authorities to teach them the truth. The world is viewed in polarities (good/bad; white/black)	Received knowledge – ideas and information from others are accepted 'as is'. 'Received knowers assume that they should devote themselves to the care and empowerment of others.... They are embedded in conformity and community'
Multiplicity – 'the student's opinion is as good as any other', no substantiation is sought	Subjective knowledge – truth is personal, informed by emotions, gut feelings, instincts, and intuition
Relativism subordinate – authority and truth are no longer considered absolute. An analytical approach is used to evaluate knowledge	Procedural knowledge – the power of reason is recognized and used in conscious, deliberate, systematic analysis in the search for truth. There are two forms of procedural knowledge: separate knowing (oriented towards impersonal rules, standards and techniques, with a sense of alienation without the personal involvement in the pursuit of knowledge) and connected knowing (concern about the objects to be understood)
Full relativism – the relativity of truth, the importance of context and individual framework, is recognized. It is understood that 'knowledge is constructed, contextual, and mutable'	Constructed knowledge – personal and learned knowledge, thinking and feeling are integrated. 'All knowledge is constructed ... the knower is an intimate part of what is known'.... Scientific theories are now seen as educated guesswork, simplified models of a complex world – rather than absolute truth

Table 4 *Two views of intellectual development*
Source: Adapted from Shepherd 1993, pp. 36–39.

press (Lewis 1994) can serve as a useful vehicle.[3] A recent study, readers were told, had found that

Female students rely almost exclusively on their teachers' assessments to measure their academic performance.... Boys, however, appear to have three sources of information about their performance:... a 'kind of in-built sense of how they were doing', ... comparisons with their peers, ... [and] teacher feedback.... The study ... suggested important implications for the ways teachers should treat boys and girls [since] critical teacher feedback is more likely to damage the self-esteem and academic performance of girls than boys. (Lewis 1994, p. 5)

Without a careful inspection of the methodology and instruments used, it is impossible to know whether this is indeed an example of research where, as Harding's argument was phrased above, the 'practices, procedures, assumptions and therefore findings of scientists are biased against females'. There is no doubt, however, that those working within a feminist perspective would not want to accept uncritically the assumption that male behavior should be accepted as the norm, and to be aimed for in general. They would wish to consider more carefully the conclusion that 'critical feedback is more likely to damage the self-esteem and academic performance of girls than boys', to examine, and possibly question, the assumptions that shaped the study, the nature of the instruments used to gather the data, the mode of response expected, other features of the classroom and school environments, peer values and expectations. Such a change in emphasis would affect the aims of the research carried out.

5. PRAGMATIC CONSIDERATIONS

Many of those engaged in research are academics employed at a university. Aitken (1990) has argued that in this setting 'only one element of life and work is organized to attract esteem – and that is research' (p. 12). In practice, university appointments and promotions are governed by research output and performance. 'It means that there are two classes of academic: those who are in a position to do their research and the ones who can't get the money or time to do theirs' (p. 12). What research is most likely to be funded?

In Australia, as well as in many other countries, it is now frequently argued that research should contribute to the country's social, economic, and cultural priorities and needs; influence and inform practice, and address the needs of policy makers, administrators, teachers, students, parents (Review Panel, Strategic Review of Research in Education 1992). There is no question such goals reflect the priorities adopted by funding bodies and influence which projects attract, or fail to attract, research funds. For example, the current (late 1994–1995) *Strategic Review of Mathematical Sciences and Advanced Mathematical Services* in Australia includes the following among its terms of reference:

A1. To determine the degree to which a strong fundamental research base is required in all branches of the mathematical sciences in Australia and recommend on future support of this research....
4. To evaluate benefits gained from participation in international mathematical research programs and recommend actions to enhance those benefits....
B1. To examine how advanced mathematical services contribute to other fields of endeavour, and to assess the national benefits from Australia's investment in the mathematical sciences.
2. To determine the areas of the mathematical sciences most used by business and industry and identify those most likely to be needed in the next decade.... (p. xii)

Thus advanced mathematical sciences in Australia, it is clearly implied, should contribute to the country's business and economic climate as well as to its research needs.

At a time when quality control, performance appraisal and designated priority areas influence the funding made available for educational research, there is an increased likelihood that research endeavors are not theory-driven but instead are heavily influenced by pragmatic, financial, and economic considerations:

Under the new regime then, academics are expected to conduct their research to schedule, offer a product for which there is an identifiable market, and compete for a buyer in that market.... The activity of research, and the ideas to which it gives rise, are no longer in this system treated as ends in themselves, but as means to basically mercenary ends. (Mathews 1990, p. 20)

Not only individuals, but institutions too, need to 'justify' their output. Lists of aims or institutional quality outcomes to be assessed frequently contain items such as the following:

- Research-trained graduates – enrollments for research-based higher degree courses, research trained graduates per annum, likely destinations in research and industrial careers
- Research publications in refereed journals
- Evaluation of significance of publications
- Patent applications
- Funding indicators – competitive peer review grant income, contract research income, consultancy income
- Reputational analysis – invitations to high level research conferences, evaluation by peers in relation to specific disciplines, assessment after site visits by review panels (Department of Employment, Education and Training 1989).

Those of us involved in educational research have come to realize that, in many cases, the transition from quality/performance indicators such as those listed above to accepted aims of research is surprisingly small.

6. CONCLUDING COMMENTS

Evidence has been presented in this chapter of the ways in which the aims of research are affected by the changing nature of research, by the diversity of individual perspectives and values, by differences in the methods and assumptions of disciplined inquiry, and by the constraints imposed by external forces and pragmatic considerations. In this climate it is imperative that curiosity-driven research not be forgotten or ignored.

NOTES

1. Reviewing earlier work to facilitate the formulation of the issues/problems to be explored is an important part of the research process. I acknowledge my debt to those whose ideas I have summarized and apologize in advance for the inevitable simplifications of the views they expressed in their lengthier responses. The complete correspondence can be found in volume 2(1) and volume 4(1) of For the Learning of Mathematics.

2. The complete correspondence can be found in volume 2(1) and volume 4(1) of For the Learning of Mathematics.

3. Also see, for example, Fennema, Damarin, Campbell, Becker & Leder (1994). Or consider material from a special half-day workshop on feminist approaches to research methods: renovation or resolution also held at the 1994 AERA annual meeting. Discussions of differences between research devised in a traditionally empirical-analytic and a feminist framework included:

● 'Generally, the typical way is to look at one perspective with the hope of repeating the study through other eyes later on ... what happens are add-ons to the breakthrough study. Those perspectives not included in the breakthrough study may not be included later' (Boodoo 1994).

● How can we do research from a feminist standpoint? 'State "problems" as women experience them', 'refer to more and varied literatures; engage in counter-readings (against the grain) deconstructive readings', use multiple methods, results 'are descriptions of "reality" as women experience it (and are often questions)', conclusions and implications are 'truth-telling and rethinking of educational practice' (Damarin 1994).

REFERENCES

Aitken, D.: 1990, 'How Research Came to Dominate Higher Education and What Ought to be Done About It', *Australian Universities' Review* **33** (1 & 2), 8–13.

Belenky, M. F., Clinchy, B. M., Goldberger, N. R. & Tarule, J. M.: 1986, *Women's Ways of Knowing,* Basic Books, New York.

Boodoo, G. M.: 1994, 'Feminist Approaches to Research Methods: Renovation or Revolution', Paper presented at the AERA workshop on 'Feminist Approaches to Research Methods', New Orleans, LA, April 6.

Damarin, S.: 1994, 'Research from Feminist Standpoint(s)', Paper presented at the AERA workshop on 'Feminist Approaches to Research Methods', New Orleans, LA, April 6.

Department of Employment, Education and Training: 1989, *Committee to Review Higher Education Report,* Australian Government Publishing Service, Canberra.

Fennema, E., Damarin, S. K., Campbell, P. B., Becker, J. R. & Leder, G. C.: 1994, 'Research on Gender and Mathematics: Perspectives and New Directions', Papers prepared for a symposium at the annual meeting of AERA, New Orleans, LA, April 7.

Gilligan, C.: 1982, *In a Different Voice,* Harvard University Press, Cambridge, Massachusetts.

Harding, S.: 1991, *Whose Science? Whose Knowledge?,* Open University Press, Milton Keynes.

Harding, S. & O'Barr, J. F.: 1987, *Sex and Scientific Inquiry,* University of Chicago Press, Chicago.

Keller, E. F.: 1983, *A Feeling for the Organism,* W. H. Freeman & Company, New York.

Kilpatrick, J.: 1992, 'A History of Research in Mathematics Education', in D. A. Grouws (ed.) *Handbook of Research on Mathematics Teaching and Learning,* Macmillan, New York, 3–38.

Lewis, J.: 1994, 'Girls Rely on Teachers' Views: Study', *The Age* (Melbourne), April 15, 5.

Mathews, F.: 1990, 'Destroying the Gift: Rationalising Research in the Humanities', *Australian Universities' Review* **33**(1&2), 19–22.

Mura, R.: 1991, *Searching for Subjectivity in the World of the Sciences: Feminist Viewpoints,* Canadian Research Institute for the Advancement of Women, Ottawa, Ontario.

Perry, W. G.: 1970, *Forms of Intellectual and Ethical Development in the College Years,* Holt, Rinehart & Winston, New York.

Poole, M. E.: 1992, 'The Patterning and Positioning of Educational Research', *Australian Educational Researcher* **19**(1), 1–18.

Review Panel, Strategic Review of Research in Education: 1992, *Educational Research in Australia,* Australian Government Publishing Service, Canberra.

Romberg, T. A.: 1992, 'Perspectives on Scholarship and Methods', in D. A. Grouws (ed.) *Handbook of Research on Mathematics Teaching and Learning,* Macmillan, New York, 49–64.

Shepherd, L. J.: 1993, *Lifting the Veil,* Shambhala Publications, Boston, MA.

Shulman, L. S.: 1986, 'Paradigms and Research Programs in the Study of Teaching: A Contemporary Perspective', in M. C. Wittrock (ed.), *Handbook of Research on Teaching*, Macmillan, New York, 3–36.

Wheeler, D.: 1981, 'A Research Program for Mathematics Education (I)', *For the Learning of Mathematics* 2(1), 27.

Wheeler, D.: 1984, 'Research Problems in Mathematics Education (I)', *For the Learning of Mathematics* 4(1), 40.

Gilah C. Leder
La Trobe University,
Bundoora,
Victoria,
Australia, 3083

JAMES HIEBERT

AIMING RESEARCH TOWARD UNDERSTANDING: LESSONS WE CAN LEARN FROM CHILDREN

The thesis of this chapter is that the primary goal of research in mathematics education should be to understand what we study. It will be argued that developing understanding satisfies both fundamental or theoretic aims and practical aims simultaneously. Research efforts that do not aim to understand satisfy neither of these goals and are not worth making.

The thesis is built on a common and simple definition of understanding that says we understand something when we can identify the important elements in a situation and describe the relationships between them. This definition, appropriate when considering issues of learning and teaching mathematics (Hiebert & Carpenter 1992), is also appropriate when doing research. That is because doing research and learning mathematics are both best viewed as sense-making activities. Both researchers and children are sense makers. The argument that research should aim to understand can be developed, in part, by considering how children learn mathematics, how they construct understandings, and how they show us that they understand.

The value of understanding important phenomena in mathematics education and the parallels between what we do to develop such understanding and what children do to learn mathematics will be elaborated by considering four features of research: the kinds of problems we investigate, the role of theories in our work, the kind of data we collect, and the coordination of theories and evidence. How we think about these four features determines what we mean by aiming research toward understanding and provides a measure of whether we are getting there.

1. THE NATURE OF THE PROBLEMS WE INVESTIGATE

It is becoming increasingly recognized that children learn mathematics by solving problems. Learning begins with problems (Brownell 1946; Davis 1992; Hiebert, Carpenter, Fennema, Fuson, Wearne, Murray, Olivier & Human 1997). Problems stimulate mathematical activity and provide the context for working out relationships between mathematical ideas. Understanding is constructed as problems are defined and methods for solution are developed and refined. That is because solving a problem usually means identifying the important elements in the situation and building relationships between them. For children (and

141

Sierpinska, A. and Kilpatrick, J. Mathematics Education as a Research Domain: A Search for Identity, 141–152.
© *1998 Kluwer Academic Publishers. Printed in Great Britain.*

adults) constructing mathematical understandings and solving mathematical problems are nearly synonymous (Brownell 1946).

A similar situation holds true for doing research. First, research begins with problems (Laudan 1977; Romberg 1992). Doing research means trying to solve a problem. Second, solving a research problem, like solving a mathematical problem, means identifying the crucial elements in a situation and seeing how they are related. As before, solving a research problem means coming to understand the problem. As Karl Popper (1972, p.166) phrased it, 'The activity of understanding is, essentially, the same as that of all problem solving'.

Consider the research problem of how young children acquire addition and subtraction skills. Partial solutions to this problem are now available because important elements have been identified, such as early counting skills (Fuson 1988; Gelman & Gallistel 1978) and the role of word problem semantics in triggering addition and subtraction behaviors (Carpenter & Moser 1984). Furthermore, well-reasoned accounts have been offered for how the elements are related (Carpenter 1986; Riley, Greeno & Heller 1983). Progress clearly is being made in solving this problem.

If research begins with problems, then important research begins with important problems. But what constitutes an important problem? Important problems are likely to share at least two features. First, they will be rich problems. Second, they probably will be related to other important problems that have been solved. Rich mathematical problems engage students' curiosities and provide opportunities to wrestle with important mathematical ideas. Such problems are often nontrivial, multifaceted, and solvable using a variety of strategies. Moreover, children's understanding grows as they engage in the full range of processes and activities that emerge as they work toward solutions to these problems.

Rich problems for researchers are, in many ways, like rich problems for children. They are nontrivial and multifaceted. They can be approached from a variety of perspectives, and the process of solving them is often filled with intermediate results and insights. New questions are often spawned during the solution process. In the end, it is not only the answer to the problem that is useful; our increased understanding is attributable to the full set of activities and processes that have been engaged in working toward a solution.

A second feature of important problems is that the problems probably will be related to other important problems that have already been solved. One of the clearest results from research on learning is that people learn by building on what they already know. Children make sense of new situations by relating them to what they already understand (Hiebert & Carpenter 1992). More specifically, children solve new problems by relating them to problems they have already solved. In fact, it appears that young students who understand arithmetic problems they have solved can solve related problems by adjusting their strategies but are essentially unable to solve new problems for which they perceive their strategic repertoire as inappropriate (Hiebert & Wearne 1992, 1996).

Building carefully on previous results and working on related problems is a useful research strategy as well. There are several reasons for this usefulness.

First, it is possible to make good progress in solving new problems by building on what we already know. Although the whole story of scientific progress does not show a smooth cumulative character, productive research programs, over shorter periods of time, do build in a cumulative way and display a chain-of-inquiry nature (Cronbach & Suppes 1969; Lakatos 1978). Given the current state of research in mathematics education, the chain-of-inquiry model seems appropriate. Second, problems that are related to others that have already been addressed are more likely to be resolved. Useful ways of conceiving the problem and productive methods of attack may have been developed from the work on related problems. These benefits should not be underestimated. Making progress on the research problems we have defined, even if such progress depends heavily on previous work, yields increased understanding. A third reason is more subtle but simple: Problems that are related to previous work are more likely to be recognized or defined as problems. Situations that are completely unrelated may not even be seen as problems (Laudan 1977).

There are many research problems in mathematics education that would qualify as rich and connected problems. For illustration purposes, consider the relationship between teaching and learning in mathematics classrooms. That is, in what ways does teaching affect learning and vice versa? How do different instructional approaches lead to different kinds of learning? These questions define a problem that is rich. It is nontrivial and multifaceted. Pursuing a solution to the problem has triggered numerous additional questions that have received attention from a variety of perspectives (Artigue & Perrin-Glorian 1991; Cobb, Wood, Yackel & McNeal 1992; Fernandez, Yoshida & Stigler 1992; Hiebert & Wearne 1993; Laborde 1989; Yoshida, Fernandez & Stigler 1993). For example, Yoshida et al. investigated whether students' active search for relevant ideas during mathematics instruction might account for differential learning. It is likely that solutions to the original problem will be the explanations that accumulate for these individual questions that arise along the way.

The problem of how teaching and learning are related also is connected to other problems of interest and research can build productively on previous work. Much research has been conducted on students' learning and on the nature of teaching. Linking the two in classroom situations can build on what we know from these previous efforts. Because we know a good deal about students' learning of arithmetic in the primary grades, it may be that initial breakthroughs in explaining teaching–learning relationships will come in these classrooms. Here researchers can take maximum advantage of what we understand about how students acquire mathematical competence.

As the problem of understanding teaching–learning relationships illustrates, important research problems also allow work that is both fundamental and practical. Theoretical developments are needed to help conceptualize the crucial issues in the relationship and research on these issues is likely to have consequences for how we design instructional environments.

2. HOW WELL WE ARTICULATE OUR THEORIES

Research that aims to understand depends on the development of well-articulated theories. I argue in this section that theories are needed for several reasons – to guide the research process, to communicate with others, and to make the results useful. Developing and refining theories also provide the best evidence that we are making progress in our understanding.

Before considering the research process, let us look again at children's mathematical activity. As children construct solutions to problems, they continue to learn as they demonstrate their solutions to others. They make claims of understanding by showing their strategies and explaining how things work. Their explanations are taken as evidence that they understand. That is true both as they respond to adult questions about their understanding and as they explain their thinking to peers. These explanations are children's local theories of how things work. Making their theories explicit keeps the classroom's attention focused on the task of understanding rather than, say, imitating a prescribed procedure.

Children's explanations or local theories also provide productive ways of communicating and interacting. Assertions and hypotheses are examined and revised as students explain, question, and defend solutions to problems (Lampert 1989). This kind of public scrutiny and debate is made possible only by making assertions and explanations explicit and open to critique.

It is important to note that children's theories that guide their problem-solving activity can be useful, even if they are wrong (Karmiloff-Smith & Inhelder 1975). What is important is that children become conscious of the theories they are using and aware of mistaken predictions when they occur. From a slightly different point of view, it is becoming clear that as mathematics classrooms become arenas for sense making (rather than only for skill acquisition), errors become much more constructive (Ball 1993; Hiebert, Wearne & Taber 1991; Schoenfeld, Smith & Arcavi 1993). Errors, made explicit, can be sites for fruitful analysis, discussion, and increased understanding.

When children can construct explanations and local theories, they can use them to guide their solution efforts on novel problems. They can use procedures acquired in other settings and adjust them to meet the demands of the new problem. Imagine 8-year-olds learning to subtract multidigit whole numbers. If they can explain the procedure they use by identifying its core features and how they are related, we say they understand the procedure. It turns out that these children can often modify and adjust their procedure to solve a related but new problem. In contrast, children who have memorized a procedure they cannot explain can seldom adjust the procedure to solve a new problem (Hiebert & Wearne 1996).

What lessons can we learn from children about articulating theories when doing research? First, we need to think of theories not as grand global theories that unify the elements in mathematics education but rather as the products of making explicit our hypotheses and hunches about how things work. Theories result from making public our private intuitions about the important elements and relationships of the situation. They provide explanatory solutions to the problems under study. They demonstrate that we are seeking to understand. In this sense, theory building is a natural aspect of all research that aims to understand.

It is useful to note that explicit theories provide clear interpretive frameworks for the research process. It is well-recognized now that all observations, all data, are theory-laden. We do not make neutral observations; our observations are biased by the theories we use. So the question is not whether we use theories but how we use them. As N. L. Gage (1963) expressed it, researchers 'differ not in whether they use theory, but in the degree to which they are aware of the theory they use' (p. 94). Making these theories explicit provides a way of dealing openly with the biases that infuse our work. At the same time, the process of making theories explicit keeps the light shining directly on activities that contribute to understanding.

A second lesson from children is that making theories explicit provides a communication bridge with other work. It is useful for the researcher because it allows the study to be situated in the context of related work. That increases the meaning and importance of the study. Communication also allows for public debate and critique. That is especially significant because, just as in the mathematics classroom, the process of public critique is critical for making progress as a research community (Lakatos 1978; Phillips 1987; Popper 1968). Developing acceptable solutions to important problems depends on expressing potential solutions as clearly and explicitly as possible and engaging colleagues in vigorous critiques of them. In Popper's (1968) words,

So my answer to the questions 'How do you know? What is the source or the basis of your assertion? What observations have led you to it?' would be: 'I do *not* know; my assertion was merely a guess. Never mind the source ... if you are interested in the problem which I tried to solve by my tentative assertion, you may help me by criticizing it as severely as you can'. (p. 27)

The point is that our theories reveal our tentative assertions and the more explicit we can be about them, the more useful the criticism will be.

A third lesson we can learn from children is that, in Popper's sense, the usefulness of theories has almost nothing to do with their correctness. Theories are useful if they are made explicit, whether or not they eventually turn out to be correct or widely accepted. Explicitness, not correctness, is needed to guide the research process and to communicate with others. Making mistakes is not such a bad thing if we can use them to make progress. Indeed, according to Popper (1968), correcting our mistakes is the only way we can make progress. Just as with children, errors can be fruitful sites for exploration and for gaining new insights.

A fourth lesson is that well-articulated theories make the data useful. Clearly expressed theories, or explanations, allow the research data to be interpreted thoughtfully and used to inform decisions in a variety of settings. As children can modify a specific procedure for a new problem, researchers can use data gathered in one setting to inform decisions in another setting. It is in that sense that research aimed at understanding is the most practical kind of research.

Consider the example of studying the effects of a teaching innovation. Without explicit hypotheses about what makes the innovation effective, the study is likely to simply test whether the innovation is more effective than a traditional method. The researcher might compare the two methods by measuring which group of students showed the largest gain on a general achievement test. The problem is that without a clearly articulated theory, these results, regardless of which method wins, are of limited use. It is not clear whether the innovative method would be effective with other topics, other groups of students, other teachers, and so on. It is impossible to use the original data to shed light on a new situation.

In contrast, if the aim of the study was to understand the effects of the innovation, the researcher would design a different kind of investigation to focus on the elements hypothesized to be crucial to the outcome and on the relationships between them. Hypotheses, or local theories, would point to the elements and relationships to consider. Checking the success of the hypotheses provides a way of explaining the outcomes. Although this kind of study is more difficult, the results can be used to suggest how the innovation might be modified to be more effective and to be effective in different contexts.

3. THE KIND OF DATA WE COLLECT

Theories are useful if they are made explicit, even if they are wrong. But the story does not end here. Theories that are wrong can interfere with progress in understanding if they are held too long in the face of conflicting evidence. That is true in both children's activities and researchers' activities. It also is true, in both arenas, that revising theories or abandoning them in light of new evidence is one of the most difficult things to do.

Karmiloff-Smith (1988) reported that children can resist so strongly giving up their theory that is guiding their problem-solving actions that they make up a new theory that handles the counterexamples and hold both theories simultaneously. Sometimes they even invent new evidence to confirm their original theory. Deanna Kuhn and her colleagues have reported a similar resistance to change in a variety of problem-solving contexts. D. Kuhn and Phelps (1982), for example, found that, when children were solving chemical mixture problems, the greatest impediment to progress was not the acquiring of new strategies but rather the abandoning of old ones. D. Kuhn, Amsel & O'Loughlin (1988) found several ways in which children justify retaining a theory that is not supported by the evidence. Several of those ways involve over-interpreting evidence as confirming. For example, children interpret information showing covariation between variables as supporting their hypothesis of cause and effect, and they interpret a single unit of supporting data as confirming their theory.

There are a number of interesting parallels between children and researchers. The history of scientific research is filled with examples of theories being retained in spite of critical disconfirming evidence (Greenwald, Pratkanis, Lieppe & Baumgardner 1986; T. S. Kuhn 1970; Lakatos 1978). There are many reasons for this phenomenon, including overly strong advocacy of particular theories (Greenwald et al. 1986) and pressures exerted by the social community of researchers (Campbell 1986). Evidence interpreted as confirming is usually given greater emphasis in public presentations than evidence that is disconfirming. Disconfirming evidence is often ignored or attributed to faulty procedures or measurement error. Small confirming data sets are weighted too heavily.

Given the tendency of people (researchers and children) to confirm theories rather than disconfirm them, what kinds of data are most useful? First, it must be remembered that errors are likely, both in theory construction and data interpretation. No sources of knowledge are completely reliable (Popper 1968). No one method of research nor type of data provides the solution. A variety of methods (e.g. qualitative and quantitative) and a variety of data are needed to provide checks on each other. All methods should be sensitive to detecting errors, to revealing disconfirming evidence.

A second consideration is that greater understanding may result from changing the research question from 'Is this hypothesis true?' to 'Under what conditions does the predicted effect occur?' Greenwald et al. (1986) argued that the first question is not very useful, because usually it will be answered yes. The second question is better because it redirects the researcher's predisposition to search for confirming evidence into a search that refines the theories' predictions. The second question also defines an important aspect of understanding a phenomenon – formulating more complete descriptions of the conditions that limit the reported findings (Greenwald et al. 1986). It is interesting that Cronbach (1986), in voicing support for this approach, cited an example from research in mathematics education. Cronbach argued that Brownell and Moser (1949) were successful in shedding light on an instructional approach because they did not try to confirm their hypothesis that 'This method of subtraction is best' but rather asked 'Under what conditions is this method better?'

To summarize, no one kind of data will provide complete solutions to the complex problems in mathematics education. So, no one method holds the answers. Evidence that both confirms and disconfirms hypotheses can be helpful, but the most useful data may be data that shed light on the conditions under which the hypotheses are appropriate. Information of this kind suggests how theories should be adjusted and hypotheses refined.

4. HOW WE COORDINATE THEORIES AND EVIDENCE

The activity of coordinating theories and evidence raises the research process to a very conscious level. It requires reflecting on alternative explanations and considering how well the evidence matches each one. Research that aims to understand phenomena necessarily involves coordination because getting better explanations, explanations that more closely match the data, is the best evidence that our understanding is increasing.

The lesson we can learn from children is that this coordination is not an easy activity and that it requires much practice. Although the seeds of the coordination may be present in young children in simple contexts (Samarapungavan 1991), most children do not coordinate theory and data in more complex situations (D. Kuhn et al. 1988; D. Kuhn, Schauble & Garcia-Mila 1992). Often they do not see coordination as an important activity, or they simply do not think to do it. At a conscious, deliberate, meta-level, the coordination of theories and evidence is difficult to apply consistently.

But coordinating theories and data is essential for deepening our understanding. It can serve two specific functions. First, the reflection required by this coordination serves as a powerful check on intuitions and speculations about what works. Education in general and mathematics education in particular are subject to wildly oscillating opinions about courses of action. Without understanding the situations in question, we have no informed basis for making decisions or for refuting the opinions of others. Fashion, rather than rational choice, is likely to determine the course of action. Research aimed toward understanding and the specific activity of coordinating theories and evidence provide the kind of reasoned and reflective check that is often needed.

A second function of coordinating theories and evidence is that it generates a cycle that continually deepens our understanding. Understanding is not an all-or-none affair. It develops along a complicated path and over a long period of time. The research activity that drives this process is one of stepping back, reflecting on the theories (explanations), and asking how well they fit the data. Are there more powerful explanations that might be formulated?

Ultimately, the process of coordinating theories and evidence is a process of warranting knowledge claims. Claims to know, within this chapter, mean claims that we understand something of importance in mathematics education. Warrant, according to *Webster's Third New International Dictionary*, means to give proof of the authenticity of something, or to give sufficient grounds or reasons for something. Dewey (1938) argued that the process of warranting knowledge claims is basic to all human intellectual activity. It is a process that plays out over and over in many ordinary daily activities.

Perhaps a mundane example would illustrate Dewey's claim. On a recent morning, while I was working on this chapter, I was listening to a sports-talk program while driving to campus. The discussion was about the appeal of sports because of their competitiveness and uncertainty of outcome. A caller claimed that the reason sports possessed these features was that all games began with the score of 0–0. The rather unforgiving host criticized the caller for not thinking carefully about the issue and pointed out instances where games could begin with scores of 0–0 and be very uncompetitive with no doubt about the outcome. Using our terms, the caller claimed to have some knowledge about sports and offered a warrant for his claim. The host disputed the warrant by pointing to some counterexamples, by presenting disconfirming evidence.

Dewey (1938) also argued that there were no differences, in principle, between warranting knowledge claims in science, in other disciplines, and in daily life. Knowledge claims and warrants must be expressed clearly, they must be connected through sound arguments, they must be matched against observations, and they must be examined carefully by peers and critics. That is true whether one is warranting a claim about the competitiveness of sports or about how we learn or teach mathematics.

However, this argument does not imply that all knowledge claims and warrants are equally valid or equally productive in helping us understand. Dewey devoted much of his work to describing the maturing process of humans along these lines, beginning with curiosities and the discovery and definitions of problems to the skillful warranting of solutions to genuinely important problems. The maturing process in warranting claims, in coordinating theories and evidence, is gradual and requires much reflective practice.

5. THE INTERNATIONAL RESEARCH ENTERPRISE

The principles that have been identified for guiding research that aims to understand educational phenomena are principles that are likely to be endorsed by researchers working within a variety of research paradigms. Given the diversity of paradigms that are popular across the international scene in mathematics education, it is important to note that there are still points of commonality and contact. Although we are inclined to emphasize differences, researchers share many basic assumptions about the research process.

In fact, I would argue that the shared assumptions are sufficiently fundamental that it is possible, in principle, to evaluate the warrants for claims across individuals and research groups, even if these groups are located within different paradigms. This possibility allows researchers working within different research cultures and traditions to communicate with each other. Although there are popular arguments to the contrary (Feyerabend 1975; T. S. Kuhn 1970), I believe the incommensurability argument often goes too far (Laudan 1990; Phillips 1987; Siegel 1980), especially as applied to the mathematics education research community. Perhaps we simply have used the paradigm vocabulary too quickly and easily, and have mistaken differences for incommensurability. Different cultures and research traditions make it more difficult to establish communication but do not, in principle, preclude it.

Based on the conversations at the 1994 International Commission on Mathematical Instruction study conference at the University of Maryland, I believe the international research community in mathematics education is in the position of working to open lines of communication that will, if pursued, be possible and productive. I do not believe that there are incommensurable differences between current paradigms of research. Evidence for this claim includes the number of meaningful disagreements (as well as agreements) that emerged, disagreements that were comprehensible to all parties. Such disagreements would not even arise if we were working within incommensurable paradigms (Siegel 1980). If that is true, then to continue communication should be especially productive because different perspectives encourage further reflection on our own interpretations. That has the potential to push our understandings further.

The process through which communication can be sustained is public conversation and debate. Points of agreement and disagreement provide bridges that can open the conversation. The goal must be to clarify claims that we understand and to examine warrants for these claims. This process need not be adversarial; it can be engaged throughout the research process (Cronbach 1986). In the end, however, our knowledge claims must be able to withstand the heat of public debate and criticism. This process is essential for making progress in our collective understanding.

REFERENCES

Artigue, M. & Perrin-Glorian, M.-J.: 1991, 'Didactic Engineering, Research & Development Tool: Some Theoretical Problems Linked to This Duality', *For the Learning of Mathematics* **11** (1), 13–18.

Ball, D. L.: 1993, 'Halves, Pieces, and Twoths: Constructing and Using Representational Contexts in Teaching Fractions', in T. P. Carpenter, E. Fennema & T. A. Romberg (eds.), *Rational Numbers: An Integration of Research*, Erlbaum, Hillsdale, NJ, 157–195.

Brownell, W. A.: 1946, 'Introduction: Purpose and Scope of the Yearbook', in N. B. Henry (ed.), *Forty-Fifth Yearbook of the National Society for the Study of Education: Part I. The Measurement of Understanding*, University of Chicago, Chicago, 1–6.

Brownell, W. A. & Moser, H. E.: 1949, 'Meaningful versus Mechanical Learning: A Study in Grade III Subtraction', *Duke University Research Studies in Education*, 8, Duke University Press, Durham, NC.

Campbell, D. T.: 1986, 'Science's Social System of Validity-Enhancing Collective Belief Change and the Problems of the Social Sciences', in D. W. Fiske & R. A. Shweder (eds.), *Metatheory in Social Science: Pluralisms & Subjectivities*, University of Chicago Press, Chicago, 108–135.

Carpenter, T. P.: 1986, 'Conceptual Knowledge as a Foundation for Procedural Knowledge: Implications from Research on the Initial Learning of Arithmetic', in J. Hiebert (ed.), *Conceptual and Procedural Knowledge: The Case of Mathematics*, Erlbaum, Hillsdale, NJ, 113–132.

Carpenter, T. P. & Moser, J. M.: 1984, 'The Acquisition of Addition and Subtraction Concepts in Grades One Through Three', *Journal for Research in Mathematics Education* **15**, 179–202.

Cobb, P., Wood, T., Yackel, E. & McNeal, B.: 1992, 'Characteristics of Classroom Mathematics Traditions: An Interactional Analysis', *American Educational Research Journal* **29**, 573–604.

Cronbach, L. J.: 1986, 'Social Inquiry by and for Earthlings', in D. W. Fiske & R. A. Shweder (eds.), *Metatheory in Social Science: Pluralisms & Subjectivities*, University of Chicago Press, Chicago, 83–107.

Cronbach, L. J. & Suppes, P.: 1969, *Research for Tomorrow's Schools: Disciplined Inquiry for Education*, Collier-Macmillan, London.

Davis, R. B.: 1992, 'Understanding "Understanding"', *Journal of Mathematical Behavior* **11**, 225–241.

Dewey, J.: 1938, *Logic: The Theory of Inquiry*, Holt, New York.

Fernandez, C., Yoshida, M. & Stigler, J. W.: 1992, 'Learning Mathematics from Classroom Instruction: On Relating Lessons to Pupils' Interpretations', *Journal of the Learning Sciences* 2, 333–365.

Feyerabend, P.: 1975, *Against Method*, Humanities Press, Atlantic Highlands, NJ.

Fuson, K. C.: 1988, *Children's Counting and Concepts of Number*, Springer-Verlag, New York.

Gage, N. L.: 1963, 'Paradigms for Research on Teaching', in N. L. Gage (ed.), *Handbook of Research on Teaching*, Rand McNally, Chicago, 94–141.

Gelman, R. & Gallistel, C. R.: 1978, *The Child's Understanding of Number*, Harvard University Press, Cambridge, MA.

Greenwald, A., Pratkanis, A., Lieppe, M. & Baumgardner, M.: 1986, 'Under What Conditions Does Theory Obstruct Research Progress?', *Psychological Review* 93, 216–229.

Hiebert, J. & Carpenter, T. P.: 1992, 'Learning and Teaching with Understanding', in D. A. Grouws (ed.), *Handbook of Research on Mathematics Teaching and Learning*, Macmillan, New York, 65–97.

Hiebert, J., Carpenter, T. P., Fennema, E., Fuson, K., Wearne, D., Murray, H., Olivier, A. & Human, P.: 1997, *Making Sense: Teaching and Learning Mathematics with Understanding*, Heinemann, Portsmouth, NH.

Hiebert, J. & Wearne, D.: 1992, 'Links between Teaching and Learning Place Value with Understanding in First Grade', *Journal for Research in Mathematics Education* 23, 98–122.

Hiebert, J. & Wearne, D.: 1993, 'Instructional Tasks, Classroom Discourse, and Students' Learning in Second-Grade Arithmetic', *American Educational Research Journal* 30, 393–425.

Hiebert, J. & Wearne, D.: 1996, 'Instruction, Understanding, and Skill in Multidigit Addition and Subtraction', *Cognition and Instruction* 14, 251–283.

Hiebert, J., Wearne, D. & Taber, S.: 1991, 'Fourth Graders' Gradual Construction of Decimal Fractions during Instruction Using Different Physical Representations', *Elementary School Journal* 91, 321–341.

Karmiloff-Smith, A.: 1988, 'The Child Is a Theoretician, Not an Inductivist', *Mind and Language* 3, 183–195.

Karmiloff-Smith, A. & Inhelder, B.: 1974, 'If You Want to Get Ahead, Get a Theory', *Cognition* 3, 192–212.

Kuhn, D., Amsel, E. & O'Loughlin, M.: 1988, *The Development of Scientific Thinking Skills*, Academic Press, San Diego, CA.

Kuhn, D. & Phelps, E.: 1982, 'The Development of Problem-Solving Strategies', in H. Reese (ed.), *Advances in Child Development and Behavior*, Academic Press, New York, 1–44.

Kuhn, D., Schauble, L. and Garcia-Mila, M.: 1992, 'Cross Domain Development of Scientific Reasoning', *Cognition and Instruction* 9, 285–327.

Kuhn, T. S.: 1970, *The Structure of Scientific Revolutions* (2nd ed.), University of Chicago Press, Chicago.

Laborde, C.: 1989, 'Audacity and Reason: French Research in Mathematics Education', *For the Learning of Mathematics* 9 (3), 31–36.

Lakatos, I.: 1978, *The Methodology of Scientific Research Programmes* (J. Worrall & G. Currie, eds.), Cambridge University Press, New York.

Lampert, M.: 1989, 'Choosing and Using Mathematical Tools in Classroom Discourse', in J. E. Brophy (ed.), *Advances in Research on Teaching*, JAI Press, Greenwich, CT, 223–264.

Laudan, L.: 1977, *Progress and Its Problems: Toward a Theory of Scientific Growth*, University of California Press, Berkeley, CA.

Laudan, L.: 1990, *Science and Relativism*, University of Chicago Press, Chicago.

Phillips, D. C.: 1987, *Philosophy, Science, and Social Inquiry*, Pergamon Press, New York.

Popper, K. R.: 1968, *Conjectures and Refutations: The Growth of Scientific Knowledge*, Harper Torchbooks, New York.

Popper, K.: 1972, *Objective Knowledge*, Oxford University Press, Oxford.

Riley, M. S., Greeno, J. G. & Heller, J. I.: 1983, 'Development of Children's Problem-Solving Ability in Arithmetic', in H. P. Ginsburg (ed.), *The Development of Mathematical Thinking*, Academic Press, New York, 153–196.

Romberg, T. A.: 1992, 'Perspectives on Scholarship and Research Methods', in D. A. Grouws (ed.), *Handbook of Research on Mathematics Teaching & Learning*, Macmillan, New York, 49–64.

Samarapungavan, A.: 1991, Children's Metajudgments in Theory Choice Tasks: Scientific Rationality in Childhood. Paper presented at the annual meeting of the American Educational Research Association, Chicago.

Schoenfeld, A. H., Smith, J. P. & Arcavi, A.: 1993, 'Learning: The Microgenetic Analysis of One Student's Evolving Understanding of a Complex Subject Matter Domain', in R. Glaser (ed.), *Advances in Instructional Psychology*, Erlbaum, Hillsdale, NJ, 55–175.

Siegel, H.: 1980, 'Objectivity, Rationality, Incommensurability, and More', *British Journal for Philosophy of Science* 31, 359–384.

Yoshida, M., Fernandez, C. & Stigler, J. W.: 1993, 'Japanese and American Students Differential, Recognition Memory for Teachers' Statements During a Mathematics Lesson', *Journal of Educational Psychology* 85, 610–617.

James Hiebert
College of Education,
University of Delaware,
Newark, DE 19716,
USA

NERIDA F. ELLERTON & M. A. ('KEN') CLEMENTS

TRANSFORMING THE INTERNATIONAL MATHEMATICS EDUCATION RESEARCH AGENDA

1. MATHEMATICS EDUCATION RESEARCH UNDER THE MICROSCOPE

During the second half of the 1980s, the international mathematics education research community was prepared to ask fundamental questions about the major issues and methods of its field of endeavor (see, e.g., Charles & Silver 1989; Grouws, Cooney & Jones 1988; Hiebert & Behr 1988; Sowder 1989; Wagner & Kieran 1989). This questioning has continued into the 1990s, with mathematics education researchers around the world scrutinizing the assumptions and methodologies associated with their work (see, e.g., Biehler, Scholz, Sträβer & Winkelmann 1994; Kilpatrick 1992, 1993; Mason, present volume; Secada, Fennema & Adajian 1995; Sierpinska 1993; Skovsmose 1994a, 1994b).

One manifestation of this interest in the foundations of mathematics education research was a symposium on 'Criteria for Scientific Quality and Relevance in the Didactics of Mathematics' held in Gilleleje, Denmark, in 1992. Another was the International Commission on Mathematical Instruction (ICMI) study conference held at the University of Maryland in May 1994 in connection with the present book.

At the Gilleleje symposium, papers by Kilpatrick (1993) and Sierpinska (1993) responded to the agenda set for the symposium by commenting on the following eight criteria, which have been used to evaluate the scientific quality of education research in general, and mathematics education research in particular: relevance, validity, objectivity, originality, rigor and precision, predictability, reproducibility, and relatedness. Both papers were concerned with discussing the meanings these eight constructs might have in the context of different kinds of research methodologies, and with whether these constructs should continue to be regarded as fundamental criteria for assessing mathematics education research.

Despite the willingness of mathematics education researchers to put their work under the microscope, we believe that the debate over the agenda and methods of mathematics education researchers has been too conservative. In the mid-1990s around the world well over one billion humans are attending formal mathematics classes – in schools and other educational institutions – in which they are being taught the internationalized version of mathematics. A massive amount of money (probably of the order of US$1000 billion annually) is spent on formal mathematics education. Yet, the international mathematics education

153

Sierpinska, A. and Kilpatrick, J. Mathematics Education as a Research Domain: A Search for Identity, 153–175.
© *1998 Kluwer Academic Publishers. Printed in Great Britain.*

research community seems to have only rarely paid attention to the fact many children around the world are still required to sit in classrooms where they understand only partially the words and symbols being used by teachers and by textbook writers.

This chapter calls for a radical reconstruction not only of the international mathematics education research agenda, but also of prevailing attitudes towards what constitutes good research. It acknowledges the tenor of Skovsmose's (1994a) view that mathematics education, as both a practice and a form of research, should discuss 'basic conditions for obtaining knowledge, ... be aware of social problems, inequalities, suppression, and ... [be] an active progressive social force' (pp. 37–38). Skovsmose maintains that critical education cannot support a prolongation of existing social inequalities.

Much of the contemporary theorizing and practice by mathematics education researchers continues to take place as if mathematics education itself is, and should be, a neutral, value-free activity. One view of mathematics education research is that it is an enterprise conducting careful studies that are *informative*, in the sense that they generate share-able knowledge that is simultaneously nontrivial, applicable, and not obvious. We would agree with that view, but would add that for us mathematics education researchers should pay more attention to their role of helping to create more *equitable* forms of mathematics education around the world. In short, mathematics education researchers need to recognize that their research ought to be both applicable and transformative.

2. KEY QUESTIONS FACED BY THE INTERNATIONAL MATHEMATICS EDUCATION RESEARCH COMMUNITY

A background paper (Balacheff, Howson, Sfard, Steinbring, Kilpatrick & Sierpinska 1993) was prepared in order to stimulate and direct the thinking of those who would contribute to the 1994 Maryland workshop. Five key questions were raised:

1. What is the specific object of study in mathematics education?
2. What are the aims of research in mathematics education?
3. What are the specific research questions or *problématiques* of research in mathematics education?
4. What are the results of research in mathematics education?
5. What criteria should be used to evaluate the results of research in mathematics education?

Although these questions by Balacheff et al. (1993) are important, they beg a number of larger questions. For example, with respect to the first question, one might ask: 'Should mathematics education have a *specific* object of study, or should it be acknowledged from the outset that there will be *many* legitimate objects of study?' Similarly, with respect to the third question, it could be argued that any list of *problématiques* of research in mathematics education would be dangerously limiting, in the sense that it would inevitably reflect the emphases of a particular group of people. The fifth question raises the issue of whether there should be clearly specified criteria for evaluating research in mathematics education that are unique to the field of mathematics education.

Bishop's (1992) approach was slightly different in that he identified the following five critical relationships in research in mathematics education:

1. 'What is' and 'what might be'.
2. Mathematics and education.
3. The problem and the research method.
4. The teacher and the researcher.
5. The researcher and the educational system.

The text that follows provides commentary on Balacheff et al.'s (1993) questions and Bishop's (1992) critical relationships.

Avoiding Another Straitjacket

Although the criteria for evaluating mathematics education research expounded by Kilpatrick and Sierpinska at the Gilleleje Symposium, and the key questions raised by Balacheff et al. (1993) and debated at the Maryland workshop, would seem to provide a fruitful basis for further discussion, it is possible that such discussion could inhibit progress towards a much-needed radical transformation of mathematics education and mathematics education research.

Our concern, expressed above, that any list of *problématiques* that might define a research agenda for mathematics education could be dangerously limiting – in the sense that such a list would inevitably reflect the emphases of a particular group of people – is given greater credence if the influence of previous sets of *problématiques* for mathematics education researchers are considered. It is our view, for example, that the widely acclaimed and influential book by Begle (1979), *Critical Variables in Mathematics Education*, and the equally influential edited collection of articles on *Research in Mathematics Education* (Shumway 1980), straitjacketed thinking within the international mathematics education research community for many years.

In the editors' preface to Begle's (1979) book the following excerpt from his address at the First International Congress on Mathematical Education, held in Lyon in 1969, is quoted:

I see little hope for any further substantial improvements in mathematics education until we turn mathematics education into an experimental science, until we abandon our reliance on philosophical discussion based on dubious assumptions, and instead follow a carefully constructed pattern of observation and speculation, the pattern so successfully employed by the physical and natural scientists.

We need to follow the procedures used by our colleagues in physics, chemistry, biology, etc., in order to build up a theory of mathematics education.... We need to start with extensive, careful, empirical observations of mathematics learning. Any regularities noted in these observations will lead to the formulation of hypotheses. These hypotheses can then be checked against further observations, and refined and sharpened, and so on. To slight either the empirical observations or the theory building would be folly. They must be intertwined at all times. (pp. x–xi)

Begle (1979) lamented the 'lack of a solid knowledge base' (p. 156) in mathematics education, and called for mathematics education researchers to develop and test theories in their domain. For him, that was the only way ahead.

Critical Variables in Mathematics Education first appeared in 1979, towards the close of a decade in which one particular form of 'scientific' research had dominated education research in general, and mathematics education research in particular. This particular form of research emphasized the need for carefully designed studies and the testing of hypotheses using statistical controls and mechanisms. Constructs like reliability, validity, replicability, and so on, were defined in largely statistical terms.

In fairness, it should be said that Begle was primarily interested in arguing the case for the need for better quality mathematics education research, for research in which scientific methods and checks were routinely employed. He certainly should not be expected to carry the blame for the tendency of mathematics educators to equate his call for better quality mathematics education research with the need to emphasize research which employed inferential statistical techniques. Indeed, he was only one of many education researchers of his time who displayed a preference for research involving experimental designs and statistical analyses. It is easy in the mid-1990s to recognize that Begle was defending a perspective on mathematics education research which would be fundamentally challenged in the 1980s, but in the late 1970s most other research methodologies, and particularly interpretive methodologies, were regarded as being likely to generate only 'soft' data for mathematics education.

Nevertheless, it is our contention that the publication of *Critical Variables in Mathematics Education* jointly by the Mathematical Association of America and the National Council of Teachers of Mathematics (NCTM) served to legitimize, in the minds of many mathematics education researchers in the United States and in other parts of the world, the notion that the 'scientific' approach, supported by 'appropriate' inferential statistics, provided the most rigorous and useful way ahead for education research. According to Begle and Gibb (1980), at the end of the 1970s the field of mathematics education resembled the state of agriculture in the United States several generations earlier. Although mathematics educators drew from 'general theories' they had 'no established general theory to provide a basis for [their] discussions' (p. 9). Begle and Gibb (1980) believed that what was needed was 'the construction of edifices of broad theoretical foundations for the teaching and learning of mathematics as well as curricula, content and organization' (p. 9).

Begle's (1979) call for a renewed emphasis on 'scientific' approaches in mathematics education research came at a time when fundamental questions were being asked about whether the designed type of research, relying heavily on the types of controls used in scientific laboratories, was, in fact, appropriate to education settings (Carver 1978; Freudenthal 1979). However, such questions were hardly, if at all, raised in Shumway's (1980) edited collection. A careful reading of this volume, which was published by NCTM, suggests that at the beginning of the 1980s many North American mathematics educators had well and truly committed themselves to a path which, they expected, would, through the application of statistical models and techniques, lead them to grand theories that would enable mathematics education to become a science.[1]

At the Eighth International Congress on Mathematical Education (ICME-8) held in Seville, Spain, in July 1996, the focus of one of the working groups was 'Criteria for Quality and Relevance in Mathematics Education Research'. A draft version of this present chapter was discussed at the first two sessions of the Working Group, and Jeremy Kilpatrick, who had been one of Begle's students, stated, in reacting to the draft chapter, that he did not accept the view that the Begle and Shumway publications were as influential as we had suggested in the chapter. He maintained that, in fact, the volumes represented the end of an era, and that this was recognized by mathematics education researchers at the time. We would disagree with this point of view, and would argue that the authority deriving from the fact that the books were published by the Mathematical Association of America and the National Council of Teachers of Mathematics gave the books large credibility. Furthermore, the fact that Begle's (1979) *Critical Variables in Mathematics Education* was reviewed at a series of special sessions at the fourth International Congress on Mathematical Education (ICME-4) held in Berkeley, California, in 1980, indicated to mathematics educators around the world that Begle's framework for mathematics education research needed to be seriously considered.

James Wilson and Jeremy Kilpatrick, both former students of Begle, became editors of the *Journal for Research in Mathematics Education (JRME)*. During the time of their editorships, many of the articles published in *JRME* were of the 'scientific' type – in which null and research hypotheses were formulated and tested using inferential statistical techniques. As Wilson (1994) has written, 'a large segment of the mathematics education community [between 1976 and 1982, when he was Editor of *JRME*] had an image of *JRME* as totally reactive, responding to the field with more concern for research rigor than for the significance of research problems' (p. 2). Wilson added that 'to many folks the image of *JRME* was quite stereotyped by statistical, methods-comparison research reports'.

An important aspect of the scientific approach to education research adopted in the 1960s and 1970s was the use of statistical significance testing. For two decades now, the legitimacy of the use of inferential statistical techniques in research in education, in general, has been regularly challenged, and in the mid-1990s there are many who are asserting, bluntly, that statistical significance testing should *never* be used in education research (Haig 1996; Menon 1993; Schmidt 1996; Shea 1996). Haig (1996) has stated that it is a major professional embarrassment that researchers continue to employ statistical significance tests 'in the face of more than three decades of damning criticism' (p. 201). In the same vein, Schmidt (1996) has challenged educators to articulate 'even one legitimate contribution that significance testing has made (or makes) to the research enterprise' (p. 116). He stated that he believed it would not be possible to find any way in which statistical significance testing had contributed to the development of cumulative scientific knowledge.

Although there are still some defenders of the faith (see, for example, Bourke 1993; McGaw 1996), it is now clear that the emphases in education research of the 1960s and 1970s on theory-*driven* analysis (as opposed to theory-*exploration* synthesis) was premature. Often, such research not only uncritically accepted the worth of largely untested theories, but it also employed methods of analysis that were based on inadequate statistical modeling and methods.

At the conclusion of *Critical Variables in Mathematics Education*, Begle (1979) included a list of goals for mathematics education. He also listed five variables for research in mathematics education to which he accorded highest priority. These were:

1. The relationship between teacher knowledge of subject matter and student achievement.
2. Drill.
3. Expository teaching of mathematical objects.
4. Acceleration.
5. Predictive tests.

Such a list reads strangely in the mid-1990s – the five 'critical variables' seem to be very narrow in scope.

Begle's (1979) list focused research attention on what many would now believe to be educational wastelands. At the same time, vitally important areas of concern, such as the influence of cultural or linguistic factors on mathematics learning, seemed to be overlooked or accorded low priority in the scheme of things. Perhaps that is why, for example, over the past 15 years there has not been, in our view, a coordinated series of investigations, among mathematics education researchers in the United States, into how language factors impinge on mathematics teaching and learning (see Ellerton & Clements 1991 for an extensive review of pertinent literature).

Undoubtedly, Begle's exhortations, and the early emphasis on statistically analyzed research by the editors of *JRME*, reflected the mood and academic culture of the time. Nevertheless, we believe that in the new millennium it is important for mathematics educators to cast off shackles of the past, while at the same time building on established strengths in their modes of operation.

3. AN EXAMPLE OF A RECENT MATHEMATICS EDUCATION RESEARCH STUDY THAT DOES NOT FIT THE MOLD

When we read Kilpatrick's (1993) and Sierpinska's (1993) papers – delivered at the Gilleleje symposium in Denmark – we felt uncomfortable about their eight criteria (relevance, validity, objectivity, originality, rigor and precision, predictability, reproducibility, and relatedness) for evaluating mathematics education research. We felt uncomfortable because we had only recently completed a major investigation which, although definitely within the category of 'mathematics education research', did not rest easily with the eight criteria.

Our research has been reported in detail in a 400-page book entitled *The National Curriculum Debacle* (Ellerton & Clements 1994). The book provides a history of an attempt, over the period 1987–1993, by Australian politicians and education bureaucrats, to establish a national curriculum. We extensively documented our argument that consultative processes adopted by the politicians and bureaucrats were inadequate. The outcomes-based curriculum framework that formed the basis of the nationally developed school mathematics documents was not acceptable to leading mathematicians, mathematics educators and mathematics teachers in Australia. In the book, we argued that outcomes-based forms of education have their origins in behaviorism, and claimed, on the basis of existing literature, that behaviorist approaches to mathematics education have rarely been associated with quality mathematics teaching and learning.

One of our concerns about the eight criteria listed by Kilpatrick and Sierpinska is it would be foolish to apply them to historical research of the kind that we conducted unless careful thought had been given to their meanings in the contexts of planning, conducting, reporting and evaluating historical research. Certainly, for example, we would hope that our research was relevant and valid, but we would find it difficult to define what the words *relevant* and *valid* mean in the context of our research. So far as 'objectivity' is concerned, in the foreword to our book, we acknowledge that there is no such thing as 'objective' history. Despite Kilpatrick's (1993) point that researchers should strive for objectivity even if they cannot achieve it, we wish to say that in the domain of investigation covered by our research it is difficult to know what 'objectivity' might look like. In addition, it would be difficult to apply the expression *rigor and precision* to our research (although we believe that our investigations were comprehensive and thorough).

Regarding reproducibility, we recognize that although someone might seek to replicate our research, the perspectives we brought to bear on the various issues raised in the book were idiosyncratic. No one else would have chosen to carry out, or could now carry out, the research in *exactly* the way we did. From that point of view the issue of 'reproducibility' would seem to be irrelevant so far as historical investigations are concerned. Certainly, the documents we examined and discussed are still available, but for most historians the idea of someone else replicating research is not regarded as sensible. Documents can be checked, but the interpretations of documents and the marshaling of evidence that might persuade a reader to accept suggested causal links renders the idea of replicating a historical study as nonsensical. That is not what good historians do.

Our last sentence ('That is not what good historians do') suggests that mathematics education research can, for example, be historical research, and that if it is historical research then criteria for evaluating the quality of historical research must apply. Of course, scholarly historians do not all agree on the criteria which apply to their domain, a fact which further suggests that any attempt to specify criteria for judging the quality of mathematics education research must inevitably result in failure. Mathematics education research can be in the realm of history, psychology, anthropology, linguistics, sociology, philosophy, mathematics, and so forth, and there is not within-domain or between-domain agreement on criteria for evaluating research. Hence, we must conclude that any attempt to introduce more than loose criteria, or guidelines, for assessing the quality of mathematics education research, would be foolish.

Probably, *The National Curriculum Debacle* would fit both Kilpatrick's and Sierpinska's criterion for 'relatedness'. But, as has been argued above, most of the eight criteria do not really match the research we carried out. Yet we would be adamant in asserting that the research has considerable relevance for mathematics education, especially (though not only) in the Australian context.

It would appear that the eight criteria echo lists of criteria found in treatises on how to conduct 'scientific' research. Our research was of an historical nature.[2]

4. TEN PROPOSITIONS FOR MATHEMATICS EDUCATION RESEARCH

Although we have argued above that it can be unhelpful to list specific criteria for assessing the quality of mathematics education research, we nevertheless agree with Balacheff et al. (1993) that some consideration should be given to describing the main *problématiques* of mathematics education. The identification of these concerns should assist mathematics education researchers to generate specific research questions.

We wish to put forward ten propositions that, if accepted, would have major implications for those seeking to identify the special *problématiques* of mathematics education research.

(1) Identifying Assumptions Influencing Current Practices in School Mathematics

The face of primary and secondary education around the world changed dramatically during the twentieth century. In 1900, less than 1% of the world's population had gained a secondary education. Indeed, despite claims by some historians that 'the nineteenth century had solidly established universal primary education' (Connell 1980, p. 4), less than 10% of the world's population in 1900 had attended elementary schools for more than 1 or 2 years. In most countries, school mathematics programs were based on rigid, externally prescribed curricula, formal textbooks, and written examinations, and it was accepted that only the best students should continue with the formal study of mathematics (Clements 1992).

As the twentieth century progressed, universal elementary education became a reality in many countries, and an increasing proportion of people gained a secondary education. Also, because of the sharp and continuous increase in world population, schools educated ever-increasing numbers of children. This phenomenon has generated important curriculum and associated teacher-training and professional development questions (Bishop 1993; Volmink 1994).

Our first proposition calls for the identification of the underlying assumptions from which current mathematics curricula and 'school mathematics cultures' have been developed. Implicit in the proposition is the question: What can be done to reduce, and ultimately eliminate, the prevalence and force of unwarranted assumptions?

Proposition 1. Many outdated assumptions influence the way school mathematics is currently practiced. The identification of those assumptions which most urgently need to be questioned represents the first, and perhaps most important, *problématique* of contemporary mathematics education research.

We believe that during the twentieth century, in most countries, mathematics curricula were unsuitable for the majority of secondary school students (Ellerton & Clements 1988). More generally, the assumptions that influenced normal patterns of behavior in mathematics classrooms (represented by teaching methods, assessment methods, and the attitudes of students, teachers, school administrators, employers, parents, etc.) were inherited from nineteenth-century patterns of schooling in leading Western nations such as the United States, England, France, and Germany. These assumptions, and corresponding expectations for what school mathematics programs should achieve, increasingly did not fit the new circumstances of the twentieth century. They are likely to be hopelessly inadequate for the twenty-first century.

To illustrate the point being made, consider the case of developing nations such as Indonesia. In the second half of the 1990s, less than half of Indonesia's school-aged children are receiving a secondary education, and only about one-tenth enroll in any form of higher education (Suprapto Brotosiswojo 1995). Under such circumstances it is legitimate to ask whether school mathematics curricula in Indonesia should be designed mainly to meet the needs of those who will go on to study mathematics and/or science in higher education institutions. What kind of mathematics curricula for the nation's schools are needed in such a situation? Given the dramatic expansion of formal education provision in Indonesia, what kinds of initial teacher-education and professional development programs are needed for primary and secondary teachers to enable them to cope with the demands of the mathematics education programs being offered in the nation's schools?

A word of warning concerning this first proposition is in order: it is unlikely that the set of unwarranted assumptions would be culture-free. In other words, findings with respect to Proposition 1 are likely not only to vary from nation to nation, but also from culture to culture.

(2) Doing More Than Preparing Students for the Next Level of Mathematics

The mathematics curriculum and standards of children at any grade level have traditionally been linked with those at the next grade level, and these expectations have not changed much over time. It is the responsibility of researchers to question whether it is reasonable for universities to continue to expect their entering students to be able to do the same kind of mathematics, at more or less the same level, as students in the past.

Many universities have introduced 'bridging programs' to bring 'deficient' students to an 'acceptable' level; others have chosen simply to fail greater proportions of students. The possibility that the old first-year tertiary mathematics programs are no longer appropriate has hardly, if ever, been taken seriously. Attempts by school systems to reform their mathematics curricula have often been opposed by mathematicians in universities, on the grounds that new courses would not prepare students as well as the old for first-year tertiary mathematics courses. This is the basis for the second proposition.

Proposition 2. Those concerned with mathematics education need to free themselves of the idea that preparation for higher level mathematics courses is the main concern of school mathematics.

In particular, the quality of mathematics programs in schools should not be judged by how many students subsequently study tertiary mathematics (or by how well those students who do proceed to study first-year tertiary mathematics perform in the mathematics examinations at a university).

(3) Making Language Factors a Central Concern

Our third proposition is in line with comments made by Secada (1988) on how cultural minorities and students in many so-called developing countries have been tacitly regarded as deviant and marginal in much mathematics education research. Secada pointed out that in many parts of the world it is normal to be bilingual, yet most mathematics education research has been with monolingual children – with bilingual children being regarded as 'handicapped', and therefore likely to need special attention.

Secada's point is well made. In many countries, including industrialized Western countries, there are children (in some cases, a majority of school children) who are required to learn mathematics in classrooms where the language of instruction is not their first language. Sometimes young children sit in mathematics classrooms for years without understanding what the teacher or textbook writer or examiner is saying. As Khisty (1995) has stated:

The teaching and learning process consists of an interaction between persons for the purpose of developing and sharing meanings. It logically follows that language is crucial if the development of meaning is to occur. Consequently, if we are to fully understand instructional dynamics – and obstacles that arise in the process and constrain minority children – we must examine not only curriculum and classroom activities, but also classroom discourse, that is, what is said and how it is said. (p. 280)

Ironically, sometimes it is claimed that in order to achieve 'equality of educational opportunity' the language of instruction and the curriculum should be the same for all, or most, children attending schools in the same country. This leads to our third proposition:

Proposition 3. The implications for mathematics education of the fact that many mathematics learners are bilingual or even multilingual urgently need to be explored.

Many linguistic factors impinge upon mathematics learning, and although much has been done in regard to identifying and relating these factors, more research is needed (Ellerton & Clements 1991).

It would be foolish to pretend that the answers to questions which must be asked will be easy to find. Take, for example, the nation in Papua New Guinea in which almost 750 indigenous languages are spoken. Until 1995, English was the language of instruction for all grades in PNG schools, which meant that young children sat in classrooms for several years not understanding of the teacher and the textbook (Clements & Jones, 1973). Now however, the official policy is that in Community Schools the language of instruction is the mother tongue of the learners. A different policy is in place in Indonesia, where 300 indigenous languages are spoken. In Indonesia the national government has decreed that the national language, *Bahasa Indonesian*, is to be the language of instruction in schools (Suprapto Brotosiswojo, 1995). On the difficult issue of language of instruction, mathematics education researchers need to carry out a great deal more sensitive and detailed research than they have, up to now, if they are to provide pertinent and helpful guidance to education policy makers.

(4) Rejecting Cultural Imperialism in Mathematics Education Policies and Practice

What we have said of language can also be said, even more generally, of 'culture'. Bishop (1990) has argued that although many cultures have contributed to the development of the internationalized version of 'Mathematics', 'Western Mathematics' has been one of the most powerful weapons used in the imposition of Western culture on 'developing' nations by colonizing powers.

Colonialism is an attitude of mind, accepted by both the leaders and representatives of the colonizing power and by those who are being colonized, that what goes on 'at home' should also take place in the colonies (Clements, Grimison & Ellerton 1989). This 'acceptance' is sometimes a conscious act, but more often it is unconscious – people behave in a colonialist way simply because that is the way they have learned to behave (Clements et al. 1989). D'Ambrosio (1985), in concluding a paper in which he presented a case for a radical revision of mathematics curricula in Third World countries, stated that 'we should not forget that colonialism grew together in a symbiotic relationship with modern science, in particular with mathematics and technology' (p. 47).

History provides a measure of support for D'Ambrosio's position. The nineteenth and twentieth centuries provided many examples of indigenous peoples uncritically accepting Western forms of mathematics as an essential component of the curricula of community schools created by their colonial supervisors. Soon after white settlement in Australia, for example, the Colonial Office established schools for 'civilizing' native Aboriginal children (Clements et al. 1989). Arithmetic was an important subject, and arithmetic textbooks from England and Ireland were used. Instruction was in the English language.

According to Bishop (1990), the reason Western Mathematics has been so powerful is that everyone has regarded it as neutral, as the least culturally-laden of all the school subjects imposed on indigenous pupils. The mystique and values of Western Mathematics have readily been accepted by most local, indigenous leaders, for there is a common belief that, somehow, economic and educational development is to be associated with these values.

Almost invariably, indigenous children in most nations where the curriculum has demanded they receive a Western version of school mathematics have struggled to cope with 'foreign' concepts inherent in arithmetic, algebra, geometry, trigonometry, calculus, Western measurement systems, and so on. A few students have succeeded, but most have become sacrificial lambs. When, finally, independence has come, the new class of indigenous rulers were often among those who had succeeded at school – and therefore they chose to continue the emphasis on Western Mathematics in their schools. In many countries they were supported in their efforts to do this by expert foreign consultants, funded through agencies such as Unesco, the World Bank, and the Asian Development Bank (Kitchen 1995). Our fourth proposition is:

Proposition 4. The assumption that it is reasonable to accept a form of mathematics education that results in a large proportion of school children learning to feel incompetent and helpless so far as 'Western' mathematics is concerned, should be rejected. Alternative forms of mathematics education, by which greater value would be accorded to the cultural and linguistic backgrounds of learners, should be explored.

In putting forward this proposition we recognize that mathematics education must always be political – in the sense that it inevitably imposes certain values on students and compels them to be involved in certain activities (Mellin-Olsen 1987; Harris 1991).

(5) Working Out the Implications of Situated Cognition Research Findings for Mathematics Education

Our fifth proposition arises from recent 'situated cognition' research in mathematics education. Researchers such as Carraher (1988), Lave (1988), and Saxe (1988), have found that people in all classes and walks of life are capable of performing quite complex 'mathematical' operations, despite the fact that many of them are not able to perform apparently similar operations under more formal circumstances − such as when asked to do so on pencil-and-paper mathematics tests. In Australia, for example, children who could easily calculate mentally how much change they should receive from $5.00 if they were to buy a chocolate bar for 45 cents were unable to find the value of 500 − 45, when this was presented as a pencil-and-paper task (Lovitt & Clarke 1988).

But the educational implications of this research are not at all clear. Suppose, for instance, there were 3 street 'candy sellers' in a mathematics class of 35 junior secondary students in a school in Brazil. How can the ordinary teacher be expected to know the special calculation abilities of these 3 children? After all, the teacher teaches quite a few different classes. And, even if the teacher did know of the special calculation abilities of the 3 children, how should that influence the teacher's normal class planning and teaching methods? There are 32 other students in the class, and no doubt many of these also have special situated mathematical knowledge (relating, for example, to sporting interests, gambling practices, etc.).

Intuitively, mathematics educators are inclined to believe that situated cognition research should send important messages to mathematics teachers and mathematics curriculum developers. But at present it is not clear what these messages are. Future research needs to be directed more at exploring the classroom implications of what is now a well-known phenomenon.

Proposition 5. The implications of situated cognition research for mathematics curricula, and for the teaching and learning of school mathematics, need to be investigated in creative ways.

With respect to Proposition 5, we do not think that controlled experiments, in which artificial education environments are created, would be helpful.

(6) Reconceptualizing the Role of Theory in Mathematics Education Research

We believe that some influential mathematics educators have been beguiled by the idea of mathematics education research becoming a science. There has been a view − often expressed publicly by powerful figures in the international mathematics education research community, and applied in the refereeing of journal articles and research proposals – that the spectacular achievements of theory-driven research in the physical sciences will be possible within the domain of mathematics education when adequate theories are developed.

It needs to be recognized, however, that the definition of the term *theory*, and the role of theory (whatever it means) in mathematics education research is not something on which all education researchers agree. Those who prefer to work from existing theories tend to justify their approach to research by referring to metaphors (such as workers 'building a wall', by continually adding new 'bricks', the final aim being to complete the 'wall' under construction). From our perspective, many of the theories – and the metaphors used to support these theories – are simplistic and have received more attention than they deserve. Furthermore, because some theories, and associated 'models' (another loosely defined term), have become so widely accepted and used, they have come to be regarded as being objectively 'true', despite there being considerable evidence that they do not apply in many contexts.

The preference for theory-driven, theory-confirming research has gone so far that funding bodies in the United States have instructed assessors of proposals to deny funding to applicants for grants who have not elaborated well-defined theoretical bases for their proposed research.

We wish to assert that the domain of mathematics education is not controlled by culture-free laws that can be progressively identified through research. Mathematics educators should emphatically reject descriptions of mathematics education as being at some point in a sequence from 'myths', to 'traditions', to 'model building', to 'paradigm selection', to 'scientific revolution' (Romberg 1981, 1983).

Unlike many others, we do not see theoretical positions such as are associated with the van Hieles levels of learning, the SOLO taxonomy, certain information-processing models for additive and subtractive task situations, and complex models for explaining rational number concept development in children, as having significantly enhanced the quality of international mathematics education research over the past two decades. However, powerful figures who have insisted that research based on such theoretical models is superior to research in which more exploratory approaches are preferred have succeeded in creating a mindset within an influential section of the international mathematics education research community.

The view that mathematics education research should resemble scientific research has resulted in theoretical structures being conjectured and much data being generated in an attempt to confirm or partly confirm proposed theories. Thus, for example, Riley, Greeno & Heller's (1983) information-processing model for explaining how children process 'change', 'combine', and 'compare' addition and subtraction word problems has been enormously influential. Yet Riley et al.'s (1983) study which supported the theoretical model involved small, fairly select samples, as have most subsequent investigations based on the same theory.

One difficulty with such an approach is that the theories can all too quickly become set in stone, and counter results are all too easily ignored. Thus, for example, although Del Campo and Clements (1987), and Lean, Clements & Del Campo (1990), produced data which demonstrated beyond doubt that the Riley et al. (1983) theory did not stand with Australian and Papua New Guinean children, that does not seem to have had any effect on subsequent writings of American researchers (Fuson 1992; Hiebert & Carpenter 1992).

According to Secada (1988), theories purporting to explain the development of addition and subtraction are 'coherent but incomplete' (p. 33). They have been based on research that attended to majority cases. By excluding minority children from its findings, such research legitimizes an ongoing view of such children as deviant.

We would go even further than Secada on this matter, by asking: Who is 'normal'? And to which populations can one generalize from a study that, for instance, involved mainly middle-class Anglo-Saxon children, somewhere in a university town in an affluent Western nation?

Proposition 6. The idea that the best mathematics education research is that which is based on a coherent theoretical framework should be subjected to careful scrutiny. Furthermore, popular existing theories for which strong counter data have been reported, should either be abandoned immediately, or substantially modified.

We are *not* saying that logical classification schemes, such as the 'Change, Combine, Compare' system (for additive and subtractive arithmetic word problems), and the Newman interview approach (for analysing written responses to mathematics tasks), have not had beneficial effects. We *are* saying that research rigidly based on generalizations from speculative theories that do not stand the test of multicultural investigations can be dangerously limiting.

Rhetorical statements such as Kurt Lewin's 'there is nothing as practical as good theory and nothing as theoretical as good practice' are often thrown at those who argue that at the present moment mathematics education needs *more* reflective, more culture-sensitive, more practice-oriented, and more theory-generating research, and *less* theory-driven research. Although we would accept the second part of Lewin's statement ('... there is nothing as theoretical as good practice' – see Proposition 8, below), we would wish to put forward an alternative to the first part of Lewin's statement. Our alternative statement would begin: 'In mathematics education there is nothing so *dangerous* as speculative theory based on data deriving from a single culture'.

(7) Developing a New Epistemological Framework for Mathematics Education Research

Our seventh proposition concerns the need to establish an epistemological framework[3] for mathematics education research. We believe that mathematics education research will be most informative if attention is given to bringing together the literature on (a) the histories of mathematics and mathematics education; (b) mathematical understanding and achievement in different cultures; (c) influences of culture on young (preschool) children's conceptions of mathematics; and (d) the impact of schooling on learners' conceptions of mathematics.

An example of the type of research we are contemplating was provided by Clements and Del Campo (1990) who, in attempting to answer the question 'How natural is fraction knowledge?', provided commentaries on three bodies of literature which had a bearing on this question. These were (a) the history of the development of fraction concepts; (b) what is known about the fraction knowledge of cultures outside the dominant Western European tradition; and (c) what is known about how very young children seem to acquire fraction concepts, and how that is influenced by culture. This approach could be extended, from merely covering fractions to every significant theme in mathematics.

Proposition 7. A suitable framework for achieving a more unified and systematic approach to mathematics education research is needed. One possible approach would focus on coordinated research programs that linked (a) the histories of mathematics and mathematics education; (b) mathematical understanding and achievement evident in different cultures; (c) influences of culture on young (preschool) children's conceptions of mathematics; and (d) the impact of schooling on learners' conceptions of mathematics.

A framework based on Proposition 7 would not be unduly restrictive, for most existing research in mathematics education could comfortably be located within it.

(8) Questioning the Basis for Assessing Achievements in Mathematics

Evaluators of mathematics education programs, and those responsible for the conduct of national and international mathematics achievement surveys and mathematics competitions, are still inclined to regard achievement on pencil-and-paper tests as fundamentally important. Thus, for example, researchers investigating the effectiveness of outcomes-based education approaches to school mathematics have used pencil-and-paper tests to define operationally the main dependent and independent variables (Sereda 1993). So too have researchers comparing the mathematics programs in schools in the United States and some Asian nations (Stigler & Baranes 1988).

Yet, research has generated data suggesting that students who give correct answers to items on so-called 'valid' and 'reliable' pencil-and-paper mathematics tests sometimes have little or no understanding of the mathematical concepts and relationships the tests were designed to measure (see, e.g., Ellerton & Clements 1995). That seems especially likely to be the case with tests consisting of multiple-choice items. A large research study into this question, jointly supervised by the authors and carried out in Thailand by Thongtawat (1992), found that the proportion of students who gave correct answers to multiple-choice mathematics items but who did not understand the mathematical concepts and relationships involved in the items, was much higher than that for corresponding short-answer *but not multiple-choice* items. Thongtawat also found that students who scored poorly on a test could sometimes have a good conceptual grasp of the material which the items covered.

Proposition 8. Closer research scrutiny needs to be given to the issue of how achievement is best measured in mathematics, and pressure should be exerted on education systems, testing authorities, and mathematics competition directors to apply the findings of this research.

In other words, mathematics education researchers should begin to pay more than lip service to the truism that, in education, what is assessed is what is valued, and what is valued is what is assessed.

The issue of equity in mathematics education is inextricably linked with methods of assessment. As Nettles and Bernstein (1995) have stated, 'the dual issues of improving the ways student performance is measured, and eliminating the racial and socio-economic gaps on whatever measures are used, are enduring ones for policy makers to address in the years ahead' (p. 6).

(9) Establishing Research Communities which Value All Participants

None of the above propositions touches on the important issue of who should be involved in mathematics education research, and in what ways. Bishop and Blane (1994) have argued that research problems and questions need to be couched in terms within the participants' sphere of knowledge, thereby offering the dimensions of generality which make the difference between research and problem solving. Theory will enter through the lenses of the participants' knowledge schema, and 'insofar as these schema involve connections with published theory, so will that theory play a part in shaping the research' (p. 63).

Action research that attempts to maximize the potential contributions of all participants in a research exercise, at all stages of the exercise – including the design and reporting stages – and that aims at achieving improvement through cooperation, has not been as widely used in mathematics education as it might. Yet reports of successful action research projects in mathematics education have been published (see, for example, Ellerton, Clements & Skehan 1989; Tan Sean Huat & Sim Jin Tan 1991). A study by Desforges and Cockburn (1987) suggests that unless we tap into the wisdom of practice, and incorporate this wisdom generously into mathematics education research, teachers will simply not listen to the admonitions and theories of remote, self-designated 'expert' researchers in mathematics education.

Proposition 9. Practising mathematics teachers need to be involved, as equal partners, in action research projects, and the theoretical assumptions and practical approaches in projects should not be predetermined by outside 'experts'.

Just as center-to-periphery models of curriculum change are likely to be actively resisted by teachers, so too are recommendations emanating from research carried out by persons who are perceived, by teachers, to be too remote to understand the pressures shaping what is possible in school mathematics programs and classrooms.

In relation to Proposition 9, it needs to be recognized that the term *action research* has been used in several ways in the literature, and that some education researchers regard it as little more than teachers reflecting on aspects of their practice. Other education researchers seem to believe that any research project which involves practicing teachers can be regarded as action research. From our perspective, both of these represent inadequate views of action research. We accept the definition of action research put forward by Kemmis and McTaggart (1988) as a form of *collective* self-inquiry undertaken by participants in social situations in order to improve the rationality and justice of their own social or educational practices, as well as their understanding of these practices and the situations in which these practices are carried out. Groups of participants can be teachers, students, principals, parents and other community members – any group with a shared concern. The approach is only action research when it is *collaborative*, though it is important to realize that the action research of the group is achieved through the *critically examined action* of individual group members.

Particular emphasis is placed on the notion of collaboration between participants, all of whom are active professional individuals who have common concerns.

(10) Making the International Mathematics Education Research Community Truly International

There is a need to end the Eurocentric/North American domination of mathematics education research which has emerged over the past three decades. For example, most of the keynote speakers at the annual conferences of the International Group for the Psychology of Mathematics Education Research (PME) have been from North America or Continental Europe, and only rarely have more than a handful of participants from Bangladesh, China, India, Indonesia, Korea, Pakistan, Vietnam, the Philippines – and indeed all other Asian nations except Japan – attended. Yet these last-named Asian nations contain half the world's population. It is simply not good enough to say that PME conferences are open to whoever wishes to attend – often those interested in mathematics education in these countries are not aware of international mathematics education conferences, and even those who are, are unable to find the money needed to go.

Proposition 10. The present international mathematics education research community needs to move proactively so that full and equal participation is possible for mathematics educators in countries that are currently underrepresented in the community.

Among the issues requiring resolution are the need to: (a) identify funding sources which will assist mathematics educators from all countries not only to attend, but also to participate actively in, important mathematics education conferences on a regular basis (Unesco, the World Bank, and the Asian Development Bank may be able to help); (b) foster quality research by teams of researchers containing a judicious mix of experienced and inexperienced mathematics education researchers from different countries; (c) develop agreements by which the reporting of this research will give local researchers full credit; and (d) expand the set of nations in which important mathematics education conferences are held.

5. CONCLUDING COMMENTS

The issues raised in this chapter need to be confronted urgently by the international mathematics education research community. A careful reading of our ten propositions will reveal that collectively they emphasize the need for the international mathematics education research community to: (a) examine fundamental assertions currently driving research activities in mathematics education; (b) give due accord to how linguistic and cultural factors influence mathematics education; (c) question whether it is helpful to work towards the developments of 'grand theories', on the assumption that mathematics education is progressing towards being a science; and (d) demonstrate a greater respect for the wisdom of practice deriving from the classroom knowledge and the action-oriented theories of practicing teachers of mathematics in different countries around the world.

ACKNOWLEDGMENTS

Although the writers accept full responsibility for the idiosyncratic views presented in this chapter, we wish to acknowledge the constructive suggestions and criticisms of an earlier version of the chapter by members of the Working Group on 'Criteria for Quality and Relevance in Mathematics Education Research'. The Working Group dedicated two sessions at the Eighth International Congress on Mathematics Education – held in Seville (Spain) in July 1996 – to discussing the chapter. In particular, the contributions of Alan Bishop, Jeremy Kilpatrick, Christine Keitel, Dick Lesh, Ramakrishnan Menon, and Ken Ruthven are acknowledged.

NOTES

1. This was the case, despite the existence of North American mathematics educators – such as Les Steffe and Robert Davis – who, in 1980, were clearly seeking to carry out scientific mathematics education research without a heavy use of statistical analyses.

2. In July, 1996, Jeremy Kilpatrick stated, in reacting to a draft of this present paper at the Working Group on 'Criteria for Quality and Relevance in Mathematics Education Research' (at ICME-8), that

there was never any intention for the eight criteria to be applied to historical research. Rather, he said, the eight criteria were meant to apply to 'scientific' approaches to mathematics education research. This comment by Kilpatrick raises the issue of the nature of 'scientific research' in the context of mathematics education.

3. At the July, 1996, Working Group on 'Criteria for Quality and Relevance in Mathematics Education Research' (at ICME-8), reactors to a draft of the present paper argued that Propositions 6 and 7 seemed to contradict each other in the sense that the 'framework' suggested in Proposition 7 was, in fact, a 'theory'. However, a framework for discussion is hardly a theory. At issue, again, is the meaning to be given to the term 'theory'.

REFERENCES

Balacheff, N., Howson, A. G., Sfard, A., Steinbring, H., Kilpatrick, J. & Sierpinska, A.: 1993, 'What Is Research in Mathematics Education, and What Are Its Results?', *ICMI Bulletin* 33, 17–23.
Begle, E. G.: 1979, *Critical Variables in Mathematics Education*, Mathematical Association of America and the National Council of Teachers of Mathematics, Washington, DC.
Begle, E. G. & Gibb, E. G.: 1980, 'Why Do Research?', in R. J. Shumway (ed.), *Research in Mathematics Education*, National Association of Teachers of Mathematics, Reston, VA, 3–19.
Biehler, R., Scholz, R. W., Sträßer, R. & Winkelmann, B. (eds.): 1994, *Didactics of Mathematics as a Scientific Discipline*, Kluwer Academic Publishers, Dordrecht.
Bishop, A. J.: 1990, 'Western Mathematics: The Secret of Cultural Imperialism', *Race and Class* 32 (2), 51–65.
Bishop, A. J.: 1992, 'International Perspectives on Research in Mathematics Education', in D. Grouws (ed.), *Handbook of Research on Mathematics Teaching and Learning*, Macmillan, New York, 710–723.
Bishop, A. J.: 1993, 'Influences from Society', in *Significant Influences on Children's Learning of Mathematics*, UNESCO Science and Technology Education (Document Series No. 47), UNESCO, Paris, 3–26.
Bishop, A. J. & Blane, D. C.: 1994 (May), 'Responding to the Pressure for "Effectiveness" in Mathematics Teaching', in *What Is Research in Mathematics Education, and What Are Its Results?* (Background papers for an ICMI Study Conference), International Commission on Mathematics Instruction, University of Maryland, College Park, MD, 52–68.
Bourke, S.: 1993, 'Babies, Bathwater, and Straw Persons', *Mathematics Education Research Journal* 5 (1), 19–22.
Carraher, T. N.: 1988, 'Street Mathematics and School Mathematics', in A. Borbas (ed.), *Proceedings of the Twelfth International Conference on the Psychology of Mathematics Education*, International Group for the Psychology of Mathematics Education, Veszprém, Hungary, 3–23.
Carver, R. P.: 1978, 'The Case against Statistical Significance Testing', *Harvard Educational Review* 48, 378–399.
Charles, R. I. & Silver, E. A. (eds.): 1989, *Research Agenda in Mathematics Education: The Teaching and Assessing of Mathematical Problem Solving*, National Council of Teachers of Mathematics, Reston, VA, and Lawrence Erlbaum, Hillsdale, NJ.
Clements, M. A.: 1992, *Mathematics for the Minority: Some Historical Perspectives of School Mathematics in Victoria*, Deakin University, Geelong.
Clements, M. A. & Del Campo, G.: 1990, 'How Natural is Fraction Knowledge?', in L. P. Steffe & T. Wood (eds.), *Transforming Children's Mathematics Education: International Perspectives*, Lawrence Erlbaum, Hillsdale, NJ, 181–188.
Clements, M. A., Grimison, L. & Ellerton, N. F.: 1989, 'Colonialism and School Mathematics in Australia', in N. F. Ellerton & M. A. Clements (eds.), *School Mathematics: The Challenge to Change*, Deakin University, Geelong, 50–78.

Clements, M. A. & Jones, P.: 1983, 'The Education of Atawe', in I. Palmer (ed.), *Melbourne Studies in Education 1983*, Melbourne University Press, Melbourne, 112–144.

Connell, W. F.: 1980, *A History of Education in the Twentieth Century World*, Curriculum Development Centre, Canberra.

D'Ambrosio, U.: 1985, 'Ethnomathematics and its Place in the History and Pedagogy of Mathematics', *For the Learning of Mathematics* 5 (1), 44–48.

Del Campo, G. & Clements, M. A.: 1987, 'Elementary School Children's Processing of "Change" Arithmetic Word Problems', in J. C. Bergeron, N. Herscovics & C. Kieran (eds.), *Proceedings of the Eleventh International Conference on the Psychology of Mathematics Education*, International Group for the Psychology of Mathematics Education, Montréal, Vol. 2, 382–388.

Desforges, C. & Cockburn, A.: 1987, *Understanding the Mathematics Teacher: A Study of Practice in First Schools*, Falmer, London.

Ellerton, N. F. & Clements, M. A.: 1988, 'Reshaping School Mathematics in Australia 1788–1988', *Australian Journal of Education* 32 (3), 387–405.

Ellerton, N. F. & Clements, M. A.: 1991, *Mathematics in Language: A Review of Language Factors in Mathematics Learning*, Deakin University, Geelong.

Ellerton, N. F. & Clements, M. A.: 1994, *The National Curriculum Debacle*, Meridian Press, Perth.

Ellerton, N. F. & Clements, M. A.: 1995, 'Challenging the Effectiveness of Pencil-and-Paper Tests in Mathematics', in J. Wakefield & L. Velardi (eds.), *Celebrating Mathematics Learning*, Mathematical Association of Victoria, Melbourne, 444–449.

Ellerton, N. F., Clements, M. A. & Skehan, S.: 1989, 'Action Research and the Ownership of Change: A Case Study', in N. F. Ellerton & M. A. Clements (eds.), *School Mathematics: The Challenge to Change*, Deakin University, Geelong, 284–302.

Freudenthal, H.: 1979, 'Ways to Report on Empirical Research in Education', *Educational Studies in Mathematics* 10 (3), 275–303.

Fuson, K. C.: 1992, 'Research on Whole Number Addition and Subtraction', in D. Grouws (ed.), *Handbook of Research on Mathematics Teaching and Learning*, Macmillan, New York, 243–275.

Grouws, D. A., Cooney, T. J. & Jones, D. (eds.): 1988, *Research Agenda in Mathematics Education: Perspectives on Research on Effective Mathematics Teaching*, National Council of Teachers of Mathematics, Reston, VA, and Lawrence Erlbaum, Hillsdale, NJ.

Haig, B. D.: 1996, 'Statistical Methods in Education and Psychology: A Critical Perspective', *Australian Journal of Education* 40 (2), 190–209.

Harris, P.: 1991, *Mathematics in a Cultural Context: Aboriginal Perspectives of Space, Time and Money*, Deakin University, Geelong.

Hiebert, J. & Behr, M. (eds.): 1988, *Research Agenda in Mathematics Education: Number Concepts and Operations in the Middle Grades*, National Council of Teachers of Mathematics, Reston, VA, and Lawrence Erlbaum, Hillsdale, NJ.

Hiebert, J. & Carpenter, T. P.: 1992, 'Learning and Teaching with Understanding', in D. Grouws (ed.), *Handbook of Research on Mathematics Teaching and Learning*, Macmillan, New York, 65–100.

Kemmis, S. & McTaggart, R. (eds.) (1988), *The Action Research Planner*, Deakin University, Geelong.

Khisty, L. L.: 1995, 'Making Inequality: Issues of Language and Meanings in Mathematics Teaching with Hispanic Students', in W. G. Secada, E. Fennema & L. B. Adajian (eds.), *New Directions for Equity in Mathematics Education*, Cambridge University Press, Cambridge, 279–297.

Kilpatrick, J.: 1992, 'A History of Research in Mathematics Education', in D. Grouws (ed.), *Handbook of Research on Mathematics Teaching and Learning*, Macmillan, New York, 3–38.

Kilpatrick, J.: 1993, 'Beyond Face Value: Assessing Research in Mathematics Education', in G. Nissen & M. Blomhøj (eds.), *Criteria for Scientific Quality and Relevance in the Didactics of Mathematics*, Danish Research Council for the Humanities, Roskilde, Denmark, 15–34.

Kitchen, R.: 1995, 'Mathematics Pedagogy in the 3rd World: The Case of a Guatemalan Teacher', *ISGEm Newsletter*, 10 (2), 1–4.

Lave, J.: 1988, *Cognition in Practice: Mind, Mathematics and Culture in Everyday Life*, Cambridge University Press, Cambridge.

Lean, G. A., Clements, M. A. & Del Campo, G.: 1990, 'Linguistic and Pedagogical Factors Affecting Children's Understanding of Arithmetic Word Problems', *Educational Studies in Mathematics* 21, 165–191.

Lovitt, C. & Clarke, D. M.: 1988, *Mathematics Curriculum and Teaching Program Activity Bank* (Vol. 1), Curriculum Development Centre, Canberra.

McGaw, B.: 1996, 'Response to Brian D. Haig', *Australian Journal of Education* 40 (2), 209–213.

Mellin-Olsen, S.: 1987, *The Politics of Mathematics Education*, Reidel, Dordrecht.

Menon, R.: 1993, 'Statistical Significance Testing Should Be Discontinued in Mathematics Education Research', *Mathematics Education Research Journal* 5 (1), 4–18.

Nettles, M. T. & Bernstein, A. R.: 1995, 'Introduction: The Pursuit of Equity in Educational Testing and Assessment', in M. T. Nettles & A. L. Nettles (eds.), *Equity and Excellence in Educational Testing and Assessment*, Kluwer Academic Publishers, Boston, MA, 3–21.

Riley, M. S., Greeno, J. & Heller, J. I.: 1983, 'Development of Children's Problem-Solving Ability in Arithmetic', in H. P. Ginsburg (ed.), *The Development of Mathematical Thinking*, Academic Press, New York, 153–196.

Romberg, T. A.: 1981, 'Towards a Research Consensus in Some Problem Areas in the Learning and Teaching of Mathematics', in *Proceedings of the Fifth International Conference on the Psychology of Mathematics Education*, International Group for the Psychology of Mathematics Education, Grenoble, Vol. 3, 31–46.

Romberg, T. A.: 1983, 'Towards "Normal Science" in Some Mathematics Education Research', *Zentralblatt für Didaktik der Mathematik* 15, 89–92.

Saxe, G.: 1988, 'Linking Language with Mathematics Achievement: Problems and Prospects', in R. Cocking & J. P. Mestre (eds.), *Linguistic and Cultural Influences on Learning Mathematics*, Lawrence Erlbaum, Hillsdale, NJ, 47–62.

Schmidt, F. L.: 1996, 'Statistical Significance Testing and Cumulative Knowledge in Psychology: Implications for Training of Researchers', *Psychological Methods* 1 (2), 115–129.

Secada, W. G.: 1988, 'Diversity, Equity, and Cognitivist Research', in E. Fennema, T. P. Carpenter & S. J. Lamon (eds.), *Integrating Research on Teaching and Learning Mathematics*, University of Wisconsin-Madison, Madison, 20–58.

Secada, W. G., Fennema, E. & Adajian, L. B. (eds.): (1995), *New Directions for Equity in Mathematics Education*, Cambridge University Press, Cambridge.

Sereda, J.: 1993, 'Educational Quality Indicators in Art and Mathematics', *Alberta Journal of Education* 39 (2), 17–233.

Shea, C.: 1996 (September 11), 'Are Stats Skewing Psychology?' *The Australian*, 26.

Shumway, R. J. (ed.): 1980, *Research in Mathematics Education*, National Association of Teachers of Mathematics, Reston, VA.

Sierpinska, A.: 1993, 'Criteria for Scientific Quality and Relevance in the Didactics of Mathematics', in G. Nissen & M. Blomhøj (eds.), *Criteria for Scientific Quality and Relevance in the Didactics of Mathematics*, Danish Research Council for the Humanities, Roskilde, Denmark, 35–74.

Skovsmose, O.: 1994a, 'Towards a Critical Mathematics Education', *Educational Studies in Mathematics Education* 27 (1), 35–57.

Skovsmose, O.: 1994b, *Towards a Philosophy of Critical Mathematics Education*, Kluwer Academic Publishers, Dordrecht.

Sowder, J. T. (ed.): 1989, *Research Agenda in Mathematics Education: Setting a Research Agenda*, National Council of Teachers of Mathematics, Reston, VA, and Lawrence Erlbaum, Hillsdale, NJ.

Stigler, J. W. & Baranes, R.: 1988, 'Culture and Mathematics Learning', in E. Z. Rothkopf (ed.), *Review of Research in Education 15, 1988–89*, American Educational Research Association, Washington, DC, 253–306.

Suprapto Brotosiswojo, B.: 1995 (November), 'An Experience in Managing a Large System of Distance Education'. Paper presented at the First International Symposium on 'Networking into the 21st Century', Jogyakarta, Indonesia.

Tan Sean Huat & Sim Jin Tan: 1991, *Thinking in Science and Mathematics Project: Interim Report*, SEAMEO-RECSAM, Penang, Malaysia.

Thongtawat, N.: 1992, *Comparing the Effectiveness of Multiple-Choice and Short-Answer Paper-and-Pencil Tests*, SEAMEO/RECSAM, Penang, Malaysia.

Volmink, J. D.: 1994, 'Mathematics by All', in S. Lerman (ed.), *Cultural Perspectives on the Mathematics Classroom*, Kluwer Academic Publishers, Dordrecht, 51–67.

Wagner, S. & Kieran, C. (eds.): 1989, *Research Agenda in Mathematics Education: The Teaching and Assessing of Mathematical Problem Solving*, National Council of Teachers of Mathematics, Reston, VA, and Lawrence Erlbaum, Hillsdale, NJ.

Wilson, J. W.: 1994, 'Guest Editorial', *Journal for Research in Mathematics Education* **25** (1), 2–3.

Nerida F. Ellerton
Faculty of Education
The University of Southern Queensland
Toowoomba
Queensland, 4350
Australia

M. A. ('Ken') Clements
Sultan Hassanal Bolkiah Institute of Education
Universiti Brunei Darussalam
Bandar Seri Begawan
Negara Brunei Darussalam

JUAN D. GODINO & CARMEN BATANERO

CLARIFYING THE MEANING OF MATHEMATICAL OBJECTS AS A PRIORITY AREA FOR RESEARCH IN MATHEMATICS EDUCATION

1. THE NATURE OF MATHEMATICAL OBJECTS: THE QUESTION OF MEANING

The specific aim of mathematics education as a research field is to study the factors that affect the teaching and learning of mathematics and to develop programs to improve the teaching of mathematics. In order to accomplish this aim mathematics education must consider the contributions of several disciplines: psychology, pedagogy, sociology, philosophy, etc. However, the use of these contributions in mathematics education must take into account and be based upon an analysis of the nature of mathematics and mathematical concepts, and their personal and cultural development. Such epistemological analysis is essential in mathematics education, for it would be very difficult to efficiently study the teaching and learning processes of undefined and vague objects.

Thus, research in mathematics education cannot ignore philosophical questions such as:

- What is the nature of mathematical objects?
- What roles are played by human activity and sociocultural processes in the development of mathematical ideas?
- Is mathematics discovered or invented?
- Do formal definitions and statements cover the full meaning of concepts and propositions?
- What is the role played, in the meaning of mathematical objects, by their relationships with other objects, the problems in which they are used and the different symbolic representations?

It must also recognize the complexity of these questions and the variety of possible answers. As A. Dou says in the preface to Cañón's book (1993), 'The ontology of mathematical entities and, even more so, their epistemology is interpreted in an incredibly disparate way and it still remains a mystery' (p. 14). Piaget (1979) also stated, 'it was never possible to agree upon what in fact mathematical entities are' (p. 147). The acknowledgment of the difficulty of the questions does not mean, however, that attempts at their clarification should be given

Sierpinska, A. and Kilpatrick, J. Mathematics Education as a Research Domain: A Search for Identity, 177–195.
© *1998 Kluwer Academic Publishers. Printed in Great Britain.*

up. We think it is important to address them if some progress is to be made in the setting of a coherent research program aimed at defining the field of research in mathematics education.

The essential role of the study of the meaning of mathematical objects for mathematics education is emphasized, amongst others, by Balacheff (1990), who stated: 'A problem belongs to a *problématique* of research of mathematics teaching if it is specifically related to the mathematical meaning of pupils' behavior in the mathematics classroom' (p. 258). Sierpinska (1994) stresses the close relationship between the notions of meaning and understanding.

However, within this area of knowledge there is a lack of explicit theories regarding the meaning and genesis of mathematical concepts and procedures according to newer tendencies in the philosophy of mathematics (Wittgenstein 1953; Lakatos 1976; Kitcher 1984; Tymoczko 1986; Ernest 1991; Dossey 1992).

In this paper we present a theory of the nature and meaning of mathematical objects (concepts, propositions, ...), which takes into account their epistemological and psychological dimensions. This theoretical framework is applied to frame certain basic research questions in mathematics education.

The theory of the meaning of mathematical objects that we present has an intrinsic kinship with Chevallard's anthropological approach to mathematical knowledge (especially his ideas of *objet* and *rapport à l'objet* (1991, 1992)) and Wittgenstein's doctrine of 'meaning as use' (1953), as interpreted by Kutschera (1971), McGinn (1984) and McDonough (1989). Our educational perspective and integrative intention lead us to complement these approaches with theoretical elements such as personal or mental objects, in line with a psychological epistemology (Kitcher 1984) and the psychological theory of situated cognition (Brown, Collins and Duguid 1989).

We highlight the view of mathematical objects as signs of cultural units, whose systemic and complex nature cannot be described merely by formal definitions when the perspective taken is that of the study of teaching and learning processes. Based on this viewpoint, we intend to explain some learning misconceptions and difficulties, not only in terms of mechanistic mental processes, but by recognizing the complexity of meaning and the necessarily incomplete teaching processes in schools.

Finally, we shall study the possible effectiveness of the presented theoretical system in formulating a *problématique* of research in mathematics education in which the center of interest in studying meaning and understanding is shifted from the mental processes to the institutional and cultural contexts. This change of perspective has been proposed in philosophy of language by Wittgenstein, in psychology – by authors following the cultural psychological trend which also emphasizes the idea of meaning (Bruner 1990), and, within mathematics education, by ethnomathematical studies (e.g. Bishop 1988; Nunes 1992; D'Ambrosio 1994).

2. INSTITUTIONAL AND PERSONAL MEANING OF MATHEMATICAL OBJECTS:
A PRAGMATIC AND RELATIVIST THEORY

According to the aforementioned tendencies in the philosophy of mathematics, the epistemological and cognitive assumptions that serve as a basis for our theory are the following:

(a) Mathematics can be seen as a human activity involving the solution of so-cially-shared problem-situations, which refer to the real or social world or which are within the realm of mathematics itself. As a response or solution to these external and internal problems, mathematical objects (concepts, procedures, theories, etc.) progressively emerge and evolve. People's acts must be considered, therefore, as the genetic source of mathematical conceptualization – in line with the Piagetian constructivist theories.
(b) Mathematical activity creates a symbolic language in which problem-situations and their solutions are expressed. The systems of symbols, as culturally embodied, have a communicative function and an instrumental role, which changes the person himself, or herself, when using the symbols as mediators (Vygotsky 1977; Rotman 1988).
(c) Mathematical activity aims, among others, at the construction of logically organized conceptual systems. The logical organization of concepts, theorems, and properties also explains the great number of problems involved in the learning of mathematics: A system cannot be simplified into a sum of isolated components because what makes it a system are exactly the inter-relationships between the components.

It is thus necessary to distinguish two interdependent dimensions in the genesis of mathematical knowledge: the personal (subjective or mental) dimension and the institutional (objective, contextual) dimension. Given that subjects grow up and live within different institutions, their knowledge is mediated by the peculiarity of the corresponding contextual knowledge. It is important to recognize that mathematics, as a cultural reality (Wilder 1981), adopts different 'ways of life and of operation' within different human groups. Nevertheless, we should recognize the dominant and controlling role of the formal and logical deductive organization adopted by mathematics in the institution of the 'producer of knowledge', mainly due to its effectiveness in setting and solving new problems and in communicating the solutions.

Hence, we recommend considering the objects and their meanings in a relativistic way, with respect to different institutions (in a sense that will be described later in the chapter). This will allow us to better appreciate the adaptation (or transposition didactique, as Chevallard puts it) and mutual influences that mathematical objects undergo as they are transmitted between people and institutions.

Below, we have defined the theoretical concepts of practice, objects (personal and institutional) and meaning by adopting, as a primitive, the notion of problem-situation, attempting to make evident and operative the aforementioned triple nature of mathematics, the personal and institutional genesis of mathematical knowledge and the mutual interdependence between the latter two. The presentation of the theoretical notions using a definition format does not intend to establish any 'axiomatics' for the complex ontological and epistemological issues that are raised. The definition format has been used in the aim of expressing our thoughts in a precise way and facilitating the analysis and debate thereof. The concept of *statistical association* has been chosen in order to illustrate the proposed theoretical model and make it less abstract and more generic. A more detailed presentation of the framework can be found in Godino and Batanero (1994), where the concept of *mean* had been used as another example. In this chapter, the implications of the theoretical framework are extended and systematized from the point of view of developing a research agenda for research in mathematics education.

Mathematical Problems, Practices and Institutions

Our theoretical system is based on the notion of *problem-situation*. We think that this notion takes into account the main components of the activity of mathematizing as described by Freudenthal (1991), and the three types of situations proposed by Brousseau (1986) (action, formulation/communication, and validation).

We shall assume that, for any given person, *a problem-situation is any type of circumstance in which mathematizing activities are needed.*

As examples of mathematizing activities we could highlight:

- building or looking for possible solutions that are not immediately accessible;
- inventing an adequate symbolization to represent the situations and the solutions found and to communicate these solutions to other people;
- producing new meaningful expressions and statements through symbolic manipulations;
- justifying (validating or arguing) the proposed solutions;
- generalizing the solution to other contexts, problem-situations and procedures.

These activities are not restricted to mathematics, but they become mathematical if mathematical objects, such as numbers, geometric figures, functions, logical reasonings, etc., take part in them.

We are not trying to *define* the notion of problem-situation here, but only to explain it. In fact, in our case, as we aim at defining the notions of object, meaning and understanding in mathematics, we need to consider the notion of a mathematical problem-situation, or mathematizing, as primitive.

The generality that we attribute to the notion of mathematical problem-situation is motivated by our desire to integrate the invention, application and diffusion contexts in the same epistemological model of mathematical knowledge.

Problem-situations do not appear in isolation, independently from one another, rather they constitute classes of interrelated problem-situations, sharing similar solutions, representations, etc., which we shall call *problem fields*.

We consider the notions of problem-situation and problem field to be very general, and dependent on the institutional context and the subjects involved. Subjects are not required to accomplish all the different types of mathematizing activities, or to completely build a mathematical model. School activities are, in fact, problem-situations if the students have no trivial and immediate answer to what they are asked to do.

Let us consider the following item as an example of a problem-situation:

Problem 1: In a medical center 250 people have been observed in order to determine whether the habit of smoking has some relationship with bronchial disease. The following results have been obtained:

	Bronchial disease	No bronchial disease	Total
Smoker	90	60	150
Non-smoker	60	40	100
Total	150	100	250

Using the information contained in this table, would you think that, for this sample of people, bronchial disease depends on smoking?

This is the simplest form of a contingency table or cross-tabulation, which is used to present the frequencies in a population or sample, classified by two statistical variables. It is a particular item of the problem field from which has originated the notion of statistical association. This concept extends the notion of functional dependence to cases in which the independent variable does not determine a unique value, but a frequency distribution for the dependent variable.

We might enunciate other similar situations by changing the context, the intensity of the dependence between the variables, the numerical values of the frequencies, etc. We could also increase the number of rows and columns or consider other types of variables, such as, for example, inquiring into the linear correlation between two quantitative variables.

The data of Problem 1 suggest the following problem from a different field:

Problem 2: Assuming that the data of Problem 1 have been drawn at random from a given population, what would be your estimate of the proportion of smokers in this population? Could you give an interval for the variation of the proportion of smokers in the population, with an error probability smaller than 5%?

This problem and other similar problems, for example, the question of deciding if the given estimation is optimal, in what sense it is optimal, or the problem of computing the sample size necessary to produce a confidence interval of a given size actually belong to a different problem field – the field of estimation of parameters.

What actions could be carried out by people without specific knowledge of statistics to solve this type of problems? In our research (Batanero et al. 1996) we proposed the contingency table problem to a sample of 213 students, without instruction in this subject. Some of them compared the ratio of bronchial disease in the smokers (90/150) with the ratio of bronchial disease in non-smokers (60/100) and, as these ratios were identical, they argued that there was no relationship between the two variables in the given sample, which is the correct answer (let us denote this 'way of solving' or *practice* by P1). Other students also obtained the correct answer by comparing the proportion of bronchial disease in smokers (90/150) with the proportion of bronchial disease in the total sample (150/250) (P2). Another practice, (P3), was to compare the ratio between the number of people with bronchial disease and the number of people with no bronchial disease in smokers (90/60) with the same ratio in non-smokers (60/40).

The processes of solving problems from this problem field require relating or operating with mathematical objects such as frequencies, ratios, totals, etc. or identifying previously built objects satisfying some given conditions. Situations and practices of this kind are essential for the building of the concept of statistical association. In general, the basic role of the activity – in its wider sense – for building mathematical objects is synthesized in Definitions 1 and 2.

Using these and the following definitions, we try to build a theoretical model that allows us to distinguish the subjective and institutional dimensions of knowledge, meaning and understanding in mathematics, as well as to point out the relationships between both dimensions. Furthermore, we propose to base our epistemological model for mathematics on the activity of subjects involved in problem-situations, mediated by semiotic instruments provided by institutional contexts.

DEFINITION 1: *Let us call practice any action or manifestation (linguistic or otherwise) carried out by somebody to solve mathematical problems, to communicate the solution to other people, so as to validate and generalize that solution to other contexts and problems.*

These different types of practices (action, formulation/communication, and validation) attempt to consider the category of situations that generate forms of mathematical knowledge as described by Brousseau (1986).

Concrete and abstract objects intervene in mathematical practices and they can be represented in textual, oral, graphical or even gestural form.

In general, rather than in one particular practice for solving a specific problem, we are interested in *types of practices*, that is to say, in the operative invariants shown by people during their actions concerning problem-situations. These invariants shall be called *prototype practices*. Generally, for each problem field and, in principle, for each person, we can identify a system of prototype practices.

The development of mathematical activity carried out by people involved in problem-solving is not usually a linear and deductive process. On the contrary, it is fraught with failed attempts, trials, errors, and unfruitful procedures that are abandoned. Thus, we consider that it is necessary to introduce the notion of meaningful personal practice:

DEFINITION 2: *We say that a practice is meaningful (or that it makes sense) for a person if, for that person, this practice fulfills a function in solving a problem, or in communicating the solution to another person, or in validating and generalizing the solution to other contexts and problems.*

Generally, problem-situations and their solutions are socially-shared, that is to say, they are linked to institutions. For example, different collectives are interested in studying statistical association problems. The contingency table problem, or some variation thereof, could be of interest to 'secondary school students', 'university students', 'medical researchers', 'public health chiefs', 'applied statisticians', etc. Each of these groups has different aims and uses different tools for solving these problems. Whilst for secondary school students the descriptive study is sufficient, at the university, students must apply the Chi-square test of independence. Practices P1 to P3 would be considered insufficient to solve the problem at the university level. Applied statisticians and researchers would have statistical packages available and would include other different variables in the analysis, to evaluate whether the empirical association could be influenced, or not, by these other variables.

These groups of people are examples of institutions in which problem-situations are dealt with using specific aims, tools and practices, and so they constitute differentiated epistemological formations. Therefore, we propose the following descriptions of the notions of *institution* and *system of social practices.*

DEFINITION 3: *An institution (I) is constituted by the people involved in the same class of problem-situations, whose solution implies the carrying out of certain shared social practices and the common use of particular instruments and tools.*

We shall use the name of *mathematical institution* (M) for people involved in solving new mathematical problems, i.e. for the producers of new mathematical knowledge. Other institutions involved in *mathematical problems* are applied mathematicians, scientists, technicians, teaching institutions, etc.

As has been shown in the example, specific practices for solving a *problem field* are carried out within different institutions. It is essential to consider the set of such practices from a systemic perspective, with the aim of inquiring into its main components and structure. In teaching institutions, this information should be used as a reference universe for selecting representative samples of teaching and assessment situations.

DEFINITION 4: *The system of institutional practices of an institution I, linked to a problem field C, is constituted by the meaningful practices to solve C, shared within I.*

The social nature of these practices implies that they are observable. As examples of social practices, we may quote: problem descriptions, symbolic representations, definitions of objects, statements of propositions and procedures characteristic of the problem field, argumentations. We shall denote the system of institutional practices by $P_I(C)$.

Institutional and Personal Objects

Examples P1 to P3 of students' actions in the solving process of the contingency table problem constitute the phenomenological substratum for students' intuitive conceptualization of statistical association. For example, when a particular student uses practice P2 it is because he or she supposes that the relative frequency distribution for the dependent variable must change in the case of dependence when we restrict the sample to a given value of the independent variable (smokers). This practice is, for different people, an operative invariant for this type of problem. Its mathematical formalization is expressed in the following definition for independence: *'Independence between two statistical variables, A and B, means the invariance of the distribution of B when conditioned by a value of A'*. The operative invariants linked to P1 and P3 lead to different characterizations for statistical association.

When widening the scope of the contingency table problem, more complete mathematical procedures would be needed. For example, testing hypotheses concerning association requires completing the Chi square test or the Fisher test. The measures of association (PHI, Contingency C and V, Goodman Lambda, etc.) have been created to assign a degree to the intensity of association. Therefore, the concept of association has emerged and evolved progressively over time and practices created to solve problems. It has also generated some related concepts, such as multiple or partial association. Moreover, it has been the basis for developing new problem fields and tools for solving them. For example, the problem of geometrical representation and reduction of dimension in multivariate data was solved using the correlation coefficient and led to factor analysis techniques.

This process has a general character: mathematical objects are abstract entities that emerge progressively from the socially-shared system of practices, linked to the activities of solving a given field of mathematical problems. According to Morin (1977), the notion of emergence means that the overall product of the activities that form a system has its own qualities which produce feedback from the activities of the system from which they cannot be separated.

Since practices may vary from one institution to another, we must give the object a relativity with respect to institutions. In our example, the concept of association is very different for applied statisticians and for secondary school students because the things that the students are able to say about association and to do in association situations are very limited. Thus, we propose the following definition:

DEFINITION 5: *An institutional object O_I is an emergent of the system of social practices linked to a problem field, that is to say, an emergent of $P_I(C)$. The elements of this system are the empirical indicators of O_I.*

The emergence of the object is progressive over time. At any given time the object is recognized as such by the institution, but even afterwards it undergoes progressive transformations as the problem field widens. The institutional objects are the constituents of the objective knowledge in the sense described by Ernest (1991).

The progressive nature of the construction of scientific objects has its parallels in the learning by the subject and in the invention of new mathematical ideas. 'Not only in its practical aspects, but also in its theoretical aspects, knowledge emerges from problems to be solved and situations to be mastered. It is true for the history of science and technologies; it is also true for the development of cognitive instruments of young children' (Vergnaud 1982, p. 31).

During the learning process, students may develop some practices that do not coincide with those considered appropriate by the teaching institution.

Practices P1 to P3 are examples of correct practices for solving the contingency table problem, from the point of view of the descriptive study of association (there is no inference from this sample to a wider population). But some of our students used procedures that were statistically incorrect, even from the descriptive point of view, as in the following cases:

- P4: Using only the cell of smokers with bronchial disease (90), to reason that there is dependence between the variables, because this frequency is maximum.
- P5: Basing a judgment only on the frequencies in one row or one column of the table; for example, reasoning that there is dependence between the variables because the number of smokers with bronchial disease (90) exceeds the number of healthy smokers (60).
- P6: Not taking into account the empirical data and basing a judgment on preconceived ideas about the association that ought to exist between the variables.

We introduce the notions of 'a system of personal practices' and 'personal object' to differentiate between objective and subjective dimensions of knowledge.

DEFINITION 6: *The system of personal practices linked to a problem field C is constituted by the prototype practices that a person carried out to solve C. This system will be denoted by $P_p(C)$.*

DEFINITION 7: *A personal object O_p is an emergent of the system of personal meaningful practices linked to a problem field, that is to say, an emergent of $P_p(C)$.*

The emergence of the object is progressive during the subject's lifetime as a consequence of the subject's experience and learning. Personal objects are constituent parts of subjective knowledge (Ernest 1991).

Institutional and Personal Meaning of an Object

Objects are named and described by means of certain practices that are usually considered as the definitions of the objects (these practices are even identified with the object through metonymy). However, Vergnaud (1990) considers that the meaning of a mathematical object, from a didactic and psychological point of view, cannot be reduced merely to its definition. We agree with this author in that the meaning of mathematical objects must refer to the actions ('interiorized' or otherwise) that the subject carries out in relation to these objects. We also think that it is necessary to distinguish between the institutional and the personal dimension of meaning. For example, the term 'association' has different meanings in different people and institutions. In secondary schools the curricula propose solving descriptive problems of contingency tables and linear bivariate correlation and regression. Only simple data sets are studied. The students compare the intensity of association using the correlation coefficient but they do not compute confidence intervals, nor do they test hypotheses concerning this coefficient. At the university level, students would perform the inferential study; they use the square correlation coefficient to decide the percentage of variability explained in the analysis of variance and to decide the order in which different independent variables are to be included in stepwise multiple regression analysis. A statistical consultant would decide how many factors should be retained using the size of multiple correlation coefficient in factor analysis or using the percentage of the Chi-square coefficient in correspondence analysis.

DEFINITION 8: *The meaning of an institutional object O_I is the system of institutional practices linked to the problem field from which O_I emerges at a given time.*

We shall denote the meaning of O_I by $S(O_I)$. This notion of meaning is a construct that depends on the institution and on time. Symbolically, $S(O_I) = P_I(C)$. If $I = M$, we talk about the mathematical meaning of an object.

The proposed notion of meaning allows us to introduce, in the didactic research program, the study of the structure of the system of social practices from which mathematical objects emerge, as well as their temporal evolution and institutional dependence. Also, the semiotic analysis of the institutional objects involves considering the situations that produce those social practices.

DEFINITION 9: *The meaning of a personal object O_p is the system of personal practices that a person p carries out to solve the problem field from which the object O_p emerges at any given time.*

Thus, this meaning depends on the subject and on time. Symbolically, $S(O_p) = P_p(C)$. Some personal practices can be observed, but not the 'interiorized' actions.

3. MEANING AND UNDERSTANDING

From the same problem field C in which an institution I produces an object O_I, with the meaning $S(O_I)$, a person could produce an object O_p with a personal meaning $S(O_p)$. The intersection of these two systems of practices is what the institution considers correct manifestations, that is to say, what the person 'knows' or 'understands' about the object O_I, from the institution's point of view. The remaining personal practices would be considered 'errors' according to the institution.

Concerning the practices that the students performed in our research to solve the contingency table problem, P1 to P3 would be considered correct from the point of view of the competence intended in secondary education. The teacher would consider that a particular student understands the idea of association if he or she shows one of these practices. On the contrary, P4 to P6 would be considered mistaken and related to a conceptual misunderstanding of association.

This situation is described in the following definition:

DEFINITION 10: *The meaning of an object O_I for a subject p, from the point of view of the institution I, is the subsystem of personal practices linked to a problem field that are considered in I as adequate and characteristic practices for solving these problems.*

In an ideal situation, and within a given institution, we would say that a subject 'understands' the meaning of the object O_I, or that he/she 'has grasped the meaning' of a concept, if he/she is able to recognize its properties and representations, to relate it to other mathematical objects and to use it in the prototype problem-situations within the institution. The understanding reached by a subject, at a given moment, will not be complete or null, but it will cover partial aspects of the different elements of meaning.

The concept of understanding that we have derived from Definition 10 is closely related to the notion of 'good understanding' described by Sierpinska (1994, chap. 4), to which this author also attributes a relative character with respect to cultural or institutional settings.

We also consider that the notions of acts and processes of understanding, in their mental or subjective dimension, could be derived from our notions of meaningful practice and personal meaning. Nevertheless, the compatibility and complementarity of our theory of mathematical object and meaning with the theory of understanding developed by Sierpinska should require further study and development.

4. A BACKGROUND TO THE PROPOSED THEORY

In this section, we shall present a brief summary of the sources that we have taken into account to support our theory and we shall mention authors and theories which agree with our viewpoints. We are aware, however, that a more in-depth study of the common ground and differences between our proposal and these theories should be carried out in the future.

The notion of meaning that we propose is inspired by Wittgenstein's ideas about meaning and understanding (Wittgenstein 1953), interpreted according to authors such as Kutschera (1971), McGinn (1984) and McDonough (1989). The doctrine of 'meaning as use' implies that the key concept is that of 'context embeddedness'. The context is understood here, not merely as the physical environment of a linguistic utterance; rather the reference is made to the institutional and cultural context. As McDonough (1989) points out, Wittgenstein's Copernican revolution in the theory of meaning, still undigested by the sciences and technologies of human cognition, describes the neural system as conceptually dependent on a new center: the institutional and the cultural contexts. In didactics of mathematics this approach is being persistently supported by Chevallard (1992), who places the study of 'The Didactic' (le didactique) within the confines of cognitive anthropology.

The notions of *institution, practice and object* are used by Chevallard (1989, 1992) to define his concept of 'rapport au savoir', although we believe that the meaning that he attributes to these notions does not completely coincide with the one we have proposed in this paper. According to our theorization, not all practices are pertinent to the emergence of objects (some practices are incorrect, inappropriate or irrelevant). Moreover, a practice should be considered to be linked to a corresponding *problem field*. The introduction of the notion of meaning (personal and institutional or, to put it in another way, psychological and epistemological) as a system of components – elements of meaning, meaningful prototype practices – focuses our attention on the systemic and complex nature of meaning.

We consider it useful to distinguish between the name of an object, the object (as a cultural and psychological entity) and the system of practices linked to the solving of problems from which this cultural unit emerges; that is to say, the meaning of the object. This formulation allows us to better conceptualize the inference processes that are needed to characterize subjects' knowledge about mathematical objects, from the empirical manifestations of this knowledge.

With our definition of an institutional object we postulate the cultural existence of different objects, according to the reference institution, in situations in which an absolutist conception of mathematics only perceives one object. This formulation is a consequence of the pragmatic assumptions which we have taken as a basis and their utility for the anthropological analysis of cognitive and didactic phenomena. Rotman (1988) has reached a similar conclusion in his semiotic analysis of mathematical activity when he asserts that the numbers studied by the Babylonians, Greeks, Romans and present-day mathematicians are different. Nevertheless, we believe that these numbers are similar because of the phenomenon of regressive appropriation.

We would refer the reader to our paper (Godino and Batanero 1994) in which we further develop the links between our idea of meaning and the ideas of authors such as Ausubel, Bunge, Douady, Putnam, Rotman, Steinbring, and particularly Vergnaud.

5. A RESEARCH AGENDA FOR MATHEMATICS EDUCATION: 'SEMIOMETRY' AND THE ECOLOGY OF MEANING

Below, we attempt to show the utility of the theorization we have presented in the setting of certain basic research questions in mathematics education. We shall also present some research examples that have been carried out from this theoretical perspective in the Department of Didactics of Mathematics at the Granada University in Spain.

We shall classify the research questions into two categories: The characterization of institutional and personal meanings – which we shall call the *semiometry problem* – and the study of the evolution and interdependence of meanings in which the ecological paradigm could be a useful model (Godino 1993).

Semiometry

The consideration of the meaning of mathematical objects as systems of practices and the discrimination between personal and institutional meaning introduces, into the didactic *'problématique'*, the study of the structure and characterization of these theoretical entities. This characterization can be conceived as a 'measurement', not in a strict psychometric or mathematical sense, but in its general sense, that is, as the categorization of quantitative or qualitative variables, including the use of nominal, ordinal, interval or ratio measurement scales. Furthermore, it highlights the sampling nature of the process of selecting teaching and evaluation situations and of the students' manifestations and behavior. Thus, it contributes to overcoming the illusion of deterministic transparency that is frequently adopted when considering these problems.

A primary class of didactic research studies must be oriented towards the determination of institutional meanings, especially the meanings within mathematical institutions. We have to research into the characteristic uses of mathematical concepts, propositions and theories and to identify their different representations. This reference meaning may be compared with the meaning of mathematical objects in teaching institutions. We can also study the conditioning factors producing the development and changes of these meanings.

This type of research was carried out by Vallecillos (1994), who analyzed the institutional meaning of statistical hypothesis testing in university teaching. In her dissertation, Vallecillos showed that the original problem field from which statistical inference has emerged refers to the search for inductive procedures of validation of empirical hypotheses. This problem has been substituted in the Neyman-Pearson theory by another related problem field for which the test of hypothesis is a satisfactory solution: obtaining an inductive rule of behavior. Nevertheless, this shift in the problems that interest applied scientists and users of statistics is, in general, not sufficiently explained in the teaching of the subject.

The experimental study proved that the personal meanings of hypothesis testing built by students did not coincide, in general, with the statistical institutional meaning. This fact caused many errors, incorrect inference applications and misconceptions. In particular, subjects conceived erroneously the level of significance (or the p-value) in a statistical test as an 'a posteriori' probability of the hypothesis, given the data obtained. Thus, the students identified testing hypotheses with an inductive procedure to compute the probability of the hypotheses.

The theoretical system we have described in this paper also allows us to study, from a new perspective, the problem of assessing mathematical knowledge. By 'assessment' we mean, following Webb (1992), 'the comprehensive accounting of an individual's or group's functioning within mathematics or in the application of mathematics' (p. 662).

According to our theory, a subject's cognitive system (his/her conceptual and procedural knowledge, his/her intuitions, representation schemas, ...), that is to say, the network of personal objects at a given time, is an organized and complex totality. The distinction we have established between the domain of ideas or abstract objects (personal and institutional) and the domain of meanings, or systems of practices from which such unobservable objects emerge, is used to clarify the problem of looking for the correspondence between both domains, i.e. the problem of assessing institutional and personal knowledge.

The assessment of subjective knowledge necessarily requires performing inference processes, from the set of observed practices in evaluation situations, whose reliability and validity must be guaranteed (Messick 1991; Feldt and Brennan 1991). The complexity of this inference process is deduced, first of all, from the interrelationships between the knowledge of different mathematical objects. Subjects' knowledge concerning a given mathematical object cannot be reduced either to a dichotomy (to know or not to know) or to a degree or unidimensional percentage (to know × %). Students' mathematical knowledge is not unidimensional; it is a complex system. Nor could it be measured on an interval or ratio scale. Assessment requires a multidimensional approach and weaker measurement scales (ordinal or nominal). Therefore, it is not appropriate to apply the classical psychometric theories of latent trait and domain mastery to the assessment of mathematical knowledge (Snow and Lohman 1991).

For example, in his dissertation, Estepa (1993) used students' strategies and judgments of association to assess their conceptions concerning statistical association. He used a written questionnaire made up of 10 descriptive association problems that included contingency tables, scatter plots and comparison of the same variable in different samples. The different signs and intensities of association and the agreement between subjects' previous theories and the empirical association in the data were also considered. *Factor analysis* of students' answers showed a multifactor structure in the judgment of association in which the influence of the aforementioned task variables was proven. *Correspondence analysis* has demonstrated the multifactor structure of students' strategies (in which we included practices P1 to P6, described in the previous sections), which varied not only according to the mathematical contents of the problem, but also depending on the students' prior beliefs regarding the association suggested by the context of the problem.

The recognition of the complexity of meanings emphasizes the problem of assessing students' knowledge. Which criteria should be chosen for selecting the system of empirical indicators which characterize the cognitive state, i.e. a subject's knowledge concerning a mathematical object?

As a consequence of our theorization: the observable nature of social practices allows us to determine the *problem field* associated with a mathematical object, as well as its institutional meaning, with the help of a *phenomenological* and *epistemological* analysis. The analysis of the task variables for this problem field provides a first criterion to structure the population of possible tasks. From this population, a representative sample could be drawn to guarantee content validity for the assessment instrument. These two elements, *problem field* and *task variables* thereof, shall provide the first reference points in the selection of relevant evaluation situations for assessing subjective knowledge.

This 'semiometric category' of research studies may be related to the 'historico-empirical' approach to understanding in mathematics described by Sierpinska (1994).

ECOLOGY OF MEANINGS

The problems involved in studying the evolution of institutional meanings of mathematical objects could be modeled with the help of the ecological metaphor (Chevallard 1989; Godino 1993): a particular object performs a function in different types of institutions and it is required to identify the necessary and/or sufficient conditions that allow this object to play its role in these institutions.

The notions of institutional object and meaning are intended to be used as conceptual instruments in this ecological and semiotic analysis of mathematical ideas.

The two types of studies described above would constitute the institutional and personal 'statics of meanings' in this ecological metaphor. Its aim would be to find the 'state and control variables' of meaning, considered as a system, at a particular moment in time. These studies of the static aspects of meaning should be supplemented with dynamic studies, which we are going to describe below.

The study of changes that the institutional meaning of a mathematical object undergoes to become knowledge to be taught in different teaching institutions (curricular design, mathematical textbooks, ...) would constitute the dynamics of institutional meanings (didactic transposition, ecology of meanings, Chevallard 1985).

We could quote, as an example of this type of research, the work by Ruiz Higueras (1993), concerning the study of students' conceptions about functions. She supported her research with a prior analysis of the evolution of the meaning of the function object throughout its historical development and of the institutional meanings presented to the students in her sample, using the analysis of official guidelines, textbooks and notes taken by students in the classroom.

Another fundamental problem in this category is the construction of adequate institutional meanings referring to a mathematical object for a specific school level, i.e. the curricular design. According to the theorization proposed, teaching should be based on the presentation of a representative sample of problems and other elements of the meaning of mathematical objects, taking into consideration the time and resources available.

This problem is tackled for combinatorial reasoning in Batanero et al. (1994) where a curriculum is presented for the teaching of combinatorics, based on a sequence of didactic situations. The selection of the situations and their sequencing was supported by a prior study of the structure of simple combinatorial problems to provide a representative sample of this problem field.

The meaningful learning (relational or significative) of the subject can be modeled as a sequence of 'acts of understanding', or acts of overcoming obstacles (Sierpinska 1990, 1994). The characterization of these acts and the identification of the mechanisms which produce the obstacles (Artigue 1990) is a central theme in the dynamics of personal meaning of mathematical objects. Metaphorically, the study of teaching and learning processes could be viewed as the study of the effects on personal meanings of 'shocks' of didactic sequences, which hold the elements of meanings.

Equally, a part of the characterization of the dynamics of personal meaning would be the study of the evolution of students' conceptions, i.e. the transformation of personal meanings as a consequence of instruction. Estepa (1993) employed systematic observation of classroom work by a pair of students and analysis of their interaction with computers. Using these data, he identified acts of understanding in relation to statistical association and assessed the over-coming of some obstacles by the students during the learning process.

6. CONCLUSION

All of us, mathematics teachers and researchers, are interested in improving students' knowledge by means of instruction. This task requires the characterization of students' knowledge and calls for the clarification of the proper nature of knowledge. As Wheeler asks (1993), 'How can we assess what we do not know?' (p. 87). One could add, how can we teach what we do not know?

The search for appropriate answers to these theoretical problems has led us to elaborate the theory that we have presented, from which we could extract some general conclusions:

1. We must postulate two dimensions of mathematical knowledge: institutional (epistemological) and personal (psychological), which are linked by complex interrelationships.
2. The meaning of mathematical objects (concepts, propositions, theories,...) should be considered from a systemic complexity paradigm, on both the epistemological and the psychological levels. We see the *meaning of a mathematical object* as an extensional entity that can play the role of a universe of reference from which to select assessment and teaching situations.
3. As students are subjects in different institutions, their knowledge is mediated by these institutions. Consequently, the characterization of institutional knowledge should be a prior step for assessing students' knowledge.
4. The phenomenological, semiotic and epistemological analysis of mathematical objects, as cultural entities, must provide criteria for the representative sampling of evaluation situations of the students' knowledge and for organizing didactic situations that favor their adequate evolution.

ACKNOWLEDGMENT

This research had been supported by the DGICYT grant PS93-0196, M.E.C., Madrid.

REFERENCES

Artigue, M.: 1990, 'Epistémologie et Didactique', *Recherches en Didactique des Mathématiques* **10** (2/3), 241–286.
Balacheff, N.: 1990, 'Towards a "Problématique" for Research on Mathematics Teaching', *Journal for Research in Mathematics Education* **21** (4), 259–272.
Batanero, C., Godino, J. D. & Navarro-Pelayo, V.: 1994, *Razonamiento Combinatorio*, Síntesis, Madrid.

Batanero, C., Estepa, A., Godino, J. D. & Green, D.: 1996, 'Intuitive Strategies and Preconceptions about Association in Contingency Tables', *Journal for Research in Mathematics Education* **27** (2), 151–169.

Bishop, A. J.: 1988, *Mathematical Enculturation: A Cultural Perspective on Mathematics Education*, Reidel, Dordrecht.

Brousseau, G.: 1986, 'Fondements et Méthodes de la Didactiques des Mathématiques', *Recherches en Didactique des Mathématiques* **7** (2), 33–115.

Brown, J. S., Collins, A. & Duguid, P.: 1989, 'Situated Cognition and the Culture of Learning', *Educational Researcher*, January–February, 32–42.

Bruner, J.: 1990, *Actos de Significado. Mas Allá de la Revolución Cognitiva*, Alianza Col. Psicología Minor, Madrid.

Cañón, C: 1993, *La Matemática: Creación o Descubrimiento*, Universidad Pontificia de Comillas, Madrid.

Chevallard, Y.: 1985, *La Transposition Didactique – Du Savoir Savant au Savoir Enseigné*, La Pensée Sauvage éditions, Grenoble.

Chevallard, Y.: 1989, Le Concept de Rapport au Savoir. Rapport Personnel, Rapport Institutionel, Rapport Officiel, Séminaire de Didactique des Mathématiques et de l'Informatique de Grenoble, IREM d'Aix de Marseille.

Chevallard, Y.: 1991, Dimension Instrumentale, Dimension Sémiotique de l'Activité Mathématique, Séminaire de Didactique des Mathématiques et de l'Informatique de Grenoble, IREM d'Aix de Marseille.

Chevallard, Y.: 1992, 'Concepts Fondamentaux de la Didactique: Perspectives Apportées par une Approche Anthropologique', *Recherches en Didactique des Mathématiques* **12** (1), 73–112.

D'Ambrosio, U.: 1994, 'Cultural Framing of Mathematics Teaching and Learning', in R. Biehler, R. W. Scholz, R. Sträßer, and B. Winkelmann (eds.), *Didactics of Mathematics as a Scientific Discipline*, Kluwer Academic Publishers, Dordrecht, 443–455.

Dossey, J. A.: 1992, 'The Nature of Mathematics: Its Role and its Influence', in D. A. Grouws (ed.), *Handbook of Research on Mathematics Teaching and Learning*, Macmillan, New York, 39–48.

Ernest, P.: 1991, *The Philosophy of Mathematics Education*, Falmer Press, London.

Estepa, A.: 1993, Concepciones Iniciales Sobre la Asociación Estadística y su Evolución como Consecuencia de una Enseñanza Basada en el uso de Ordenadores. Ph.D. Dissertation, Departamento de Didáctica de la Matemática, Universidad de Granada.

Feldt, L. S. & Brennan, R. L.: 1991, 'Reliability', in R. L. Linn (ed.), *Educational Measurement* (3rd ed.), American Council on Education and Macmillan, New York, 263–331.

Freudenthal, H.: 1991, *Revisiting Mathematics Education*, Kluwer Academic Publishers, Dordrecht.

Godino, J. D.: 1993, 'La Metáfora Ecológica en el Estudio de la Noosfera Matemática', *Quadrante* **2** (2), 69–79.

Godino, J. D. & Batanero, C.: (1994), 'Significado Institucional y Personal de los Objetos Matemáticos', *Recherches en Didactique des Mathématiques* **14** (3), 325–355.

Kitcher, P.: 1984, *The Nature of Mathematical Knowledge*, Oxford University Press, New York.

Kutschera, F. von: 1971, *Filosofía del Lenguaje*, Gredos, Madrid.

Lakatos, I.: 1976, 'A Renaissance of Empiricism in the Recent Philosophy of Mathematics', in I. Lakatos (ed.), *Philosophical Papers*, Cambridge University Press, Cambridge.

McDonough, R.: 1989, 'Towards a Non-Mechanistic Theory of Meaning', *Mind* **xcviii** (389), 1–21.

McGinn, C.: 1984, *Wittgenstein on Meaning*, Basil Blackwell, Oxford.

Messick, S.: 1991, 'Validity', in R. L. Linn (ed.), *Educational Measurement* (3rd ed.), American Council on Education and Macmillan, New York, 13–104.

Morin, E.: 1977, *El Método I; la Naturaleza de la Naturaleza*, Cátedra, Madrid.

Nunes, T.: 1992, 'Ethnomathematics and Everyday Cognition', in D. A. Grouws (ed.), *Handbook of Research on Mathematics Teaching and Learning*, Macmillan, New York, 557–574.

Piaget, J.: 1979, 'Los Problemas Principales de la Epistemología de la Matemática', in J. Piaget (ed.), *Tratado de Lógica y Conocimiento Científico. 3*: Epistemología de la Matemática, Paidós, Buenos Aires.

Rotman, B.: 1988, 'Toward a Semiotics of Mathematics', *Semiotica* **72** (1/2), 1–35.
Ruiz Higueras, L.: 1993, Concepciones de los Alumnos de Secundaria sobre la Noción de Función: Análisis Epistemológico y Didáctico. Ph.D. Dissertation, Departamento de Didáctica de la Matemática, Universidad de Granada.
Sierpinska, A.: 1990, 'Some Remarks on Understanding in Mathematics', *For the Learning of Mathematics* **10** (3), 24–36.
Sierpinska, A: 1994, *Understanding in Mathematics*, The Falmer Press, London.
Snow, R. E. & Lohman, D. R.: 1991, 'Implications of Cognitive Psychology for Educational Measurement', in R. L. Linn (ed.), *Educational Measurement* (3rd ed.), American Council on Education and Macmillan, New York, 263–331.
Tymoczko, T.: (ed.), 1986, *New Directions in the Philosophy of Mathematics*, Birkhauser, Boston.
Vallecillos, A.: 1994, Estudio Teórico-Experimental de Errores y Concepciones sobre el Contraste Estadístico de Hipótesis en Estudiantes Universitarios. Ph.D. Dissertation, Departamento de Didáctica de la Matemática, Universidad de Granada.
Vergnaud, G.: 1982, 'Cognitive and Developmental Psychology and Research in Mathematics Education: Some Theoretical and Methodological Issues', *For the Learning of Mathematics* **3** (2), 31–41.
Vergnaud, G.: 1990, 'La Théorie des Champs Conceptuels', *Recherches en Didactiques des Mathématiques* **10** (2/3), 133–170.
Vygotsky, L. S.: 1977, *Pensamiento y Lenguaje*, La Pléyade, Buenos Aires.
Webb, N. L.: 1992, 'Assessment of Students' Knowledge of Mathematics: A Step toward a Theory', in D. A. Grouws (ed.), *Handbook of Research on Mathematics Teaching and Learning*, Macmillan, New York, 661–683.
Wheeler, D.: 1993, 'Epistemological Issues and Challenges to Assessment: What is Mathematical Knowledge? ', in M. Niss (ed.), *Investigations into Assessment in Mathematics Education: An ICMI Study*, Kluwer Academic Publishers, Dordrecht.
Wilder, R.: 1981, *Mathematics as a Cultural System*, Pergamon Press, Oxford.
Wittgenstein, L.: 1953, *Investigaciones Filosóficas*, Crítica, Barcelona.

Juan D. Godino and Carmen Batanero
Departamento de Didáctica de la Matemática,
Facultad de Ciencias de la Educación,
18071 Granada,
Spain

PAOLO BOERO & JULIANNA RADNAI SZENDREI

RESEARCH AND RESULTS IN MATHEMATICS EDUCATION: SOME CONTRADICTORY ASPECTS

The idea of this joint contribution arose from some similarities between traditions of research in mathematics education in Hungary and Italy, as well as from
some common problems in establishing mathematics education as a specific,
recognized field of research in our academic institutions.

Concerning the first issue, we should bear in mind the engagement of some
eminent mathematicians of the past in problems concerning mathematics education; the fact that at present most researchers in mathematics education come
from mathematics and work in mathematics departments; the close collaboration
existing between university researchers in mathematics education and mathematics teachers; the involvement of many researchers in innovative processes in
school (cf. Barra et al. 1992). Moreover, in the past, some Hungarian and Italian
researchers were involved in joint efforts – as, for instance, during the 70s, the
translation into Italian and adaptation for Italian schools of part of a project for
primary school mathematics education, directed by T. Varga.

Concerning the issue of the present status of research in mathematics education at the academic level, the situation is not easy, neither in Hungary nor in
Italy (and we suppose that this can also be the case for other countries).

In Hungary, for mathematicians, 'research' means only research in pure or
applied mathematics, as refereed in standard international journals such as
Mathematical Reviews. Papers of research in mathematics education published
in outstanding international journals, when taken into account by Hungarian
mathematicians, are quoted as concerning 'experiments', or 'innovation', or
'observations', or 'case studies', or 'essays on classroom situations' – not 'research'. And, as a consequence, people who do 'such things' are not considered
'researchers'.

In Italy the situation is better in some respects. The academic structure is such
that evaluation of research in mathematics education, and in the history of mathematics as well, is performed at all levels: for promotion of people or for
funding, by researchers in mathematics education or in the history of mathematics, or at least by mathematicians with some competence in those fields. The
situation is very difficult for other reasons: for instance, it seems very difficult at
present to create a doctorate in 'mathematics education and history of mathematics'. Many mathematicians have the same attitude towards 'research' in
mathematics education as already reported for Hungary.

*Sierpinska, A. and Kilpatrick, J. Mathematics Education as a Research Domain: A Search for
Identity, 197–212.*
© *1998 Kluwer Academic Publishers. Printed in Great Britain.*

In Hungary, as in Italy, 'research' exists (sometimes concerning teaching and learning of mathematics) in the domain of methodology of education. But it is mainly carried out by researchers belonging to the academic field of 'sciences of education' – not particularly expert in the content domain – and results are evaluated according to standards which do not include the specific interest for mathematics education.

The opinions reported in this paper are strictly personal – even if they rely in part upon some ideas which are largely shared by our colleagues in our two countries.

1. INTRODUCTION

Researchers in mathematics education usually consider mathematicians and different categories of people engaged in the school system (not only mathematics teachers, but also curriculum advisers, administrators, etc.) as natural and unavoidable interlocutors for their research works. Usually, these interlocutors are primarily interested in 'products' they think are suitable to improve mathematics education: such as, for instance, better methodologies to prepare mathematics teachers; productive innovations in the domain of mathematics education; new devices suitable to help teaching of more advanced and 'modern' mathematics. Frequently they also acknowledge tools and results for the quantitative assessment of mathematics curricula, or quantitative and comparative studies concerning different curricula, mathematics education in different countries, etc.

Many researchers in mathematics education nowadays are aware of the limited value of many 'products' of these kinds, both from a scientific and a pragmatic point of view. They would like to offer different tools and results, allowing people to better tackle the teaching and learning of mathematics as a very complex problem. In connection with this aim, many researchers in mathematics education are presently engaged in the effort of establishing mathematics education as a specific field of research, with its own methodologies and autonomous criteria to establish whether a scientific result is pertinent and valid.

This situation produces a lot of tension between the community of mathematics educators, the school system and the mathematicians' community. It is caused by certain contradictions: between the researchers' aims and the present situation of scientific work in mathematics education; between 'products' expected by the outside community and research results offered internally by mathematics educators.

In our opinion it is necessary to take a closer look at such contradictions in order to understand whether it will be possible to overcome them in the future, and, if so, how. This paper tries to make a contribution in this direction.

We will propose (in the next section) a classification of results in mathematics education, suitable (in our opinion) to better analyze some of the present internal and external difficulties and contradictions. Next, some contradictions deriving from the expectations of the natural interlocutors of mathematics educators will be discussed. Also contradictions inherent in the effort of establishing mathematics education as a specific field of research will be considered. Finally, we will try to establish some connections between the needs emerging from the preceding investigation and the problem of mathematics teachers' preparation, in relationship with present reality and current mathematicians' ideas about this problem.

2. RESULTS IN MATHEMATICS EDUCATION

Taking into account the existing tension between results needed by the school system (teachers, administrators, etc.), results recognized as useful by mathematicians, and results nowadays discussed and offered by researchers, we think that it is useful to propose a specific classification of results in mathematics education suitable to better clarify the problems inherent in this tension. The categories used are different in some aspects from those implicitly suggested in the ICMI Study Discussion Document (Balacheff et al. 1993):

1. 'Innovative patterns' to teach a specific subject (old or new for school mathematics), or to develop some mathematical skills; or, more generally, innovative methodologies, curricula, projects, etc. Results may consist in innovative teaching material, 'proposals', or reports about innovations or projects that have been experimented. An important variable is the dimension of innovation both in terms of time – a short sequence vs. an innovative five year curriculum, and in terms of content – a specific subject vs. an integrated system of topics and methodologies.

2. 'Quantitative information' about the consequences of: educational choices concerning the teaching of a specific mathematical subject; general methodologies; curricular choices (including comparative and quantitative studies). Or, quantitative information about general or specific difficulties regarding learning mathematics, and their possible correlation with factors influencing the learning process. Information is based on quantitative data, collected and analyzed according to standard or ad hoc statistical methods. The level of statistical treatment may be elementary (only percentages and histograms) or quite sophisticated.

3. 'Qualitative information' about the consequences of some methodological or content innovation, or some general or specific difficulties concerning mathematics, etc. In this case, information is based on careful analysis of pupils' protocols, of recorded teacher–pupil interactions, of recorded group or classroom discussions, etc. Frequently, these analyses implicitly or explicitly refer to general educational or psychological or didactic theories.

4. 'Theoretical perspectives' regarding: what is the relationship in the classroom between 'teacher', 'pupils' and 'mathematical knowledge'; the role of the mathematics teacher in the classroom; what happens in the relationships between school mathematics and mathematicians' mathematics; topics to be taught; the relationships between research results and classroom practice in mathematics education, etc. These results may concern descriptions and classifications of 'phenomena', interpretations of 'phenomena', 'models', historical or epistemological analyses (oriented towards educational aims), etc.

In mathematics education, the same study may utilize and produce different types of results. For instance, an 'innovative pattern' experimented in the school system may be the object of a study producing 'quantitative' and/or 'qualitative information' and may raise questions about specific phenomena needing 'theoretical perspectives' in order to interpret them (see, for example, Brousseau 1980, 1981).

Referring to the Discussion Document, we find that:

- most of the results of the 'innovative patterns' type (but also some results of the other types) belong to the 'energizers of practice' category;
- many results of the other types belong to the 'demolishers of illusions' category; but some results of the 'theoretical perspectives' category belong to the 'economizers of thought' category;
- concerning the distinction between 'pragmatic' and 'fundamental scientific' aims, we find that today most research works with 'pragmatic' aims produce results of the 'innovative patterns' and 'quantitative information' types and (in some cases) of the 'qualitative information' type; most research works with 'fundamental scientific aims' produce results of the 'quantitative information', 'qualitative information' and 'theoretical perspectives' types. Some researchers consider results of the 'qualitative information' and especially 'theoretical perspectives' types relevant only to researchers (Figure 1).

We may also observe that, at present, results of the 'innovative patterns' type, and, in part, results of the 'quantitative information' type, are the most accessible for teachers (and the most popular amongst them). Results of the 'quantitative information' type and partly of the 'qualitative information' type, and sometimes of the 'theoretical perspectives' type (depending on local traditions and orientations of research) are the most popular amongst researchers.

Different reasons may explain why this happens; in our opinion, one of them is the present preparation of teachers and researchers. Most researchers in mathematics education come from mathematics, or from experimental sciences of education; almost all mathematics teachers have a 'scientific' background (mathematics plus, possibly, other experimental sciences). On the other hand, research and related results of the 'qualitative information' and 'theoretical perspectives' types need a specific 'human sciences' vocabulary, and are neces-

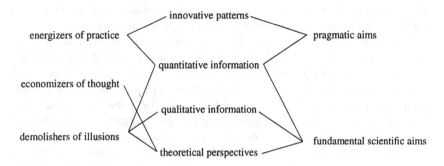

Figure 1

sarily more overtly grounded in philosophical assumptions. Researchers producing results of the 'qualitative information' and 'theoretical perspectives' types need both backgrounds. Furthermore, teachers and researchers often do not understand each other and we feel that in many cases a 'translation' is necessary.

The main concern of this paper will be to try to suggest that results of the 'qualitative information' and 'theoretical perspectives' types are important, not only in themselves, but also because they allow teachers and researchers to get the other kinds of results under control – while the specificity of research in the domain of mathematics education depends especially on results of the 'innovative patterns' and 'theoretical perspectives' types and related methodologies.

3. SOME CONTRADICTIONS DERIVING FROM EXTERNAL REQUESTS

In order to support the preceding statements, we will focus on some contradictions underlying the results of the 'innovative patterns' and 'quantitative information' types and current problems of mathematics education, and will try to show how these contradictions may be tackled with suitable results of the 'qualitative information' and 'theoretical perspectives' types.

Useful and successful innovations (as results of the 'innovative patterns' type) are very important in justifying research in the mathematicians' and teachers' eyes and in ensuring 'specificity' to research in mathematics education. Innovations should be replicated from the level of proposals and prototypes to the level of widespread diffusion; but it is well-known that reproducing innovations on a large scale without substantial degeneration and loss of effectiveness is very difficult.[1]

As to this issue, some phenomena should be quoted. We will only consider some examples of innovations concerning specific mathematical subjects. In the past twenty years, mathematics educators have been concerned with the distance between proposals and prototypical innovation, and widespread innovation about subjects like the 'set' approach to natural numbers, or 'ratio, proportion and linear functions'. A comprehensive research perspective is lacking.

Other phenomena refer to 'popular' and 'not popular' innovations. In Italy, the introduction of substantial topics of elementary probability theory in the comprehensive school (even if many prototypes are available) is confronted with many obstacles. On the other hand, the introduction of elements of analytic geometry in the comprehensive school has taken only a few years to succeed. Concerning probability, the situation in the UK and Hungary is quite different! It is not easy to tackle these phenomena in a research perspective.

As researchers, we think that intercultural differences should be taken into account because frequently the situation is not the same in different countries. Other aspects should also be considered: in-service teacher education; the institutional aspects (official programs, relationships with the preceding and following levels of schooling, structure and content of final examinations); specific 'didactic transposition' (Chevallard 1991) problems; difficulties in integrating new subjects in the old curriculum and/or changing some parts of it in the perspective of new subjects. But there are few research articles and surveys available on these issues.

In order to begin to understand which 'variables' may determine the success (or failure or degeneration) of innovations and of the large-scale reproduction of innovations in the field of mathematics education, we would need wide-ranging and careful descriptions and analyses of:

- classroom teaching–learning processes, and (possibly) their interpretations and modelizations;
- students' long-term learning processes – in order to better understand many aspects of the development of pupils' mathematical knowledge, depending on individual and social factors, cultural influences and affective constraints (see Bishop 1988; Carraher 1988; Boero & Szendrei 1992), etc.;
- teachers' conceptions and beliefs, in different countries.

Results of the 'qualitative information' and 'theoretical perspectives' types are needed in order to go beyond our present, very limited, knowledge of some of these aspects. In particular, teachers' conceptions and students' mathematical experiences in everyday life should be carefully investigated because they may have deep effects on classroom teaching–learning of mathematics; we may also observe that this is the reason why, frequently, 'quantitative' information provided by international comparative studies is biased due to the lack of knowledge about these aspects.

As an example, we may consider the different situations in Italy and in Hungary concerning the learning of natural numbers. In the Italian language the names of natural numbers (*'one, two, three, four...'*) are also used to indicate the days of the month (only the first day is commonly named *'first of...'*), whereas in the Hungarian language all the days of the month are named with the ordinal adjectives *'first, second, third, ...'* ; so in the two countries the relationships between 'cardinal' and 'ordinal' aspects of natural numbers are different in the first mathematical experiences of pupils. These differences may affect both the cognitive hierarchy between different aspects of the number concept and the opportunities offered to the teacher by real-life situations to approach these aspects in the classroom.

If we agree that research in mathematics education must produce results of the 'innovative patterns' type, these results must be circulated and compared through international journals and meetings of mathematics educators; but if we analyze outstanding journals and take part in specialized international meetings we find less and less 'didactic proposals' or reports about innovations.

For instance, we may estimate that in the period from 1970 to 1974 over 50% of the papers published in *Educational Studies in Mathematics* dealt with 'innovative patterns' type of results; in the period from 1987 to 1992 less than 15% of the papers dealt with 'innovative patterns' type of results. Less and less results of the 'innovative patterns' type are considered as outstanding 'research products' worth publishing in important journals. Indeed, nationally-circulated journals continue to publish a good number of papers of that kind, but they are not considered (in most countries) as research journals. And also, in many rather specialized international meetings of researchers (like the annual conferences of the *International Group for the Psychology of Mathematics Education*) the presentation of 'innovations' takes place essentially in poster sessions.

On the other hand, if we consider most of the results of the 'innovative patterns' type, presented at congresses or in local journals, we see that they are limited to teaching materials and/or descriptions of proposed (or experimented) short-term or long-term didactic sequences concerning narrow or wide subjects, and do not try to thoroughly analyze educational, epistemological or psychological problems. The consequence of this fact is that conditions under which innovations can be reproduced, variables conditioning effectiveness, etc., are not known.

Taking these remarks into account, we need a proper style of presentation of the results of the 'innovative patterns' type, supported and framed by already existing results of the other types (especially of the 'qualitative information' and 'theoretical perspectives' type), in order to make them valuable as research results. This is true both for the fundamental research perspective and the pragmatic perspective.

If an innovation has undergone a small-scale (or even a large-scale) successful experiment, a careful analysis of the conditions which allowed such success (for instance: motivation of teachers, cultural background of teachers and pupils, school traditions, etc.) should be provided, and possible limitations in the 'reproducibility' of the innovation should be pointed out. In an experimental plan concerning innovation, phenomena such as 'obsolescence of innovation' (Brousseau 1986) – with consequent effects on the teaching–learning process – should be taken into account.

Another example: the epistemological analysis of mathematical content and the analysis of the relationships between current and historical points of view in mathematics, and in the school mathematics concerning it, seems to be necessary in order to situate an educational proposal concerning that content from the cultural point of view. A teacher must therefore be encouraged to take some distance from his cultural background and possible 'epistemological obstacles' inherent in that content must be predicted (see: Bartolini Bussi & Pergola 1992; Boero 1986, 1989b; Brousseau 1983, 1986; Chevallard 1991; Laborde 1988; Sierpinska 1985, 1987).

If we agree that research in mathematics education must keep a close contact with today's mathematics and help teachers to teach contents which are relevant to modern views about mathematics – or take into account the opportunities offered by new technologies – research should then create connections between these modern views and these opportunities. But, if we read articles published in leading international journals of mathematics education, we will observe how most of them are concerned with 'paradigmatic' contents and problems (i.e., contents and problems which allow for the comparison of new research results and perspectives with previous ones) regardless of the importance of the content or problems in mathematics education – while many traditional and new fields are not covered. Also, results of the 'innovative patterns' type are concentrated in few directions (regardless of the importance and difficulty of the subject).

In the last twenty years many studies have been reported in mathematics education journals concerning fractions, ratio and proportion, additive problems and multiplicative problems; few research studies have been reported in the same period about percentages, parallelism and perpendicularity, absolute value, integrals, discrete mathematics, graphs, etc.

Another example: today, 'rational numbers, decimal numbers and approximations of numbers in a calculator or computer environment' form a very important topic in mathematics education which, however, is almost neglected in the main journals.

As regards the results of the 'quantitative information' type, they are very popular amongst school teachers and administrators; some mathematicians recognize them as the only 'scientific' results in mathematics education.[2] We think that these results must be seriously considered by researchers in mathematics education. Indeed, as we observed earlier, they provide information which frequently seems 'objective', 'scientific', and easily intelligible to teachers (and also to parents, school administrators, etc.). Taking this aspect into account, the quantitative evaluation of pupils, teachers, school systems, projects and innovations needs special attention as it may cause serious damage: for instance, it may orient the teachers' work towards preparing pupils to be successful in assessment tests.

On the one hand, researchers should endeavor to find better tools than the existing ones in order to reduce the potential for damage. In particular, there is a need for additional statistics in order to better exploit, where possible, the multidimensional information of quantitative assessment – going far beyond the percentages of 'good' answers. On the other hand, researchers (and teachers) should use results of the 'qualitative information' and 'theoretical perspectives' types to get the methodologies and the results of the 'quantitative information' type under control. For instance, it is crucial to understand whether it is possible (and, in this case, *how* it is possible) to keep 'constant' mathematics education variables in order to create effective control groups (cf. Artigue 1990).

More generally, results of the 'qualitative information' and 'theoretical perspectives' types (some as 'demolishers of illusion', others as 'energizers of practice' of research) are necessary to get under control, in the field of mathematics education, the usage of experimental and statistical methodologies borrowed from educational sciences. In particular, it should be important to clarify why some 'laboratory' results of the 'quantitative information' type, which may be interesting from a psychological or sociological point of view, may be scarcely relevant to mathematics education.

4. CONTRADICTIONS INHERENT IN THE AIM OF ESTABLISHING MATHEMATICS EDUCATION AS A SPECIFIC FIELD OF RESEARCH

If we claim that experimental research in mathematics education (like experimental research in physics, biology, psychology) must produce data which can be reproduced under similar conditions, and whose interpretations may be partly or totally falsified by further research, then researchers must isolate variables, keep experimental conditions under control, and communicate this in their papers. But, mathematics education is a complex process: research results are interesting (for their consequences in the school system, and also by themselves, as scientific results concerning 'mathematics education') when this complexity is taken into account. 'Complexity' concerns mainly the fact that almost every meaningful teaching–learning process appears as a long-term process involving a large number of interrelated variables; moreover, at present, to isolate and measure some of the relevant variables does not appear easy.

Some examples concern:

1. Deep cultural and cognitive aspects of 'context sensitivity' in mathematical problem-solving: These aspects may be revealed (in the school setting) only by long-term exploratory studies involving many variables in real classroom situations managed by a teacher inducing (and giving importance to) those aspects. The classroom must be involved for a long period in meaningful problem-solving activities concerning a particular context in order to change the traditional relationship between 'context' mathematics and learning processes. Usually, a real life 'context' is evoked by the text of an isolated word problem and does not become the subject of a long-term investigation involving many problem situations. (Concerning this issue, mainly related to mathematics education as *a long-term process*, see: Artigue 1990; Bartolini Bussi 1991; Boero 1989a, 1989b, 1990, 1992; Carraher 1988; Douady 1984; Lesh 1985; Robert 1992).

2. Research about the effects of LOGO on learning mathematics. Separating variables and reducing the complexity of tasks, researchers have shown good quantitative effects of LOGO-based activities regarding the learning of some geometric concepts (some examples concern 'measure of angles', 'orientation of angles', 'polygons'). Comparing performances of groups of students working in a 'paper-and-pencil' environment and a LOGO environment, on very simple tasks, better results were found in the case of the LOGO environment. But Hillel et al. (1989), and other researchers' qualitative results concerning more 'open' problem situations, point to the dangers of a particular usage of the LOGO environment on the development of conceptual geometric thinking: case studies were realized through qualitative analysis of strategies produced to solve 'complex' problems taking into account the learning environment (including the teacher!). (Concerning this issue, mainly related to mathematics education as a *process depending on many interrelated variables*, see: Hillel, Kieran & Gurtner 1989; Hughes, MacLeod & Potts 1985; Noss 1987; Olson, Kieren & Ludwig 1987; Simmons & Cope 1990, 1993).

In general, mathematics educators at present tend to agree with the statement that the study of many aspects of mathematics education cannot be reduced to the study of a finite number of independent, measurable variables; but this statement should be supported by a suitable analysis of long-term classroom processes and theoretical considerations.

If we claim that research in mathematics education must be similar to research in any 'normal' science, 'cumulation' and 'universality' of research results are needed, and the existence of the progress must be evaluated comparing new results with previous ones.

But, if we read most research articles on mathematics education published in leading international journals, we see that most of the references point to the research school to which the author belongs (the barrier of languages acts as a deterrent, but even in articles written in English, 'substantial' quotations concerning other articles written in English are frequently limited to a particular audience or research school ignoring contributions from other research schools, also written in English).

As an example, if we consider articles written in English and published in *Educational Studies in Mathematics* and *For the Learning of Mathematics*, in the last four years in some specific domains (problem solving, classroom interaction and social construction of mathematical knowledge and ethnomathematics), we find that only less than half of the articles make substantial references to papers issued from different research schools. The 'classroom interaction and social construction of mathematical knowledge' and 'ethnomathematics' topics involve rather precise fields of research: in the last four years 18 articles were published in these domains, with only 8 of them making substantial references to different research schools.

We think that this is not due to lack of information, or to misjudgment of the results of others; the problem is that frequently results obtained by different research schools are very difficult to compare, and researchers prefer to stay within a strictly homogeneous reference system.

This happens for various reasons: there is the possibility of the existence of hidden variables; the same technical words may have different meanings for different schools, frequently referring to different epistemological, philosophical or psychological frameworks. For instance, 'interiorization', with reference to a Piagetian versus a Vygotskian framework (see Mugny 1984; Perret-Clermont 1980; Piaget & Inhelder 1947; Vygotskij 1967, 1978, 1990). Different words may be used for partially similar phenomena: considering an example related to specific research problems in mathematics education, in 'ratio and proportion research', three different definitions (*'internal ratios, external ratios'; 'within-state ratios, between-state ratios'; 'function, scalar'*) were proposed by Freudenthal, Noelting and Vergnaud in order to distinguish between two different kinds of ratio. These were used in proportionality problems pointing out different aspects: epistemological aspects in Freudenthal 1978; psychological aspects in Noelting 1980; structural and psychological aspects in Vergnaud 1981, 1983.

An example related to more general learning problems concerns the words *'scheme', 'script', 'frame'* in Piaget's, Minsky's, Shank & Abelson's, and Vergnaud's papers: these words point out different aspects of the same psychological phenomena which are also relevant to mathematics education (see: Minsky 1986; Piaget & Inhelder 1959; Shank & Abelson 1977; Vergnaud 1990). Also, different theoretical frameworks are frequently used to plan experiments: some relevant differences concern the role of the teacher ('observed' actor in the classroom, like in Brousseau 1986, or 'participant observer', like in Eisenhart 1988); selection and isolation of variables in short-term (like in many 'laboratory studies'), or long-term classroom studies involving many variables; observations centered on individual pupils, or on classroom interactions. (For further discussion, see: Artigue 1990; Balacheff 1990; Bartolini Bussi 1991; Boero 1994; Brousseau 1986; Chevallard 1992; Kilpatrick, 1992; Robert 1992; Steiner 1984).

As to the problem of the 'cumulative' character of research in mathematics education, the best solution is not, in our opinion, to refuse articles which do not make substantial quotations from all the previous, outstanding results concerning the same topics but rather that the community of researchers should help the individual and orient him or her in this work. We need to invent new kinds of scientific meetings where different schools compare their results – especially their vocabulary and methodology regarding the same subject. This might also produce the possibility of revealing hidden cultural, linguistic, and other variables.

Also, renowned journals might contribute to the enhancement and deepening of the exchange between different schools and traditions of research in mathematics education: for instance, it might be useful to publish special survey articles seeking to point out subjects worthy of closer investigation ('white spots' of our knowledge); or, to publish short, preliminary information about ongoing research.

And, more generally, we also need theoretical tools of the 'theoretical perspectives' type concerning comparison of different types and methodologies of research in mathematics education; i.e. going beyond descriptions (see Boero 1994). Classifications, and models – if possible – are necessary. In this sense, we share the idea that:

...this ICMI study does not seek to describe the state of the art. Nor does it intend to tell anyone what research in mathematics education is or is not. Instead, the organizers of this study propose to clarify the different meanings these ideas have for mathematics educators.... (Balacheff et al. 1993)

5. SOME LINKS WITH THE PROBLEM OF TEACHERS' PREPARATION

As we have tried to prove, results of the 'qualitative information' and 'theoretical perspectives' type are crucial for practitioners and researchers alike, but usually those results need a specific vocabulary and precise references to human sciences (anthropology, philosophy, sociology, psychology). This may be hard for mathematics teachers and researchers in mathematics education.

From this perspective, it appears evident that changes are needed in the preparation of teachers and researchers. But what changes? How does one negotiate these changes with mathematicians? How does one put them into practice?

Concerning teachers, in order for them to become able to use more research results of the 'qualitative information' and 'theoretical perspectives' types in the life of ordinary schools, we think that extensive research should be carried out on teacher training and in-service training. Furthermore, researchers must make a greater effort to make their work understandable to ordinary teachers. If we look at the reality of present day teachers' and researchers' preparation, we perceive a large gap in this regard.

Although mathematics educators think that teachers' preparation must be based on good mathematics preparation and good preparation in the educational field, we observe that in almost all countries, most current university courses of mathematics attended by future teachers provide a deep and up-to-date specialized insight into many topics recognized as 'basic contents' for advanced applications and research in mathematics. But school mathematics (especially in elementary and secondary schools) deals with other topics which are not normally taught or taken into consideration at university level. On the other hand, most current university courses attended by future teachers neither have connections with school mathematics, nor cover in depth the relationships with other domains of mathematics, or give any idea of the cultural 'meaning' of what is being taught. We may add that most university courses in the domain of Education attended by future teachers are 'academic' courses detached from their cultural and professional needs. Moreover, unfortunately, the teacher training process is not a popular research domain.

The problem of the relationships between school mathematics and mathematicians' mathematics (as well as 'street' mathematics, and social needs) should be dealt with in-depth to seriously tackle the difficult problem of the mathematics curriculum for future teachers, and the connections between their 'professional' and their 'cultural' preparation in mathematics.

Concerning the preparation of teachers, it is also necessary to overcome the false idea, so frequently shared in the mathematicians' community, that effective teaching of mathematics is essentially based on good 'technical' knowledge of the topics to be taught, and on the quality of the teacher as a self-made 'artist'. (This idea has well-known, important consequences for the preparation of teachers in many countries.)

Again, results of the 'qualitative information' and 'theoretical perspectives' types might be useful to tackle these problems. But stressing the importance of these types of results may create friction within the mathematicians' community due to their lack of background in human sciences: there are mathematicians who, as 'scientists', are suspicious about 'humanities': philosophy, anthropology, sociology, psychology, etc. Many mathematicians may take into consideration (at least as 'useful' or 'interesting') the results of the 'innovative patterns' and 'quantitative information' type: few of them are already open-minded towards results of the 'qualitative information' and 'theoretical perspectives' types.

As a key issue in this discussion, we think that the 'explanatory' power of the results of the 'qualitative information' and 'theoretical perspectives' types must be proved on common, recognized meaningful grounds; for instance, advanced mathematics teaching, or phenomena concerning dissemination of innovations, and in close connection with 'innovative patterns' type results. Also, problems which are generally considered important in this historical period (like multicultural teaching of mathematics: classrooms with mixed mother tongues and cultural traditions) might be tackled to show the effectiveness of those results.

Results of the 'qualitative information' type and (at least for the 'description' part) 'theoretical perspectives' type require the involvement of teachers in the research and collaboration between teachers and researchers. The same thing is needed during the preparation of teachers and researchers to exploit/produce such results. But this is not possible in some countries, especially where innovations are involved: 'innovation' implies very complex procedures for permission, since researchers (or students) cannot enter classrooms without special permissions which are rather difficult to obtain, etc. We think that experimental (and observation) schools or sections must be created (especially in the countries where it is difficult to perform experiments and long-term observations in ordinary classes) and existing experiences about 'observation' schools should be analyzed and compared.

6. CONCLUSION

Results of the 'innovative patterns' type (for obvious reasons, connected to pragmatic aims, and for other reasons, connected to the specificity of research in mathematics education) and results of the 'theoretical perspectives' type (as a consequence of the preceding analyses) are needed to characterize 'mathematics education' as a specific field of research. But, in our opinion, results of the 'innovative patterns' type alone are not consistent by themselves as research results, and results of the 'theoretical perspectives' type alone are not sufficient to justify research in mathematics education in the teachers' and mathematicians' eyes.

Results of the 'qualitative information' and 'theoretical perspectives' types are important not only in themselves, but also in allowing teachers and researchers to get the other kinds of results under control: for this reason, they should gradually enter into teacher training. Unfortunately results of types 'qualitative information' and 'theoretical perspectives' nowadays are not popular among mathematics teachers and mathematicians: we think that their 'explanatory' power must be proved on common and recognized meaningful grounds.

NOTES

1. Concerning the idea of contradiction between the notions of innovation and dissemination of innovation, see also the discussion by G. Brousseau (1992, 'Didactique: What it can do for the Teacher', in R. Douady & A. Mercier (eds.), *Research in Didactique of Mathematics, Selected Papers*, La Pensée Sauvage éditions, Grenoble, 7–40). (Editors' note).

2. A quantitative information about the distribution of different types of thinking in mathematics in a given population of school children was found, by S. A. Amitsur, a mathematician interviewed by A. Sfard for our book, to be the most important result to be provided by mathematics education research to mathematicians who are participating in curriculum design (Editors' note).

REFERENCES

Artigue, M.: 1990, 'Ingénierie Didactique', *Recherches en Didactique des Mathématiques* **9**, 281–308.

Balacheff, N.: 1990, 'Future Perspectives for Research in the Psychology of Mathematics Education', in P. Nesher & J. Kilpatrick (eds.), *Mathematics and Cognition*, Cambridge University Press, Cambridge, 135–147.

Balacheff, N. et al.: 1993, 'What is Research in Mathematics Education and What are its Results? Discussion Document for an ICMI Study', *Zentralblatt für Didaktik der Mathematik* **93** (3), 114–116.

Barra, M., Ferrari, M., Furinghetti, F., Malara, N. & Speranza, F. (eds.): 1992, *The Italian Research in Mathematics Education: Common Roots and Present Trends*, T.I.D.-C.N.R., Quad. 12.

Bartolini Bussi, M.:1991, 'Social Interaction and Mathematical Knowledge', *Proceedings of PME-XV* (Assisi), **I**, 1–16.

Bartolini Bussi, M. & Pergola, M.: (1992), 'History in the Mathematics Classroom: Examples from a Research Project for High School', in H. N. Jahnke, N. Knoche & M. Otte (eds.), *Interaction between the History of Mathematics and Mathematics Learning*, Vandenhoek & Ruprecht, Göttingen.

Bishop, A.: 1988, *Mathematical Enculturation*, Kluwer Academic Publishers, Dordrecht.

Boero, P.: 1986, 'Usage de l'Histoire des Mathématiques dans la Production de Séquences d'Enseignement et de Situations Didactiques', *Comptes Rendus IV-ème Ecole d'Eté de Didactique des Mathématiques* (Orleans), 140–156.

Boero, P.: 1989a, 'Semantic Fields Suggested by History: Their Function in the Acquisition of Mathematical Concepts', *Zentralblatt für Didaktik der Mathematik* **20**, 128–133.

Boero, P.: 1989b, 'Mathematical Literacy for All: Experiences and Problems', in *Proceedings of PME-XIII* (Paris), Vol. I, 62–76.

Boero, P.: 1990, 'On Long-Term Development of Some General Skills in Problem Solving: A Longitudinal Comparative Study', *Proceedings of PME-XIV* (Oaxtpec, Mexico), Vol. 2, 169–176.

Boero, P.: 1992, 'The Crucial Role of Semantic Fields in the Development of Problem Solving Skills in the School Environment', in J. Ponte, et al. (eds.), *Mathematical Problem Solving and New Information Technologies*, Springer-Verlag, Berlin, 77–91.

Boero, P.: 1994, 'Situations Didactiques et Problèmes d'Apprentissage: Convergences et Divergences dans les Perspectives de Recherche', in M. Artigue, R. Gras, C. Laborde & P. Tavignot (eds.), *Vingt Ans de Didactique des Mathématiques en France*, La Pensée Sauvage, Grenoble, 17–50.

Boero, P. & Szendrei, J.: 1992, 'The Problem of Motivating Pupils to Learn Mathematics', *Actes de la 42è Rencontre de la Commission Internationale pour l'Etude et l'Amélioration de l'Enseignement Mathématique*, Chicago.

Brousseau, G.: 1980, 'Problèmes de l'Enseignement des Décimaux', *Recherches en Didactique des Mathématiques* **1** (1), 11–59.

Brousseau, G.: 1981, 'Problèmes de Didactique des Décimaux', *Recherches en Didactique des Mathématiques* **2** (1), 37–127.

Brousseau, G.: 1983, 'Les Obstacles Epistémologiques et les Problèmes en Mathématiques', *Recherches en Didactique des Mathématiques* **4**, 165–198.

Brousseau, G.: 1986, *Théorisation des Phénomenes d'Enseignement des Mathématiques*. Thèse d'Etat, Université de Bordeaux I.

Carraher, T.: 1988, 'Street Mathematics and School Mathematics', *Proceedings of the 12th Conference of the International Group for the Psychology of Mathematics Education*, OOK-Veszprem, Vol. I, 1–23.

Chevallard, Y.: 1991, *La Transposition Didactique* (2nd ed.), La Pensée Sauvage éditions, Grenoble.

Chevallard, Y.: 1992, 'Concepts Fondamentaux de la Didactique: Perspectives Apportées par une Approche Anthropologique', *Recherches en Didactique des Mathématiques* **12**, 73–112.

Douady, R.: 1984, *Jeux de Cadres et Dialectique Outil-Objet*. Thèse d'Etat, Université Paris VII.

Eisenhart, M. A.: 1988, 'The Ethnographic Research Tradition and the Mathematics Education Research', *Journal for Research in Mathematics Education* **19**, 99–114.

Freudenthal, H.: 1978, *Weeding and Sowing* , D.Reidel, Dordrecht.

Hillel, J., Kieran, C. & Gurtner, J. L.: 1989, 'Solving Structured Geometric Tasks on the Computer: The Role of Feedback in Generating Strategies', *Educational Studies in Mathematics* **20**, 1–39.

Hughes, M., MacLeod, H. & Potts, C.: 1985, 'Using Logo with Infant School Children', *Educational Psychology* **5**, 287–301.

Kilpatrick, J.: 1992, 'A History of Research in Mathematics Education', in D. A. Grouws (ed.), *Handbook of Research on Mathematics Teaching and Learning*, Macmillan, New York.

Laborde, C.: 1988, 'L'Enseignement de la Géométrie en tant que Terrain d'Exploration de Phénomènes Didactiques', *Recherches en Didactique des Mathématiques* **9**, 337–364

Lesh, R.: 1985, 'Conceptual Analysis of Mathematical Ideas and Problem Solving Processes', *Proceedings of the 9th Conference of the International Group for the Psychology of Mathematics Education*, OW & OC, Utrecht, Vol. 2, 73–96.

Minsky, M.: 1986, *The Society of Mind*, Simon and Schuster, New York.

Mugny, G. : 1984, *Psychologie Sociale du Développement Cognitif* , Peter Lang, Berne.

Noelting, G.: 1980, 'The Development of Proportional Reasoning and the Ratio Concept, Part II', *Educational Studies in Mathematics* **11**, 331–363.

Noss, R.: 1987, 'Children's Learning of Geometrical Concepts through Logo', *Journal for Research in Mathematics Education* **18**, 342–362.

Olson, A. T., Kieren T. E. & Ludwig, S.: 1987, 'Linking Logo, Levels and Language in Mathematics', *Educational Studies in Mathematics* **18**, 359–370.

Perret-Clermont, A.N.: 1980, *Social Interaction and Cognitive Development in Children* , Academic Press, London.

Piaget, J. & Inhelder, B.: 1947, *La Représentation de l'Espace chez l'Enfant*, Presses Universitaires de France, Paris.

Piaget, J. & Inhelder, B.: 1959, *La Genèse des Structures Logiques Elémentaires*, Delachaux & Niestl, Neuchâtel-Paris.

Robert, A.: 1992, 'Problèmes Méthodologiques en Didactique des Mathématiques', *Recherches en Didactique des Mathématiques* **12**, 33–58.

Shank, R. & Abelson, R.: 1977, *Scripts, Plans, Goals and Understanding*, Lawrence Erlbaum Associates, Hillsdale, NJ.

Sierpinska, A.: 1985, 'Obstacles Epistémologiques Relatifs à la Notion de Limite', *Recherches en Didactique des Mathématiques* **6**, 5–67.

Sierpinska, A.: 1987, 'Humanities Students and Epistemological Obstacles Relative to Limits', *Educational Studies in Mathematics* **18**, 371–397.

Simmons, M. & Cope, P.: 1990, 'Fragile Knowledge of Angle in Turtle Geometry', *Educational Studies in Mathematics* **21**, 375–382.

Simmons, M. & Cope, P.: 1993, 'Angle and Rotation: Effects of Different Types of Feedback on the Quality of Response', *Educational Studies in Mathematics* **24**, 163–176.

Steiner, H. G.: 1984, *Theory of Mathematics Education*, Occasional Paper 54, Institut für Didaktik der Mathematik, Universität Bielefeld.

Vergnaud, G.: 1981, *L'Enfant, la Mathématique et la Réalité*, Peter Lang, Berne.

Vergnaud, G.: 1983, 'Multiplicative Structures', in R. Lesh & M. Landau (eds.), *Acquisition of Mathematical Concepts and Processes*, Lawrence Erlbaum Associates, Hillsdale, NJ.

Vergnaud, G.: 1990, 'La Théorie des Champs Conceptuels', *Recherches en Didactique des Mathématiques* **10**, 133–170.

Vygotskij, L. S.: 1967, *Gondolkodas és beszéd*, Akadémiai Kiado, Budapest.

Vygotskij, L. S.: 1978, *The Mind in Society*, Harvard University Press, Cambridge, MA.

Vygotskij, L. S.: 1990, *Pensiero e Linguaggio* (ed. L. Mecacci), Laterza, Bari.

Paolo Boero
Dipartimento di Matematica,
Università di Genova,
Via Dodecaneso, 35,
16146 – Genova – Italy,

Julianna Radnai Szendrei
Mathematics Department,
Teachers' Training College,
Kiss Janos 40,
1126 – Budapest – Hungary

CAROLYN KIERAN

MODELS IN MATHEMATICS EDUCATION RESEARCH:
A BROADER VIEW OF RESEARCH RESULTS

1. INTRODUCTION

It may seem strange to some to imagine models as research results. How are research results defined? Narrowly, in the sense of empirical findings, or more broadly as the products of research endeavors that are in some way linked to empirical work. At international conferences of mathematics education researchers, for example, the International Group for the Psychology of Mathematics Education (PME), one finds in the guidelines for conference proposals two categories of research reports: empirical and theoretical. But the separation between empirical and theoretical work in mathematics education research may, in fact, be a false one – especially with respect to models. At the least, the dividing line between the two is blurring.

What are the aims of research in mathematics education? Clearly one of them is the development of a better understanding of the ways in which students learn mathematics and the ways in which teachers teach mathematics. Another is the design of innovative learning environments with the potential of changing for the better the ways in which students come to understand their mathematics. But whether one conducts research in the spirit of the former or the latter aim – and these are not the only aims of mathematics education research (see the discussion document of this ICMI Study) – the current reporting of research suggests that both involve the description of observed phenomena by means of models. Sometimes these descriptions might be based on *implicit* models or theories that are held intuitively by the researcher. But as is often the case, the models that are either being used as tools or being created in order to explain the data are *explicitly* formulated. The point is that the way in which research results are presently being reported frequently takes the form of descriptive or explanatory models. This observation is not meant to suggest that all empirical work that is reported includes explanatory models, but rather that most of the work involving theoretical models does incorporate empirical results (which does not exclude the possibility of referring to someone else's empirical findings). That there is an inescapable synergy between the two – empirical results and explanatory models – is clear; I return to this aspect below. So, we are beginning to see more and more often that the reporting of research results is not simply the enumeration of the observed empirical facts but also the description of a model that has been developed (or an existing model that has been further refined) to explain what

213

Sierpinska, A. and Kilpatrick, J. Mathematics Education as a Research Domain: A Search for Identity, 213–225.
© *1998 Kluwer Academic Publishers. Printed in Great Britain.*

has been identified; that is, that models are as much the results of doing research as are the empirical observations.

However, there is a great deal of variation in the way in which the term *model* (or theory or principle or ...) is used in the field of mathematics education research. One person's model can be another's theory. The aim of this chapter is: first, to illustrate some of the various ways in which mathematics education researchers have in the past few years been using the terms *models* and *theories*; and second, to provide a more detailed example of a model that has emerged in the past decade and that illustrates clearly the inseparable interaction between empirical and theoretical work in mathematics education research.

2. USES OF THE TERMS: MODELS, THEORIES, ...

The first example of some of the various ways in which the mathematics education research community has been using the terms *models* and *theories* is drawn from a monograph reporting the discussions of the Research Workshop on Learning Models held in Durham, New Hampshire, 1977. John Richards (1979), one of the participants, claimed that 'in the literature in mathematics education, models are usually not distinguished from theories' (p. 6). In attempting to provide a distinction between a theory and a model, he argued the following:

> The relationship between a model and a formal theory is one of the open questions in the philosophy of science. ... I propose that theories are complex conceptual systems comprising explanations, problems, methods, technological devices, and so on. Theories, in a non-trivial sense, determine the data to be analyzed and structured. ... If we view theories as including the larger context of scientific enquiry, then clearly the use of models is an intrinsic part of that undertaking. Theoretical considerations are therefore necessarily prior to building or using a model. ... In contrast to a theory, a model is not even intended to be complete, or final. It is not intended to capture or explain a situation totally. ... A model simplifies, represents, visualizes, preserves structure (mainly), and finally disappears. ... It is presented as a working hypothesis. ... In fact, within a theory it is *expected* that various divergent models will be proposed. Each may emphasize different aspects of the theory. ... The ramifications of this framework for the researcher interested in models are fairly straightforward. First, the researcher is always operating within a tradition (a methodological research programme) and the researcher must be cognizant of this. For mathematics education the programme may be Piagetian, behavioral, logical, linguistic, and so on. Each of these has a fairly well developed metaphysical and epistemological outlook. Each determines data which must be understood, and each delineates general problems to be examined and resolved. Second, within a programme there is a more or less well-accepted, up-to-date, theory which is adopted by the researcher. This provides specific problems and perspectives. Mathematics education may still be in the stage of generating theories. It is far from being a mature science. Nevertheless, there are still well-defined programmes determining the research. A proposed model functions differently than a theory within these programmes. That is, even if the programme itself is general, or borrowed from other disciplines, a theory within the programme still determines the data and the problems. Models may then be proposed to resolve these problems. (pp. 17–23)

Richards provided a specific example of the distinctions he was making by referring to the work of Steffe, Richards & von Glasersfeld (1979): 'The models presented there relate the learning of whole number concepts to the Piagetian tradition, and more specifically, to a theory within this tradition (i.e., constructivism)' (p. 23). And the nature of these models has been described by Leslie Steffe (1983) as follows:

A constructivist research program in mathematics education has, as its central problem, the explanation of the process of construction of mathematical objects as it actually occurs in children. ... The explanations we formulate in our research constitute models. It must be understood that these models are formulated in the context of intensive observation of children's construction of mathematical objects. But while they are based on actual human behavior, they also are based on our interpretation of the meaning that the children attribute to their behavior or, in other words, why they exhibit the observed behavior. ... The model, as we use it, can be understood using the metaphor of a black box ... ; unlike behaviorists, however, who are not interested in the machinery inside of the box, a constructivist attempts to design models which ideally will, given an 'input', produce the same output as the black box. One never speculates as to the function of the black box, but only attempts to design models which seem to be a viable explanation of the input–output relation. (pp. 469–470)

The kind of model being discussed is the implicit model that is guiding the child's behavior. Much of the research in our field that involves model construction is based on the search for implicit models.

Guy Brousseau (1972) was one of the first, I believe, to use the term *implicit models*. In his descriptions of the processes by which the child constructs mathematical knowledge, he noted that when children are placed before certain objects or elements that are related, they form mental models of the situation that control their actions:

[Lorsqu'un enfant] dans une suite de situations comparables (qui réalisent une même structure) a une suite de comportements comparables (qui relèvent d'une même conduite), on est fondé à estimer qu'il a perçu un certain nombre d'éléments et de relations de cette structure. Il a donc au moins un certain *modèle mental* de cette situation qui règle son action. (p. 57)

The implicit model thus defined by Brousseau introduces the idea of a certain mental representation of the relations between the givens, which modulates the actions of the child in the presence of comparable situations. This notion of implicit mental models has been further developed by Brousseau and other researchers of the French community of mathematics education researchers into the Theory of Didactic Situations.

Rouse and Morris (1985), in their synthesis of the various definitions, uses, and functions of mental models, point out that 'the common themes are describing, explaining, and predicting, regardless of whether the human is performing internal experiments, scanning displays, or executing control actions' (p. 11). Other aspects of mental models, which have been emphasized by Fischbein, Tirosh, Stavy & Oster (1990), are their autonomy with respect to the original and their stability:

Being structurally unitary and autonomous, the model often imposes its constraints on the original and not vice versa! Consequently a model is not simply a substitute, an auxiliary device (more simple, more familiar, more accessible). (p. 24)

The autonomy and stability of mental models seem to suggest that they are not mere products, mere reflections of the originals. They belong to the mental structure of the individual, well integrated into this structure, reflecting its requirements, its particularities, its schemata, its laws. (p. 29)

Efraim Fischbein (1994) gives much priority to the intuitive character of implicit models, as seen in the following excerpt aimed at explaining a number of misconceptions held by elementary school teachers concerning the mathematical concept of set:

A very simple interpretation may account for all these misconceptions. If the model one has in mind, when considering the concept of set, is that of a *collection of objects*, all these misconceptions are predictable. ... I do not affirm that students identify, explicitly and consciously, the mathematical concept of set with the notion of a collection of concrete objects. What I affirm is that, while considering the mathematical concept of set, what they have in mind – *implicitly* but *effectively* – is the idea of a collection of objects with *all* its connotations. There is no subjective conflict here. The intuitive model manipulates from *behind the scenes* the meaning, the use, and the properties of the formally established concept. The intuitive model seems to be stronger than the formal concept. (p. 236)

Fischbein, Deri, Nello & Marino (1985) have been able to verify the presence of the implicit models that they have posited by constructing situations that introduce constraints for which the model would no longer be able to function. They have observed that, every time the problem givens did not respect the constraints imposed by the model, the pupils were led to make predictable errors. This approach permits not only the formulation of hypotheses with respect to the models used by pupils but also the confirmation of their existence.

The preceding brief discussion has focused on implicit models, that is, those whose non-explicit character makes it necessary for the researcher to try and find ways to induce their presence. But not all modeling in mathematics education research is directly oriented toward the uncovering of a subject's implicit models. (Please note that the examples mentioned in this section of the chapter are not intended to be comprehensive, but merely illustrative. As well, there has been no attempt to include teaching models, that is, models used by teachers in order to present certain mathematical concepts – such as, the various teaching models for negative numbers.)

In a somewhat different perspective, Margaret Brown (this volume) describes research that is aimed at 'reducing the complexity of an enormous variety of individual behaviors or beliefs by modeling the field in terms of a limited number of categorizations or "ideal types"' (p. 265). Such research, according to Brown, results in models that attempt to describe and explain understanding, attainment, beliefs and/or practice. She cites as an example the Concepts in Secondary Mathematics and Science study (Hart 1981), an assessment project that involved data from thousands of students and which led to the development of models of progression/levels in mathematical thinking. Brown (this volume) argues:

The act of identifying common features in different groups of individuals and using these to describe a number of 'ideal types' is a form of psychological modeling akin to mathematical modeling.

In each case the model presents a general way of thinking about the behavior of specifics (objects or people). This act of identifying common features means that other more idiosyncratic features have been ignored. Such models are only valid if they are useful, which in general means that those features ignored can be demonstrated or agreed to be of secondary importance for a particular purpose (p. 272).

Eugenio Filloy (1990) has described a kind of modeling that results in what he calls *local theoretical models*:

The stability of the observed phenomena and the well-established replicability of the experimental designs that were used in our studies confronted us with the need to propose a theoretical component to deal with different types of (1) algebra teaching models for the teaching–learning processes together with (2) models for the cognitive processes involved, both of which are related to (3) formal competency models to simulate the competent performance of an ideal user of elementary algebraic language. It was necessary to concentrate on *local theoretical models* appropriate to specific phenomena, which were nevertheless able to take account of all these components; we also proposed ad hoc experimental designs to throw light on the interrelationships and oppositions arising during the development of all the processes relevant to each of these three components. (p. PII.19)

In their descriptions of the mathematical models constructed by students and teachers – another use of the term *model* – Lesh and Kelly (1994) emphasize the representational aspect of models, in particular the way in which these models are subject to modification:

In the tutoring study, for example, students proposed a variety of different ways to think about a problem. In the early stages, they suggested several models based on additive relationships, subtractive relationships, fractions, or proportions. These models were expressed in a variety of different ways: as numbers, as verbal arguments, as graphs, as sketches, and so forth. As the students explored a relationship through a given representation, they oftentimes pursued features of the representation that, in turn, suggested the pursuit of an alternative relationship. In this way, the models were dynamic, unstable, and subject to mutation. (p. 278)

Implicit in the mathematical modeling engaged in by the students described by Lesh and Kelly is the students' underlying use of their own mental models. The latter are also subject to gradual change, as has been shown by the work of Fischbein and his colleagues and students. But just as children's models of the world evolve, so too do researchers' models of the phenomena they are studying; and this evolution of researchers' models sometimes occurs over a period of several years. For example, the long-term research program of Jacques Bergeron and Nicolas Herscovics led to several versions (e.g., Bergeron & Herscovics 1989; Herscovics & Bergeron 1983) of their two-tiered model of the understanding of early number. Basically, the two tiers are related to the acquisition of certain physical concepts, followed by the corresponding mathematical concepts. But rather than providing a detailed picture of the dynamics of movement through the two tiers, this model has been used to suggest guidelines to teachers about the conditions that should be satisfied in order for a pupil to construct an enriched concept of early number.

Another model of the growth of mathematical understanding is the one that has been developed by Susan Pirie and Tom Kieren (1992) – also the result of several years of classroom observation and case study analyses. This model, with its eight modes of understanding depicted by concentric circles tangent at a point, attempts to capture the nonlinear, dynamic growth of mathematical understanding at any level in any topic.

As has been seen from the two examples just mentioned, a model is often not the result of a single study; its formulation can be based on the analysis of data from several studies. Its development might also be subject to input from additional sources, such as other researchers' models or the history of mathematics. For example, the Sfard (1991) model of mathematical conceptual development, which is discussed at greater length in the next section, has its basis in both psychological-empirical and historical work. Thus, the construction of models can reflect varying degrees of contribution from empirical data, over varying time periods.

In addition to the evolving form of models, there is another aspect to be noted here. That is the fact that they tend to be situated within the context of a more global, overarching theory – as was pointed out earlier by Richards. For example, Lesh and Kelly (1994), in their paper on student and teacher models, emphasize that 'we begin with the assumption that students actively construct meaning ... thus, we are in general accord with the precepts of what has become known as constructivism' (p. 277). And Pirie and Kieren (1992) state that 'our theory is constructivist in its roots, elaborating the nature of understanding as the personal building and re-organization of one's knowledge structures' (p. 243).

But it is important to note that in the Pirie and Kieren description of their model, there appears to be more to the term *theory* than simply that the model has its roots in the theory of constructivism, for they refer to their model as 'a *model* for the *theory* of growth of mathematical understanding' (p. 245). Thus, they seem also to be developing a theory of mathematical understanding for which their model might be but the current representation or metaphor. So we see here an example of not only the development of a model within a theory, but also the development of a theory within a theory.

This kind of modeling, which is closely and explicitly tied to the development of a theory, reminds us of the earlier mentioned theoretical work of Brousseau and colleagues (i.e., the development of the Theory of Didactic Situations that was based on his research involving children's implicit mental models). Yet even these broader theories, which are intended to be more complete and of wider application than their partial dynamic models, are seen by some as still being local models:

As often happens in the development of science, the selection of 'narrow' pieces of reality to be modeled can solve the problem of both acceptable modeling and theoretical coherence: 'Narrowness' could result, in turn, in a limitation of either the number of subjects involved, the duration of observation, or the items of knowledge. A good example is the *theory of didactical situations* (Brousseau 1986), which is successful for microdidactical studies, in which a given item of knowledge and a given problem situation is considered. (Bartolini Bussi 1994, p. 127)

Presmeg (this volume) sums up this viewpoint when she says: 'The dilemma of constructing theory is that any theory, by its nature, is a simplified model which facilitates understanding of some elements of a phenomenon but excludes others; thus no theory can be complete or final (p. 63).'

The phenomena that are being modeled in our field are basically teaching and learning phenomena. Because mathematics education is an applied discipline, the models we create do not look like, say, models of the atom. They are models of what people actually do when they learn or teach mathematics. The goal of this section of the chapter has been to show, directly, some of the various ways in which mathematics education researchers use the terms *models* and *theory* and, indirectly, how closely model development in the discipline is tied to the conduct of empirical work and ought thus to be considered a *result* of research. In other words, *results* ought not to be limited to the products of empirical work; separating the theoretical work of modeling from empirical work would seem to be artificial.

In the following section, I take one particular model – the process-object model of mathematical conceptual development – and in showing the path followed by its evolution illustrate the inseparable interplay between theoretical and empirical work. The model that is featured is one for which different versions exist, but all share certain basic theoretical assumptions regarding the duality of mathematical thinking, situated within the larger theoretical framework of constructivism.

3. PROCESS-OBJECT MODELS OF MATHEMATICAL CONCEPTUAL DEVELOPMENT

In 1985 a PME working group on Advanced Mathematical Thinking was established, and soon after began a series of discussions that was to continue over the course of several years on the learning of certain mathematical concepts in terms of process-object conceptualizations (e.g., Dreyfus 1990; Dubinsky 1991; Gray & Tall 1994; Harel & Kaput 1991; Thompson 1985). At the same time, other researchers not directly involved with the Working Group were doing related work on model development. For example, Sfard (1987, 1991) was gathering evidence for a historical-psychological model of the operational-structural duality of mathematical thinking; Douady (1985) was conducting empirical work associated with the tool-object model that she was developing. Thus, several individuals were working in parallel on the construction of similar models.

When these various versions are compared, it becomes clear that, despite their obvious resemblances, there are also certain differences. For example, the nature of the movement through the models, the degrees of groundedness in past empirical research findings, as well as the nature of the other components influencing the theorizing process, all provide means of differentiating the assorted renderings of the models. For these reasons, the path taken in the development of the process-object model varies according to the version being described. A large part of the discussion below will center on the Sfard version.

Robert B. Davis (1975) was one of the first mathematics educators to point to the 'name-process' dilemma facing mathematics students, that is, seeing an expression such as 3 + 5 as both a name for a number (in other words, an object) and as a process to be calculated. More recently, Davis (1992) commented on the notion that children learn many parts of mathematics as 'operations', and only later, as a result of reflecting upon their actions, do they come to see that each process can be seen as a thing in itself:

This, of course, is not a new idea; indeed, it lies at the heart of the misconception that the equals sign means '... and the answer is ...', which makes it difficult for many children to deal with 7 = 3 + 4. Fractions are clearly, at first, operators, a point used by Max Beberman three decades ago in his middle school mathematics curriculum: students can deal with 'one half *of*' something-or-other, long before they can come to see 'one half' as a thing in itself (and of course an abstract thing, not tied to any concrete realization). In Davis (1984) this is discussed as acquiring first 'verb' status, and only later acquiring 'noun' status. (p. 234)

There was a gap of several years between the bringing of this cognitive fact to light over 20 years ago and the developing of a research model based on the ubiquitousness of the phenomenon.

Anna Sfard's (1991) historical-psychological analysis of different mathematical definitions and representations has showed that abstract notions such as number and function can be conceived in two fundamentally different ways: structurally (as objects) or operationally (as processes). She has claimed that the operational conception is, for most people, the first step in the acquisition of new mathematical notions. The transition from a 'process' conception to an 'object' conception is accomplished neither quickly nor without great difficulty. After they are fully developed, both the process and the object conceptions are said to play important roles in mathematical activity.

Some of the historical examples on which Sfard (1991) has drawn in arriving at these conclusions are as follows:

For long periods did mathematicians perform some special manipulations with already acknowledged kinds of numbers before they were able to sever an abstract product from these new processes and to accept the resulting entities as a new kind of mathematical object. For instance, a ratio of two integers was initially regarded as a short description of a measuring process rather than as a number. ... For a long time the term 'number' appeared mainly in the context of measuring processes. The Pythagorean discovery that in certain squares the usual procedure for finding the length of the diagonal cannot be described in terms of integers and their ratios (because the diagonal and the sides have no common measure) was greeted with astonishment and bewilderment. ... Much time elapsed before mathematicians were able to separate the notion of number from measuring processes and to acknowledge the fact that the length of any segment represents a number even if it cannot be found in the 'usual' way. Eventually, the set of numbers was broadened again, to include positive irrationals along with integers and fractions. This enlarged set, in its turn, gave birth to new kinds of computational processes, and then to new kinds of numbers. ... The term 'negative number' and the symbol $\sqrt{-1}$ were initially considered nothing more than abbreviations for certain 'meaningless' numerical operations. They came to designate a fully-fledged mathematical object only after mathematicians got accustomed to these strange but useful kinds of computation. (pp. 11–12)

While in the process of developing her model, Sfard reports that she constantly moved back and forth between her conjectures derived from the historical examples of mathematical development and the search for supporting evidence from cognitive studies (see Sfard 1991 for illustrations); as well, she conducted some studies of her own. In 1987, she attempted to find out whether sixty 16- and 18-year-olds, who were well-acquainted with the notion of function and with its formal structural definition, conceived of functions operationally or structurally (Sfard 1987). The majority of the students were found to view functions as a process for computing one magnitude by means of another, rather than as a correspondence between two sets. In a second phase of the study involving ninety-six 14- to 17-year-olds, students were asked to translate four simple word problems into equations and also to provide verbal prescriptions for calculating the solutions to similar problems. They succeeded much better with the verbal prescriptions than with the construction of equations. This evidence suggests a predominance of operational conceptions among Sfard's algebra students. The empirical foundations of Sfard's model also derived support from the results of a previous study (Soloway, Lochhead & Clement 1982) that showed that students can cope with translating a word problem into an equation when that equation is in the form of a short computer program specifying how to compute the value of one variable based on another.

The way in which mathematical concepts evolved historically, supported by the available psychological-empirical research, led Sfard to elaborate a parallel model of mathematical conceptual development (see Sfard 1991 for details). She hypothesized three phases in the evolution of the process-object continuum: interiorization, condensation and reification. During the first phase, called *interiorization,* some process is performed on already familiar mathematical objects. The second phase, called *condensation,* is one in which the operation or process is squeezed into more manageable units. The condensation phase lasts as long as a new entity is conceived only operationally. The third phase, *reification,* involves the sudden ability to see something familiar in a new light. Whereas interiorization and condensation are lengthy sequences of gradual, quantitative rather than qualitative changes, reification seems to be a leap: A process solidifies into an object, into a static structure. The new entity is detached from the process that produced it.

In a subsequent study designed to follow, at close range, the movement of computer-programming students through the various phases in progressing towards a structural notion of function, Sfard (1989) reported that 'our attempt to promote the structural conception cannot be regarded as fully successful' and conjectured that 'reification is inherently so difficult that there may be students for whom the structural conception will remain practically out of reach whatever the teaching method' (p. 158).

Further evidence from other areas of mathematics learning research, such as that on rational number, illustrates the difficulty many students appear to have in learning to interpret certain mathematical entities as objects. Behr, Wachsmuth & Post (1985) reported that, for a particular task that they presented to fifth grade students, 'the [cognitive] load would appear to be particularly heavy for a child who deals with a fraction as [the division of] two whole numbers and is unable to perceive it as a conceptual unit' (p. 129). Examples of a similar phenomenon in the area of linear algebra have been provided by Harel (1985). And another case in point, which bears on a study carried out by Kaput (1991) in which students were determining functions from numerical data, has been described as follows by Harel and Kaput (1991):

There were two types of students: One type were essentially 'pre-algebraic' in their thinking, and treated every potential rule that they inferred from their numerical data in a table (which they generated) as a natural language-based rule. That is, they thought of $2x + 1$ as doubling and adding one. ... The latter [type] were looking for growth rates, which they interpreted as the first parameter's value, etc. For them, a linear function was experienced as a 'thing', a conceptual entity, whose identity is determined by the two parameters. The other students were looking for a way to translate from their natural language-based encoding of an unencapsulated process to algebra. (p. 90)

In contrast to the combined historical and psychological roots of Sfard's process-object model, Dubinsky (1991) has pointed out that his version of the process-object model has its basis primarily in Piaget's theory of Reflective Abstraction. But he has emphasized that this theory is only one of the bases of his 'genetic decomposition' version of the process-object model:

The details [of the model]... come from three sources. First, there is the psychological data that we have gathered through observations of students in the midst of trying to learn these concepts. ... This data, along with the ideas of Piaget formed the basis for the derivation of our theory, which is the second source of the genetic decompositions. That is, for each phenomenon that was observed, we tried to use our theory to describe it, adjusting the theory when necessary. (As the necessity for adjustment occurs less often, our confidence in the theory increases.) The third source of the descriptions is our mathematical understanding of the concepts in question. It seems important that a genetic decomposition should make sense from a mathematical point of view, although it might not be exactly how the mathematician might have analyzed the subject in thinking about how to teach it. (p. 110)

In addition, Dubinsky has insisted that, 'some of the statements we make are based on observations of students and others are only suppositions, derived as a preliminary to observations, from the general theory and our knowledge of the mathematics' (p. 104). This comment draws our attention to an aspect of model building that is sometimes used to distinguish theoretical work from empirical research, that is, that model construction and/or modification/extension almost always involves partial supposition. However, even the 'straightforward' analysis of empirical data can be said to involve a certain amount of conjectural interpretation. As Richards (1979) has pointed out, no data analysis is theory-free; but, in our field, as we have seen from the several examples provided, the converse is also the case: No theorizing is data-free.

4. CONCLUDING REMARKS

I have shown how one particular model has evolved in the field of mathematics education research, as well as the relation of the development of this model with empirical results. I have claimed that models such as the ones I have illustrated ought to be included under the heading, *Results of Research*, because they have been both developed and refined on the basis of empirical work. There is no escaping the fact that, in mathematics education, theory building and empirical studies form the vicious circle of research; each requires the other. Neither is truly possible without the other in a field such as ours. Margaret Brown (this volume) has argued that that is the case for all types of research, even large-scale assessment studies:

> Data alone do not help us to interpret and understand the situation, although they may stimulate us to attempt to do so. The results of SIMS attainment tests alone do not shed much light on the nature of and the reasons for the difference between English pupils and, say, Belgian pupils, although they suggest some hypotheses. One of the reasons for the lack of insight from SIMS test results alone was the relative lack of theorization in their design and analysis (p. 264).

James Hiebert reminded us at the conference that gave rise to this volume that 'we do not yet have theories in mathematics education that unify the field and generate specific hypotheses'. It is not even clear that such grand theories are realistic goals for mathematics education. Gerald Goldin (1992), in exploring what should be involved in developing a broader, unified psychological model of mathematical learning and problem solving, discusses many of the diverse, yet pertinent, empirical results that have been obtained. But taking into account all of these results could lead to a model that would be so complex as to be unusable or so general as to be of limited value. Robert B. Davis (1990) has pointed out one of the pitfalls of attempting to produce a theory that would be widely applicable: 'Unfortunately, that which is true for most people (or even true for *all* people) is nearly always trivial' (p. PI.14).

But on the basis of our empirical work, we in the mathematics education research community are succeeding at developing non-trivial, locally applicable, theoretical models. A search of current research journals in our field, as well as the proceedings of research conferences, attests to the growing number of researchers whose *empirical research* and *development of related theoretical models* are closely interconnected. Nevertheless, models are often the product of several years of work by dedicated researchers in a given research program. Thus, the perspective that model development is a *result* of research is indeed a long-term view of what constitutes research results.

ACKNOWLEDGMENTS

My thanks to Anna Sierpinska for her helpful comments regarding an earlier version of this chapter, and to Anna Sfard for our many conversations on process-object models.

REFERENCES

Bartolini Bussi, M. G.: 1994, 'Theoretical and Empirical Approaches to Classroom Interaction', in R. Biehler, R. W. Scholz, R. Sträßer & B. Winkelmann (eds.), *Didactics of Mathematics as a Scientific Discipline*, Kluwer Academic Publishers, Dordrecht, 121–132.

Behr, M. J., Wachsmuth, I. & Post, T. R.: 1985, 'Construct a Sum: A Measure of Children's Understanding of Fraction Size', *Journal for Research in Mathematics Education* 16, 120–131.

Bergeron, J. C. & Herscovics, N.: 1989, 'A Model to Describe the Construction of Mathematical Concepts from an Epistemological Perspective', in L. Pereira-Mendoza & M. Quigley (eds.), *Proceedings of the 1989 Annual Meeting of the Canadian Mathematics Education Study Group*, Memorial University of Newfoundland, St. John's, 99–114.

Brousseau, G.: 1972 (February), 'Processus de Mathématisation', *Bulletin de l'association des professeurs de mathématiques de l'enseignement public*, 282, 57–84.

Brousseau, G.: 1986, *Théorisation des Phénomenes d'Enseignement des Mathématiques*. Unpublished postdoctoral dissertation, University of Bordeaux, France.

Davis, R. B.: 1975, 'Cognitive Processes Involved in Solving Simple Algebraic Equations', *Journal of Children's Mathematical Behavior* 1 (3), 7–35.

Davis, R. B.: 1984, *Learning Mathematics: The Cognitive Science Approach to Mathematics Education*, Routledge, London.

Davis, R. B.: 1990, 'The Knowledge of Cats: Epistemological Foundations of Mathematics Education', in G. Booker, P. Cobb & T. N. de Mendicuti (eds.), *Proceedings of the 14th International Conference for the Psychology of Mathematics Education*, PME Program Committee, Oaxtepec, Mexico, PI.1–PI.24.

Davis, R. B.: 1992, 'Understanding "Understanding"', *Journal of Mathematical Behavior* 11, 225–241.

Douady, R.: 1985, 'The Interplay between Different Settings: Tool–Object Dialectic in the Extension of Mathematical Ability – Examples from Elementary School Teaching', in L. Streefland (ed.), *Proceedings of the Ninth International Conference for the Psychology of Mathematics Education*, State University of Utrecht, The Netherlands, Vol. 2, 33–52.

Dreyfus, T.: 1990, 'Advanced Mathematical Thinking', in P. Nesher & J. Kilpatrick (eds.), *Mathematics and Cognition: A Research Synthesis by the International Group for the Psychology of Mathematics Education*, Cambridge University Press, Cambridge, 113–134.

Dubinsky, E.: 1991, 'Reflective Abstraction in Advanced Mathematical Thinking', in D. Tall (ed.), *Advanced Mathematical Thinking*, Kluwer Academic Publishers, Dordrecht, 95–123.

Filloy, E.: 1990, 'PME Algebra Research. A Working Perspective', in G. Booker, P. Cobb & T. N. de Mendicuti (eds.), *Proceedings of the 14th International Conference for the Psychology of Mathematics Education*, PME Program Committee, Oaxtepec, Mexico, Vol. I, PII.1–PII.33.

Fischbein, E.: 1994, 'The Interaction Between the Formal, the Algorithmic, and the Intuitive Components in a Mathematical Activity', in R. Biehler, R. W. Scholz, R. Sträßer & B. Winkelmann (eds.), *Didactics of Mathematics as a Scientific Discipline*, Kluwer Academic Publishers, Dordrecht, 231–245.

Fischbein, E., Deri, M., Nello, M. S. & Marino, M. S.: 1985, 'The Role of Implicit Models in Solving Problems in Multiplication and Division', *Journal for Research in Mathematics Education* 16, 3–17.

Fischbein, E., Tirosh, D., Stavy, R. & Oster, A.: 1990, 'The Autonomy of Mental Models', *For the Learning of Mathematics* 10 (1), 23–29.

Goldin, G. A.: 1992, 'On Developing a Unified Model for the Psychology of Mathematical Learning and Problem Solving', in W. Geeslin & K. Graham (eds.), *Proceedings of the 16th International Conference for the Psychology of Mathematics Education*, PME Program Committee, Durham, NH, Vol. III, 235–261.

Gray, E. M. & Tall, D. O.: 1994, 'Duality, Ambiguity, and Flexibility: A "Proceptual" View of Simple Arithmetic', *Journal for Research in Mathematics Education* 25, 116–140.

Harel, G.: 1985, *Teaching Linear Algebra in High School*. Unpublished doctoral dissertation, Ben-Gurion University of the Negev, Beer-Sheva, Israel.

Harel, G. & Kaput, J.: 1991, 'The Role of Conceptual Entities and their Symbols in Building Advanced Mathematical Concepts', in D. Tall (ed.), *Advanced Mathematical Thinking*, Kluwer Academic Publishers, Dordrecht, 82–94.

Hart, K. (ed.): 1981, *Children's Understanding of Mathematics 11–16*, John Murray, London.

Herscovics, N. & Bergeron, J. C.: 1983, 'Models of Understanding', *Zentralblatt für Didaktik der Mathematik* 15, 75–83.

Kaput, J. J.: 1991, 'Notations and Representations as Mediators of Constructive Processes', in E. von Glasersfeld (ed.), *Radical Constructivism in Mathematics Education*, Kluwer Academic Publishers, Dordrecht, 53–74.

Lesh, R. & Kelly A. E.: 1994, 'Action-Theoretic and Phenomenological Approaches to Research in Mathematics Education: Studies of Continually Developing Experts', in R. Biehler, R. W. Scholz, R. Sträßer, & B. Winkelmann (eds.), *Didactics of Mathematics as a Scientific Discipline*, Kluwer, Dordrecht, 277–286.

Pirie, S. E. B. & Kieren, T. E.: 1992, 'Watching Sandy's Understanding Grow', *Journal of Mathematical Behavior* 11, 243–257.

Richards, J.: 1979, 'Modeling and Theorizing in Mathematics Education', in K. C. Fuson & W. E. Geeslin (eds.), *Explorations in the Modeling of the Learning of Mathematics*, ERIC Clearinghouse for Science, Mathematics, and Environmental Education, Ohio State University, Columbus, 5–26.

Rouse, W. G. & Morris, N. M.: 1985, *On Looking into the Black Box: Prospects and Limits in the Search for Mental Models* (Research report), Georgia Institute of Technology and Systems Engineering (ERIC Document Reproduction Service No ED 268 131)

Sfard, A.: 1987, 'Two Conceptions of Mathematical Notions: Operational and Structural', in J. C. Bergeron, N. Herscovics & C. Kieran (eds.), *Proceedings of the 11th International Conference for the Psychology of Mathematics Education*, Université de Montréal, Canada, Vol. III, 162–169.

Sfard, A.: 1989, 'Transition from Operational to Structural Conception: The Notion of Function Revisited', in G. Vergnaud, J. Rogalski & M. Artigue (eds.), *Proceedings of the 13th International Conference for the Psychology of Mathematics Education*, G. R. Didactique, CNRS, Paris, Vol. 3, 151–158.

Sfard, A.: 1991, 'On the Dual Nature of Mathematical Conceptions: Reflections on Processes and Objects as Different Sides of the Same Coin', *Educational Studies in Mathematics* 22, 1–36.

Soloway, E., Lochhead, J. & Clement, J.: 1982, 'Does Computer Programming Enhance Problem Solving Ability? Some Positive Evidence on Algebra Word Problems', in R. J. Seidel, R. E. Anderson & B. Hunter (eds.), *Computer Literacy*, Academic Press, New York, 171–185.

Steffe, L. P.: 1983, 'The Teaching Experiment Methodology in a Constructivist Research Program', in M. Zweng, T. Green, J. Kilpatrick, H. Pollak & M. Suydam (eds.), *Proceedings of the Fourth International Congress on Mathematical Education*, Birkhäuser, Boston, 469–471.

Steffe, L. P., Richards, J. & von Glasersfeld, E.: 1979, 'Experimental Models for the Child's Acquisition of Counting and of Addition and Subtraction', in K. C. Fuson & W. E. Geeslin (eds.), *Explorations in the Modeling of the Learning of Mathematics*, ERIC Clearinghouse for Science, Mathematics, and Environmental Education, Ohio State University, Columbus, Ohio, 27–44.

Thompson, P. W.: 1985, 'Experience, Problem Solving, and Learning Mathematics: Considerations in Developing Mathematics Curricula', in E. A. Silver (ed.), *Teaching and Learning Mathematical Problem Solving: Multiple Research Perspectives*, Lawrence Erlbaum, Hillsdale, NJ, 189–236.

Carolyn Kieran
Département de Mathématiques,
Université du Québec à Montréal,
Montréal, QC H3C 3P8
Canada

GÉRARD VERGNAUD

TOWARDS A COGNITIVE THEORY OF PRACTICE

As I view it, the topic of the relations between practice and theory in mathematics education conveys two main problems. The first problem concerns the relationship between mathematics as a practical activity, and mathematics as a theoretical body of knowledge. The second problem concerns mathematics education, and the relationship between the teacher's activity in the classroom and research in mathematics education and psychology. I think that, for both problems, we need a theory of practice and of articulation between theory and practice. This theory must be quite general since it does not concern mathematics or mathematics education alone, but all human activities. To do this, we also need some practice of theory, or theorizing. And we need examples, since theory without examples is usually incomprehensible.

Let me start with two examples outside mathematics (I will come later to more specific examples in mathematics education). My first example is the satellite expert's. The satellite expert has books on his shelves, reports in his drawers, and fifteen or twenty years' experience in satellite conception. No satellite is exactly the same as the former ones; therefore one has to reinvent something each time. There is no individual able to conceive, alone, of a satellite. Satellite conception is the task of a whole community of people; sometimes several hundreds. And yet, in this community, one individual may appear, on some occasion, as irreplaceable. For instance, one may have to solve, in 1996, a technical problem that is similar to a problem that had been dealt with six or seven years ago. But the person who did it then has now retired. He or she had formed a specific competence which is not in books and reports. This competence may be technical. It may also have social components since the solution may have involved a whole network of human and technical resources. Finally this person will be called back from his or her retirement place, if it is possible. There is no expert-system to replace this person; there is not even any younger engineer that would have been trained in time for the purpose of inheriting and capitalizing on that critical knowledge.

Analyses of people's activity at work show that we develop many personal competences. Some of them are critical, in the sense that the functional entity in which they take place (workshop, laboratory, or else) cannot survive without them; also in the sense that they make a difference between those who have formed such competences, and those who have not.

My second example concerns the swineherd who guides pigs with a stick at the entrance of a slaughterhouse: His job is low-qualified and the director of the

227

Sierpinska, A. and Kilpatrick, J. Mathematics Education as a Research Domain: A Search for Identity, 227–240.
© *1998 Kluwer Academic Publishers. Printed in Great Britain.*

slaughterhouse sees it as a very simple and unimportant job which does not require any ergonomic analysis. Yet a group of ergonomists, who had to do some expertise in this factory, insisted upon observing him and analyzing his activity. They discovered that he was likely to be the most important person in that part of the factory. The pigs may have developed heart disease; they may have strong emotions and even die before being killed. When this happens the loss of money is important, but even when they are only emotionally stressed, the quality of the meat is low. This man's competence was critical and had been formed through experience: he was able to guide the pigs as smoothly as possible, detect those of them likely to die, and get them to the 'sacrifice place' before it was too late. He had great diagnostic competences and had categories for this: these categories were cognitive, if not theoretical, and very efficient.

There are many other examples. For instance, in a cement factory, one man was able to repair water-pumps. When he fell ill, nobody was able to repair these pumps for three months.

Competences formed in practice concern all levels of children's development and professional experience; at all levels of qualification. This is true even for scientists. Most of our knowledge consists of competences. We have thousands of them. They are organized in hierarchical systems and develop through life. These competences are not made of a-conceptual procedures as some psychologists pretend them to be when they contrast procedural knowledge with conceptual knowledge, or instrumental understanding with relational understanding, or analogical processes with propositional processes. On the contrary, these competences rely heavily upon efficient concepts-in-action and theorems-in-action; that is to say, upon relevant categories from which to select the information available, deal with it, and generate from it plausible goals, subgoals, actions and expectations. The concept of scheme is essential to the understanding of the cognitive structure of competences. This concept is not very well known, even though Piaget introduced it sixty years ago – borrowing it from Kant. It has occasioned many misunderstandings due to the confusion with other ideas, like those of schema, script, pattern, frame, scenario, and others. Therefore I will define and analyze the concept of scheme as precisely as possible.

1. THE CONCEPT OF SCHEME

Originally, Piaget (1980) used it to analyze what he called – like most psychologists of that period – 'sensory-motor activity'. He should have called it 'perceptual and gestural activity', since its main characteristic is its organization of perceptions and gestures, which is not simply sensorial and motoric but strongly structured. The main role of the schemes, according to Piaget, is to generalize efficient behavior to new objects. For Piaget, schemes are *functional dynamic totalities*. We need a more precise and more general definition. I propose the following one:

Definition:

Scheme: Invariant organization of behavior for a certain class of situations.

This definition is very useful in mathematics education since it covers a wider range of behaviors than the concept of algorithm, for example, which is also an invariant organization of behavior for a certain class of situations. Both concepts are closely connected. But algorithms are schemes, whereas not all schemes are algorithms. Moreover, algorithms are *effective* (they provide the solution in a finite number of steps, if a solution exists), whereas most schemes are only *efficient* (and are only *likely* to lead to success). They may even be wrong. Even when students and adults are supposed to learn and use algorithms, they develop personal schemes: for calculations, for geometry, for algebra, for reasoning. The invariant character of a scheme for an individual does not mean that schemes are stereotypes. What is invariant is the organization of behavior, not the behavior itself.

What does this organization consist of?

- Goals, subgoals and expectations.
- Rules of action: They can be considered as the generative part of schemes, the part that generates behavior as a function of some situation variables.
- Operational invariants: They consist mainly in concepts-in-action (to categorize and select information), and theorems-in-action (to infer, from the available and relevant information, appropriate goals and rules).
- Possibilities of inference: These possibilities are essential since there is always some inference and computation in any activity.

Definition:

Theorems-in-action are propositions held to be true by the subject when he or she acts.

Counting a set of objects involves important concepts-in-action like those of one-to-one correspondence and cardinality. It may also involve additive or multiplicative decompositions, commutativity, distributivity, etc. Solving ordinary arithmetic problems involves important and non-trivial theorems-in-action, as I have demonstrated in my contributions on additive and multiplicative structures (Vergnaud 1983). Even algebra, which is supposed to trace operations and rules most explicitly, relies upon strong non-explicit theorems-in-action. They concern, for instance, conservation of equality, conservation of solution, equivalence, etc.

The concept of scheme is very general:

- We have schemes to catch something with the hand, sit down, walk, jump and dance, fix a nail or use a square rule, etc.

- We have schemes to speak: phonological ones (quite different for English and French); syntactico-lexical ones to produce understandable sentences; and also schemes to dialogue, argue, get into a conflict and cooperate.
- We have schemes to organize sequences of physical or intellectual operations in order to face some task, e.g. estimating, buying, cutting and laying linoleum in a room, or analyzing a problem in physics and translating it into a system of equations.

The best way to understand the process of thinking is to view it as a gesture, that is to say, as an activity that takes place over some period of time and follows some organization; with rules to modify external behavior and internal activity as a function of specific circumstances. To understand the relationship between practice and theory, we need to analyze representation as an intricate combination of both a set of competences and a set of conceptions.

2. CONCEPTIONS AND COMPETENCES

What are conceptions composed of? They are composed of objects, properties, relationships, transformations and processes. There are objects at quite different levels: ordinary physical objects, sets, numbers, functions, graphs, groups, differential equations, etc. Part of mathematics consists of producing sentences, true or likely to be true, about these objects. Mathematics is composed of propositions and texts; so too are physics, chemistry and technology. This is the predicative analysis of scientific knowledge.

What are competences composed of? They are composed of schemes aimed at facing situations: They are not made of texts. Schemes are the operational side of knowledge.

If one does not want to fall into a dualistic and schizophrenic view of cognition, one must analyze the representational ingredients (objects, properties, and propositions) contained in schemes, so as to make the connection between science as a practical and rational activity, and science as a theoretical and textual enterprise. The concept of operational invariant (concepts-in-action and theorems-in-action) is the keystone that makes the connection between practice and theory. Why are operational invariants so crucial? Because practice is action; because action is always efficient under certain conditions; because action is driven by rules of the form 'if C_1, C_2, ... then A_k, ... , A_j'; and finally, because the possibility for such rules to emerge would not be understandable if there were no cognitive categories to analyze these conditions, to analyze the components of action, and to analyze the relationships between goals, conditions and actions. For instance, the efficacy of algorithms is due to the bonds of necessity that connect the properties of the relevant variables in the set of situations with the properties of the operations.

The part of conceptualization is therefore essential in practice, even though it may not be explicit. Most researchers' theoretical views today, both in psychology and in mathematics education, minimize the role of conceptualization. But conceptualization is crucial in the development of human competences. This is true, not only for mathematical competences but for all human competences: professional, social, linguistic, physical.

The fact that this process of conceptualization is not directly observable, and has to be reconstructed from observed behaviors, makes it a challenge for research, as is also the fact that schemes are the psychological and subjective counterparts of situations. The analysis of schemes makes it necessary to analyze and classify situations carefully so as to understand what is essential to be conceptualized in them; even if this conceptualization remains totally or partially implicit.

These ideas are not specifically mine. They can be traced back to the work of several other authors, like Piaget (1971, 1980) of course, but also Vygotsky (1962). Vygotsky's stress on the role of language in teaching and learning makes one sometimes forget that he was also interested in the formation of concepts in daily life, through experience. He even mentioned, in his paper on everyday and scientific concepts, that some concepts are 'unconscious'.

Having stressed the role of situations and schemes in the formation of knowledge, I must also stress the idea that the status of knowledge changes when it is made explicit, put into words and symbols, and communicated to others. Explicit mathematics does not have the same cognitive status as implicit mathematics contained in schemes. It is only explicit knowledge that can be discussed, argued, proved or disproved. Therefore, in the teaching of mathematics, one should find not only interesting and challenging situations for action but also situations for formulation and proof, as Brousseau has so convincingly shown in his work (see, for example, Brousseau 1981).

Operational invariants are the source of concepts and propositions, but their scope of availability and validity is usually very limited and local, whereas scientific concepts have a much wider scope and are organized in integrative systems. Science is systemic or it is not science. Our intuitive grasp of situations in ordinary life is not usually systemic.

For instance, primary school children can grasp some aspects of negative and positive numbers, and some aspects of the linear function:

- a win of 4 and a loss of 6 is a loss of 2
- more of something makes it more expensive; twice as much makes it twice as expensive.

But the integration of directed numbers into a system which makes it possible to solve any equation $a + x = b$, whatever the value of a and b may be, and whatever a, b and x may represent (transformation, state, part, whole, abscissa or difference...) gives directed numbers a much more general cognitive status. Similarly, the concepts of function and variable, the writing of such linear properties as $f(nx) = nf(x)$ and $f(x) = ax$, or the plotting of different numerical values on a graph changes the cognitive status of the covariation of two magnitudes, which students can grasp a long time before they can be taught functions explicitly.

It is important that we sort out what can be used from the work of such great authors as Piaget and Vygotsky and what cannot be used, and what may even be an obstacle to new and fruitful ideas. A science which is not cumulative is not a science, but just a [passing] fad. This is certainly a problem today for cognitive psychology and for mathematics education. From my point of view, one should pay more attention to Piaget and Vygotsky, even though their theories can be criticized on many points. We need a more precise mathematical description of the knowledge involved in new competences.

3. CONCEPTS-AND-THEOREMS-IN-ACTION

Let us take the example of the developmental move from the counting-all procedure (1, 2, 3, 4, 5 for the first set, then 1, 2, 3 for the second set, then 1, 2, 3, 4, 5, 6, 7, 8 for the union), to the counting-on procedure (count from the cardinal of the first set, as many digits as there are elements in the second set: 5..., 6, 7, 8) or to the number fact (5 + 3 = 8). A crucial question is the following :

'What is the implicit mathematical knowledge in that move that characterizes the critical move from the counting-all procedure to the counting-on procedure?'

Let me word it as a theorem of equivalence :

'It is equivalent to count each set first and add their cardinals afterwards, or to make the union set first and state its cardinality afterwards':
$$card(A \cup B) = card(A) + card(B) \text{ provided } A \cap B = \varnothing.$$

This is one of the axioms of the theory of measure. Five- or six-year-olds are, of course, unable to put such a homomorphic property in words, but they show in action that they understand it for small numbers.

Similarly, when they have to calculate 3 + 6, young children usually count from the first set as many steps as there are elements in the second set (3...4, 5, 6, 7, 8, 9), with the help of their fingers, eventually. Then they move to a new procedure: counting-on-from-the-larger-set (6...7, 8, 9). This commutativity-in-action emerges from the need and the possibility to control more easily the double counting (7, 8, 9 like 1, 2, 3 instead of 4, 5, 6, 7, 8, 9 like 1, 2, 3, 4, 5, 6). But this need would have no effect if there were no recognition of the equivalence: 3 + 6 = 6 + 3. This is a typical theorem-in-action; its scope of validity is, of course, very small.

The concept of scheme covers not only the overt behavior, but also the goals, operational invariants and rules that underlie implicit choices and decisions. Consider, for example, the following problem:

Suppose a student starts by dividing 972,000 by 120 (this procedure is frequently observed). There are two different schemes in it, illustrated in Figure 1.
The first one generates the choice to divide 972,000 by 120, while the second
consists of the division algorithm.

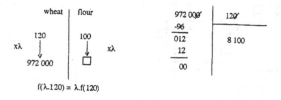

Figure 1

*How much flour can one make with 972,000 kg of wheat, if 120 kg of wheat
give 100 kg of flour?*

The conceptual contents of each scheme are very different.

The first one has to do with proportion: The student tries to calculate the
scalar ratio between 120 kg and 972,000 kg of wheat to apply it later to the flour,
which is another kind of magnitude. It is impossible to understand this choice if
the student does not understand and believe that this ratio is conserved when one
transforms wheat into flour.

The second scheme has to do with the polynomial decomposition of numbers
and the base-ten system of numeration. The rules are complex but have nothing
to do with the proportionality between the wheat and the flour any longer.

A proof of the difference between these two schemes is evident from the fact
that the difficulties met by students can be quite different. In the first case they
may multiply 972,000 by 100, or divide 120 by 100 and then be unable to go
any further. If one replaces the data 972,000 by 96, many students think it is impossible to divide 96 by 120 because 'one cannot divide a smaller number by a
larger one'. In the second scheme, students' difficulties occur due to a wrong understanding of numeration, and to the presence of zeros in the dividend, or the
divider, or the quotient. There are epistemological obstacles likely to interfere
with both schemes, but the schemes are different.

This example also shows that procedures which are taught, and thereby supposed to be algorithms, often degenerate in personal schemes: most students,
and even most mathematicians, do not follow the rules of the algorithm strictly.
They have shortcuts – for better or for worse!

It is worth analyzing, in some detail, the knowledge that underlies human activity. In mathematics, this analysis must be made in mathematical terms, even
though the mathematics underlying schemes may be totally implicit, or even
wrong.

My further examples are taken from geometry (Figure 2). The first task (the
'fortress') can be offered to 9-year-old students. The scheme which is necessary
to meet this particular task has a certain range of operationality; for all drawings

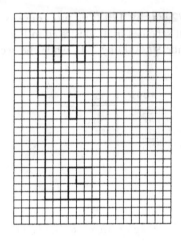

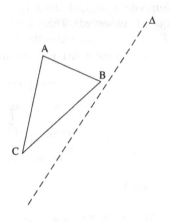

Draw the symmetrical figure   Draw the symmetrical figure

Figure 2

that follow the lines of the squared sheet. In that sense it is general. But students do not have to use the concept of angle, because there are only right angles, nor do they have to use complex instruments: a ruler will suffice. All they have to do is conserve or inverse orientations, and count square-sides, or half-sides.

The second task is much more complex and can hardly be proposed to students before secondary school. The conservation of angles and lengths is more problematic; new instruments are needed; the diagonal orientation of the axis may be misleading; if the axis crosses the triangle it is even worse. The scheme required is much more sophisticated both from a gestural point of view, and from a mathematical point of view, although it obviously belongs to the same conceptual field as the scheme needed for the fortress.

4. LANGUAGE AND SYMBOLS

Suppose now that teachers require students to put into words what they see, or what they have done, using such words as: 'angle, length, triangle, symmetrical, symmetry, conservation, isometry'. Here are four possible sentences:

1. The fortress is symmetrical.
2. Triangle A'B'C' is symmetrical to triangle ABC in relation to line Δ.
3. Symmetry conserves lengths and angles.
4. Symmetry is an isometry.

One can easily imagine that these sentences are not equally difficult. Understanding the first one is easy, even if it may be difficult for young students to produce it. But the main point to be stressed is the change of status of the concept of symmetry throughout these four sentences.

In sentence 1, symmetrical is a one-place predicate and the argument is the whole figure, whereas in sentence 2, it is a three-place predicate with three arguments: A′B′C′, ABC, Δ. The move from sentences 1 and 2 to sentence 3 is even more important since symmetry becomes an object which has its own properties and relationships. The linguistic operation of substantiation is essential in this transformation of a predicate into an object. In sentence 4, isometry becomes an object. The singular for symmetry and isometry actually covers the whole class of symmetries and isometries, and sentence 4 expresses the inclusion relationship of one class in the other. The singular for triangle ABC, or line Δ in sentence 2 is a true singular.

The production and the understanding of such sentences requires schemes which are both linguistic and conceptual. We have schemes to produce sentences for mathematics, as well as for other domains of human activity: we have schemes to discuss and argue, dialogue with others, give lectures, or write texts. Schemes have physical, linguistic and social components. Their main characteristic is their operationality: they operate on situations and deal with them in order to overcome the difficulties, and organize progress in the managing of these situations. When our schemes fail, we develop cognitive activities to accommodate them to the properties of the situations that may have caused trouble. Even emotion is made of schemes. Emotional schemes are often seen as negative, but they may also be very positive and drive us to new ways of doing and representing.

Algebra is a semiotic system that covers a small part of the objects and relationships that one might be willing to represent. It is not a powerful way to communicate with other people. Therefore it is not, properly speaking, a language. Its importance in mathematics comes from its computable character and from the fact that equations usually represent just the necessary and sufficient characteristics of the problem we want to solve. Too little work has been devoted to the 'put-into-equation' task in the learning of algebra. Although algebra is not a language, the put-into-equation task also raises problems of expression: not only does the writing of equations request the identification of complex relationships and functions, but also the way these relationships are worded in natural language contaminates the use of algebraic symbols.

Da Falcao Rochas (1992) provides us with good examples of such contamination.

The students (16-year-olds) had to deal with different problems concerning friends working part-time at different travel agencies. Their salaries were composed of three parts: a part proportional to the number of hours worked, a part proportional to the number of tickets sold, and a constant part. All three parameters (salary for one hour, salary for one ticket, constant part) could vary from one agency to the next, but were held constant for each travel agency.

Different formulas could be elaborated, depending on the unknown, for example :

$$S = Hh + Bb + C$$

     S salary
     H number of hours worked

$$H = \frac{S - (Bb + C)}{h}$$

     h salary per hour
     B number of tickets sold

$$B = \frac{S - (Hh + C)}{b}$$

     b salary per ticket
     C constant part

Different tasks could be proposed:

- referring to one agency only
- referring to two given agencies
- referring to one or several unknown agencies; i.e., the parameters become unknowns
- a system of several equations may become necessary.

Here are two symbolic expressions, which are wrong from an algebraic point of view, but perfectly intelligible for the students who produced them.

    case 1 $S = H + B + constant\ part$

This expression is a mixture of algebraic symbols and natural language. Moreover H does not hold for the number of hours worked but for the whole part, Hh; B also holds for the whole part, Bb.

    case 2 $(xH \times 1H) + (xB \times 1B) + constant\ part = S$

Here, the first parenthesis holds for: unknown number of hours (xH) multiplied by the salary per hour $(1H)$. The second one for: unknown number of tickets (xB) multiplied by the salary per ticket $(1B)$.

This writing is totally wrong, from an algebraic point of view. The first H holds for the type of unit (hour) and the unknown number of hours is represented by x; the expression $1H$ holds for an amount of money. The reading is the same for $xB \times 1B$.

In spite of this ambiguity the student was able to produce correct calculations for the salary, knowing all the relevant information. When he had to find the number of tickets sold, he moved to a different writing:

$$\frac{S - (xh \times 1H + C)}{b} = B$$

One can notice that he now uses different types of letters for the unit 'hour' (h) and the salary per hour (H), and even destroys his strange and awkward syntax for the tickets: 'B' now represents the number of tickets and 'b' the salary per ticket. This way of writing equations is totally opportunistic and idiosyncratic; it is nevertheless transparent for the student who uses it.

The production of mathematical expressions has therefore something in common with the production of natural language, and can be strongly contaminated by it.

5. A THEORY OF REPRESENTATION

Logical positivism has played a very perverse part in science, especially in the philosophy of mathematics. Knowledge is not a language, and mathematics cannot be reduced to a system of symbols, whatever the role of symbols may be in mathematics. Not only this: one should also get rid of the simplistic view of the relationship between reality, representation and symbolism, as it is commonly conveyed by the triangle (Figure 3).[1]

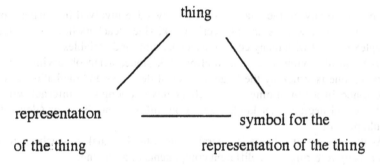

Figure 3

First point: Reality is not only made of physical objects, but also of predicates of different complexity (properties and several-argument relationships). Objects can exist at very different levels of abstraction: a vector space is an object for the mathematician. *Second point*: Reality is also made of situations (i.e. problems to be dealt with).

Therefore we need to look at reality as both a set of situations of different complexity and a set of objects at different levels. This dualistic analysis of the reference of mathematical knowledge is made necessary by the fact that situations and schemes are essential in the development of knowledge.

When using natural language or symbolic representation like algebra, graphs, tables and diagrams, we have to do with conventional systems that do not reflect exactly what we think. Therefore, one cannot just identify the signified, as it is conveyed by the linguistic or symbolic signifiers, with the operational invariants that we develop or have developed in dealing with situations. It is a nominalistic deviation to do so. I have summarized this analysis in Figure 4.

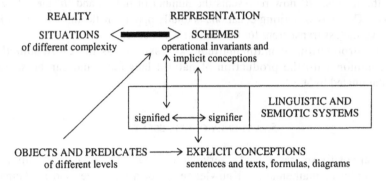

Figure 4

6. TEACHERS AS MEDIATORS

Let me come now to the analysis of the knowledge involved in teachers' practice. One can view it as an engineer's work, since teachers have to manage a complex process with many conditions, constraints and variables.

One can also view it as a mediation task since, when observing teachers working, one is struck by the large number of decisions and mediation acts that take place in a few moments. Teachers have schemes to interact with the students and decide what kind of question, information, help, or silence they should provide.

As the concept of scheme is central to me, I tend to analyze teachers' mediation acts by referring to the different components of a scheme.

According to Vygotsky, there are two different components of mediation. The first one is the others-component: Mediation consists in the help brought by somebody else (parent, teacher or peer). The second one is the linguistic-semiotic component: Mediation consists in the part played in thinking and learning by symbols of all kinds. Since teachers talk a lot to students, one tends to confuse both ideas. This is wrong: Teacher's mediation schemes cannot be reduced to dialogues.

The first important mediation act consists in offering situations and activities aimed at helping students to learn. But once a situation has been offered to the students, the teacher has several other tasks:

- he or she may have to clarify the goal to be reached, or the subgoals, or the expectations;
- he or she may have to show how to do something, or at least scaffold and monitor the different steps to face the situation. This kind of help is usually different for different students (which makes life difficult for teachers);

● he or she may have to help a student by selecting and identifying the relevant pieces of information; also by inferring, from that information, what subgoal to reach, what events to expect, what sequencing of operations and controls to generate.

Not only do teachers talk a lot to students, they also smile, make gestures of all kinds, especially with their face and their hands. It is sometimes sufficient help for a student when the teacher points with his finger at a certain piece of physical or written information.

Figure 5 symbolizes different mediation acts of teachers in relation to the theoretical analysis of a scheme I have provided earlier. One should study the teacher's behavior in the classroom more than one does today, since mediation is probably his or her most important job.

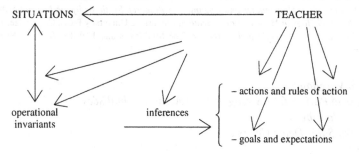

Figure 5

Of course, words are important, as are symbolic representations. However we must not forget that verbal communication conveys much less information than would be relevant. Teachers always leave an important burden to students: to reconstruct, from what is said or expressed, the meaning to the task. Action schemes are very selective; communication schemes even more so.

Teachers make their own decisions all the time. It is unavoidable. And there is no engineering that could offer students the possibility to learn alone. The best way to help teachers improve their practice is to offer them better representations of mathematical knowledge, of children, of learning and development, of action and language. Teachers improve their representations mainly through experience; this is also unavoidable. But training, research, theory and examples necessarily help teachers in interpreting their own experience.

NOTE

1. The 'triangle' referred to here by Vergnaud is not representing exactly the same idea as the so called 'epistemological triangle', proposed in 1923 by C. K. Ogden & I. A. Richards (1946, *The Meaning of Meaning*, Harcourt, Brace and Company, New York) and adapted in mathematics education research for analyzing the contents of mathematics classes by, mainly, H. Steinbring (see his chapter in this volume) (Editors' note).

REFERENCES

Brousseau, G.: 1981, 'Problèmes de Didactique des Décimaux', *Recherches en Didactique des Mathématiques* **2** (1), 37–128.

Da Rocha Falcao, J.: 1992, Représentation du Problème, Ecriture de Formules et Guidage dans le Passage de l'Arithmétique à l'Algèbre. Thèse, Université Paris V.

Piaget, J.: 1971, *Genetic Epistemology*, Norton, New York.

Piaget, J.: 1980, *Adaptation and Intelligence*, University of Chicago Press, Chicago.

Vergnaud, G.: 1981, *L'Enfant, la Mathématique et la Réalité*, Peter Lang, Berne.

Vergnaud, G.: 1982, 'A Classification of Cognitive Tasks and Operations of Thought Involved in Addition and Subtraction Problems', in T. P. Carpenter, J. M. Moser, & T. A. Romberg (eds.), *Addition and Subtraction: A Cognitive Perspective*, Lawrence Erlbaum Associates, Hillsdale, New Jersey.

Vergnaud, G.: 1983, 'Multiplicative Structures', in R. Lesh and M. Landau (eds.), *Acquisition of Mathematics Concepts and Processes*, Academic Press, New York, 128–203.

Vergnaud G.: 1990, 'Epistemology and Psychology of Mathematics Education', in J. Kilpatrick & P. Nesher (eds.), *Mathematics and Cognition, A Research Synthesis by the International Group for the Psychology of Mathematics Education*, Cambridge University Press, Cambridge, 14–30.

Vygotsky, L.: 1962, *Thought and Language*, The MIT Press and J. Wiley & Sons, New York, London.

Gérard Vergnaud
Université Paris 8 CNRS, Cognition et Activités Finalisées
2, rue de la Liberté
F-93 526 Saint-Denis Cedex 2
France

New ICMI Studies Series

Volume 4

Published under the auspices of The International Commission on Mathematical Instruction under the general editorship of

Miguel de Guzmán, President   Mogens Niss, Secretary

The titles published in this series are listed at the end of Book 2.

Mathematics Education as a Research Domain: A Search for Identity

An ICMI Study

Book 2

Edited by:

Anna Sierpinska

Concordia University, Montreal, Quebec, Canada

and

Jeremy Kilpatrick

University of Georgia, Athens, Georgia, USA

Kluwer Academic Publishers

Dordrecht / Boston / London

Library of Congress Cataloging-in-Publication Data

Mathematics education as a research domain : a search for identity :
    an ICMI study / edited by Jeremy Kilpatrick and Anna Sierpinska.
        p.   cm. — (New ICMI studies series ; v. 4)
    Includes index.
    ISBN 0-7923-4599-1 (hardcover : alk. paper)
    1. Mathematics—Study and teaching—Research. I. Kilpatrick,
Jeremy. II. Sierpinska, Anna. III. Series.
QA11.M3757   1997
510'.71—dc21   97–20240

ISBN (set) 0-7923-4599-1 (hardback)   ISBN (Book 2) 0-7923-4946-6 (hardback)
ISBN (set) 0-7923-4600-9 (paperback)   ISBN (Book 2) 0-7923-4948-2 (paperback)

Published by Kluwer Academic Publishers,
P.O. Box 17, 3300 AA Dordrecht, The Netherlands.

Sold and distributed in the U.S.A. and Canada
by Kluwer Academic Publishers,
101 Philip Drive, Norwell, MA 02061, U.S.A.

In all other countries, sold and distributed
by Kluwer Academic Publishers,
P.O. Box 322, 3300 AH Dordrecht, The Netherlands.

Printed on acid-free paper

Printed in Great Britain

TABLE OF CONTENTS

BOOK 1

Part VI: Mathematics Education and Mathematics

PART IV

DIFFERENT RESEARCH PARADIGMS IN
MATHEMATICS EDUCATION

FERDINANDO ARZARELLO & MARIA G. BARTOLINI BUSSI

ITALIAN TRENDS IN RESEARCH IN MATHEMATICAL EDUCATION: A NATIONAL CASE STUDY FROM AN INTERNATIONAL PERSPECTIVE

> Researchers in didactics of mathematics
> run the risk of being like spiders which
> produce shining but brittle webs;
> or like ants,
> which accumulate blindly grains for winter.
> Instead they must be like bees
> which produce honey.
> (*adapted from Francis Bacon*)

Every discussion about research in mathematics education (referred to as 'RME' from now on) that is carried on at the international level emphasizes contrasting and even competing approaches: the increasing number of monographs that review national contributions to RME in a self-contained way (e.g. Barra et al. 1992; Blum et al. 1992; Douady & Mercier 1992; Kieran & Dawson 1992) seems to emphasize the existence of different research traditions that are developed locally with their own store of epistemological debates, institutional constraints, research questions, methods, results and criteria, leading even to the birth of paradigms (e.g. the French *didactique des mathématiques*). Despite this increasing volume of information at the international level, the discussion of research questions related to local projects seems to be difficult (see Silver 1994). Two different yet related levels of problems are involved:

1. *communication*: How is it possible to convey to the international professional community of researchers in mathematics education meaningful information about the context of a local research project?
2. *relevance*: (a) What are the individual aspects of context-bound research that are supposed to be relevant to and influential for the international community, and (b) What are the general aspects of international research that are supposed to be relevant to and influential for any local research project?

This paper aims to address both problems by starting from an analysis of the Italian situation. In the first part of the paper, the authors present some elements of a national case study, in order to communicate information about the roots and the present state of the core of Italian research in mathematics education. It

Sierpinska, A. and Kilpatrick, J. Mathematics Education as a Research Domain: A Search for Identity, 241–262.
© *1998 Kluwer Academic Publishers. Printed in Great Britain.*

is not a detailed presentation of the whole history (for which, see Bazzini & Steiner 1989; Barra et al. 1992; Malara & Rico 1994), but a retrospective reconstruction of the research scene in Italy since the middle of the sixties, which seeks to identify the reasons that make research in Italy different from research in other countries. As with every reconstruction, also this one suffers from the bias of the authors' scientific and professional interests and from the focus chosen for this presentation. In particular, the authors point to the roles played by the internal roots of Italian research arising from its traditions and by the external influences emanating from the increasingly frequent contacts between Italian researchers and researchers from other countries. In the second part of the paper, the authors reflect on the significance of the Italian experience for the international professional community of researchers in mathematics education; they find it in its contribution to the emergence of an original paradigm that has *research for innovation* at its core.

1. ELEMENTS FOR A NATIONAL CASE STUDY

The First Period of the Italian RME

For the purpose of this paper, we have distinguished two periods, the first dating from the mid-sixties up to the mid-eighties and the second concerning the last decade. We shall analyze these periods by means of a conceptual structure quite similar to that introduced by Bishop (1992), who distinguished three different research traditions: the Pedagogue tradition, the Empirical Scientist tradition and the Scholastic Philosopher tradition. The two conceptual structures have been generated independently (Arzarello 1992): the fact of their similarity can suggest that the hope, expressed by Bishop, that such a structure can help understanding some of the developments in different countries, is not unjustified.

Two internal trends: concept-based didactics and innovation in the classroom

In the first period, a large amount of research was produced in the universities by professional mathematicians who followed the tradition of the far past (e.g. Veronese, Cremona, Peano, Enriques, Vailati, quoted in Barra et al. 1992). They aimed at producing ideas for improving mathematics teaching in 'generic' situations and focused on the logical organization of concepts, as considered from inside mathematics. Specific problems were determined by cultural issues (such as, *what is really important about the concept of limit?*). Disciplinary contents were elaborated in a form which was correct from a mathematical point of view and with a strict adherence to the language of professional mathematicians. Due to exclusive emphasis on the disciplinary contents, the problems of the didactic transposition (Chevallard 1980), i.e. the relationships between the knowledge to be taught and scientific knowledge that legitimizes it, were not explicitly addressed. The didactic actions in the classroom were designed according to conceptual difficulties and complexities; possible psychological difficulties were not ignored but rather reduced to their mathematical counterparts. When classroom experiments were carried out, a considerable didactic ingenuity was used to make up for the little if any reference to research on learning processes. Consequently, when the same experiments were analyzed, practice was explained in terms of concepts. Thus attention was focused on products (e.g. students' performances and attitudes) rather than on the processes of mathematics teaching and learning.

The best example of such an analysis is the so-called Syllabus produced in the late seventies on behalf of the Unione Matematica Italiana (UMI 1980), that is the official association of Italian mathematicians, where the main mathematical abilities and knowledge required of students at the end of high school (at least for those intending to start a scientific course in university) were described in a conceptual hierarchy, which put forward the possible main difficulties and pitfalls of students by means of challenging questions and intriguing examples.

Even if different positions can be detected by more detailed analysis, many contributions of professional mathematicians in this first period can be clustered within the trend, whose main features have been described above and which will be referred to in the sequel as *concept-based didactics* or, briefly, as trend A.

However, in the same period from the sixties, a movement for innovation was generated in schools, with the strong involvement of teachers and teacher associations. This movement aimed at producing paradigmatic examples of improvement in mathematics teaching in 'specific' situations and focused on concrete problems of mathematics education. Specific problems were determined by practical issues (such as *how to teach area in 6th grade?*). As these problems were embedded in a wider environment and were analyzed also from a sociological point of view, the analysis required a language distinguished from and more complex than the disciplinary one. The didactic actions in the classroom were designed pragmatically. When classroom experiments were carried out, an explicit recourse to the professional competence of the teacher was often claimed to change the design according to the classroom needs. Analysis of classroom experiments was usually rather limited, evidence being the fact that the experiment had worked in the specific school situation where it had taken place. However, even if only at a descriptive level, in this movement for innovation, the processes of teaching and learning did enjoy some attention.

As a meaningful example, we quote a paper by Emma Castelnuovo (1965), whose work deeply influenced much subsequent research in Italy (Barra et al. 1992). After a detailed description of activities about the concept of area and surface with 11–12 years old pupils, she concluded:

'If one wishes for the teaching of intuitive geometry to have constructive and hence formative features the recourse to both object and action is necessary; this is our conclusion and, we hope, also the reader's conclusion. Object and action must not follow a predetermined design but must always be inspired by the actual needs of the classroom: the teacher must be sufficiently sensitive to grasp them. The examples given have arisen exactly from such needs. The concrete materials to realize experiences are not so important: they can be models, devices, experiences, made up by means of real stuff or purely imagined; they can be lights or shadows. And perhaps it is this very freedom in designing and interpreting, which both the teacher and the pupils have, is one of the main features of the constructive method' (Castelnuovo 1965).

Even if the roots of the innovation movement were in schools, some professional mathematicians were also engaged in the activity. Moreover, their meetings with groups of motivated teachers gave rise to the first Nuclei di Ricerca Didattica (Educational Research Teams) which played a very important role in the next period. Also in this case, a detailed analysis would suggest different positions: however, the increasing number of papers related to the innovation movement can be clustered around a trend that will be referred to as *innovation in the classroom* or, briefly, as trend B.

Up to now, we have suggested a clustering of research contributions around two trends: the concept based didactics and innovation in the classroom, according to the main aims and foci of research. We are well aware that, in some cases, classifying the contribution of an author into one trend or another may be difficult and even misleading, because of the integration of different trends. Moreover both types of produced results (e.g. conceptual networks for the former; patterns of experiments for the latter) can be considered as *energizers of practice* (Balacheff et al., 1993), still to be elaborated before being used in the classroom.

However, some major differences are evident as far as the image of research and the influence of research on the educational system are concerned. In concept based didactics, a top-down model is assumed: the starting point is conceptual analysis by means of which practice is determined and described. In innovation in the classroom, action research is assumed: the starting point is the specific teaching problem as it is perceived by the teachers, while practice is determined by the concrete condition of action in the classroom. In the former case, the influence on the educational system is mainly played at the level of intended curriculum (e.g. school programs and textbooks, Barra et al. 1992), while in the latter it is played at the level of implemented curriculum (concrete realizations in the classroom). In both cases dissemination is based on an optimistic faith in teachers. In the former, the teachers, provided with better pre- and in-service education are supposed to be able to realize change (hence many efforts in improving teacher training); in the latter, the teachers, provided with good opportunities for collaborative work, are supposed to be able to pass professional competence on to each other.

In other countries (e.g. in France, see Rouchier 1994; in Germany, see Griesel & Steiner 1992) in spite of the presence of similar trends, the pressure of academic institutionalization of didactics as a scientific discipline has led to a complete separation between forms of theoretical research (closer to concept based didactics) and forms of action research (closer to innovation in the classroom), to the extent that even communication between them appears not to be easy.

In Italy, a different pathway was followed, namely the progressive integration of both trends, under pressure to support effective innovation in school, in a delicate period when the programs in most school grades were being fundamentally changed after decades of 'wait and see' policy (Barra et al. 1992; Malara & Rico 1994). Embryos of this attitude, realized in a very pragmatic way, have been present since the early seventies. To quote but a couple of examples, the textbooks written by professional mathematicians for high school (namely F. Speranza, G. Prodi, E. Magenes, L. Lombardo Radice, V. Villani and others, see Barra et al. 1992, p. 35) are the products of cooperation between academics and groups of motivated teachers. At primary school level, an emblematic example is represented by the RICME Project (Pellerey 1980), where an explicit model of action-research is stated. The importance of the cooperation between professional mathematicians and school teachers (with all the implications that will be stated in the following sections) can hardly be overestimated and is crucial to understand all the subsequent developments in Italian research (further details can be found in Boero 1994).

An external trend: laboratory observation of processes

In the late seventies, several Nuclei were working in the departments of mathematics at various universities on projects for curricular innovation, with the financial support of the CNR (National Council for Research). They consisted of mixed groups of professional mathematicians and school teachers at all levels, from primary to high school (Prodi 1992; Barra et al. 1992; Malara & Rico 1994).

In the progressive development of research a need grew up around the necessity to make more precise the didactic dimension of the projects in order to understand, for instance, why some innovations would work in some classrooms and would not in others. This necessity was similar to that expressed by Douady (1988) in her review of the IOWO Project (Treffers 1987). But the tools for meeting such a need were not available in the Italian tradition of RME. Actually, cooperation between researchers in didactics of mathematics and researchers in education was not very widespread in Italy, due to an inheritance of the idealistic positions of the whole Italian culture and school (see a discussion in Barra et al. 1992). This fact is confirmed by the scarce presence of Italian researchers in the international forum of mathematics educators (for example, with a few exceptions, the scientific presence of Italian researchers in PME dates back to only 1988).

Some crucial events made new tools and methods of research in mathematical education available to Italian scholars. In fact, both the Trento Conferences (1980, 1983, 1984, 1991), where many prominent international researchers were invited to lecture, and the XXXIII-CIEAEM Conference in Pallanza (Pellerey 1981) put Italian researchers in contact with different research traditions and mainly with the methods of what we call *laboratory observation of processes*, which will be illustrated later (we shall refer to it briefly as trend C). By trend C we mean the constellation of research projects which aim to increase knowledge of classroom processes (mainly of a short and middle term nature), which can be used to design and implement classroom experiments. Because of the focus on knowledge, the specific problems were usually inspired by research issues. Most of PME research reports clustered around this trend. Practice was designed, interpreted and/or modeled by looking at cognitive difficulties, patterns of interaction and so on. Instruments were borrowed from psychology, sociology, pedagogy and applied to experiments (in real laboratories or in classrooms), concerning specific mathematical concepts to be taught or processes to be enhanced. The types of results produced (e.g. taxonomies, patterns of interaction) can be conceived as *economizers* and *demolishers of illusion* (Balacheff et al. 1993). The impact on the educational system was not addressed as a main issue, as the focus was on observation of classroom processes. As this research claimed to be basic in the sense of classical natural science, it left the problems of wider application to others.

The introduction to this research tradition and the awareness that it could meet the need of making the didactic dimension of individual projects more precise caused a deep restructuring of activity amongst Italian researchers, which coincided with their increasing participation in international conferences. Looking at the national development from an international perspective makes it clear that the three trends are the local mirror images of more general trends in the international community (see the distinction between the *Pedagogue tradition* – our trend B, the *Empirical Scientist tradition* – our trend C, and the *Scholastic Philosopher tradition* – our trend A, introduced by Bishop 1992).

The Second Period of the Italian RME

The result of the reflective process in the Italian community described above was stated officially in the Eighth Session of the National Seminar in Didactics of Mathematics (*What is Research in Mathematics Education?* Pisa 1991; Arzarello 1992; Boero 1992; Malara 1992; Prodi 1992). It was pointed out that there had been a progressive and deep mixing of the trend B (innovation in the classroom) with the essential elements of the other two trends. A new trend, which concerns innovation as research and not only as action in the classroom (henceforth called *research for innovation* or trend D) is now becoming consolidated through the efforts of most Italian researchers to such an extent that we can say that it now constitutes the core of Italian RME.

This trend is still in progress and so a complete definition of its features is not yet possible, but a short tentative description of its main elements may be given.

The object of research is the teaching–learning of mathematics, concerning either specific classroom situations or throughout the wider educational system. The main aim of research is twofold:

(a) to produce paradigmatic examples of improvement in mathematics teaching (in the form of projects for curricular innovations concerning either the whole mathematics curriculum or some special parts of it);

(b) to study the conditions for their realization as well as the possible factors underlying their ineffectiveness.

While the former is present also in the trend A, the latter is related to the study of classroom processes as in trends B and C. The didactic actions in the classroom are designed according to cultural and epistemological choices (trend A), pedagogical reasons (trend B) and conceptual (trend A) or cognitive (trend C) difficulties with attention also to the quality of interaction (trend C). The influence on the educational system is played at different levels:

(i) the gradual expansion of projects for curricular innovation into larger and larger groups of teachers;
(ii) the recourse in teacher training to the progressively produced knowledge about complex classroom processes, in order to increase teachers' awareness of their own work in the classroom (Malara & Rico 1994).

The specific problems of research are manifold: a typical problem is to produce knowledge about the relationships between proposals for innovation (grounded in epistemological, cultural and cognitive hypotheses) and their implementation in the classroom (analyzed by means of different instruments) (Boero 1994). A strong experimental component (teaching experiments in the classroom) is present, an inheritance of trends B and C. Because of the involvement of teachers in all the phases of research, *participant observation* (Eisenhart 1988) is the most common way of collecting and interpreting data from the classroom.

Typically the products of such a research are projects for curricular innovation concerning either the whole mathematics curriculum or some special part of it (e.g. concerning the methods or the contents), or examples of careful modeling of classroom processes (concerning, for example, the teacher's role). However, such products, unlike those of trend B, are usually given within a theoretically-based frame (with local validity as we shall argue in the following), which itself has been originated by the same research and is rooted in its methodology. Hence the same objects and data are not only practical products produced in some precise environment, but are also basic results insofar as the variables themselves which feature the given environment are made explicit while doing the research. Thus basic and practical features sustain each other and are obtained because of the 'mixed' approach to didactic problems including the deep cooperation between teachers and researchers, both as designers and as participant observers. Thus, theory and practice are generated together, neither one preceding the other during the research itself, except for a subsequent careful analysis of the teaching–learning processes and of their effects. This point is crucial, insofar as it distinguishes this type of research from those where the project is designed according to a previously articulated theoretical framework and the methodology of research separates the roles of the (detached) observer and the observed.

Another feature of this trend of research is that no global coherent theoretical framework is assumed, because of its complexity and the number of interrelated variables to take into account. Rather, instruments are borrowed from various theoretical approaches or produced inside, and applied as elements of a kit of tools. Local coherence of the framework is necessary, but global coherence is considered impossible or at least irrelevant. We shall elaborate further on this point later, discussing the notion of complementarity.

Briefly, the core of Italian RME is trying to overcome the distinction between theoretical and pragmatic relevance (addressed by Sierpinska 1993) by means of developing their mutual relationships from the very beginning. This attempt is made initially in a very pragmatic way, by means of the joint work of researchers and teachers, as in other countries (see Kilpatrick 1992). Yet, it becomes clearer and clearer over time that a very strong epistemological choice is involved in it. Renouncing the classical distinction between observed and observer (represented in educational research respectively by the classroom, including a teacher, and a researcher) means a shift from the classical positivistic view of natural sciences that has also influenced research in education.

In fact, observation of classroom processes is a relevant common issue of the trends B, C and D. However a deep difference exists. We can interpret the three positions according to a model proposed by Raeithel (1990) to describe three modes of relationships between actor and observer in the inquiring activity. Trend B corresponds to the naïve problem solver who considers the meaning of objects (classroom processes, in our case) to be inherent to them and who is not able to build a symbolic structure inseparable from perceived reality. Trend C corresponds to the detached observer who aims at understanding the flow of activity by means of modeling the process to cope with its complexity – the development of classical science is a good example of this attitude. Trend D corresponds to the participant observer, who develops a split between observing and observed subjects in a dialogical relation. By accepting participant observation, Italian researchers have been coping with the paradoxes of actor and observer pointed out by Brousseau (1986).

Consequently, in every feature of trend D we find the presence of two types of contrasting issues: the first are more empirical, pragmatic, concrete and specific, whilst the second are more speculative, theoretical, abstract and general. They correspond to two different historical periods and to the two different poles of the dialectic observed/observer, in the process of teaching–learning. That is, the general issues are the most recent while the more specific ones are older and rooted in trend B (sometimes in trend A). The former are rooted in the reflections made by the observer on the processes lived by the observed.

Hence, the big problem today is to elaborate tools adequate for analyzing such a twofold nature, namely grasping deeply the connections between the (practical) products and the (basic) results of the research.

We believe that existing theories are not enough to grasp properly such new tendencies with their consequent products and results. In fact, the description of the core of Italian RME given insofar does not elicit something very important: the dynamism between the components. In the following section we shall try to fill this gap and to elaborate a dynamic analysis of trend D.

2. BEYOND THE NATIONAL CASE STUDY

A Dynamic Analysis: Adequacy Tests

The contradictions in trend D concern mainly the relationships between the components inherited from the other trends (trend A: concept based didactics; trend B: innovation in the classroom; trend C: laboratory observation of processes). The Italian core of RME must be described as a living process, not only as a list of chosen categories. In this sense we need what might be called *second-order properties*. Terminology is borrowed from logic and denotes relationships of relations. As far as we know, in the context of mathematics education it has been used in CSMS papers (Hart 1981) to indicate the grasping of inter-functional dependence between variables, i.e. when a pupil understands a rule from a table. Here we mean the necessity to look at mutual dynamic relationships among the trends we have described: RME, not only in Italy, can be described properly only by considering such second-order properties in a systemic and dialectic way. (The main features of trends A, B, C, D are compared and summarized in a table at the end of the chapter.)

Trend D is an example of a second-order approach to the other trends. We can say that many recent innovative research projects in Italy are looking for such second-order properties. Even if single researchers do not use this terminology, many of them are concerned in research involving specific second-order issues.

Minimality test

The analysis of recent tendencies and changes in the core of Italian RME makes us ready to make a challenging claim; namely, we can state two *adequacy tests for the research in mathematical education.*

TEST 1 (the minimality test): *In a research study in mathematics education at least two of the three components that correspond to trends A, B and C are considered. Relevance requires that all three of the components be considered.*

Some comments are necessary to avoid misunderstandings. The above components are based on an analysis of the prevailing situation in Italy. Many Italian research studies (which, of course, does not mean all) are in fact respecting such a test; hence the test, thus formulated, applies only to Italy. However, we propose it as a test having possibly a more general validity, provided one introduces different components. We believe that much of RME nowadays, explicitly or implicitly, is working in order to integrate, at the second order, the main components that characterize research in the different specific histories of its development (for such a tendency see the analysis of research on problem solving instruction in Lester & Charles (1992) and the characterization of recent developments in research in mathematics education in Bishop (1992).

Let us sketch some examples.

The first examples have a general character and are taken from the paper of Boero and Szendrei (this volume). They point out, for instance, the necessity of considering not only quantitative aspects but also 'qualitative information about the consequences of some methodological or content innovation'. This is a typical way to look for second-order properties. Moreover, the authors' theoretical perspective focuses on 'what is the relationship in the classroom between "teacher", "pupils" and "mathematical knowledge"; what happens in the relationship between school mathematics and mathematicians' mathematics; the relationship between research results and classroom practice in mathematical education'. These are all second-order issues.

The second example is taken from French research. Perrin-Glorian (1994) reconstructs the developments of research activity in France, referring to the 'integration of external contributions', such as: the didactic transposition, the tool–object dialectic, and, more recently, the concepts introduced by Chevallard (1994) in his anthropological approach, and into the theory of didactic situations, originally developed by Guy Brousseau (1986). This last has for a long time been identified as the core of the French RME. However the recourse to concepts elaborated in different frameworks and originating from different initial problems (e.g. the gap between scholarly knowledge and knowledge to be taught, as studied in the theory of didactic transposition; and the shift from knowledge to be taught to taught knowledge in the classroom, as studied in the theory of situations) makes it necessary to consider second-order relationships between them.

Second-order approaches are also suggested by Anglo-Saxon researchers. See, for an example, the analysis of Cobb (1994) on interaction and learning in mathematics classroom situations where the initial problem is 'that of clarifying the relationship between students' situated conceptual capabilities and their small group interaction'. This is a typical second-order problem.

Dynamic inter-functionality test

In a research study in mathematics education relevant to this analysis, the components considered are not combined together as in a puzzle, but rather the relationship between them has a systemic character which can be described as *dynamic inter-functionality*. We have borrowed this idea from the discussion in chapter one of *Thinking and Speech* (Vygotskij 1990): while discussing the development of the study of thinking and speech, the author distances himself from classical psychology, which has decomposed the study into separate elements, that of thinking and speech. He claims that there is a need to carry out an analysis that decomposes the unitary set into *component units*, meaning 'the products of the analysis that, even in the difference of elements, have the fundamental properties of the set and that are living parts (no more decomposable) of the above global unit'. The metaphorical example he suggests is the following: it makes no sense to study separately hydrogen and oxygen in order to study water since they do not have the properties of water; rather it is necessary to study molecules and molecular movement. Applying this idea to RME, one can say that it makes no sense to study separately conceptual hierarchies in mathematics and social interaction in the mathematics classroom in order to analyze the process of teaching–learning mathematics.

The above discussion can be summed up by means of a second test. Suppose that a research study has passed Test 1, then at least two components are taken into account.

TEST 2 (dynamic inter-functionality): *In a research study in mathematics education the dynamic inter-functionality between the components is respected, i.e. analysis is focused on the mutual relationships between the components rather than on the components themselves.*

Such an approach in RME is a further elaboration of the notion of *complementarity,* as discussed by Steiner (1985), in order to avoid its reduction to a phenomenological level. In his paper, Steiner himself proposes activity theory, that has been developed from the seminal work of Vygotskij, as a source of conceptual tools to cope with deep inter-functionality.

From this point of view a research study is *deeper* – the more it enters into these problems – and *wider* – the more it involves the different components in a global and inter-functional way.

Many researchers nowadays in Italy agree that doing RME means, above all, studying processes of teaching–learning mathematics in all their complexity as dynamical systems. This means that they are analyzed (or at least the purpose is to analyze them) as processes where different components live together. Even if this new trend is clear enough and generally accepted by most of Italian researchers, the concrete results are still provisional, at least as far as methodology is concerned.

Two Examples

To give a more specific idea of research inspired by these new issues, which can possibly pass both tests, we sketch two typical examples. These come from the authors' personal research and appear deeply interconnected so that it is possible to identify a common 'second-order' development. More cases and details are reported in the paper by Boero (1994), while the booklet by Barra et al. (1992) has a more official and complete list of Italian papers.

Example 1: Perspective drawing and mathematical discussion
(M. G. Bartolini Bussi)

The basic idea is to design, implement and analyze long-term teaching experiments (grades 3–5 or 6–8) about the representation of the visible world by means of perspective drawing, visualized at that level, alongside the systematic introduction of both individual problems and mathematical discussions orchestrated by the teacher. In this project, the historical and epistemological analysis of the birth of perspective (trend A) is strictly intertwined with an analysis of the function of geometry in school programs that determines the choices concerning the teaching–learning process (trend B).

Explicit assumptions about the teaching–learning process are made with reference to Vygotskij (1990) and most of the further work is based on the development of Vygotskian hypotheses, concerning mainly internalization and semiotic mediation. The study of the relationships between short-term and long-term processes as well as between the personal constructions of the pupils and existing mathematical knowledge (trend B) is made by means of concepts borrowed from activity theory (Leont'ev 1978); namely, the distinction between activity, actions and operations and between meaning and personal sense.

The modeling of the teacher's role in the discussion (trend C), as well as of the subsequent processes in the classroom, is one of the main issues addressed by this research. The analysis of the teaching experiment as a paradigmatic example of the component unit type is in progress (see Bartolini Bussi in press a, in press b).

Example 2: The problem of reflexive interrelation between syntax and
semantics (F. Arzarello, G. P. Chiappini, L. Bazzini)

The main item in this research is the interplay between concepts, notations and mediators in the processes of teaching–learning mathematics. The question which Arzarello and his colleagues are now studying can be summarized as follows : 'What are the main features of component units where the pupil is supposed to grasp a rule in a system of notations compatible with those of wider society?' (Cobb et al. 1992; Bakhurst 1988; Geertz 1983; Kaput 1991 for terminology).

The problem entails three components:

(i)   an epistemological analysis of mathematical concepts evoked by notations (trend A);
(ii) an analysis of the cognitive pitfalls (trend C) that are related to the so-called Wittgenstein's paradox of the understanding of a rule (Kripke 1982; Arzarello et al. 1994);
(iii) an analysis of the social mediation in the classroom (trend B), which is deemed essential in order to overcome the pitfalls of trend C and to regen-

erate a consensual mathematical reality – which constrains pupils' ways of thinking and notating (Cobb et al. 1992, p. 116).

The ongoing research by Arzarello, Bazzini, and Chiappini attempts to integrate Vygotskij's and Leont'ev's activity theory with an approach to the theory of meaning based on intentional semantics (see Arzarello et al. 1993, 1994, 1995) and this seems to point to a new kind of component unit.

Some Consequences

The first consequence of the approach to RME represented by the above examples, and characterized by the above tests, concerns the units of analysis of every research study aimed to pass both tests. One of the strategic problems encountered here is the choice among innumerable possibilities suggested by concrete teaching problems in school (e.g. sharing the features of trends A, B and C) and paradigmatic examples to be developed in depth and analyzed from a second-order point of view (trend D). Paradigmatic examples are to be considered as component units of the global unit, that is the activity of teaching and learning mathematics in school. In this sense they have to retain the fundamental properties of the activity: hence, they cannot be too small and they cannot avoid complexity. The determination of such units as well as their further analysis (and, later, the statement of the criteria for determining them) seems to be, now, one of the major problems of the Italian RME.

The long-term development of paradigmatic examples has a number of implications itself. The considerable influence that the teaching experiment is supposed to have on students (to the extent of determining, in some cases, all the curriculum) implies the need for a careful analysis, before implementation, to avoid, as far as possible, the undesired effects. Besides, the critical role played by the teachers in all the phases of the research study requires a long-term training of the teachers before, and during the implementation of the experiment. Hence, even if specific issues of teacher training are not theoretically addressed in the research study, they are always in the background as a strong pragmatic component. These two elements are present in the development of every teaching experiment, even if they may not always be addressed in the reports – in order to meet the space and time constraints of either international journals or international conferences.

The second consequence concerns the conceptual instruments to be used in analysis. Long-term processes cannot be analyzed by means of short-term processes and macrodidactic studies are not the juxtaposition of a series of microdidactic studies. Yet most of the literature in mathematics education (a relevant exception being that of the anthropological theory of Chevallard 1994) consists of microdidactic studies where only short-term processes are analyzed in detail. So the development of conceptual instruments for long-term analysis is urgently needed by the Italian community of researchers.

The relationship between paradigmatic cases and conceptual instruments is assumed to be dialectic. On the one hand, the production of useful scientific conceptual instruments in RME, with at least a local validity, is dependent on the study of paradigmatic cases. On the other hand, the choice of paradigmatic cases is presupposed to be driven by previous choices about what to design, what to observe, etc. Whenever a theoretical framework is assumed, it can at least partially determine the choices. But, in our case, no theoretical framework as yet exists (and perhaps cannot exist globally).

The point is to elaborate a (second-order) epistemological frame, concerning the didactics of mathematics, that copes with all or most of the components typical of trend D. To do this, we need specific observation instruments (for example, for long-term processes) and concepts (for example, a more careful description of component units and their relationship with global units). Some attempts are now being made in Italy, by means of concepts borrowed from activity theory (Leont'ev 1978), in order to relate, on the one hand, the study of long- and short-term processes and, on the other, the analysis of personal constructions made by learners and the historical collective development of mathematics. The two examples quoted above are representative of this effort.

3. CONCLUSION

In this chapter we have shown how the Italian community of researchers in mathematics education has developed a specific approach, *research for innovation*, from the basis of previous trends internal to Italy, namely *concept-based didactics* and *innovation in the classroom*, alongside the external influence of research into *classroom observations of processes*. As a result, research for innovation had to face *second-order problems* and had to develop second-order instruments for analysis. The paradigmatic examples that are being developed in recent years by many Italian researchers are studied as dynamical systems of teaching–learning, using vestiges of all three components. However, the systemic approach seems to be shared by others in the literature of mathematics education, even if the components to be considered are different from those considered in the Italian literature, because of the strong inheritance of the pragmatic component of innovation in the classroom. Thus the international community can be offered contributions of two kinds:

(a) paradigmatic examples of research for innovation analyzed as component units of the global unit of the activity of teaching–learning mathematics in school;
(b) conceptual instruments to cope with an analysis of paradigmatic examples taking into account the dynamic inter-functionality between the components.

This seems to answer a need expressed some years ago by Freudenthal (1981, p. 135), in a different language. One of the aims of RME is to undertake a research that is neither empty nor blind but produces useful theories for concrete mathematical education, following the motto, inspired by Francis Bacon, that we have put at the beginning of our paper.

ACKNOWLEDGMENTS

In working on this chapter, the authors had the financial support of CNR (Consiglio Nazionale delle Ricerche) and MURST (Ministero dell'Università e della Ricerca Scientifica e Tecnologica). Hints for this contribution came from different persons, whom we wish to thank here. The teachers of our research groups gave a substantial contribution to the development of our ideas; several Italian colleagues took part in fruitful discussions during the national meetings (especially we recall the National Seminars in Pisa on the Research in the Mathematics Education, November 1991, and on the Learning of Geometry, November 1993 – both organized with funds from CNR and MURST). Special thanks are due to Michele Pellerey, for continuous discussions and suggestions on the theoretical and methodological issues and, last but not least, to Giovanni Prodi, one of the most prominent fathers of the current trend of the Italian RME; he gave all of us concrete and magnificent examples of RME, by working for many years with pupils and teachers of high schools in their classrooms and, as a professional mathematician and member of official institutions, he did everything in his power to create the cultural and material conditions for the development of this new subject of inquiry. Without his efforts and encouragement we could not even write a report on Italian RME today. Hence we wish to dedicate to him our joint paper on the occasion of his 70th birthday.

REFERENCES

Arzarello, F.: 1992, 'La Ricerca in Didattica della Matematica', *L'Insegnamento della Matematica e delle Scienze Integrate* **15** (4), 345–356.

Arzarello, F., Bazzini, L. & Chiappini, G. P.: 1993, 'Cognitive Processes in Algebraic Thinking: Towards a Theoretical Framework', in I. Hirabayashi, N. Nohda, K. Shigematsu and F. L. Lin (eds.), *Proceedings of the Seventeenth International Conference for the Psychology of Mathematics Education 1, Tsukuba, Japan*, 138–145.

Arzarello, F., Bazzini, L. & Chiappini, G. P.: 1994, 'Intensional Semantics as a Tool to Analyze Algebraic Thinking', in F. Arzarello and E. Gallo (eds.), *Proceedings of the International Workshop on Algebraic Learning*, Turin, October 1992, (WALT), Rendiconti dell'Università e del Politecnico di Torino, **52** (1), 105–125.

Arzarello, F., Bazzini, L. & Chiappini, G. P.: 1995, 'The Construction of Algebraic Knowledge: Towards a Socio-Cultural Theory and Practice', in L. Meira L. and D. Carraher (eds.), *Proceedings of the Nineteenth International Conference for the Psychology of Mathematics Education 1, Recife, Brazil*, 119–134.

Bakhurst, D.: 1988, 'Activity, Consciousness and Communication', *Quarterly Newsletter of the Laboratory of Comparative Human Cognition* 10, 31–39.

Balacheff, N., Howson, A. G., Sfard, A., Steinbring, H., Kilpatrick, J. & Sierpinska, A.: 1993, 'What is Research in Mathematics Education, and What are its Results? Discussion Document for an ICMI Study', *Zentralblatt für Didaktik der Mathematik* 25 (3), 114–116.

Barra, M., Ferrari, M., Furinghetti, F., Malara, N. A. & Speranza, F. (eds.): 1992, *The Italian Research In Mathematics Education: Common Roots and Present Trends*, Progetto Strategico del CNR – Tecnologie e Innovazioni Didattiche 12.

Bartolini Bussi, M.: in press a, 'Coordination of Spatial Perspectives: An Illustrative Example of Internalisation of Strategies in Real Life Drawing', *Journal of Mathematical Behavior*.

Bartolini Bussi, M.: in press b, 'Mathematical Discussion and Perspective Drawing in Primary School', *Educational Studies in Mathematics*.

Bazzini, L. and Steiner, H. G. (eds.): 1989, *Proceedings of the First Italian–German Bilateral Symposium on Didactics of Mathematics* (Pavia 1988), Progetto Strategico del CNR – Tecnologie e Innovazioni Didattiche 5.

Bishop, A. J.: 1992, 'International Perspectives on Research in Mathematics Education', in D. A. Grouws (ed.), *Handbook of Research on Mathematics Teaching and Learning*, Macmillan Publishing Company, New York, 710–723.

Blum, W., Keitel, C., Schupp, H., Steiner, H. G., Sträßer, R. & Vollrath, H. J. (eds.): 1992, 'Mathematics Education in the Federal Republic of Germany', *Zentralblatt für Didaktik der Mathematik* 24 (7) (Special Issue).

Boero, P.: 1992, 'Sulla Specificita' delle Ricerche in Matematica: il Caso del Formalismo Algebrico', *L'Insegnamento della Matematica e delle Scienze Integrate* 15 (10), 963–986.

Boero, P.: 1994, 'Situations Didactiques et Problèmes d'Apprentissage: Convergences et Divergences dans les Perspectives de Recherche', in M. Artigue, R. Gras, C. Laborde, and P. Tavignot (eds.), *Vingt Ans de Didactique des Mathématiques en France: Hommage à Guy Brousseau et Gérard Vergnaud*, La Pensée Sauvage, Grenoble, 17–50.

Boero, P. & Radnai Szendrei, J.: this volume, 'Research and Results in Mathematics Education: Some Contradictory Aspects'.

Brousseau, G.: 1986, *Théorisation des Phénomenes d'Einsegnement des Mathématiques*, Thése d'Etat, Université de Bordeaux I.

Castelnuovo, E.: 1965, 'L'Oggetto e l'Azione nell'Insegnamento della Geometria Intuitiva', in *Il Materiale per l'Insegnamento della Matematica*, Firenze, La Nuova, Italia, 41–65.

Chevallard, Y.: 1980, *La Transposition Didactique*, La Pensée Sauvage éditions, Grenoble.

Chevallard, Y.: 1994, 'Nouveaux Objets, Nouveaux Problèmes en Didactique des Mathématiques', in M. Artigue, R. Gras, C. Laborde, and P. Tavignot (eds.), *Vingt ans de Didactique des Mathématiques en France: Hommage à Guy Brousseau et Gérard Vergnaud*, La Pensée Sauvage éditions, Grenoble, 313–320.

Cobb, P.: 1994, 'A Summary of Four Case Studies of Mathematics Learning and Small-Group Interaction', in J. P. Ponte and J. F. Matos (eds.), *Proceedings of the Eighteenth International Conference for the Psychology of Mathematics Education* 2, Lisbon, Portugal, 201–208.

Cobb, P., Yackel, E. & Wood, T.: 1992, 'Interaction and Learning in Mathematics Classroom Situations', *Educational Studies in Mathematics* 23 (1), 99–122.

Douady, R.: 1988, 'Book review: Treffers A.: "Three Dimensions. A Model of Goal and Theory Description in Mathematics Instruction. The Wiskobas Project"', *Educational Studies in Mathematics* 19 (3), 411–417.

Douady, R. & Mercier, A. (eds.): 1992, *Research in Didactique of Mathematics*, La Pensée Sauvage éditions, Grenoble.

Eisenhart, M. A.: 1988, 'The Ethnographic Research Tradition and the Mathematics Education Research', *Journal for Research in Mathematics Education* 19 (2), 99–114.

Freudenthal, H.: 1981, 'Major Problems of Mathematics Education', *Educational Studies in Mathematics* 12 (2), 133–150.

Geertz, C.: 1983, *Local Knowledge*, Basic Books, New York.

Griesel, H. & Steiner H. G.: 1992, 'The Organization of Didactics of Mathematics as a Professional Field', in W. Blum, C. Keitel, H. Schupp, H. G. Steiner, R. Sträßer and H. J. Vollrath (eds.), Mathematics Education in the Federal Republic of Germany, Zentralblatt für Didaktik der Mathematik 24 (7) (Special Issue), 287–295.

Hart, K.: 1981, Children's Understanding of Mathematics: 11–16, The CSM Mathematics Team, John Murray, London.

Kaput, J.: 1991, 'Notation and Representations as Mediators of Constructive Processes', in E. von Glasersfeld (ed.), Constructivism in Mathematics Education, Kluwer Academic Publishers, Dordrecht, 53–74.

Kieran, C. & Dawson, A. J.: 1992, Current Research on the Teaching and Learning of Mathematics in Canada, Canadian Mathematics Education Study Group, Université du Québec à Montreal, Canada.

Kilpatrick, J.: 1992, 'A History of Research in Mathematics Education', in D. A. Grouws (ed.), Handbook of Research on Mathematics Teaching and Learning, Macmillan Publishing Company, New York, 3–38.

Kripke, S., 1982, Wittgenstein on Rules and Private Language, Basil Blackwell, Oxford.

Leont'ev, A. N.: 1978, Activity, Consciousness and Personality, Prentice-Hall, Englewood Cliffs, NJ.

Lester, F. K. & Charles, I. R.: 1992, 'A Framework for Research on Problem-Solving Instruction', in J. P. Ponte, J. F. Matos, J. M. Matos and D. Fernandes (eds.), Mathematical Problem Solving and New Information Technologies. Research in Contexts of Practice, NATO ASI series, Springer-Verlag.

Malara, N. A.: 1992, 'Ricerca Didattica e Insegnamento', L'Insegnamento della Matematica e delle Scienze Integrate 15 (2), 107–136.

Malara, N. A. & Rico, L. (eds.): 1994, Proceedings of the First Italian–Spanish Research Simposium in Mathematics Education, Modena.

Pellerey, M.: 1980, 'Il Metodo della Ricerca-Azione di K. Lewin nei suoi più Recenti Sviluppi e Applicazioni', Orientamenti Pedagogici 27, 449–463.

Pellerey, M. (ed.): 1981, Processes of Geometrisation and Visualisation, Proceedings of the 33th CIEAEM's Meeting, Pallanza, Italy.

Perrin-Glorian, M. J.: 1994, 'Théorie des Situations Didactiques: Naissance, Développement, Perspectives', in M. Artigue, R. Gras, C. Laborde and P. Tavignot (eds.), Vingt ans de Didactique des Mathématiques en France: Hommage à Guy Brousseau et Gérard Vergnaud, La Pensée Sauvage, Grenoble, 97–147.

Prodi, G.: 1992, 'Ricerca in Didattica della Matematica', Notiziario dell'Unione Matematica Italiana 19 (1–2), 146–150.

Raeithel, A.: 1990, 'Production of Reality and Construction of Possibilities: Activity Theoretical Answers to the Challenge of Radical Constructivism', Multidisciplinary Newsletter for Activity Theory 5–6, 30–43.

Rouchier, A.: 1994, 'Naissance et Développement de la Didactique des Mathématiques', in M. Artigue, R. Gras, C. Laborde and P. Tavignot (eds.), Vingt ans de Didactique des Mathématiques en France: Hommage à Guy Brousseau et Gérard Vergnaud, La Pensée Sauvage, Grenoble, 148–160.

Sierpinska, A.: 1993, 'Criteria for Scientific Quality and Relevance in the Didactics of Mathematics', in G. Nissen & M. Blomhoj (eds.), Criteria for Scientific Quality and Relevance in the Didactics of Mathematics, Danish Research Council for the Humanities: The Initiative 'Mathematics Teaching and Democracy, Roskilde, 35–74.

Silver, E. A.: 1994 (May), 'Research Questions in Mathematics Education: Opportunities and Constraints within an International Community of Researchers', in What Is Research in Mathematics Education and What are its Results? (Background papers for an ICMI Study Conference), International Commission on Mathematical Instruction, University of Maryland, College Park, 52–68.

Steiner, H. G.:1985, 'Theory of Mathematics Education: An Introduction', For the Learning of Mathematics 5 (2), 11–17.

Treffers, A.: 1987, *Three Dimensions. A Model of Goal and Theory Description in Mathematics Instruction.* The Wiskobas Project, Kluwer Academic Publishers, Dordrecht.

Unione Matematica Italiana: 1980, Syllabus, *Notiziario dell'Unione Matematica Italiana* **7** (3), 5–16.

Vygotskij, L. S.: 1990, *Pensiero e Linguaggio*, edited and translated by L. Mecacci, Laterza, Bari (first ed. in Russian 1934).

Ferdinando Arzarello
Dipartimento di Matematica,
Università di Torino,
via Carlo Alberto 10,
I 10123 Torino, Italy

Maria G. Bartolini Bussi
Dipartimento di Matematica,
Università di Modena,
via Campi 213/B,
I 41100 Modena, Italy

APPENDIX

This table is constructed from the conceptual structure of the ICMI document (Balacheff et al. 1993). It summarizes the main features of the trends A, B, C and D discussed in the paper on the basis of the analysis of two periods of Italian research in mathematics education.

	First order trends			*Second order trend*
TREND	A CONCEPT BASED DIDACTICS	B INNOVATION IN THE CLASSROOM	C LABORATORY OBSERVATION OF PROCESSES	D RESEARCH FOR INNOVATION
WHAT Object	The teaching of Maths in 'generic' situations	The teaching of Maths in 'specific' situations	The teaching–learning of Maths in 'specific' situations	The teaching–learning of Maths
WHY Aims	To produce ideas for improving Maths teaching	To produce improvement in Maths teaching	To produce instruments for improving Maths teaching–learning	To produce improvement in Maths teaching and instruments for improving Maths teaching–learning

	First order trends			*Second order trend*
TREND	**A** CONCEPT BASED DIDACTICS	**B** INNOVATION IN THE CLASSROOM	**C** LABORATORY OBSERVATION OF PROCESSES	**D** RESEARCH FOR INNOVATION
Rationale	Practice is explained/designed in terms of concepts	Practice is explained/designed in terms of paradigm examples	Practice is explained/designed in terms of cognitive difficulties or patterns of interaction	Practice is explained/designed in terms of concepts and cognitive difficulties and patterns of interaction studied in paradigmatic cases
HOW Choices	Determined by cultural issues	Determined by practical issues	Determined by research issues	Determined by cultural practical and research issues
Specific problems	Products of Maths teaching (performances and attitudes)	Innovation in the classroom (short and long term processes)	Models of teaching–learning (short and middle term processes)	Innovation in the classroom and products of Maths teaching (performances and attitudes) and models of teaching and learning (short and long term processes)
Localization of research	At one's desk	Everyday life in the classroom	Specially mounted experiments	At one's desk and specially mounted experiments (teaching experiments) in the classroom '
Methodologies	Pragmatic	Pragmatic experimental Action-research	Experimental induction Detached observation	Experimental Participant observation
Influence on the educational system	Intended curriculum	Implemented curriculum	Attained curriculum	Relationships between intended, implemented and attained curriculum

	First order trends			*Second order trend*
	A	B	C	D
TREND	CONCEPT BASED DIDACTICS	INNOVATION IN THE CLASSROOM	LABORATORY OBSERVATION OF PROCESSES	RESEARCH FOR INNOVATION
Based on	Teacher education	Gradual expansion	Further technology	Gradual expansion and teacher education
RESULTS	Textbooks grids for analysis tests and syllabuses	Projects for curricular innovation	Models of teaching (taxonomies, hierarchies, patterns of interaction)	Projects for curricular innovation and models of teaching–learning
CONTROL Theoretical framework	Inside Maths	Strong social features (rejection of positivism)	Paradigm of cognitive sciences	Kit of tools (local validity)
	Low attention to psychology	Low attention to theoretical control	Low attention to Maths content	Complementarity

MARGARET BROWN

THE PARADIGM OF MODELING BY ITERATIVE CONCEPTUALIZATION IN MATHEMATICS EDUCATION RESEARCH

1. INTRODUCTION

The breadth of research in mathematics education has expanded relatively rapidly. From one area of specialization within psychometrics in the 1950s, it has grown into a field of inquiry that draws eclectically on the theories and methodologies of a broad span of science, social science and the humanities. As this field of inquiry has become more diffuse, both its rationale and its scope of potential utility have become less clear, both for educational practitioners and for those potential customers who fund the activity; in particular, those in central or local government or educational charities. An earlier paper that describes, explains and evaluates the development of research in mathematics education in the UK (Brown 1992a) concludes that the activities of the mathematics education community in that country are seen by such interests to be at best marginal, and at worst subversive, in relation to public concerns.

In contrast, here I want to be more positive, arguing that the results of mathematics education research can, and indeed have, substantially benefited both practitioners and national educational policy. I will restrict my attention to an area of research with which I have most experience, the development of models of pupil and teacher behavior, and examine its nature and its potential strengths and weaknesses. Because this is necessarily a subjective view, much of it reflecting on my own experiences as a researcher, I will adopt a personal style of writing. Before proceeding, I need to describe as a background my view of the significant features of academic research.

2. DISTINCTIVE FEATURES OF ACADEMIC RESEARCH

The hallmark of great research is that it fundamentally changes the way that we construe the world. This is not to say that, for example, publication of new data of a significant kind cannot shift our awareness of the state of the world and lead to widespread change. For example, knowing that our own country performs poorly in an international comparison such as the Third International Mathematics and Science Study (Beaton et al. 1996) can make us more aware that our own national quality of mathematics education falls seriously short, lead

263

Sierpinska, A. and Kilpatrick, J. Mathematics Education as a Research Domain: A Search for Identity, 263–276.
© *1998 Kluwer Academic Publishers. Printed in Great Britain.*

us to analyze the situation further and to institute remedial action. In England it is possible to argue that it was the publication of the international comparative data (Robitaille & Garden 1989) that finally triggered legal enforcement of a national curriculum and national testing across all subjects.

Data alone, however, do not help us interpret and understand the situation, although they may stimulate us to attempt to do so. The results of SIMS attainment tests alone do not shed much light on the nature of and the reasons for the differences between English pupils and, say, Belgian pupils, although they suggest some hypotheses. One of the reasons for the lack of insight from SIMS test results alone was the relative lack of theorization in their design and analysis; to fulfill the SIMS aim of providing valid international comparisons, the main criterion for selection of test items presumably had to be that, within a sampling frame of syllabus topics and cognitive behaviors, the items had to be accepted by all participating countries as valid representations of their syllabus. In a sense such international surveys are, like the production of curriculum materials, more in the realm of design and development than academic research exercises; there is an overwhelming requirement to abide by external constraints that limit the coherence and consistency it is possible to achieve.

In contrast, academic research is first and foremost a *scholarly* and *systematic* inquiry. To fulfill the requirements of scholarship, such research has to be within an academic tradition, which means it must be planned and carried out in awareness of relevant theory and, where appropriate, of relevant research results. Its aim is to extend the relevant corpus of knowledge, either by adding new results, or new theory, or both. Relevant theory and, in some cases, knowledge of results are almost certain to draw from the contributory disciplines, going beyond the boundaries of the specific field of mathematics education, and, in many cases, beyond the boundaries of the field of education. For the inquiry to be systematic, the methodology used must be actually or potentially recognized as valid within one or more of the contributory disciplines.

That is not to deny the validity or importance of design exercises, including design of courses, of curriculum or assessment materials, or of surveys. Having spent six years directing the development of a continuous assessment program to be carried out by teachers (Brown 1992a) and having also been involved in the development of a national mathematics curriculum and national tests, I am only too aware that the political, practical and financial constraints of such exercises can make them considerably more complex and difficult to carry out with success than is the case with academic research. I am also convinced that such exercises benefit from being carried out from within a theoretical framework rather than simply as pragmatic undertakings.

Nor is it to deny the importance of action research carried out by and with teachers, with the aim of understanding and improving facets of their own teaching. As with the design exercises, the objectives of the exercise are rather different from those of academic research; in particular, the need to be publicly accountable to the academic community in terms of building on previous research and theory, and using methods of inquiry agreed to be valid is not present in action research, where benefit to the practitioner is paramount.

In all these cases, there are of course many boundary problems. Action research can develop into academic research if it meets the necessary criteria; well-theorized and systematically evaluated development can likewise generate valid public knowledge. Equally, some research in the academic domain is so narrowly informed and the methods so insufficiently systematic that it becomes at best action research and at worst of little use to anyone.

My own experience has led me to the belief that, as with technology and science, educational development and academic research are symbiotic; development suggests fields of inquiry that lead to new theory, which in turn feeds back to better development.

3. MODELING BY ITERATIVE CONCEPTUALIZATION

In the previous section, I referred to great research as changing the way we construe the world. One way in which it is possible to do this is to reduce the complexity of an enormous variety of individual behaviors or beliefs by modeling the field in terms of a limited number of categorizations or 'ideal types'. Many of the products of mathematics education research are in the form of models of this sort that attempt to describe and explain the beliefs or behavior of pupils and/or teachers. For example, models of mathematical development have been generated most famously by Piaget (e.g., Piaget & Inhelder 1958), and also by, among others, Biggs and Collis (1982) and Case (1985).

Categories or stages of thinking in particular mathematical topics have been generated by many researchers; for example, in geometry by the van Hieles (Fuys, Geddes & Tischler 1988), in ratio by Karplus (e.g., Karplus, Karplus, Formisano & Paulsen 1975) and Hart (1981a, 1981b), in algebra by Kuchemann (1981) and Kieran (1988), and so on. (The work referred to by Hart and Kuchemann was part of a wider study in the UK in 1975–1979, *Concepts in Secondary Mathematics and Science* (CSMS), of which I was the initial leader of the mathematics team and to which I refer later.)

Although much early work in the area of modeling behavior was related to pupils, recently there has been more emphasis on the knowledge, beliefs and practice of teachers of mathematics. For example, Thompson (1984) has illuminated how teachers' conceptions of mathematics relate to their classroom practice. Others such as Pinner and Shuard (1985) have proposed models of the professional development of mathematics teachers. Goffree (1985) describes different modes of textbook use among teachers. I have recently been involved in two studies that have presented models of teacher behavior, one in relation to use of mathematics curriculum schemes (Johnson & Millett 1996) and one in relation to methods of ongoing assessment (Gipps, Brown, McCallum & McAlister 1995).

This type of modeling is clearly similar to the modeling that occurs in the physical sciences, but the complexity of the social sciences renders the exercise more problematic. What is the methodology which relates to the formulation of such models?

Important in model building is the notion of 'grounded theory' (e.g., Glaser & Strauss 1967), arising from an ethnographic methodology rooted in sociology, which deflects the focus from the researcher onto the data. Ethnographic data collection involves amassing data from a variety of sources and using open coding methods to introduce an analytical framework that purports to reflect primarily the story in the data rather than the prior hypotheses of the researcher. Nevertheless, even here the researcher introduces a personal perspective on what is significant, in selecting the data elements on which to focus and the codes to use, and critically in evolving an analytical framework.

This necessary involvement of the researcher's perspectives raises the problem of the relationship between the 'researcher's country' and the 'pupil's country' or the 'teacher's country'. Black & Simon (1992) discuss the same point with respect to research in science education, underlining the fact that investigations of pupils' thinking are bound to be set within a concept framework evolved by those who are themselves to some extent 'experts'. Inevitably the steps towards understanding the ideas of the 'novice', the 'pupil's country', will be filtered through those who occupy the 'scientist's country'. This seems an obvious point but is rarely acknowledged by those who have investigated alternative conceptions in science as if they are objectively reporting phenomena.

I would prefer to openly acknowledge the limitations of starting in the 'researcher's country', but to accept that these are inevitable and to find ways of minimizing the effect of these limitations in obtaining as faithful a picture as possible of the behaviors or beliefs of the pupil or teacher.

It might be attractive to conclude that an open-minded researcher in mathematics education can best prepare an investigation by coming fresh to the data, remaining unprejudiced by any knowledge of mathematics or of any earlier educational theory and research findings. However, that not only contradicts the *scholarly* criterion for academic research proposed in the previous section, it also begs the question of the design of the collection of the data and leaves the researcher in attempting an analysis at the mercy of ignorance and confusion, and implicit prejudices and beliefs.

Clearly the best preparation is not just a knowledge of both the relevant theory and results in mathematics, in education and in related disciplines, but a critical appreciation of how these relate, including where they are in conflict, and the limitations arising from the contexts in which they have been generated. The researcher, in other words, must be as aware as possible of the knowledge and prejudices he or she brings to the investigation, and of its relative fragility.

My own experience is mainly of conducting research in teams, rather than individually, although I have also supervised many students working individually on Ph.D. theses as well as those attached to research teams. This experience suggests strongly that it is important to minimize the interference of the researcher's own perspectives in gaining insight into the worlds of the pupil or the teacher by having a group of people involved who are prepared to separately trial data-gathering methods, to analyze data independently and to argue constructively as to what is significant. Although there are times when researchers can find so little common ground that progress becomes impossible, in general healthy argument makes for more robust and significant models.

In collecting data, I believe that the steps needed to minimize the limitation of having to start out from the 'researcher's country' are of two kinds; first, the data from pupils or teachers, at least in the early stages, need to be gathered in classrooms or in interviews using skilled and open-minded researchers. If they are reacting directly with pupils or teachers, they need to be able to identify and probe strategies or beliefs in a non-evaluative way, where necessary being ready to change the whole course of the encounter to follow up interesting responses.

Second, the stimuli used need to be as open as possible, at least at the start, in order not to constrain responses too closely. If the design is too narrow, in order to elicit particular responses expected by the researcher, then key ideas that pupils or teachers have, but which have no counterpart in the land of the mathematics education expert, may be missed.

Linked with these facets of the data gathering is the need for the method of analysis also to be responsive enough to take account of unexpected features in the data.

These requirements can be summed up by the notion of respect for those involved in the investigation, allowing them the best opportunity to reveal and explain their worlds. That is not to confuse the roles of researchers and researched; while I am keen to consult as much as possible the pupils or teachers involved, the responsibility for the investigation remains ultimately with the researcher, who in order to gain the confidence of teachers and pupils should be as professional in her or his research activity as teachers aspire to be themselves in their teaching role. Nor is it to deny teachers the right to conduct research, but only to stress that the teacher will only attain quality in research if the roles of teacher and researcher can be clearly distinguished.

In practice, the negotiated aims of a study, the need for comparability in the data, and the time scale and resources, may put some limits on the openness and sensitivity that are possible. Yet without at least a period of openness in this sort of study, I believe that the quality and thus the usefulness of the outcome, in terms of the insight obtained into the worlds of the pupil or teacher, become seriously diminished.

I do not wish to go into the strengths and weaknesses of different methods of analyzing data and of arriving at categories. In general, I believe in trying as many different methods as possible in order to see which seem to impose some clear picture on the data. The criteria for the dimensions of the model must be partly subjective (involving what John Mason refers to as *resonance*) and partly objective in that they can be demonstrated to arise from the data. As with data collection, that puts a premium on the experience and knowledge of the researcher, or preferably researchers, in terms of methods of analysis, and is not easily done quickly.

What is especially important, I believe, is the opportunity to make a preliminary analysis of initial data and to use that as the basis for further rounds of data collection and analysis. This continual refining of the analysis by iterating between the 'pupil's country' or the 'teacher's country' and the 'researcher's country' is a key process that I will call 'iterative conceptualization'.

Iterative conceptualization has a close parallel in the ethnographic technique of 'progressive focusing'; again using the ethnographic paradigm, it allows the theory to be more closely 'grounded'. Despite the reservations expressed earlier about the misleading interpretations of the term 'grounded theory', it is clearly important to achieve a conceptualization that is as closely grounded in the data as possible.

The exact nature of these iterative steps will vary from study to study. In some cases, they may involve the same subjects, with respondent validation and/or extension of data already collected; in others, they may involve testing of draft categorizations on new subjects. In one of the best studies I have been involved in, the whole preliminary analytic structure was scrapped after a first round of data collection, when it became clear that in this case the distance between the 'researcher's country' and the 'pupil's country' was unexpectedly large. Whatever the precise nature of the iterative steps, the more extended they are in scope and number, the greater I believe is the likelihood of a robust, valid and useful model as an outcome.

Some stage or stages of validation of the final models with those involved in the process of teaching and learning seem very important in education. In proposing models of pupil understanding, the researcher should discuss these at various stages with teachers, who in my experience will almost certainly add some new insights into the analysis. Having arrived at such a model after many stages of iteration, I have interviewed a new set of pupils to see if the final outcome seems to be helpful in describing their responses, and have discussed the results with their teachers to see if these match and/or extend their impressions. However, I now regret that I have never shown the details of a proposed research model to pupils, since my experiences with the graded assessment scheme referred to earlier (Brown 1992b) have demonstrated the importance for all concerned of negotiating with pupils the description of how far their mathematical knowledge and skills extend.

In the recent study of teachers' assessment methods (Gipps et al. 1995), we have refined our models by discussing them both with the teachers who were involved in generating them and with other teachers who were first encouraged to think about whether they could use the categories to analyze their own methods. We have also tried them out at different times with several large groups of school inspectors and advisers, in mathematics and in assessment, who were asked to apply them to teachers they knew and then comment on whether they found them sufficiently clear, helpful and accurate. Again, the wider the scope of validation within the iterative conceptualization, the more valid the result.

What I have referred to here is a model of research that is neither quick nor cheap. For quality outcomes, it requires teams containing a core of experienced and knowledgeable researchers alongside others who are being trained by apprenticeship methods. It also requires time, to iterate and validate, to discuss and revise.

Unfortunately, the funding methods in the UK and no doubt in other countries militate against such endeavors. In an atmosphere of continual reduction of government and other funds, the premium has tended to go towards atheoretical consumer-related design activity, or towards accountability, in which funding bodies attempt to spread money thinly between many different small projects so as to take few risks, to show equity between different institutions and researchers, and to appear to demonstrate how much they have achieved in return for their ever-declining budgets. The result is small pieces of research, often marginal in relevance and sometimes of doubtful quality. The pressure on and uncertainty of funds also means that young and promising researchers are exploited and insecure, and likely to exit from full-time research as soon as they can. We are therefore seeing a spiraling down effect, which it is the responsibility of ourselves as leaders of the profession to combat, by demonstrating and arguing the value of good research.

4. THE USEFULNESS OF MODELS OF BEHAVIOR AND BELIEF

The value of research that provides models of behavior and beliefs is that it can radically change the way we construe the world. For example, I believe the result of the Piagetian corpus and the constructivism that derives from it is that educationists can no longer entertain a common-sense transmission model of education with the pupil as *tabula rasa*. Indeed, as I have argued in detail elsewhere (Brown 1994), the root of current arguments about curriculum and assessment between educationists and some teachers, on one side, and the politicians and the public, together with other teachers, on the other, is that the two groups have incompatible notions of knowledge and of learning.

The benefit of a model of behavior, belief or understanding is that it provides a conceptualization of an area of which researchers, teachers and others may only intuitively have been aware, or which was previously regarded as complex. Without such models, we would be confined to either particulars or to very vague generalities.

For example, most people think of teachers' attitudes and competence in rather vague and diffuse ways. Models of teacher development such as that used by Pinner and Shuard (1985) can assist headteachers, school inspectors, providers of school support and inservice courses, as well as teachers themselves, with a vocabulary and classification system that helps them to think in a more focused way about progress in professional competence.

Similarly, many people think of algebra rather broadly as something that involves using letters, and can produce a few detailed examples of bits of algebra such as simultaneous equations, or finding the subject of a formula. Not only is a model of children's different interpretations of the use of letters such as that produced by Kuchemann (1981) helpful in considering in a more focused way what progress has been made by individual pupils; it also provides a different way of conceptualizing school algebra that can be used by curriculum developers and teachers in formulating courses or communicating with pupils.

The benefits for the work of teachers of a structure of pupil development are summarized in a quotation from a teacher I have interviewed recently about the framework for national assessment:

'I think the National Curriculum has specified particular things that we needed to look for and that has really changed the way that we look at children's work. I think it's given focusing that we didn't have before.'

This comment demonstrates how a structure enables not only thought, discussion and further theory-building, but also action of a more focused kind. Such sentiments have been offered in different contexts by many of the teachers I have worked with over the years. In particular, many indicate that they can only talk about mathematics in relation to exercises or chapters in textbooks, having no clear alternative conceptual framework for the subject.

A model provides not only a classification but also a vocabulary, which allows discussion and further elaboration. The process of modeling may sometimes be limited, and be regarded as theorizing only on a very local level, but it may still help in laying a basis for informing, validating and triggering theory formation at a later date. This gain in mapping parts of an area in a more structured way is similar to modeling in science, in which for example early 'naturalist' attempts at classification of substances and living things led to later more extended theorizations.

In some cases, the influence of the findings of this type of research can spread widely, and even beyond the boundaries of mathematics education. To illustrate this observation, I will cite the example of a project I have been involved in to which I referred earlier, the Concepts in Secondary Mathematics and Science (CSMS) project.

The aim of the mathematics side of the CSMS work was to provide a picture of the development of mathematical concepts in pupils aged 11–16. Frameworks were published for ten areas of the mathematics curriculum (Hart 1981a, 1981b), each providing models of progression through levels of development (between three and six levels in each topic). These frameworks had been derived partly from the results of a large number of pupil interviews and partly from survey data, both cross-sectional data in the age groups 11 to 16, and longitudinal data over two years, using tests derived from the interviews. The results suggested that within each topic, the range of attainment at any particular age in terms of the distribution of the levels was very broad, especially when compared with the rate of progress in any child's understanding over a two-year period.

The virtue of referring to a project completed 15 years ago is that it is possible to report and evaluate the long-term effects of describing such models of progression and using them to survey the distribution of levels reached:

1. The progressions described in the CSMS results provided much of the theoretical framework for a set of curriculum materials (SMP 11–16) that were adopted by more than two thirds of British schools. The fact that SMP 11–16 was so successful clearly cannot be credited only to the fact that it was based on CSMS, but teachers do claim that one of its strengths is that there is a much better match between pupils' attainment and the progression in the scheme than has ever been the case before.

2. The Cockcroft Report (Department of Education and Science 1982) accepted the implications of the CSMS results. In particular, it reformulated the survey results in terms of a notional '7 year gap' between the ages when higher attaining and lower attaining pupils might be expected to first grasp a particular mathematical idea. The Report therefore encouraged greater differentiation and sensitivity in teaching and assessment to each pupil's current attainments. Its recommendations also took account of the CSMS results that indicated very modest attainments by large numbers of pupils. Adequate differentiation in teaching to reflect this spread, along with appropriate progression for individual pupils, are now two of the most important criteria for all subjects in the four-yearly statutory inspections of both primary and secondary schools in the UK. Sir Wilfred Cockcroft was appointed to head the Secondary Examinations Council after chairing the committee that produced the Cockcroft Report, and hence differentiation also became a guiding principle in the new General Certificate of Secondary Education (GCSE) examination introduced in 1988.

3. As the original CSMS leader, I was approached in 1982 by the largest local education authority in the UK and by one of the examination boards to use the CSMS results as the basis of an ongoing 'authentic' classroom-based assessment scheme for all secondary pupils. The scheme would be graded in the sense that it would be organized into a series of levels through which pupils could progress during their secondary school careers. We decided to focus the scheme on 80 open-ended activities relating either to real-life problem-solving or to mathematical investigation, allowing pupils at very different levels to demonstrate their attainments within the same activity, and to thus build up their own portfolios. These activities were assessed in relation to criteria assigned to the various levels of progression, the criteria having been derived partly from the CSMS results. As referred to earlier, a full set of materials has been published recently as the Graded Assessment in Mathematics Project (GAIM) (Brown 1992a); it relates closely to many of the moves now taking place to broaden assessment in the United States and in other countries.

4. The development of GAIM, a project in which the level criteria were carefully trialled over a five-year period, was overtaken by the rapid introduction

of the untrialled UK National Curriculum. The level structure of GAIM was chosen as the basis for the 10-level structure for national assessment across all ages and all major subjects (Department of Education and Science 1988), and the data gathered by CSMS were used to convince the then Secretary of State for Education of the correctness of implementing this framework rather than an age-based structure. Major sections of the content of the National Curriculum were adapted from the level criteria in GAIM, which in turn derive from the CSMS results.

5. Outside the UK, the CSMS tests have been replicated in more than 20 different countries including some in Africa and Asia. Although particular items often differ considerably in difficulty according to different curricular emphases, the distribution of pupils across the levels is often remarkably stable. This stability suggests that curricula worldwide might eventually become less dominated by age-based textbooks and more sensitive to pupil's levels of understanding.

Hence, the effect of describing a model of progression in mathematical thinking determined on the basis of iterative conceptualization, moving between the 'pupils' country' and the 'researcher's country', has been fundamental and far-reaching in English education, and may even have some marginal influence in other countries.

5. LIMITATIONS OF MODELS

So far, the terms 'pupils' country' and 'teachers' country' have been used without discussion of their ambiguity; there is of course not a single 'country' but a different 'country' for each pupil and for each teacher. It would be difficult to conceive that any pair of pupils or any pair of teachers would have identical positions on every facet of understanding or belief. This diversity might suggest that research aiming at modeling of this sort is not a worthwhile endeavor; that the most one could do would be to develop some insight into the thinking of a particular child or a particular teacher at a particular time.

Luckily, however, there is considerable evidence that there are many features in common to the different 'countries' inhabited by different pupils and teachers. The act of identifying common features in different groups of individuals and using these to describe a number of 'ideal types' is a form of psychological modeling akin to mathematical modeling. In each case, the model presents a general way of thinking about the behavior of specifics (objects or people).

This act of identifying common features means that other more idiosyncratic features have been ignored. Such models are only valid if they are useful, which in general means that those features ignored can be demonstrated or agreed to be of secondary importance for a particular purpose.

For example, using the equations for uniformly accelerated motion will give a reasonably close prediction for falling stones but not for falling leaves. In the former case, the prediction is not completely accurate but close enough to be useful for some practical purposes; in the latter case, further factors like air resistance and wind speed need to be built in before the model is usable.

The more features that are incorporated into the model, clearly the greater the possible degree of match to individual cases. Nevertheless, except for specific and well-defined purposes, simple models are often more useful than complex ones since they are more easily appreciated, internalized and applied by researchers and users.

I believe that although models of understanding, behavior or belief can be helpful and indeed influential, as described in the previous section, their limitations must be borne in mind constantly. For example those models of learning previously referred to that incorporate levels (e.g., those of Piaget, Biggs and Collis, the van Hieles, the CSMS project and the UK National Assessment) can be useful since, by conceptualizing in a detailed way the meaning of attainment or development, they provide a way of thinking more precisely about learning in general, as well as about the attainment of particular pupils. The description that can be given of a pupil's attainment using such a framework is surely a great improvement, in terms of information to teachers, children and parents, and guidance for curriculum planning, on the vague notion that a child is good or average or weak in the subject, or that he or she scored 55% or is judged grade B.

Problems occur when a model, which is a form of general hypothesis, is confused with the complexity of reality, as O'Reilly (1990) points out in his criticisms of the CSMS project. For example, when applying a model that uses levels, it is misleading to suggest either that each pupil can be exactly described by one of the levels, or that the description of a level can be used for predicting pupils' behavior in a range of contexts of differing complexity. Models, whether categorical, ordinal or linear, can readily lead to inflexible labeling, whether of a teacher as a 'restricted professional', or of a student as 'Level 2' or 'IQ 98'. Labeling can readily lead to discrimination; for example, discrimination against a child by providing a limited and inflexible curriculum, as pointed out by Noss, Goldstein & Hoyles (1989).

In order to avoid such simplistic judgments, it is often helpful to think of the 'ideal types' as not being mutually exclusive. The most helpful description of an individual is likely to be in terms of some combination of different stages or types, rather than as uniquely matching one alone.

The inherent dangers of labeling are associated with the nature of models, and should not be confused with the further difficulties introduced by the necessary unreliability of assessment tools, whether qualitative or quantitative.

There is a school of thought in education that suggests that the only fair way to treat students, or indeed teachers, is without differentiation, and that any individual information based on a model of learning, behavior or belief is discriminatory and hence harmful. I believe, in contrast, that the professionalism of doctors is demonstrated in their ability to make judgments based on and allowing for the varied, partial and often unreliable nature of information. Clearly in medical situations, mistakes can be made, but negligence arises from failure to use or properly evaluate information, or ignorance of available theory. In the same way as for doctors and, as pointed out earlier, for researchers, it is the duty of teachers and those who work with them to make sure they have access to a variety of the best theoretical models and to the best assessment tools available, and to have an awareness of their limitations. As much information and insight as possible is needed in understanding the position of a student or a teacher, and in planning their future development with them.

To summarize, models can be unjustly criticized on the basis that they fail to take account of the fact that every individual is unique and each context different, and that they lead to discrimination through labeling. Clearly, we may feel that some models are so crude and misguided that they should be rejected as on balance more likely to lead to inappropriate than appropriate interpretation and action. While accepting the limitations of models, I nevertheless believe that some models can assist in enabling us to think in a more focused way about an area of human attainment, behavior or belief, to analyze and better understand ourselves and other people, and to plan for future development.

6. CONCLUSION

In this chapter I have described what I have called a process of iterative conceptualization, which has as its product a model of behavior, belief or understanding. The iterations have been between the world of the researcher and the world of the pupil or teacher, with a sequence of successive analysis and data gathering. The result in each case has been a conceptualization of an area that researchers, teachers and others have found useful in that it has enabled them to think more clearly about something that was previously ill-focused. This process can involve theorizing only on a very local level, but as laying a basis for theory formation at a later date.

A caveat has, however, been given that such models have the potential to be over-crude and misused. A worthwhile model has to be shown to be both useful and robust, and should be used with awareness of its limitations.

The most successful examples of research that has had the most effect on education have been those in which both a relatively large team (three or more) could share and argue out ideas, and where the time scale allowed several iterations, some very exploratory in nature.

I believe for these reasons that good research that has the potential to change people's views of the world demands reasonable resources of time and people. I am concerned about the situation in educational research in England and Wales, which perhaps pertains in other countries, in which a small financial cake is spread ever more thinly, producing quantities of rather shallow research, and giving rise to inconsequential publications.

As in other aspects of education, equality may prove to be the enemy of quality.

REFERENCES

Beaton, A. E., Mullis, I. V. S., Martin, M. O., Gonzalez, E. J., Kelly, D. L. & Smith, T.A.: 1996, *Mathematics Achievement in the Middle School Years: I.E.A.'s Third International Mathematics and Science Study*, Center for the Study of Testing, Evaluation, and Educational Policy. Boston College, Boston.

Biggs, J. B. & Collis, K.: 1982, *Evaluating the Quality of Learning: The SOLO Taxonomy*, Academic Press, New York.

Black, P. J. & Simon, S. A.: 1992, 'Progression in Learning Science', *Research in Science Education* 22, 45–54.

Brown, M.: 1992a, 'Desenvolvimentos em investigação em educação matemática no Reino Unido', in J. P. Ponte (ed.), *Educação Matemática*, Instituto de Inovação Educacional, Lisbon, 15–44.

Brown, M. (ed.): 1992b, *Graded Assessment in Mathematics*, Thomas Nelson, Basingstoke.

Brown, M.: 1994, 'Clashing Epistemologies: The Battle for Control of the National Curriculum and Its Assessment', *Teaching Mathematics and Its Applications* 12 (3), 97–112.

Case, R.: 1985, *Intellectual Development: Birth to Adulthood*, Academic Press, New York.

Department of Education and Science, Committee of Inquiry into the Teaching of Mathematics in Schools: 1982, *Mathematics Counts* ('The Cockcroft Report'), Her Majesty's Stationery Office, London.

Department of Education and Science, Task Group on Assessment and Testing: 1988, *A Report*, Her Majesty's Stationery Office, London.

Fuys, D., Geddes, D. & Tischler, R.: 1988, *The van Hiele Levels of Thinking among Adolescents*, (Journal for Research in Mathematics Education Monograph No. 3), National Council of Teachers of Mathematics, Reston, VA.

Gipps, C., Brown, M., McCallum, B. & McAlister, S.: 1995, *Intuition or Evidence?: Teachers' and National Assessment of Seven-Year-Olds*, Open University Press, Buckingham.

Glaser, B. G. & Strauss, A. L.: 1967, *The Discovery of Grounded Theory*, Aldine, Chicago.

Goffree, F.: 1985, 'The Teacher and Curriculum Development', *For the Learning of Mathematics* 5 (2), 26–27.

Hart, K.: 1981a, 'Ratio and Proportion', in K. Hart (ed.), *Children's Understanding of Mathematics 11–16*, John Murray, London, 88–101.

Hart, K. (ed.): 1981b, *Children's Understanding of Mathematics 11–16*, John Murray, London.

Johnson, D.C. & Millet, A.: 1996, *Implementing the Mathematics National Curriculum: Policy, Politics and Practice*, Chapman Hall, London.

Karplus, R., Karplus, E., Formisano, M. & Paulsen, A.-C.: 1975, *Proportional Reasoning and Control of Variables in Seven Countries* (Advancing Education through Science Orientated Programs, Report ID-65), University of California, Berkeley, CA.

Kieran, C.: 1988, 'Two Different Approaches Among Algebra Learners', in A. Coxford (ed.), *The Ideas of Algebra K-12* (1988 Yearbook), National Council of Teachers of Mathematics, Reston, VA, 91–96.

Kuchemann, D. E.: 1981, 'Algebra', in K. Hart (ed.), *Children's Understanding of Mathematics 11–16*, John Murray, London, 102–119.

Noss, R., Goldstein, H. & Hoyles, C.: 1989, 'Graded Assessment and Learning Hierarchies in Mathematics', *British Journal of Education Research* 15 (2), 109–120.

O'Reilly, D.: 1990, 'Hierarchies in Mathematics: A Critique of the CSMS Study', in P. Dowling & R. Noss (eds.), *Mathematics versus the National Curriculum,* Falmer Press, Basingstoke, 77–97.

Piaget, J. & Inhelder, B: 1958, *The Growth of Logical Thinking from Childhood to Adolescence,* Routledge & Kegan Paul, London.

Pinner, M. T. & Shuard, H.: 1985, *In-service Education in Primary Mathematics,* Open University Press, Milton Keynes.

Robitaille, D. F. & Garden, R. A.: 1989, *The IEA Study of Mathematics II: Contexts and Outcomes of School Mathematics,* Pergamon, Oxford.

Thompson, A. G.: 1984, 'The Relationship of Teachers' Conceptions of Mathematics and Mathematics Teaching to Instructional Practice', *Educational Studies in Mathematics* **15**, 105–127.

Margaret Brown
School of Education,
King's College London,
London SE1 8WA,
UK

KOENO GRAVEMEIJER

DEVELOPMENTAL RESEARCH AS A RESEARCH METHOD

With Walker (1992) we may observe that a multitude of research approaches is emerging. Many of the present approaches can be captured under one of the two headings: 'explanation' or 'understanding' (Bruner 1994). Here one may think of explanation in terms of causal relations between dependent and independent variables, and of understanding as making sense of what is going on. Apart from these two main perspectives we can discern a third, which might be labeled 'transformational research', as suggested by the NCTM Research Advisory Committee (1988). 'Transformational research' means research that does not focus on 'what is' but that deals more broadly with 'what ought to be'. This involves, for instance, research that addresses the question of how to constitute education that meets certain pre-given standards or ideals.

The idea of 'developmental research', which is the topic of this contribution, falls into the latter category. Developmental research consists of a mixture of curriculum development and educational research in which the development of instructional activities is used as a means to elaborate and test an instructional theory. However, what is meant here is not the kind of symbiosis between research and development in which the research takes the shape of formative evaluation in the service of curriculum development. Instead, developmental research is seen as a form of basic research that lays the foundations for the work of professional curriculum developers.

For over two decades now, developmental research has been the driving force behind the innovation of mathematics education in The Netherlands. The innovation process started in the early 70s with the work of the Institute for the Development of Mathematics Education (IOWO), and is continued by its successor, the Research Group on Mathematics Education and Computing Science (OW&OC) – now called the Freudenthal Institute. The main goal is the development of a mathematics education that corresponds to Freudenthal's (1973, 1991) idea of 'mathematics as an human activity'. That is to say, students should be given the opportunity to reinvent mathematics by mathematizing – mathematizing subject matter from reality and mathematizing mathematical matter. In both cases, the subject matter that is to be mathematized should be experientially real for the students. That is why the envisioned education is called 'realistic mathematics education'. Furthermore, the idea of mathematizing implies a high autonomy of the students. Or, in other words, the core principle is that *mathematics can and should be learned on one's own authority and through one's own mental activities.*

Sierpinska, A. and Kilpatrick, J. Mathematics Education as a Research Domain: A Search for Identity, 277–295.
© *1998 Kluwer Academic Publishers. Printed in Great Britain.*

This contribution focuses on a description of developmental research as a research method whose aim is the constitution of a domain-specific instructional theory for realistic mathematics education; that is, an instructional theory that does justice to the above-mentioned principles. The construction of this instructional theory will be the end result of a series of research projects. These research projects can be laid out in two phases: a first phase concerning the constitution of an initial design of a course, and a second phase concerning the elaboration of that course in interaction with teaching experiments. A discussion of the methods that constitute these two phases forms the main part of this contribution. To compensate for the distant and abstract character of this discussion, the discourse will be interspersed with examples concerning the teaching of decimal numbers.

We start by giving a global sketch of developmental research and its characteristics.

1. DEVELOPMENTAL RESEARCH

As indicated above, developmental research is part of a large scale, long-term innovation project in The Netherlands. Curriculum development is one of the core activities in this innovation project. However, no so-called first-generation instructional design theories (Merrill, Li & Jones 1990) were used. In this aspect, the Dutch approach is not different from many other innovative curriculum projects – like The Madison Project, the University of Chicago School Mathematics Project (UCSMP) in the United States, and the work of the Shell Centre in England.

Traditional instructional design theories (Gagné 1977; Gagné & Briggs 1979; Romiszowski 1981) may be quite useful in cases where the curriculum goals can be easily specified and ample empirical knowledge is available. However, if the curriculum objectives are still vague and there is no sharp image of the intended instructional processes – let alone experience with it – instructional design theories that are objective-oriented and that capitalize on empirical knowledge about the intended instructional processes are of little use.

In most innovative curriculum projects the production of curriculum materials is a top priority, for it is with the help of these curriculum materials that the innovation has to be shaped. In practice, this means that there is not enough time to take standard methodological measures to ensure scientific justifications of the theories developed in the project. As a consequence, the research aspect of many curriculum projects remains in the background. However, of course, exceptions exist (see for instance Romberg 1973).

In the Dutch situation, the research aspect has been at the forefront from early on. Since there was no elaborated theoretical framework to start with, the theory had to be construed in the process. Therefore, the developmental research had not only to produce curriculum products, but also instructional theory and a justification of that theory. In this sense, the notion of developmental research is cognate to Steffe's concept of teaching experiments (Steffe 1983). It is also cognate to Kamii's work (Kamii, Lewis & Livingston Jones 1993), to the work of Lampert (1989, 1990) and to the work of Cobb, Yackel & Wood (1992).

It should be noted that the question with which the Dutch developmental researchers are struggling is very similar to the questions constructivists have to solve. For them too, the question arises: How to design instructional activities that (a) link up with the informal situated knowledge of the students, and (b) enable them to develop more sophisticated, abstract, formal knowledge, while (c) complying with the basic principle of intellectual autonomy.

In short, the Dutch researchers try to answer this question by designing and testing instructional sequences that fulfill these requirements. In reflecting upon the design process, and upon the instructional sequences themselves, the researchers try to abstract an instructional theory for 'realistic mathematics education'. The construction of this instructional theory is a long-term process; the theory is slowly emerging from a large set of individual research projects.

2. DEVELOPING A PROTOTYPICAL COURSE

In these individual projects, the question, 'What constitutes a mathematics instruction which is consonant with the basic principles of realistic mathematics education?' is answered locally – primarily at a concrete level. That is to say, the answer is sought for a specific topic, namely by developing a prototypical course for that topic. Broadly speaking, the researcher construes a provisional set of instructional activities which is worked out in an iterative process of (re)designing and testing. To construe a provisional design, the researcher may take ideas from different sources and make them fit the new course: these sources may be curricula, texts on mathematics education, research reports, etc. It should be noted, however, that adopting an activity that stems from an existing course or some experimental approach does not imply that the underlying instructional theory will be adopted as well. In that sense, instructional activities are being detached from their original context. The way in which this is done is determined by the researcher's overall vision of the end product.

This approach is called 'theory-guided bricolage' (Gravemeijer 1994a) since it resembles the manner of working of what the French call a 'bricoleur'. A bricoleur might be thought of in English as a person who tinkers and fixes things by using, as much as possible, those materials that happen to be available. To do so, many materials may have to be adapted. Moreover, the bricoleur may have to invent new applications that differ from those for which the materials were designed. The developmental researcher follows a similar approach but the way in which selections and adaptations are made will be guided by a theory; i.e., a theory for 'realistic mathematics education'. That is why we speak of theory-guided bricolage. At the beginning of the innovation of mathematics education in The Netherlands, this theory was more a global philosophy of mathematics education than a theory. Gradually, however, a domain-specific instructional theory evolved. This theory serves as a guide not only for the construction of the provisional design, but also for the further elaboration of the course which is executed in an iterative process of experimentation and adaptation.

The bricolage metaphor is taken from Levi-Strauss (cited by Lawler 1985) who uses it to describe the human thinking process. Jacob (1982) uses a similar metaphor, 'tinkering', to describe evolution. Both interpretations fit developmental research, since the first characterization corresponds fairly well with the basic design process, while the second one describes the long-term perspective of developmental research. The latter, in particular, denotes the cumulative effect of a long-term process of adapting, adjusting, improving and elaborating. This process not only concerns a product, it also includes the growth of the instructional theory: the curriculum and the theory are developed together. In developmental research these are two sides of the same coin, although the ultimate goal is theory development.

3. THEORY DEVELOPMENT

While developing a curriculum, the research question ('What constitutes a mathematics instruction which is consonant with the basic principles of realistic mathematics education?') is initially answered on a concrete level: in the form of a curriculum. At the same time, the question is also answered on a more theoretical level: by means of the learning process of the researcher. The second step in developmental research consists of explicating the result of that learning process. This is done by reflecting upon the considerations, deliberations and experiences that lie at the base of the final course design. The aim is to uncover the 'local instructional theory' that is implied in the instructional sequence.

Such a theory is local, in the sense that it describes how that specific topic should be taught to fit the basic principles. The local instructional theory differs from the instructional sequence itself, in the sense that it focuses on a rationale for the choice of the instructional activities. This rationale has to explain how these instructional activities comply with the intention to give the students the opportunity to reinvent mathematics.

Key ingredients of such local instructional theory are, among others:

- informal knowledge and strategies of the students on which the instruction can be built;
- contextual problems that can be used to evoke informal knowledge and strategies;
- instructional activities that can foster reflective processes which support curtailment, schematization and abstraction.

Moreover, the local instructional theories should contain the germs of theoretical notions that exceed the specific topic that is addressed. In the local instructional theory, these and other ingredients are to be justified, both in terms of the envisioned learning process of mathematics as a human activity, and in terms of empirical findings about the actual behavior of the students. Elaborated examples of such local instruction theories can be found in Streefland (1990b) and in van den Brink (1989).

Thus, the base level of developmental research is concrete. By formulating a local instructional theory, a shift is made towards a theoretical level. However, the theory is still local. The next step, therefore, is to move from the local to the general; although 'general' in this context is still limited. The instructional theory that is being developed is confined to mathematics education and it is linked to the realistic approach. That is why one speaks of a 'domain-specific instructional theory for realistic mathematics education'. This domain-specific instructional theory can be construed by searching the various local instructional theories for common characteristics. In this manner, Treffers (1987) construed a framework for an instructional theory.

Note that the theory is construed as an a posteriori theory. Nevertheless, the theory also reflects the basic tenets of the theoretical notions that lie at the base of the whole innovation. The three basic ingredients that Treffers uses to erect a theoretical framework consist of the theory of levels of van Hiele (1973), Freudenthal's didactic phenomenology (1983), and the idea of 'progressive mathematization'.

Treffers uses van Hiele's theory of levels to describe the global structure of a course. He labels the levels in this context as 'intuitive phenomenological level', 'locally descriptive level' and 'level of subject matter systematics', where the didactic phenomenology is cast as an expedient that helps to fill in the global framework. Progressive mathematization serves as a means to facilitate the transition between the three levels. Treffers discerns five characteristics of progressive mathematization:

1. The central place of contextual problems; as starting points, and as applications.
2. The use of schemes and models, including model-situations.
3. The input of the students in terms of their own constructions and free productions.
4. The interactive character of the instructional process.
5. The intertwinement of the various learning strands.

It shows that the reconstruction of the instructional theory is more descriptive than prescriptive in character. More recently, however, the theory has been recast in a more prescriptive version (Gravemeijer 1994b). This will be elucidated later.

4. CHARACTERISTICS

The above description shows the remarkable character of this type of research. In summary, we can say it is evolutionary, stratified and reflexive.

Evolutionary

Developmental research is *evolutionary* in the sense that theory development is gradual, iterative and cumulative. There is no detailed theory with which to start. The initial, global theory is elaborated, refined and explicated during the process of designing and testing. The theory grows out of the process, and characteristically, the overall theory is explicated in retrospect.

Stratified

Developmental research is *stratified* in the sense that theory development takes place at different levels:

- at the level of the instructional activities (micro theories)
- at the level of the course (local instruction theories)
- at the level of the domain-specific instructional theory.

These three levels can be illustrated with an example concerning decimals.

The key features of a *local instructional theory* for decimals are: 'changing the measurement unit', 'benchmark fractions', 'the double number line', and 'progressive refinement'. The idea of 'changing the unit of measurement' may be illustrated with the way students might deal with decimal numbers in the context of money. Here, the thesis is that students routinely interpret decimal prices in dollars as prices in cents, or prices in dollars and cents: e.g. $2.45 is interpreted as '245 cents', or as '2 dollars and 45 cents'. With the first interpretation, the students get rid of the decimal point, this enables them to work with these numbers as if they were integers! The second interpretation may support the students' notion of magnitude; knowing that 100 cents equals one dollar, $1.98 will easily be seen as 'almost 2 dollars'. In a similar way, $2.45 can be thought of as 'about two and a half dollars'. Other benchmarks can be created by thinking of 'quarters': e.g. $0.75 is three quarters, which is 'three quarters of a dollar'. In The Netherlands this idea of 'changing the measurement unit' can be extended to measurement since we use a metric system. This leads to the number line as a model to represent decimal numbers. A metric ruler can be converted into a 'double number line', where, for instance, kilometers can be written above the line, while meters are written beneath the line (see Figure 1). It

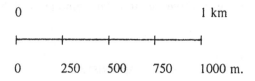

Figure 1 *Double number line*

is not possible within the scope of this paper to discuss all the characteristics of the local instructional theory. 'Progressive refinement' will be elaborated later.

In the *micro theories* that underlie the instructional activities, informal knowledge will play a central role. On a given task, the researchers may expect the students to use certain informal strategies instead of an instrumental procedure, based on a literal, formal interpretation of decimals as a set of decimal fractions (like $0.123 = 1/10 + 2/100 + 3/1000$). The micro theory might hypothesize that students will use the connections between decimals and fractions if they have to calculate the total contents of 4 bottles of 0.75 liter ($4 \times 3/4 = 3$).

On a more global level, both phenomenological embeddedness – in the context of measurement – and the way the number line is used (first as a model of repeated pacing of a measurement unit, and later as a model for reasoning about decimal numbers) come to the fore as characteristics of the *domain-specific instructional theory* for realistic mathematics education.

At each level, cyclic processes can be discerned. At the level of the micro theories there is the cyclic process in which micro theories are tested and adapted. At the level of the local theories subsequent versions of an instructional course are tested and adapted in cyclic processes. At the level of the domain-specific instructional theory more indirect cyclic processes can be discerned; the domain-specific instructional theory functions as input and output of course development.

Reflexive

Developmental research is *reflexive* in the sense that theory development is fostered by reflexive relations between the aforementioned levels. On the one hand, the domain-specific theory, for instance, is the result of a generalization over local instruction theories; while, on the other hand, the domain-specific instructional theory will influence the course designs, and thus, the local theories.

To some extent, developmental research has the characteristics of a research program.[1] The envisioned constitution and growth of the domain-specific instructional theory is to be achieved via a series of research projects. As a consequence, each individual research project will have only a limited significance. Or, in other words, the individual projects derive their significance from being embedded in the overall structure of the research program. Viewed on its own, the theoretical contribution of each research project will be small, and apart from its significance in the larger picture, an individual project might not seem to be more than a well-considered developmental project. Nevertheless, the progression of the research program depends on the execution of the smaller projects as research projects.

In the following paragraphs, the design of the smaller research projects will be discussed, then the focus will shift to the long-term perspective. The design of a developmental research project can be split into two phases: the phase of the preliminary design, and the phase of the elaboration of the course design.

The result of such a research project will not be final. New versions of the same course will be worked out in new projects. New insights may even result in the development of alternative courses. The research program, therefore, encompasses a vivid complex of research projects on various topics that is evolutionary in character and not just cumulative.

5. PRELIMINARY DESIGN

Before a research project is actually begun, the researcher will make an analysis of the situation based on the question: Why are the existing curricula unsatisfactory?

With decimals, lack of understanding comes to the fore in problems with place value, weak notions of the magnitude of a decimal number and low performance on estimation tasks. Also well-known are the problems students have with understanding that multiplication with a number smaller than 1 gives a smaller result – a result which runs counter the common belief that multiplication enlarges (Bell, Swan & Taylor 1981). Realistic mathematics education, of course, asks for understanding.

In such an analysis, the demands which the new course is expected to meet will become visible. It is expected that the new course will make contributions where the old ones were remiss. Alongside this analysis, a general concept of a course must develop before the actual experiments can begin. Moreover, as a start, such a general concept has to be explicated and justified. Both the problem analysis and the preliminary course design rest to a great degree on the domain knowledge of the researcher. Over and above that, the researcher will use the theory of realistic mathematics education by applying the central principles of this theory heuristically.

In doing so, theory-guided bricolage will not have to be confined to selection and adaptation of available instructional activities – as a minimal interpretation of the bricolage metaphor might suggest.

6. HEURISTICS

The domain-specific instructional theory, as it is now known, can be formulated in terms of heuristics for instructional design (Gravemeijer 1994a). These heuristics contain the reinvention principle, the didactic phenomenology and the mediating (or emergent) models.

The reinvention principle (Freudenthal 1973) offers the guideline: 'Think how you might have figured it out yourself.' This is basically the Socratic method, although the role of the student is more active in realistic mathematics education. As the term reinvention suggests, the original invention process can be used as a model. An analysis of the history of mathematics will bring the developer much information that can be used in the design of a course. But there is another access route. Developmental research shows that children's spontaneous solution strategies can also be used to put the developer on the track of a possible reinvention-route (Streefland 1985; Gravemeijer 1991). Note that such a reinvention route demands the planning of long-term learning processes. The focus is not the subsequent mastery of small units of subject matter, as in a learning hierarchy according to Gagné, but rather a guided reinvention entailing a gradual elaboration and sophistication of the subject matter – an approach that is to some extent similar to that which Greeno (1991) suggests for developing number sense, but more directed.

We can take decimals as an example again. It may be too much to ask of the students to expect them to reinvent decimal fractions the way Stevin did originally. Nevertheless, we may try to let them experience some of this invention process. The most salient characteristic of decimal numbers is their similarity with the decimal position system for integers. However, there is another underlying characteristic; i.e. the notion of repeated refinement. This is clear in metric measurement where the precision can be improved endlessly by introducing ever smaller measurement units – by dividing the smallest unit by ten. This idea of unlimited refinement by repeated division was already known to the ancient Egyptians, although they worked with repeated division by two. This seems to correspond to a natural approach of children, who, when in need of a smaller measurement unit for linear measurement, tend to introduce halves and 'halves of halves'. Therefore, linear measurement may be a good context for developing the notion of repeated halving. This idea of repeated halving then can be compared to the decimal approach of repeated 'decimating'.

Didactic phenomenology (Freudenthal 1983) points to applications as a possible source of information for the developer. Following from the idea that mathematics developed as increased mathematization of what were originally solutions to practical problems it may be concluded that the starting points for the reinvention process can be found in current applications. The developer should therefore analyze application situations with an eye to their didactic use in the reinvention process (see also Streefland 1993). Moreover, a didactic phenomenological analysis gives an overview of the area of applications that has to be taken into account when developing a course.

Concerning decimals, several important phenomenological aspects have already been mentioned: the relation with measurement and, especially, with money. When one makes calculations with prices, one easily shifts from dollars to cents and back, while avoiding formal arithmetic with decimal numbers. A didactic phenomenological analysis shows the potential of the implicit strategy of changing the unit of measurement. Further, the relation with benchmark fractions as points of reference comes to the fore. Also other didactic issues, like repeated halving and the modeling of linear measurement, are revealed by such an analysis.

Another feature that becomes evident is the support that contexts can offer. When multiplying 1.20×0.75 in the context of prices and weights, for instance, the students may use all sorts of informal strategies based on their understanding of the problem. Starting with situated problems, and maintaining the connection between these situations and the more formal procedures that are developed later, may help the students realize what is going on when solving more formal problems.

It should be noted that the phenomenological exploration and the search for informal strategies can be combined. Presenting applied problems to students, who have not yet learned the standard procedure or concept, appears to be a fruitful situation for observing informal strategies. One might investigate, for instance, how children solve a problem that asks for a refined measurement unit if they have not yet learned the decimal system.

Mediating, or emergent, models are deployed in realistic mathematics education in order to connect informal and formal knowledge with one another. However, the aim is not just to connect these two. The ideal is that the models emerge from the activities of the students. Subsequently, mediating models should serve as a catalyst for a growth process in which the formal knowledge evolves from the informal knowledge. The developer must therefore search for ways to model the students' informal strategies so that models, diagrams and manners of notation evolve which can then be used to generalize and formalize the informal knowledge and strategies.

In this plan, the models are first linked to contexts. That is to say, the models model contextual problems. The reference to the context situation gives them meaning. Next, solving a variety of similar contextual problems will aid towards the generalization and independence of the model. This will lead to a certain reification, so that the same models can then function as a basis for further formalization.[2] In this way, the reinvention process is structured along four levels, which have to do not only with actual models but also with concepts and strategies. Or, in other words, the term 'model' should not be taken too literally. It can concern a model situation, a scheme, a description, or a way of notating.

The four levels may be outlined as follows (see Figure 2):

1. *the level of situations*, where domain-specific, situational knowledge and strategies are used within the context of the situation
2. *a referential level*, where models and strategies refer to the situation described in the problem

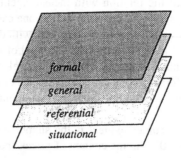

Figure 2 *Levels*

3. *a general level*, where a mathematical focus on strategies dominates over the reference to the context
4. *the level of formal mathematics*, where one works with conventional procedures and notations.

The concrete level may require some clarification since the reality outside school can hardly be brought into the classroom. Furthermore, the reality of a realistic context should not be identified with everyday reality outside school. Realistic mathematics education has to do with situations that are experientially real to the students. These may be everyday situations, but they may also be fantasy worlds in which the students can immerse themselves. And – last but not least – it may be mathematics that is experientially real (see also Davis & Hersh 1981). The objective of realistic mathematics education is that the mathematics the students develop by themselves is experienced as a developing 'common sense' Freudenthal (1991).

Looking at decimals once more, it can be seen that the number line is an important model in the realistic approach. Other instructional approaches often use base ten blocks to model decimal numbers. Or, more precisely, the blocks are used to represent decimal numbers. The researchers use manipulatives to represent their understanding of decimal numbers. In a realistic approach, the starting point is the knowledge of the students. In the context of measurement, a ruler is introduced in a natural way. Several ways of refinement can be worked out, particularly repeated halving and repeated dividing by ten. The activity of pacing the measurement units is modeled with marked segments on a ruler. Later on, this segmented number line becomes a model for the relations between the units of a different rank. Then, the segmented number line can be replaced by a double number line that integrates two scale lines: one with the larger units and one with the smaller units. It enables the students to shift between larger and smaller units, to represent the magnitude of a number, to make estimations or to identify a domain (see Figure 3).

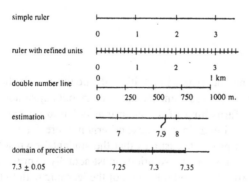

Figure 3   *Various number lines*

7. ELABORATION OF A COURSE

Even if the preliminary course design is detailed, well-considered, and based on sound knowledge, this does not mean that an elaboration in a classroom experiment is redundant. A sound development of a course is only possible in conjunction with classroom experiments. Moreover, the overall design has the character of a collection of hypotheses that are tested in the educational experiment.

In the classroom experiment the overall preliminary design is expanded and adapted in a cyclic process of design and experiment. Freudenthal (1988) speaks in this context of 'thought experiments' and 'educational experiments'. When designing an instructional activity, the researcher executes a thought experiment in which he or she envisions how the teaching–learning process will proceed. Afterwards, the instructional activity is tried out in the classroom. Then the researcher goes in search of signs that confirm or reject the expectations of the thought experiment. Moreover, the researcher keeps his or her eyes open for new possibilities.

The feedback of these empirical data into new thought experiments fuels the cyclic process of deliberating and testing. The short cycles lead to what could be called 'feed-forward'. Whenever the development of educational material and its testing in the classroom follow upon one another's heels, it is possible to react immediately to the classroom experiences. In the material still to be developed, one can take previous successes and failures into account.

The central activity that guides the bricolage process on a micro-level is the micro-didactic deliberation on the learning process. The thought experiments are to be executed in terms of 'micro theories' about the mechanism of the invented instructional activities. A term such as 'instructional theory' could leave the impression that education is viewed solely from the supply side. Nothing is further from the truth. The whole point of education is the student's own mental activities; these are central to the basic principles of realistic mathematics education. Therefore, micro theories have to be brought to the fore that describe how the instructional activities provoke the envisioned mental activities of the students, and how these mental activities contribute to the presumed growth in mathematical ability.

8. CRITERIA

The aforementioned heuristics implicitly indicate the criteria that will be used by the researchers to make assessments and to carry out adjustments since they lead to a preliminary furnishing of a course that is based on expectations derived from the heuristics. Let us look at these criteria in more detail.

The reinvention principle assumes that the students' own solutions will pave the way. That is, the students' solutions must actually express a variety of solution levels that provide a good reflection of the learning path to be followed.

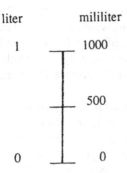

Figure 4 *A double scale line*

For example, students have to list 0.75 liter, 0.8 liter, 0.99 liter, and 0.109 liter from the smallest to the largest amount. This problem can be solved on a formal level; for instance, by comparing the first decimal place of each number ($1 < 7 < 8 < 9$), or by writing each number with three decimals ($0.109 < 0.750 < 0.800 < 0.990$). But the student can also use a number line. This number line can be segmented or it could have a double scale, representing liters and milliliters as well (see Figure 4). In the latter case, the student can write all amounts in milliliters to get rid of the decimals. In reverse order, these solution procedures signify a possible learning route.

This requirement of a dispersion of solution levels is not only significant as a formal characteristic of the reinvention concept, but guided reinvention assumes that the teacher will find a modus operandi for reconciling guidance and self-reliant invention with one another.[3] However, a productive classroom discussion is only possible when there is a dispersion both in the solutions of various students and in solution levels. If the students all use the same solution procedure then there can be no discussion; likewise if all solutions are on the same level. In other words, the availability of adequate instructional activities is essential for a productive teaching–learning process in which the students can have a substantial constructive input.

The didactic phenomenology requires the researcher to go in search of appropriate context problems. The suitability of the context problems is thereby automatically a criterion. This has to do primarily with the relation between the context and the students' spontaneous solution strategies. Do the students indeed make use of the footholds offered by the context? Do they apply their own domain-specific knowledge? And, equally important, do the solutions they come up with offer possibilities for vertical mathematization?

An example of a situation that is expected to offer the students footholds could be:

Swiss Cheese costs $1.20 per lb. How much does 0.75 lb cost?

The students might, for instance, interpret $1.20 as four quarters and four nickels. Three quarters of this would be three quarters and three nickels. Other strategies might involve doubling or halving, while keeping in mind that the prices are proportional to the quantity. The students might realize, say, that 0.75 lb is half of 1.5 lb, and therefore the price is half of $(1.20 + 0.60) (Gravemeijer 1992).

Another aspect of the phenomenology involves the applicability. Solutions will, in the first phase, be local. Each problem is approached as a new problem and the solutions will contain clear context-specific elements. After a while, however, the communal must begin to prevail and a more broadly applicable piece of mathematics must develop. Whether the knowledge developed in this manner can truly be broadly applied is an important criterion for the researcher. As indicated earlier, the didactic phenomenology outlines the area of application to be considered.

The level structure and the related role of emergent models also provide the necessary criteria. The criteria for emergent models can be captured as: 'naturalness', 'vertical power' and 'breadth of application'.

Characteristic of the realistic mathematics education approach is that the models spring from the students' own activities. It is in this sense that there must be *'naturalness'*. Ideally, the students (re)invent the models on their own. In cases where a model is presented, a requirement is that it fits in with the informal strategies demonstrated by the students. The idea is that the model offers the students the opportunity to be faithful to their own solution procedures. In other words, the students should not adapt their solution procedures to the model; the model must adapt to the students' thought processes (Gravemeijer 1993). This should become evident through, among other things, flexible use of the models.

'Vertical power' may be deemed to be present if the students spontaneously abbreviate and schematize their way of working. A level shift from 'referenced' to 'general' is crucial here. This becomes evident when the students break the correspondence between the semantic structure of a problem and the mathematical structure of the solution procedure. Another indication of shift in levels is when the students themselves bring up the matter of the efficiency of an approach.

This shift can also come to the fore in the way the students work with a model. With decimals, the number line first functions as a ruler with precise and fixed segments. A vertical shift would mean that the number line becomes a thinking model. This would be revealed through the students' drawings when they do no longer need precise and complete segmentation.

The matter of applicability was already raised by the didactic phenomenology. In relation to models, the criterion of *'breadth of application'* asks that students be able to deal with a model in a variety of applied situations.

This overview of criteria is, of course, incomplete in the sense that it is not exhaustive. Moreover, the categories are not mutually exclusive. And yet, it does offer a representative picture of the criteria used by developmental researchers in the field of realistic mathematics education. For the sake of completeness it should be pointed out that these criteria not only serve as a gauge of the expectations arising from the thought experiment, but that they also act as a searchlight for discovering the students' insightful approaches.

9. JUSTIFICATION

Being a combination of development and research, developmental research has the dual function of production and justification. The result of developmental work is a prototypical course. The result of research is a description and rationale of the course on a meta-level (a local instructional theory) and a justification. The justification will focus on why the course satisfies the core principle of realistic mathematics education. The justification will consist of an argumentation, supported by a theoretical analysis, empirical data and the interpretation of this data. A justification of this nature will contain: an analysis of the area of subject matter, an intrinsically substantiated characterization of the structure and content of the course, paradigmatic examples (of student work and interaction) and a reflection on the realistic caliber of the whole. A balanced reflection will mention any negative characteristics in addition to positive ones.

If the developer fails to note any negative aspects, it is likely that others will do so. In the realistic research community it is common practice for the prototype designs to be brought to the attention of interested parties at an early stage. This will thereby significantly expand the subjective research experience of the developmental researcher.

10. INTERSUBJECTIVE AGREEMENT

The new prototype will become the subject of discussion among the experts in the domain of mathematics education. This group consists, not only of developers and researchers, but also of school counselors, textbook authors, teacher trainers and teachers. Discussion will include topics such as effectiveness, feasibility and theoretical quality in the light of the realistic goals. Eventually, an inter-subjective agreement regarding the value of the prototype will come forth.

A characteristic and essential facet of this process is a certain homogeneity of the group in question. In order to carry on a respectable discussion and to arrive at a consensus it is necessary to have a communal frame of reference. This also demonstrates, at the same time, the limitations of the value judgments and claims. Because of the communal frame of reference, the various members of the group will interpret the same phenomenon in (more or less) the same way, giving rise to the danger of systematic errors. Not that this is anything new – every research group must contend with this problem. Blind spots may appear but remain unnoticed by the members of any research community. On the other hand, it is also true that such a research community is the basis for growth. This is, after all, the way in which research programs work (Kuhn 1970; Lakatos 1978).

Such a closed research community can be very helpful when establishing the theory. Later on, however, the aim will have to be to inform and convince the educational research community at large. To do so, the researchers have to explain and justify their findings in such a way that both the theory and the research process that led to these findings is accessible for relative outsiders. This is what Streefland (1990b) did for his local instructional theory on fractions. Other, sometimes less detailed, examples can be found in de Lange (1987), van den Brink (1989), Treffers (1987) and Gravemeijer (1994b). It should be noted that this kind of justification is congenial to the way Cobb, Perlwitz & Underwood (1996) justify an instructional sequence for multi-digit addition and subtraction, and how Lampert (1989) justifies her instructional approach for decimals. Characteristic of this kind of justification is the intertwinement of empirical and theoretical justifications.

11. CONCLUSION

The developmental research discussed in the preceding pages describes a research method that is at the heart of an innovation process; an innovation process whose aim is to establish an instructional practice consonant with the ideal of realistic mathematics education. The goal of the research efforts is to develop a domain-specific instructional theory for realistic mathematics education. It is based on a method of elaborating an instructional theory in a cyclic process of developing and testing instructional activities. In fact, several cyclic processes are involved. These concern instruction theories at three levels (micro theory, local instructional theory and domain-specific instructional theory) which are reflexively related.

The complete experiment can be seen as a research program consisting of a series of research projects. In an individual research project, the researcher makes it his or her object to find out what an instructional course for a given topic should look like to make it fit the general points of departure. In the developmental process, three key heuristics can be employed: the reinvention principle, the didactic phenomenology and the concept of emergent models. Each of these heuristics gives rise to criteria which the researcher will use in the evaluation of the instructional activities.

The theory gain is in the learning process of the researcher and the underlying process of deliberation and testing entails the justification of that theory gain. Crucial for this method is the explication of the learning process. The researcher has to explain and justify his findings in such a way that it enables outsiders to retrace this process. In this way a basis is laid for a scientific discussion that may lead to intersubjective agreement.

The manner of working of a developmental researcher was typecast as theory-guided bricolage. Looking back at the elaboration on the previous pages, we may discern three types of theory-guided bricolage. First, there is the bricolage that aims at a well-considered preliminary design. Second, there is the theory-guided bricolage process that takes care of the elaboration of a course: here the bricolage concerns the elaboration, adaptation and constitution of instructional activities in response to the experiences with trials in educational experiments. The third type concerns theory-guided bricolage as the long-term process of theory development. This applies to the evolutionary development of the domain-specific instructional theory. One may speak of 'A Never Ending Story' because the instructional theory is elaborated, refined and adapted in an endless bricolage process that interacts with changing social demands and developing technologies.

The developmental-research paradigm differs from mainstream curriculum approaches in that the idea of putting ready-made courses to the test is abandoned in favor of an experiment in which the 'treatment' is open to adaptation and elaboration. This results in a learning process for the developmental researcher. The explication of this learning process then offers a rationale and a justification of the course which is critical for the understanding, acceptance and idea-consistent adaptation of the course by textbook authors and teachers.

The theoretical yield of the research within this paradigm has a long-term effect. It took several decades for the domain-specific instructional theory for realistic mathematics education to develop. In this period, the reinvention principle and the didactic phenomenological analysis were elaborated, and the notion of emergent models was construed, together with other insights: for example, the distinction between horizontal and vertical mathematizing (Treffers 1987), the potential of 'free productions' (Streefland 1990a), the intertwinement of learning strands (Streefland 1990b; Treffers 1987), and the necessity of a dispersion of solution procedures, as mentioned above.

This rich harvest was only possible thanks to many years of developmental research and the Freudenthal Institute was very fortunate to be able to work within the same research tradition for such a long time.

ACKNOWLEDGEMENT

The preparation of this paper was in part supported by the National Science Foundation under grant number RED-9353587. All opinions expressed are solely those of the author.

NOTES

1. The term 'research program' suggests a more formal structure and more planning than is actually true in the Dutch situation.

2. Both Sfard (1991) and Ernest (1991) refer to a similar role of reification in the development of mathematics itself.

3. This tension is also stressed by Lampert (1985).

REFERENCES

Bell, A., Swan, M. & Taylor, G.: 1981, 'Choice of Operation in Verbal Problems with Decimal Numbers', *Educational Studies in Mathematics* 12 (4), 399–420.

Bruner, J.: 1994, Four Ways to Make Meaning. Invited Address at the AERA Conference 1994, New Orleans.

Cobb, P., Perlwitz, M. & Underwood, D.: (1996), 'Constructivism and Activity Theory: A Consideration of Their Similarities and Differences as they Relate to Mathematics Education', in H. Mansfield, N. Pateman & N. Bednarz (eds.), *Mathematics for Tomorrow's Young Children: International Perspectives on Curriculum,* Kluwer Academic Publishers, Dordrecht.

Cobb, P., Yackel, E. & Wood, T.: 1992, 'A Constructivist Alternative to the Representational View of Mind in Mathematics Education', *Journal for Research in Mathematics Education* 23 (1), 2–33.

Davis, P. J. & Hersh, R.: 1981, *The Mathematical Experience,* Houghton Mifflin, Boston, MA.

De Lange Jzn., J.: 1987, *Mathematics, Insight and Meaning,* OW&OC, Utrecht.

Ernest, P.: 1991, *The Philosophy of Mathematics Education,* The Falmer Press, London.

Freudenthal, H.: 1973, *Mathematics as an Educational Task,* Reidel, Dordrecht.

Freudenthal, H.: 1983, *Didactical Phenomenology of Mathematical Structures,* Kluwer Academic Publishers, Dordrecht.

Freudenthal, H.: 1988, 'Ontwikkelingsonderzoek', in K. Gravemeijer & K. Koster (eds.), *Onderzoek, Ontwikkeling en Ontwikkelingsonderzoek,* OW&OC, Utrecht.

Freudenthal, H.: 1991, *Revisiting Mathematics Education,* Kluwer Academic Publishers, Dordrecht.

Gagné, R. M.: 1977, *The Conditions of Learning* (3rd ed.), Holt, Rinehart & Winston Inc., New York.

Gagné, R. M. & Briggs, L. J.: 1979, *Principles of Instructional Design* (2nd ed.), Holt, Rinehart & Winston Inc., New York.

Gravemeijer, K.: 1991, 'An Instruction-Theoretical Reflection on the Use of Manipulatives', in L. Streefland (ed.), *Realistic Mathematics Education in Primary School,* CD-ß Press, Utrecht, 57–76.

Gravemeijer, K.: 1992, Micro-Didactics in Mathematics Education. Paper presented at the AERA Conference 1992, San Francisco, CA.

Gravemeijer, K.: 1993, 'The Empty Number Line as an Alternative Means of Representation for Addition and Subtraction', in J. de Lange, I. Huntly, C. Keitel & M. Niss (eds.), *Innovation in Mathematics Education by Modeling and Applications,* Ellis Horwood Ltd., Chichester, 141–149.

Gravemeijer, K.: 1994a, 'Educational Development and Developmental Research in Mathematics Education', *Journal for Research in Mathematics Education* 25 (5), 443–471.

Gravemeijer, K.: 1994b, *Developing Realistic Mathematics Education,* Cdß-Press, Utrecht.

Greeno, J. G.: 1991, 'Number Sense as Situated Knowing in a Conceptual Domain', *Journal for Research in Mathematics Education* 22, 170–218.

Jacob, F.: 1982, 'Evolution and Tinkering', *Science* 196, 1161–1166, 10 June 1977; republished in *The Possible and the Actual,* Pantheon Books, New York, 1982.

Kamii, C., Lewis, B. A. & Livingstone Jones, S.: 1993, 'Primary Arithmetic: Children Inventing Their Own Procedures', *Arithmetic Teacher,* Vol. 41, no. 4, 200–203.

Kuhn, T. S.: 1970, *The Structure of Scientific Revolutions* (2nd ed.), Chicago University Press, Chicago.

Lakatos, I.: 1978, *Mathematics, Science and Epistemology,* Cambridge University Press, Cambridge.

Lampert, M.: 1985, 'How do Teachers Manage to Teach? Perspectives on the Problems of Practice', *Harvard Educational Review* 55, 178–194.

Lampert, M.: 1989, 'Choosing and Using Mathematical Tools in Classroom Discourse', in J. Brophy (ed.), *Advances in Research on Teaching,* Vol. 1, JAI Press Inc., Greenwich, CT, 233–264.

Lampert, M.: 1990, 'When the Problem is not the Question and the Solution is not the Answer: Mathematical Knowing and Teaching', *American Educational Research Journal* 27, 29–63.

Lawler, R. W.: 1985, *Computer Experience and Cognitive Development: A Child's Learning in a Computer Culture,* Ellis Horwood Ltd, Chichester, and John Wiley & Sons, New York.

Merrill, M. D., Li, Z. & Jones, M. K.: 1990, 'Limitations of First Generation Instructional Design', *Educational Technology* XXX (1), 7–11.

NCTM Research Advisory Committee: 1988, 'NCTM Curriculum and Evaluation Standards for School Mathematics: Responses from the Research Community', *Journal for Research in Mathematics Education* 19, 338–344.

Romberg, T. A.: 1973, *Development Research: An Overview of How Development-Based Research Works in Practice,* Wisconsin Research and Development Center for Cognitive Learning, University of Wisconsin-Madison, Madison, WI.

Romiszowski, A. J.: 1981, *Designing Instructional Systems,* London, Kogan Page.

Sfard, A.: 1991, 'On the Dual Nature of Mathematical Conceptions: Reflections on Processes and Objects as Different Sides of the Same Coin', *Educational Studies in Mathematics* 22, 1–36.

Steffe, L. P.: 1983, 'The Teaching Experiment Methodology in a Constructivist Research Program', in M. Zweng, T. Green, J. Kilpatrick, H. Pollak & M. Suydam (eds.), *Proceedings of the Fourth International Congress on Mathematical Education,* Birkhäuser Inc., Boston, MA, 469–471.

Streefland, L.: 1985, 'Vorgreifendes Lernen zum Steuern langfristiger Lernprozesse', in W. Dörfler & R. Fischer (eds.), *Empirische Untersuchungen zum Lehren und Lernen van Mathematik,* Hölder-Pichier-Tempsky, Vienna, 271–287.

Streefland, L.: 1990a, 'Free Productions on Teaching and Learning Mathematics', in K. Gravemeijer, M. van den Heuvel-Panhuizen & L. Streefland (eds.), *Contexts, Free Productions, Tests and Geometry in Mathematics,* OW&OC, Utrecht, 33–52.

Streefland, L.: 1990b, *Fractions in Realistic Mathematics Education, a Paradigm of Developmental Research,* Kluwer Academic Publishers, Dordrecht.

Streefland, L.: 1993, 'The Design of a Mathematical Course: A Theoretical Reflection', *Educational Studies in Mathematics* 25 (1/2), 109–136.

Treffers, A.: 1987, *Three Dimensions. A Model of Goal and Theory Description in Mathematics Education: The Wiskobas Project,* Reidel, Dordrecht.

Van den Brink, J.: 1989, *Realistisch Rekenonderwijs aan Jonge Kinderen,* Freudenthal Institute, Utrecht.

Van Hiele, P. M.: 1973, *Begrip en Inzicht,* Muusses, Purmerend.

Walker, D. F.: 1992, 'Methodological Issues in Curriculum Research', in P. Jackson (ed.), *Handbook of Research on Curriculum,* Macmillan, New York, 89–118.

Koeno Gravemeijer
Freudenthal Institute,
Tiberdreef 4,
3561 GG Utrecht,
The Netherlands

GILLIAN HATCH & CHRISTINE SHIU

PRACTITIONER RESEARCH AND THE CONSTRUCTION OF KNOWLEDGE IN MATHEMATICS EDUCATION

1. INTRODUCTION

What is research in mathematics education? In this chapter we start from the position that it is

> *the intentionally controlled examination of issues, within and related to the learning and teaching of mathematics, through a process of inquiry that leads to the production of (provisional) knowledge both about the objects of the inquiry and the means of carrying out that inquiry.*

We further suggest that research in mathematics education is of limited value unless it affects classroom practice and experience. Thus a criterion for *important* research is the applicability of outcomes of the research to the classroom. In other words, the most important location for constructed knowledge of mathematics education is within the practitioner, and its most important manifestation is the knowledge-in-action displayed by the teacher in the classroom.

A related question then arises. *Who* does research in mathematics education? If all research is carried out by a dedicated body of mathematics educators privileged as researchers, then there is a danger of the isolation of research from teachers and its value cannot be realized. Our claim is that the need is for the teacher to be part of the process of research and to have direct access to the outcomes. An essential version of this claim is the 'teacher-as-researcher' empowered to examine immediately the questions of practice and meaning that arise in every mathematics classroom.

This is not to claim that all research in mathematics education can be carried out by practitioners, nor to deny the importance of those projects that can only be carried out by dedicated researchers. Rather we are saying that teachers of mathematics are uniquely placed to investigate – and record – aspects of their teaching, their classroom and their students that are hidden from others. Our aim is to describe and illustrate such investigation, first to show that this process does meet the definition offered above, and secondly to consider some of the questions of objectivity, ethics, interpretation and dissemination that arise when the process becomes overt.

Sierpinska, A. and Kilpatrick, J. Mathematics Education as a Research Domain: A Search for Identity, 297–315.

First, we consider what kinds of questions a teacher is likely to formulate explicitly both 'in-the-moment' within her classroom, and as she reflects on a lesson 'after-the-event'. These are frequently questions of the forms:

- What actually happened when ...?
- Why did that happen?
- How did my teaching actions affect what happened?
- What would happen if ...?

The first three invite the practices of what John Mason has called 'giving an account of' and 'accounting for'; the fourth leads into the areas of prediction and hypothesis testing. A mode of addressing such questions is often called *reflection,* in the sense used in the following quotation:

Reflection is the process of learning from experience. It is what a teacher does as he or she looks back at the teaching and learning that has occurred and reconstructs the events, the emotions, and the accomplishments. (Wilson, Shulman & Richert, 1987, p. 120)

Such reflective practice is the hallmark of the practitioner as professional, described by Donald Schön (1983) in his seminal work *The Reflective Practitioner*:

When someone reflects in action he becomes a researcher in the practice context. He is not dependent on the categories of established theory and technique, but constructs a new theory of the unique case. His inquiry is not limited to a deliberation about means which depends on a prior agreement about ends. He does not keep ends and means separate, but defines them interactively as he frames a problematic situation. He does not separate thinking from doing, ratiocinating his way to a decision which he must later convert to action. Because his experimenting is a kind of action, implementation is built into his inquiry. (p. 68)

This is not to say that all classroom research should be done by practitioners. Mathematics education, like medicine, is a field of practice. In medicine, both dedicated researchers and practitioners, general and specialist, can and do contribute to the body of clinical knowledge that is derived from accumulated case studies. Similarly, in mathematics education the observer-researcher and the reflective practitioner both have valid and important perspectives within the classroom. The observer-researcher can collect data from a variety of classrooms and is likely to possess greater awareness of the theoretical backgrounds developed by earlier researchers, and so the work produced offers great potential for generalization. By contrast, the reflective practitioner has extensive and intimate contact with a particular classroom, and the directness of her observations within her experience of the classroom means that the constructs of her research can be more readily shared with other teachers. Each researcher role gives access to information that is denied to the other, and even the pooled knowledge can tell only a selective part of the story of any classroom. Nevertheless, the knowledge gained by the practitioner-researcher within her own classroom is likely to be the most readily usable knowledge available and thus fulfills our essential criterion for important research.

Unlike medicine, mathematics education has few formal mechanisms for collating and disseminating practitioner research. We believe that the development of a system to assist the research community (in the most inclusive sense) to access evidence of this research genre would be of great value. It is our awareness of the existence of extensive but scattered classroom data that has motivated us to speculate on the possibility of creating a national database of such studies that would allow further use to be made of the results. This possibility is discussed in more detail in the final section of this chapter.

2. ACTION RESEARCH IN THE UK

The action research movement is often traced from the work of Kurt Lewin and his associates. In talking about this work, Adelman (1993), has quoted Marrow (1969), who describes the classification of action research into four types, namely: *diagnostic, participant, empirical,* and *experimental.*

The teacher-researcher can undertake research of any of these types (see, e.g., some of the accounts in Brubacher, Case & Reagan 1994), but it is the third – *empirical* – that is the major concern of this chapter, and the one that offers the closest analogy with the development of professional knowledge in medicine.

In the UK, there is now a robust tradition of action research in education, often inspired by work at the Universities of East Anglia and Bath. See, for example, Stenhouse (1975) and Nixon (1981). Jean McNiff (1993), who has worked in this tradition for some time, writes:

An educational epistemology of practice is one which allows teachers to undertake an enquiry that is focused on the development of personal understanding. The aim and the process of practice becomes the object of the enquiry; practice becomes enquiry. The practice of teaching others becomes the process of learning about oneself. The process of learning about oneself becomes the object of the research....

We need to see research as practice, and that pedagogic practice should be viewed as a constant process of enquiry. If we view teaching in this light, it is no longer an activity geared toward passing on information; it becomes a shared communicative exercise which is focused on generating intersubjective agreements about the nature of being. (p. 59)

In this tradition of action research, teachers of a variety of subjects offer each other mutual support in setting their own agendas, refining their research questions and developing the means of pursuing them in a whole variety of educational issues and contexts. One such group – of teachers of mathematics – published a document under the aegis of the Association of Teachers of Mathematics (1987) in which they declared:

This book is intended to stimulate discussion. We hope that it will encourage teachers and others to reflect on their interpretation of the word research and to discuss its place in their professional life. In particular this document aims to provide support for teachers in recognising aspects of their work as research and in extending and sharing their research experiences. (p. 1)

There have been some national initiatives within the UK in which an academic institution or grouping has received funding to promote, support and coordinate teachers' own research into mathematics teaching and learning, notably the Low Attainers in Mathematics Project (1987) and the Primary Initiatives in Mathematics Education project (1991). By contrast, the examples of practitioner research that we examine in the next section are instances of where individual teachers or student teachers have engaged in very specific personal research, often without any defined possibility of significant dissemination of their work, though usually in a context which allows local or peer support.

But, beyond our examples, there is also much local small-scale personal inquiry in mathematics education that involves the practitioner-inquirer in the personal construction of knowledge. Such practitioners have varied degrees of support and sharing within their peer community and opportunities for sharing and negotiating their knowledge within the wider mathematics education community. We suspect that our instances represent the tip of a very large iceberg of practitioner research into mathematics education.

3. CONTEXTS FOR PRACTITIONER RESEARCH IN MATHEMATICS EDUCATION

Personal Research in Initial Teacher Education

We believe that reflective practice is essential to effective and developing teaching, and that this implies that a research-oriented approach needs to be established in the repertoire of the beginning teacher. In accordance with this belief, it has long been the practice at Manchester Metropolitan University (MMU) to embed such an approach in the initial training of teachers on the four-year Bachelor of Education degree. In the early stages of the course, students are encouraged to identify and make explicit problems that they encounter in their classroom placements. They then investigate possible solutions through reading and peer and tutor discussion, and overtly try out some of the suggested possibilities in order to evaluate the best solution for them at that stage in their development as a teacher. Later periods of time in the classroom involve the development of broader observational skills to support the identification of 'key features' of the teaching context in which they find themselves. This implicit personal research now begins to encompass interaction between their own developing skills as teachers of mathematics and the features of the particular context in which the skills are being exercised. It attempts to draw more and more on the children's responses to teaching actions in order to evaluate practice. Throughout the course, students are required to keep an up-to-date teaching file that details classroom plans, their rationale, observed outcomes and subsequent evaluation. This file is a mechanism that both encourages a reflective and analytic approach and contributes to the assessment of the teaching practice at honors level. The best students soon manifest in their writing an attitude that suggests that indeed pedagogic practice *is* 'viewed as a constant process of enquiry', and many are able to move onto higher degree work of the kind described in our third example, at an early stage in their teaching career.

In their final year, all students undertake an extended piece of explicit action research into a question that they identify from their own teaching. They examine relevant literature and consider the forms of data collection that are open to them, as student-teachers within a school. They then create a research design and implement this through to the analysis of outcomes. The studies they produce are essentially small scale, usually involving working with a small group of children and drawing on very limited experiences of teaching, but nevertheless they produce some interesting transcripts. Collectively, they form a resource of potential richness, which might easily be of use to more experienced researchers.

Personal Research as Professional Development

Many teachers in the UK elect to further their professional development by studying part-time courses as associate students with the Open University. A general description of such courses and their genesis is given by Pimm (1993). Here we focus on one such course designed for teachers of mathematics, namely, EM236: Learning and Teaching Mathematics (Open University 1992). A declared aim of the course is as follows:

As part of this course we hope you will come to recognize or identify more clearly the current characteristics of your own practice as a teacher of mathematics. All practice is continually changing: sometimes this is obvious and in response to readily identifiable forces such as curriculum or assessment requirements but sometimes change is more slow and subtle and the causes are largely hidden. By analysing your own practice, the forces for change may be exposed and evaluated and your chance of taking control of that change – and helping others to do the same – will be increased. (EM236, Unit 1, p. 5)

Students are asked to keep a reflective diary from the beginning of the course; using techniques that aid the deliberate invocation of 'stressing and ignoring', that is, becoming aware of what has been stressed and what ignored in both practice and observation. For a fuller account of such 'researching from inside', see Mason (1994). The 'accounts of' elicited by course tasks and prompted by course materials such as classroom videotape include reports of work on 'own mathematics', 'the mathematical activity of others' and 'own teaching'. As skill is gained in separating 'accounts of' from 'accounting for', the diary becomes the record of field notes of research into practice. Almost all students on the course report enhanced insight into their own context and practice. What the course does at its most successful is develop in teachers an overt methodology for further inquiry into practice, and the skill to communicate their findings, thus potentially linking them to the research community.

Towards a Higher Qualification

One of the higher degrees offered at MMU is a modular M.Sc. in Mathematics Education that further develops the kind of approach started there in initial teacher education. The required 'teaching file' of initial training is now replaced by a diary-type file in which students on the course – who are all teachers – record those observations and reflections to do with mathematics education that *they* perceive as salient.

Some of the first-year modules involve students in studying mathematics at their own level, in part 'for its own sake', but mainly and overtly, in order to study their personal learning processes – to discover how they respond to difficulties, what ways of working aid their understanding and what feelings they experience in relation to those processes. This mathematical learning is carried out within study groups, and they are asked to address explicitly issues to do with the experience of learning in a group, including their awareness of the nature of the learning of other members of the group. They relate these experiences to theoretical accounts of learning and translate the awareness they develop back into attitude and action in their classrooms. Through their writing and the reading to which it leads, they clarify notions and questions they wish to pursue further. Examples of issues that have developed in this way include 'formal and informal language in the classroom', 'emotional conviction' – a concept related to *intuition* (as defined by Fischbein 1982) – 'a learner's perception of his success in a group and its effect on learning' and 'children's responses to mathematical games'.

The second part of the M.Sc. course involves an extended practitioner inquiry. Typically, the students select a topic that draws together a classroom concern and the issues about the nature of learning they have identified as significant in the earlier modules. Many of these pieces of work are of wide relevance to the schools or departments in which they work. Indeed, one of the assessment requirements of the course is that students involve a 'critical community' who will respond to the work that they do both as critical friends during the process and as validators of the products. Once again, students build up *particular* accounts of mathematical learning in their own classrooms, which, if linked, might enable the construction of 'meta-classroom accounts' to describe revealed *generalities*, thus adding to our knowledge of mathematics education.

Research Degrees in Mathematics Education

Like many institutions of higher education the Open University enrolls students for higher degrees by research. The Centre for Mathematics Education (CME) has an established and growing tradition in catering for students who wish to use their own practice as a major object of their research. Such students have included teachers from a variety of institutions and of a variety of age groups from early years through to adult learners. Some examples of their work include inquiries into the roles of experiential learning techniques (Chatley 1992), reflection (Billington 1993) and context (Williams 1989) in mathematics education.

Those who are privileged to work with these practitioner-researchers as supervisors or critical friends frequently observe that the techniques used in carrying out their research often mirror those used in the phenomenon to be examined and documented, and that the means of validating the findings echo methods used in their classrooms for the validation of shared knowledge in mathematics. Thus the practitioner-inquirers' research biographies are an essential part of their theses, and their developed methodologies offer a potential starting point for the inquiries of others.

Another recurring theme arises from the frequency with which these higher degree students change their professional role during the course of their studies – from classteacher to headteacher or advisory teacher or teacher educator. In such cases, research has often become a vehicle for examining that change and has enabled the more reflective – and effective – support of colleagues or students. A serendipitous but valuable test of the generalizability of the research findings then lies in their successful applicability across roles and contexts.

Professional Associations and Members' Knowledge

The role of professional associations in linking people with common interests can be crucial in enabling the creation of meta-classroom accounts. In the UK, two professional associations for teachers of mathematics are instrumental in supporting such links, namely the Association of Teachers of Mathematics (ATM), and the Mathematical Association (MA).

The ATM has been concerned for many years with the support of working groups, usually formed at the instigation of a number of members interested in pursuing a particular common theme or interest in mathematics education. These groups typically meet for residential weekends in order to share insights gained through individuals' classroom experience and to plan further cooperative or mutually supported inquiry. They are usually open to any member who wishes to be involved. Recent themes have included 'the problem-solving school', 'algebraic imagery', 'teacher intervention', and 'the use of LOGO in the classroom'. Contact between face-to-face meetings is often maintained through a process of shared reflective writing, in which writing produced by one or more members of the group is circulated and may be developed or commented on by others. Sometimes this process leads to articles that are published externally; for example, the article by Brown, Hardy and Wilson (1993), which emanated from the work of a group that is exploring the implications of constructivism and psychoanalysis for the teaching and learning of mathematics.

The Mathematical Association also supports groups who are studying topics of mutual interest and is often proactive in setting up such groups, leading to the production of MA reports. A group of practitioners is currently concerned with investigating the teaching and learning of undergraduate mathematics in the UK. It is both surveying current practice and looking to formulate some agreed conceptual stances.

We note that groups of this kind are made possible and fruitful by the networks already existing within an association, and by the opportunity for dissemination offered by the journals of the association and its other publishing ventures (see, e.g., ATM 1986, 1988a, 1988b, 1989). For classroom teachers, this activity often works better for the specialist secondary teacher than for the primary teacher, who only rarely is attached to a single subject association.

In the five examples of practitioner research we have described, it is possible to perceive a movement from contexts that provide a strong pre-existing framework for personal research to those that require a self-created or chosen framework. These contexts give rise to different degrees of constraints on the questions asked, the methods used, and the time scales available for the inquiry. These differences coincide broadly with a general growth of relevant experience and a particular movement from contexts where accreditation is needed – for example, students in initial training need to acquire qualified teaching status (QTS) – to those where accreditation is irrelevant and the creation of (at least) personal knowledge and improvement of practice is the paramount motivation for research.

4. A CLASSROOM CASE STUDY

In order to examine how data generated by teacher-researchers can contribute to both personal and professional knowledge in mathematics education, we focus on one particular transcript produced from an audio recording made by a primary teacher in her own classroom, while on an in-service course (Hallworth 1993). We suggest that this is an example of the sort of material that might profitably be held in a database indexed by various keywords.

The teacher had been studying a module on number theory in which the students were required to reflect on their own experience as learners and to look at their pupils working in related areas. We examine it, first, from her own perspective to see what she might discover about her practice and its effects on the children she teaches, and secondly, from the perspective of an 'external' researcher who serves to identify possible keywords to be associated with the transcript. We show how each can construe different but often complementary aspects of the objects of inquiry.

The children featured in the transcript had been playing a game that generated a distinct 2-digit number for each child. They each receive a score by counting the number of divisors of their number, so in the first part of the conversation they are focusing on finding the divisors of a number.

Excerpt 1

Tim: They are even – so they can be divided by 2.

Mark: Yeah, but you can divide some even numbers by other tables as well.

Tim: I know, but we only need to do it by 2.

Ian: No, you mix that up with that Greek man and his sieve.

Tim: I don't! If you can divide a number by 2, then you can divide it by 4, 6 and 8 as well *(momentary silence)*.

Susie: No, you can't. You *are* mixing up what we did with that sieve!

Tim: I don't!

The *teacher* knows the class and the children but was not present at this stage of the conversation. From the tape and its transcript, she may learn something of the dynamics of the group and whether it will be a productive grouping to use in the future. She can make deductions about the current state of mathematical knowing of the children in the group from the point of view of assessment. Her planning, both for these particular children and for future classes, can be informed.

She seems to have most evidence about Tim; this excerpt is almost a conversation between Tim and the rest of the group. In each of his first three statements, he makes an assertion that may be conjectural. The first two statements appear to indicate a degree of understanding. He seems to see the equivalence between 'even numbers' and divisibility by 2, and he is not distracted by Mark's correct but unhelpful intervention. Tim's third statement, however, seems to indicate a misunderstanding, and the momentary pause on the tape suggests that perhaps the rest of the group are worried by this formulation. The logical form of his statement is sound, 'if x then y', but unfortunately his inference is wrong.

In the same statement, the teacher may hear echoes of the words she had used when she worked with the children on the Sieve of Eratosthenes, deleting multiples of 2 from a 100 square. The transcript offers her a chance to learn about misleading conceptions that her pupils have developed in this context and to take decisions about future teaching approaches with this group and others.

The *researcher* accessing this transcript on a database would perhaps be informed that the children are nine-year-olds. We consider what keywords might have been used to allow her to access this particular script, and suggest the following:

- *Generalization.* There is certainly evidence here about the ability of children of this age to make generalized mathematical statements, such as 'they are even, they can be divided by 2', 'you can divide some even numbers by other tables as well'.
- *Discussion.* There is also evidence of a group of young children maintaining a focused and on-task discussion that could feed in to a study of small group discussions in the mathematics classroom.
- *Cognitive conflict.* In discussion the children have exposed an inherent conflict in their views. Part of the conflict is related to their ability to transfer information from one area of discourse to another. In some cases, this transfer has been achieved correctly, but the Sieve of Eratosthenes has become the site of inappropriate transference.

The transcript continues.

Susie: Well, then explain that. 2 goes into 10, 5 times, but 4 doesn't go into 10 at all, does it!

Ian: It does, but there's two left over.

Susie: Exactly, so 4 is not a factor of 10.

Tim: So, why did we say, if we have crossed out the 2s in that sieve, we don't have to the 4s, 6s, and 8s?

Peter: I think I know! When you crossed out the 2s, you crossed out all the numbers 4 goes into as well. Like 8 – you crossed out because 2 goes into it – but 4 goes also into it and 8.

The *teacher* may note that both Peter and Susie can give very coherent accounts of the problem that has disturbed Tim and may witness the way in which they are using their insights to endeavor to assist Tim to sort out his ideas. She may interpret this exchange as evidence that this grouping is working together in a supportive way.

She may also notice another instance of the possibility that the children are echoing her words from previous discussions. It could be that Ian, in saying 'but there's two left over', is recalling the words she used when working on a division algorithm and is now reproducing this phrase inappropriately. This comment could trigger her awareness of her own speech patterns and enable her to listen to herself for future instances. Such awareness could link to reading she has done about the use of language in the classroom, or prompt future reading.

The *researcher's* account might add new issues or revisit old ones:

- *Peer tuition*. Both Susie and Peter are able to produce ideas that are potentially useful to the others in the group – a counterexample in one case, and a clear explanation in the other.
- *Discussion*. There is further evidence of issues concerning discussion. These children can maintain a point of view against counter argument. They seem able to listen carefully even when not participating verbally. They are able to modify statements made by each other, as when Ian amends Susie's statement '4 doesn't go into 10 at all'.

The transcript again continues:

Tim: So, what about six?

Susie: It's the same!

Tim: How?

Susie: Well, look! It's like that. Say we take the 2 times table – 2, 4, 6, 8, 10, 12,...

Tim: What's that to do with anything?

Susie: Just listen. All those numbers have 2 in them, haven't they?

Tim: Yeah, I guess so!

Susie: But look, 6 is also in the 6 times table, and 6 is in the numbers 12 and 18, and all the other 6 times-table numbers. And all those numbers have the 2 times table in them as well.

(Tim looks totally confused and expresses that feeling.)

Tim: You've lost me!

Susie: Then I can't explain it!

(I came in because Tim has now lost all confidence in his understanding of this concept.)

The teacher has added her own comments in brackets that indicate she was now close enough to the group to be able to take an informed decision on intervention. The consequences of that intervention are not in the transcript, but might allow her to decide whether she intervened at the right point. She may deduce that Susie has worked through to an extremely coherent explanation and judge that she, not being a teacher, has reached the end of her surprisingly extensive resources. Tim, however, seems to have got even more confused. There are therefore issues concerned with classroom organization such as how she tries to ensure that such moments are picked up reliably. She can reflect on whether it was pure luck or trained awareness that caused her to be within earshot at a crucial moment. For any teacher using group work, this is a significant issue for study.

From the point of view of the *researcher,* some more potential issues have arisen:

- *Teacher intervention.* For the researcher, too, the whole question of effective teacher intervention is raised. A premature intervention might have deprived the children of a valuable opportunity to talk about mathematics. That is true both for those able to articulate their insights and for Tim in becoming convinced that there was a complexity involved that he had not recognized.
- *Informal language.* One might examine the relatively informal use of language by children in explaining mathematics, as, for example, when Susie says, 'And all those numbers have the 2 times table in them as well'. It is hardly precise mathematics but could be more helpful than its formal counterpart in assisting understanding and is worthy of study in its own right.
- *Mathematical environment.* A question that is suggested by the transcript as a whole, and for which evidence might be found, is that of how any teacher (or any school) creates a mathematical environment in which pupils are able to sustain this level of discourse.

5. THE PROCESS OF PRACTITIONER RESEARCH

At what point do the actions and reflections of a teacher become identifiable as practitioner research, rather than part of the normal process of teaching? Any answer to this question lies, in part, in the extent to which the research produced can be validated and shared, and involves consideration of the methods available to the practitioner and the ethical constraints under which she must operate.

First, however, we start with a brief consideration of what is researched.

Objects

In the case study, we drew attention to those aspects of the transcript that are of immediate use to the teacher. She now has additional recorded access to the speech and interactions of her students, which may be interpreted in terms of:

- their particular understanding of the mathematical topic,
- the possible effects of her own past speech and behavior, and
- the impact of the way she organizes her classroom on pupil behavior and potential learning.

She discovers what is particular to her own classroom and has the opportunity to compare this with the average or typical case, revealed by large-scale external research.

We also indicated the potential value of her transcript to other researchers seeking evidence of a variety of phenomena as they arise naturally in a mathematics classroom.

Methods

The methodology available to the teacher researching her own practice is essentially that of participant observer. This role may involve conscious and deliberate moves from the everyday pressures of the classroom into reflective mode. We shall call this process *distancing*. Thus a basic need is often to produce a 'record' that becomes exterior to the teacher and is separated from some of the involvement that makes abstract analysis difficult. This record, together with the creation of external data, is what moves the activity from teaching to research. In this way, an increase in objectivity is achieved that allows the researcher to declare her relationship to the data, so that the emotional involvement is not lost but is acknowledged as part of the 'event' being researched. This record could be a transcript (as in our case study), a videotape, a sequence of carefully related anecdotes, or simply perhaps a reflective journal of classroom happenings kept by the teacher. The teacher can then use the record of the events as a prompt in accounting for her observations. The 'hard copy' is also crucial in that it creates 'data' that can then be discussed by others, compared with other accounts in a variety of forms, or used directly by the teacher herself to form meta-accounts of her classroom over time. Once the account has been created, it can be revisited, and the passage of time may effect a further distancing, allowing the teacher to give an account of her position relative to the event then and now.

The diagram in the next section shows our view of the teacher-researcher process. It indicates the first step as being distancing, that is, moving from the act of classroom teaching to the research process. It also represents the ways in which validation may be achieved.

Objectivity and Validation

Questions of objectivity arise in all educational research. The methods employed, the means of gathering data, the questions asked are all the product of personally, historically and culturally-determined specific choices made by a researcher or group of researchers. The data thereby produced are necessarily constrained by these choices and need to be interpreted in the light of the probable impact of those choices, and of the context in which the research is carried out.

When qualitative approaches are used, what is observed and recorded depends crucially on what the observer notices, whether or not that observer is also the teacher. Information about context and the role of the observer and about her relationship to what is observed is needed for assessing the validity of data and for interpretation. The objectivity of an external observer has to be balanced against the fact that any extra person in the classroom inevitably affects the classroom, making it to that extent different from what it would have been if the teacher were there alone.

We see at least three possible stages of validation for practitioner research. All of these are indicated in the diagram. First, the teacher examines her own experience for face validity as she accounts for what has happened. She can then extend the validation process by sharing both accounts and analyses with other practitioners, who become a critical community for her research. Together they can consider whether conjectures created to explain the happenings in one classroom have resonance for other practitioners analyzing their own classrooms. Through this, often cyclic, process of peer validation, the teacher-researcher can develop her own grounded constructs, which can be taken directly back into her own classroom for further refinement. Professional (as opposed to academic) journals can provide access to a wider critical community. In publishing teachers' reflective accounts of classroom happenings, they provide a source for comparison with events in the researcher's own classroom. From both commonalities and dissonances that she perceives, she can elaborate and refine the personal constructs, and formulate new conjectures to test in her classroom (Figure 1).

Peer validation was manifest in each of the contexts we discussed earlier. In all five cases, the role of the *other* – tutor, supervisor, colleagues, the wider mathematics education community – in the initiation and validation of inquiry is worthy of examination. Initiation is concerned with confirmation that the inquiry is worthwhile – here a critical friend is useful in testing whether the initial constructs make sense to another practitioner. As the inquiry progresses, the validity of the methodology and the findings can be tested against tutor or peer experience. As a practitioner-researcher, *I* may try to replicate in my teaching context what *you* have done in yours, and hence explore the particularity of your investigation and the limits of its generalizability. At other times, I may be

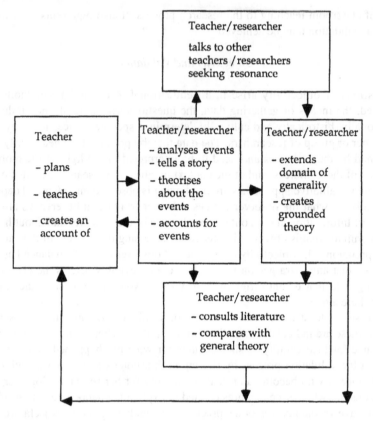

Figure 1   *Diagram for practitioner research*

stimulated by the findings of public large-scale research to discover whether its claims can be demonstrated in my own teaching context. The personal knowledge gained thereby becomes an important means of critiquing these claims.

Sometimes it will become appropriate to undertake the further validation of comparison with published theories in the academic literature. The teacher-researcher can identify themes arising from her analysis that have been widely studied elsewhere, and consider the relationship between her own observations and these theories. She can extract implications or develop modifications that have relevance to her situation. This undertaking is more likely to lead to the overt sharing and public acceptance of her findings, but it should be noted that whether research is carried out by a practitioner or an outside observer, validation and generalization arise from comparing and contrasting across different groups of pupils and different classroom situations. The difference lies only in limitations on the domain of comparison. In short, we suggest that the move from teacher to researcher can be perceived as a move along a continuum rather than a dichotomy; further, the move from face validation through peer validation to theoretical validation is also a continuum, which is often reflected by different forms of publication and dissemination.

In comparing the contribution to professional knowledge of practitioner research with that of 'dedicated' research, we may perceive different strengths and weaknesses. The generality of 'external' research and its careful embedding in the work of previous researchers should ensure that it can be applied to a wide variety of classrooms. This very generality, however, and its formal reporting may make it problematic for the practitioner to apply it in a specific situation. Practitioner research, grounded as it is in vivid accounts of the particular, clearly loses immediate generalizability, but can often speak more directly to a practitioner, who, by testing its findings in her own classroom, undertakes empirical generalization.

Ethical Considerations

All teacher-researchers, whether they are teachers who research their own practice or researchers who assume the role of teacher as part of a research project, face similar moral choices. When they are actually teaching, they must give the immediate educational needs of the pupils priority over their own research concerns. We offer an example from the experience of one of the authors who is currently acting 'as a teacher' while researching the use of games to teach the early stages of algebra (see Hatch, in preparation). When working with a small group, she sometimes has to decide whether it would be justifiable in the interests of her research to leave them to struggle with the ideas of the game with minimal teacher support. Having withdrawn the group from their normal lessons, she feels ethically constrained to offer the resource of her availability as a teacher if this seems likely to maximize immediate learning.

However, choices of this nature are rarely simple, as there is always a balance to achieve between the long-term and short-term needs of the pupils. The researcher must also decide whether to work in a way that would be applicable to a whole class or to employ techniques only available when working with a small group. Finally, for the teacher researching her own practice, there is the recurrent dilemma of balancing the gains for the current class against the potential gains of (possibly many) future classes.

A further issue arises when the practitioner-researcher moves to publication. If the identity of the teacher is declared, the identity of school and pupils can be only thinly concealed, and any interested reader could penetrate this disguise with ease. There are times, therefore, when external publication of results that could be read as being critical of a school or potentially damaging to individual pupils must be eschewed.

These ethical dilemmas are not exclusive to the teacher-researcher, but are faced by anyone engaged in qualitative research. However, they will often manifest themselves in particularly acute form where the researcher is regarded as teacher by the pupils. Sowder (this volume) has discussed such issues and their philosophical origins. Following her analysis, the conscientious researcher will always consider the costs and benefits of her actions to the pupils, and treat them as an end rather than as means. We suggest that the teacher-researcher is in any case disposed to do this by the very motivation that led her to the profession of teaching.

Dissemination of Practitioner Research

From the five examples of practitioner-researcher given earlier, the role of writing in the research process can be identified, whether this is manifest as a teaching file, a reflective diary of practice, formal field notes or shared reflective writing within a working group. We illustrated how the discipline of making observations, interpretations and insights explicit is frequently experienced as a powerful means of allowing validation and prompting further development. All this writing is essentially private, confined, at most, to a small group of others, and often seen only by the practitioner herself. Often it is quite appropriate that written dissemination should go no further – the constructed knowledge is immediately available to the practitioner, who benefits personally and professionally and uses that knowledge to the benefit of pupils and colleagues. Just as at the beginning of the article we acknowledged that not all research into mathematics education should be carried out by practitioners, now we acknowledge that not all practitioner research can, or indeed should, be widely available in written or other forms. However, as we suggested in our classroom case study, data generated by the teacher can serve two purposes: first, in its contribution to her own growing personal knowledge; and second, in adding evidence to a growing body of professional knowledge. We conclude this chapter by considering the possibilities for the further dissemination of the findings of practitioner research where that *is* appropriate.

6. NETWORKING

Research degree theses are available through institutional libraries and may often lead to publications in conventional academic journals. The products of working groups in professional associations may lead to articles in the journals of that association or to special reports or other published materials. By contrast, as we have observed, the writing produced in the course of personal inquiry in initial teacher education or an in-service course involving professional development, however potentially valuable, may have no obvious channel for wider dissemination.

We believe that accounts of small-scale action research have the potential to contribute more generally to professional knowledge of the teaching and learning of mathematics. However, in our experience it is rarely possible, and never easy, to amalgamate data from a variety of sources, representing evidence gathered by different members of the same sample set, in order to create neat classroom accounts. We therefore ask what is necessary to make such data accessible in a way that allows the accumulation of evidence with respect to particular research inquiries. We suggest that the more formal work done in contexts equivalent to any of our first three examples could be collated to form the basis of a national (or even international) database. This database could contain both:

- Directly accessible materials of several kinds; for example, transcripts giving evidence of classroom utterances, reflective writing giving evidence of practitioner philosophy and priorities and limited quantitative data collected in clearly described contexts. All of these could be referenced by keywords to permit access and analysis by theme.
- Reference to more extensive 'accounts of' and 'accounting for' produced by students and practitioners, held in a variety of locations but that could be made available to the bona fide external researcher.

Three things. are needed to realize this possibility. First is the belief by the research committee in the value and the validity of such small-scale inquiry. The second is the will in the mathematics education community to identify those pieces of student and teacher work that seem to have value outside their original context. The last is the resource to permit the establishment and support of appropriate storage and retrieval systems.

If such a project could be realized, then a number of benefits could follow. First and foremost there would be increased communication between different groupings in the mathematics education community. This step in turn might lead to more fruitful dialogue, including perhaps our suggested 'meta-classroom accounts' based on a wide and robust foundation of evidence and expressed in terms supporting further negotiation of common meaning. In addition, groups and individuals working independently of institutional support might offer their data to increase the representativeness of the collection. Practitioner-researchers whose work was included in the database would find an extra encouragement to persevere with what is often a lonely task. Others might find inspiration and motivation in their own research from the contributions of others. All these benefits would encourage the further development of practitioner research, and thence, more effective professional practice through the construction of useful and usable knowledge in mathematics education.

7. SUMMARY

The following definition of research in mathematics education is proposed:

the intentionally controlled examination of issues, within and related to the learning and teaching of mathematics, through a process of inquiry that leads to the production of (provisional) knowledge both about the objects of the inquiry and the means of carrying out that inquiry.

We have illustrated how the classroom teacher can, through developing reflective practice and by capturing recordings of classroom events, engage in that process of inquiry. In so doing, she contributes not only to her own developing knowledge of her own practice and classroom but potentially to the accumulated knowledge of the research community. Further, in writing about what she does, she offers fellow teachers a possible means of carrying out their own inquiry. Provided that writing is in some sense in the public domain, she offers it to the research community for replication or response. Finally, if mathematics education is a professional enterprise, then it is essential that those engaged in its practice, in the full range and variety of roles, contribute to the growth of professional knowledge through their own practitioner research.

REFERENCES

Adelman, C.: 1993, 'Kurt Lewin and the Origins of Action Research', *Educational Action Research* **1** (1), 7–23.

Association of Teachers of Mathematics: 1986, *The Problem Solving School*, ATM, Derby.

Association of Teachers of Mathematics: 1987, *Teacher Is Researcher*, ATM, Derby.

Association of Teachers of Mathematics: 1988a, *Reflections on Teacher Intervention*, ATM, Derby.

Association of Teachers of Mathematics: 1988b, *Mathematical Images*, ATM, Derby.

Association of Teachers of Mathematics: 1989, *Logo Microworlds*, ATM, Derby.

Billington, E. M.: 1993, Beyond Mere Experience: The Role of Reflection in the Learning of Mathematics. Unpublished M.Phil. thesis, Open University.

Brown, T., Hardy, T. & Wilson, D.: 1993, 'Mathematics on Lacan's Couch', *For the Learning of Mathematics* **13** (1), 11–14.

Brubacher, J. W., Case, C. W. & Reagan, T. G.: 1994, *Becoming a Reflective Educator: How to Build a Culture of Inquiry in the Schools*, Corwin Press, Thousand Oaks, CA.

Chatley, J.: 1992, The Use of Experiential Learning Techniques in Mathematics Education: A Personal Evaluation. Unpublished M.Phil. thesis, Open University.

Fischbein E.: 1982, 'Intuition and Proof', *For the Learning of Mathematics* **3** (2), 9–18.

Hallworth, U.: 1993, Unpublished classroom transcript. Manchester Metropolitan University.

Hatch, G.: in preparation, *Using Games in the Teaching of Early Algebra.*

Low Attainers in Mathematics Project: 1987, *Better Mathematics*, HMSO, London.

McNiff, J.: 1993, *Teaching as Learning – An Action Research Approach*, Routledge, London.

Marrow, A. J.: 1969, *The Practical Theorist: The Life and Work of Kurt Lewin*, Basic Books, New York.

Mason, J. H.: 1994, 'Researching from the Inside in Mathematics Education – Locating an I-You Relationship', *Proceedings of the Eighteenth International Conference for the Psychology of Mathematics Education*, Lisbon, 176–194.

Nixon, J.: 1981, *A Teachers' Guide to Action Research: Evaluation, Enquiry and Development in the Classroom*, Grant McIntyre, London.

Open University: 1992, *EM236: Learning and Teaching Mathematics*, Open University, Milton Keynes.

Pimm, D.: 1993, 'From Should to Could: Reflections on Possibilities of Mathematics Teacher Education', *For the Learning of Mathematics* **13** (2), 27–32.

Primary Initiatives in Mathematics Education (PrIME) Project: 1991, *Calculators, Children and Mathematics*, Simon and Schuster, New York.

Schön, D. A.: 1983, *The Reflective Practitioner*, Temple Smith, London.

Stenhouse, L.: 1975, *An Introduction to Curriculum Research and Development*, Heinemann, London.

Williams, H. J.: 1989, Tuning in to Young Children: An Exploration of Contexts for Learning Mathematics. Unpublished M.Phil. thesis, Open University.

Wilson, S. M., Shulman, L. S. & Richert, A. E.: 1987, '150 Different Ways of Knowing: Representations of Knowledge in Teaching', in J. Calderhead (ed.), *Exploring Teachers' Thinking*, Cassell, London, 104–124.

Gillian Hatch
Manchester Metropolitan University,
Didsbury School of Education,
Manchester M20 2RR,
UK

Christine Shiu
Open University,
Centre for Mathematics Education,
Walton Hall,
Milton Keynes MK7 6AA,
UK

RUDOLF VOM HOFE

ON THE GENERATION OF BASIC IDEAS AND INDIVIDUAL IMAGES: NORMATIVE, DESCRIPTIVE AND CONSTRUCTIVE ASPECTS

1. CONCEPTS ABOUT BASIC IDEAS

According to present-day knowledge, the process of mathematical problem solving, even at the university level, is always combined with intuitive images and assumptions which unconsciously affect the outcome. This has been shown, in particular, by Fischbein, in several investigations (1979, 1983, 1987, 1989, 1990). There seems to be no thinking without images, in this sense. At best these images can affect mathematical thinking positively. But they can also lead us astray – as Fischbein has shown as well – if inadequate images become fixed as unconsciously effective 'tacit models'. Therefore the question arises how to deal with this intuitive level. For instance, does one assume that adequate images develop automatically with an appropriate formal contact with mathematics, or does one consciously follow and support the generation of adequate images and ideas?

There is every reason to believe that we should deal with the generation of the intuitive level more thoroughly than we have done so far. Experience with both lessons and empirical research shows that the prime reasons for some serious problems of understanding and communication found in mathematics lessons are based on conflicts concerning the intuitive level. To put it more concretely, problems occur because of the fact that concepts and symbols which are used in the teaching of mathematics are often understood by students with a totally different meaning from what would be adequate from the teacher's point of view.

For about two centuries there have been approaches in the German tradition of mathematics education which have tried to counteract these problems with concepts of 'basic ideas' ('Grundvorstellungen'). Partly with these concepts, or others with similar names, these approaches have been developed to emphasize the constitution of meaning as a central aim of mathematics education. The remarks which follow in this section are concerned with a description of this German tradition. They are not necessary for the understanding of sections 2–4, in which the concept of 'basic ideas' is applied to general problems of misunderstanding in current mathematics education.[1]

Pestalozzi (1803), and several other nineteenth-century educators who were influenced by him (see especially, Herbart 1804, Hentschel 1842 and Diesterweg 1850), tried to base the learning of mathematics on the generation of

317

Sierpinska, A. and Kilpatrick, J. Mathematics Education as a Research Domain: A Search for Identity, 317–331.
© *1998 Kluwer Academic Publishers. Printed in Great Britain.*

'Anschauungen' – i.e. on the creation of visual mental models representing mathematical contents or structures. To realize this aim, teaching methods for *visual instruction* were created. This was considered to be an indispensable premise for comprehensible mathematical thinking and acting, which would otherwise degenerate into senseless manipulation, based merely on learning by heart – like the 'verbalisms' of the Middle Ages. According to the psychology of the nineteenth century, as influenced by Herbart (1804), the generation of 'Anschauungen' was assumed to be a passive static-depictive process based on laws of association, which did not depend on the conditions of the learning individual.

After the turn of the century, under the influence of Wundt (1907), who emphasized the active concept of 'apperception' in contrast to the passive concept of 'perception' in the Herbartian sense, the individual learning process was seen as being something more active. Recognizing the importance of the affective and motivational aspects of the student, Kühnel (1916) stated that the learning of mathematics must be based on *personal activity and insight*. With reference to Wundt, he developed the first psychology-based course of mathematics instruction in German-speaking countries, emphasizing the generation of 'Stellvertretervorstellungen' (which can be translated as 'representation images' or 'representation ideas'). They were supposed to represent abstract mathematical contents by visual or auditory mental images. Kühnel described these ideas as objects of transition; as a kind of bridge between the world of abstract mathematical thinking and the world of real-life experience.

'Stellvertretervorstellungen':

• were developed by activity in familiar contexts,
• were the mental models which make vivid thinking possible,
• allowed one to apply mathematical concepts to real life situations.

Kühnel's ideas exerted a great influence on the further development of mathematics education in the German primary school. This was especially true for his concept of the generation of 'Stellvertretervorstellungen', which was inherent in nearly every important later approach dealing with elementary concept development. Thus we also find this idea in J. Wittmann's (1929) holistic teaching method ('Ganzheitlicher Unterricht'), where the generation of 'Anschauungen' is emphasized.[2] Wittmann's approach was, besides Kühnel's concept, one of the most important in the first half of the twentieth century in German mathematics education. In contrast to his predecessors, Wittmann was both a psychologist and a mathematician and based his ideas about concept development on insights from both psychology and mathematics; the latter influenced by Cantor's set theory.

While the above-mentioned concepts in the nineteenth and in the early twentieth century were influenced mostly by pedagogical and psychological ideas, reflection about mathematical subject matter has been playing an increasingly important role since the middle of the twentieth century. Accordingly, *didactic analyses of mathematical subject matter* ('Didaktische Sachanalysen') characterized concepts of the generation of basic ideas to an increasing degree, besides psychological and pedagogical reflections. Breidenbach (1957) was one of the first important advocates of the idea that didactic analyses of mathematical subject matter must take priority in mathematics education.[3] His contention was as follows: Because the teacher has been familiar with elementary calculation techniques for a long time, he has difficulty comprehending the structural problems students have with new mathematical contents, for example, with algorithms in arithmetic operations. Therefore, the structural connections of the contents and its application to real life contexts must be thought out by the teacher. Thus grasping mathematical content means acquiring insight into its characteristic structure. On the basis of didactic analyses, Breidenbach constructed concrete or graphic models – which he called 'image foundations' ('Vorstellungsgrundlagen') – to represent the characteristic structure of mathematical content in accordance with the psychological state of the student's development. Based on these 'image foundations', corresponding teaching units were developed to introduce the respective mathematical content and to support the generation of basic ideas.

Oehl (1962) picked up the above-mentioned approaches and formed them into an integrated concept. He was the first to use the term basic idea ('Grundvorstellung') in an explicit and systematic way. Taking up the principles of *visual instruction* (Pestalozzi) and *self-doing and insight* (Kühnel), he merged the knowledge of traditional lesson practice, didactic analyses of mathematical subject matter and the knowledge gained from psychological approaches – especially the cognitive psychology of Piaget. Oehl used the term 'basic idea' (Grundvorstellung) – like his predecessors – in a twofold way: on the one hand, characterizing psychological mental models which students are supposed to generate and, on the other, describing concrete or graphic models or schemes of action which are intended to initiate and support the corresponding learning process. The second meaning was put into concrete form in his widely-circulated courses for mathematics instruction in Germany. In these courses, basic ideas were used as *prescriptive notions* which were supposed to characterize an adequate meaning of mathematical content by giving verbal or graphic descriptions. To choose a simple example: a 'basic idea' of adding positive integer numbers is the idea of 'putting together'. 'Basic ideas' of dividing are, for instance, the ideas of 'splitting up' or 'sharing out'. Under the influence of Piaget, Oehl interpreted the generation of basic ideas in a dynamic way, assuming learning and understanding as a process of change, modification and reinterpretation. In this (psychological) connection, 'basic ideas' are not regarded as isolated fixed entities, but as variable parts of a growing and changing system.

Referring to Oehl, one can speak of *the* (one integrating) *concept* of generation of basic ideas. His approach had much influence on later concepts; it was continued and enlarged to include mathematical content at the secondary school level by Griesel (1971), Blum & Kirsch (1979), Bender (1991), and vom Hofe (1992), among others. Basic ideas were used as didactic categories for structuring curricula in a *prescriptive* way; in particular, in the lesson units developed by authors who advocated the tradition of 'Stoffdidaktik'.[4]

All in all, basic ideas can be used to describe *relations between mathematical contents and the phenomenon of the individual generation of concepts*, characterizing three aspects of this phenomenon:

1. *constitution of meaning of mathematical concepts* based on familiar contexts and experiences,
2. *generation of corresponding (visual) representations* which make operative thinking (in the Piagetian sense) possible,
3. *ability to apply a concept to reality* by recognizing corresponding structures in real-life contexts or by modeling a real-life situation with the aid of the mathematical structure.

The main *didactic* task of 'basic ideas' is to describe adequate real-life contexts which represent the 'heart' (or 'essence') of the respective mathematical contents in a way which is understandable for the student. From a *psychological* point of view, basic ideas can be interpreted as elements of connection or as objects of transition between the world of mathematics and the individual world of thinking. Therefore, basic ideas describe relations between mathematical structures, individual-psychological processes and real-life contexts, or, more simply, as *relations between mathematics, the individual and reality*. Looking, with Blum (1991), at the didactics of mathematics as the science of examining and constructively structuring the relations between mathematics, the individual and reality in teaching–learning contexts, one can understand the term 'basic idea' as a *genuine category of mathematics education* at the center of this discipline.

In the following sections, the concept of generation of basic ideas (in the sense of Oehl) will be stated more precisely in three steps. First, a distinction between basic ideas and individual images – which is not made in the approaches mentioned above – will be introduced. This will be done in order to deal more specifically with the ambiguity of the term, i.e., its didactic-instructional and its psychological sense. Then, basic ideas and individual images will be seen in concrete forms through a paradigmatic case study. Finally, the method of the case study will be generalized to an enlarged model about the generation of basic ideas.

2. BASIC IDEAS AND INDIVIDUAL IMAGES

While concepts of 'basic ideas' have almost become universal in primary school practice they are more or less unknown in many other mathematical areas. One reason for this could be that the scope of these concepts has so far been restricted by the fact that they have been used only in a *prescriptive way;* i.e., 'basic ideas' have been postulated as normative categories which should train the students for an adequate understanding of concepts. The question of which individual images and ideas the students generate *in actual fact* and therefore the question for possibly *inadequate* images was not addressed in this context. This is the reason why, up until now, concepts of 'basic ideas' have not played a major role as far as *descriptive didactics* is concerned. But looking at students' ideas and images by means of descriptive methods is desirable for several reasons.

From the present point of view, the generation of 'basic ideas' is not – as it was assumed in the nineteenth century, under the influence of the psychology of association – a passive static-depictive process but an active dynamic-operative process which takes place in a complicated network of relations over a long period of time. Therefore, one can assume that the individual images of the students may differ from the 'basic ideas' intended by the teacher or they may even lead to inadequate images, even though they were thoroughly introduced by the teacher. Accordingly, it is desirable that the teacher not only try to teach *basic ideas* in a *prescriptive* way but also that he or she develop a purposeful sensitivity for the *individual images* the students may in fact have. Therefore, *descriptive* working methods are necessary to analyze the mathematical activities of the students, especially their mistakes. Methodical help to discover divergences between the teacher's adequate intended basic ideas and the students' actual ascertainable individual images is therefore desirable. Furthermore, constructive solving methods of these conflicts of understanding and communication would be helpful both from the point of view of lesson practice and research methods.

From these considerations, two questions arise: How can the generation of basic ideas be supported? and, How can it be prevented that inadequate individual images become fixed in the students' minds? In accordance with the insights of the former German mathematics education of the elementary school – which are more or less forgotten today (see for example Kühnel 1916, p. 193) – Fischbein points to the problems which arise during the introduction of new mathematical content because of contradictions between the formal and the intuitive level (Fischbein 1983). The robustness of old mental models can obstruct the generation of new basic ideas. Kirsch (1987) argues in a similar way. In the face of these problems, he suggests taking care of the structuring of school curricula to avoid situations in which too many new basic ideas must be generated at the same time. For example, in the German school curriculum, an accumulation of such new basic ideas is at present combined with the introduction of negative integers in the second semester of the seventh school year. An avoidance of such problems could possibly be achieved by beginning the teaching of negative integers in elementary school. This would extend the generation of the negative integers and their respective basic ideas over a longer period of time, thus reducing the number of above-mentioned problems with their resulting mistakes.

An imperative condition for the realization of such suggestions concerning the school curriculum in a successful way is a *purposeful sensitivity for a reflective dealing with the individual understanding process of the students*. For this, a methodical help to *analyze students' strategies and mistakes resulting from these strategies in a constructive way* may be useful. By means of the following case study the extent to which the concept of basic ideas and individual images can provide elements for this methodical help will be discussed.

This will be done by combining prescriptive and descriptive working methods. On the one hand, the *basic ideas* which are adequate to a specific situation, from the mathematical point of view, will be examined. On the other hand, the *individual images* of the students will be analyzed. This will be attempted by using psychological tools taken from the theory of 'Micro Worlds' or 'Subjektive Erfahrungsbereeiche' (Lawler 1981; Minsky 1982; Bauersfeld 1983). A comparison between the basic ideas involved and their respective individual images and the question of the contradictions which possibly exist between the prescriptive and the descriptive level will then lead to an interpretation and explanation of the misunderstandings and the disruption of communication. Furthermore, this will lead to reflections on a constructive way of overcoming these misunderstandings. Let us look, then, at the following case study.

Ingo and the Temperature

The following task and its accompanying interview are taken from a learning sequence carried out and documented by Malle (1988). It deals with an introduction to negative integers in the context of temperatures. Malle's cognitive theoretical approach is not meant to be evaluated or criticized in this paper. Rather, this example was chosen to show how a comparison between normative and descriptive thoughts could lead to explanations of students' strategies which are not discovered, if one – like Malle – measures the students' abilities basically with normative criteria which are theoretically deduced:

    1. *Problem*: It is 5 degrees below zero in the evening. Because a warm wind is moving inland the temperature rises by 12 degrees through the night. What's the temperature in the morning? (T = Tester, I = Ingo)
        T: What's the temperature in the morning?
    5. I: Yes, probably plus degrees.[5]
        T: Can you answer more exactly?
        I: The warm wind is normally warm air, isn't it? Well, when the warm wind is blowing it is always plus degrees. Because it is already plus 12 degrees in the night, well, in the night, because the warm wind is
    10.   coming.... First in the evening – well, evening that is about 8 o'clock, around 5 o'clock, during dinner time. At that time it was 5 degrees, minus. And then there came, let's say around 12 o'clock, the warm wind. It brought warm air and then it was plus 12 degrees and then it can only be plus degrees in the morning.
        T: And how many?
    15. I: 7 degrees. Because the air mixes somehow, doesn't it?

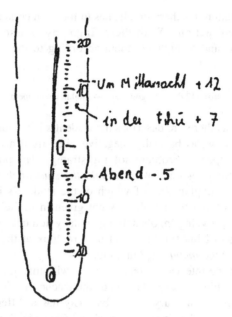

Figure 1 *Ingo's drawing*

> T: Can you draw this somehow?
> I: Must I draw three thermometers?
> T: Try to draw it within one thermometer.
> I: (*draws as shown in Figure 1.*)
>
> 20. T: What does this mean: at midnight plus 12 degrees? What do you mean
>   by this?
>   I: That the warm wind came at that time.
>   T: Is that the 12 degrees that the temperature rose because of the warm
>   wind moving inland?
> 24. I: Yes! Then it rose. Well, first in the evening it was minus 5 degrees.
>   Then came the warm wind moving inland. Then the thermometer came
>   to plus 12 degrees and in the morning it went down to plus 7.

Malle interprets this transcript basically as follows:

- Ingo has difficulties identifying the things that are important to solve the task,
- he is uneasy about differentiating between states and changes,
- Ingo tells fairy tales or invents information in addition to the given text (p. 263).

In Malle's expected problem solving strategy the movement of the mercury is stressed as the basic image of action. To be able to solve the task the student has to – as Malle says – structure the image of the movement of the mercury with adequate punctuation. Furthermore, he has to focus on certain states and sections of the (imagined) actions. With these, Ingo has to associate the quantities given in the text and to organize them according to the following scheme of assignment:

(5 degrees; below zero), (12 degrees; going up) → (unknown temperature; unknown level)

Malle's analysis therefore describes the student's behavior, based on his theoretical point of view, as basically 'negative', i.e., from his failures and deviations from the expected problem solving strategy. His point of view can be described with the questions: *What should the student think?* and *What is he doing wrong?* An explanation of which of the student's images leads to his chosen way of solving the problem is not given in Malle's interpretation. To what extent can analyzing Ingo's solving strategy at a descriptive level enhance our interpretation of his solution? Let us now look at the transcript with the question *What is Ingo imagining?* in mind.

A first look at the interview already leads – with this question – to an important indication: While the expectations of the teacher are fixed on the image of the movement of the mercury, Ingo obviously has a different image of what happens, namely that 'the air mixes somehow' (line 15). A more precise picture arises, if one specifies the question mentioned above referring to the theories of Lawler (1981), Minsky (1982) or Bauersfeld (1983): *Which 'micro worlds' are activated in the student's mind, and with which individual images does the student try to solve the problem?*

If we look at the interview from this point of view, we realize Ingo's orientation and attachment to his experiences: The warm wind coming inland, which in the teacher's eyes is obviously just used as an illustration to explain the rising of the temperature, is for Ingo something physical and vividly imagined from his own experiences. A closer look at the text shows how meaningful the warm wind image is for Ingo's problem solving strategy. It dominates the whole attempt at solving the problem. Ingo thinks of 'warm air' which comes (line 7) and that 'the air mixes somehow' with the previously existing cold air (line 15). He connects the given times of day in the task ('in the evening', 'through the night', 'in the morning') with his own experiences: 'evening that is about 8 o'clock, around 5 o'clock, during dinner time' (line 11). These are the times with which he tries to orient himself. While the teacher sees the movement of the mercury as the adequate image of action – therefore a measurable phenomenon caused by primary natural events – Ingo obviously thinks about *how* the temperature can rise at all. Ingo therefore asks for the reason of the movement of the mercury, which means that he asks for the genuine natural event. This leads him to an individual image of movement which this question explains for him: *the mixture of two air masses.*

Obviously there are misunderstandings which result from the differences between the solving activities that the teacher expects and the student's chosen attempt at solving the problem. An extensive interpretation and explanation of this misunderstanding gives us an analysis of the involved *basic ideas* and *individual images* with the following leading questions:

- Which basic ideas are adequate to solve the problem from the teacher's point of view? *(normative level)*
- Which individual images can be recognized in the student's attempt at solving the problem? *(descriptive level)*
- What are the reasons for possible divergences and how can they be removed? *(analytical and constructive level)*

First of all let us deal with the analysis of the adequate basic ideas in this case. Meaningful interpretations of the addition of integers are, for instance, following Kirsch (1987):

- The fitting together of two states to one new state (S – S – S).
- The changing of one state into a new state (S – C – S).
- The carrying out of two changes – one after the other – with the result of an overall change (C – C – C).

The temperature is an intensive quantity where changes cannot be realized by material adding or taking away. For measuring the changes of temperature, symbolic scale-like graphic representations are needed. Concerning the above-mentioned interpretations, the last two basic ideas which are connected with operator aspects are adequate: the basic idea (S – C – S) or (operand – operator – operand) is applicable to the given context. The material basic idea of fitting together two states (S – S – S) – with which the student is familiar because of dealing with extensive quantities like length or cash value (e.g., combining two bank accounts) – cannot be meaningful when put into concrete terms in this case.

But it is precisely this interpretation which seems to be suggested from Ingo's point of view when he tries to explain what happens with the help of the image of mixing two air masses. The foundation for this is the basic idea that adding is fitting together two quantities, or as Griesel (1973, p. 70) says more accurately: *fitting together two representatives of quantities to one new representative.* Obviously, for Ingo, the evening air and the warm wind coming overnight are two air masses, two representatives of the 'quantity of temperature' with the features 5 degrees below zero and 12 degrees above zero, so to speak. He combines these two air masses to one new representative with the feature 7 degrees above zero; namely, the mixed morning air masses. Accordingly, Ingo tries to draw three thermometers to explain his way of solving the problem which means for him to measure each air mass separately (line 17). This attempt is stopped by the teacher who interrupts him (line 18).

All in all, this comparison of normative and descriptive thoughts leads to a conclusive interpretation of Ingo's solution strategy: An explanation of the inadequate strategy can be seen in the fact that the student is unable to develop the basic idea intended by the teacher (S – C – S) because he is fixed on the familiar and dominant idea of fitting together (S – S – S). This fixation is stressed by his experiences (micro worlds) which are activated because of the context of the task. This is manifested in a provable interference of understanding and therefore in an interference of the communication between student and teacher (lines 20–29).

Along with an explanation of the student's strategy, the analysis also leads to indications for a constructive repairing of the interference. The repairing, in this case, could possibly be realized by changing the context of the example to activate more suitable visual experiences.

3. GENERATION OF BASIC IDEAS

The way of interpreting this case study is to look at the basic ideas involved and the individual images and to compare them. In this connection, the terms *basic idea* and *individual image* are used to describe different aspects of the teaching–learning process:

1. The term *basic idea* is used as a prescriptive notion. It is a didactic category of the teacher which is specified from relevant mathematical thoughts intended to be transposed into adequate learning contexts.
2. The term *individual image* is used as a descriptive notion. Analyzing individual images gives information about the individual explanation models of the student which are integrated into the system of his experiences ('society of mind', Minsky 1982) and which, therefore, can be activated.

The basic assumption of the concept supported in this paper is that the process of teaching–learning with the *interplay between individual images and basic ideas* can lead to the constitution of 'basic ideas of the student' (meant here in a psychological sense), or briefly: *basic ideas can be generated.* To put it in a different way: By realizing the didactic category, corresponding individual explanation models can be generated which have – in spite of all the individual differences – a common core.

A graphic model might show how to imagine this (see Figure 2): The left side tries to illustrate the didactic decisions of the teacher, the right side – the activities of the student which naturally are not 'triggered' by the didactic endeavors but which could be supported by them. The first didactic decision the teacher has to make is reflection on the respective mathematical content or procedure in order to *specify* an adequate basic idea. This step cannot be done just by analyzing the mathematical structure; it requires, at the same time, a consideration of psychological and pedagogical aspects of the respective class. In particular, it requires a reflection on what the mathematical content could mean to the students – considering their individual experiences. For example, the interpretation of the

Generation of Basic Ideas

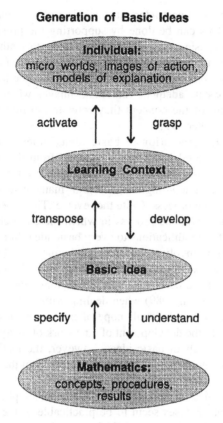

Figure 2   *Generation of Basic Ideas*

'operator' as a 'machine' – one of many possible basic ideas of the function concept – could be such a step which refers to mathematics on the one hand and to the students' experience on the other.

The *transposition* of the basic idea; i.e., the construction or identification of an adequate learning context, should represent the core of the concept which is to be generated. The learning context should be able to *activate* the *individual images* and explanation models of the student. Under ideal conditions the student would then be able to *grasp* the learning contexts from the perspective of his or her explanation models and experiences. This can lead, in the long run, to a *development* of the intended basic idea by the student, who integrates this idea into his own system of explanation models. In this way he can finally – accordingly to his individual conditions – participate in the core of the mathematical concept, which is to '*understand*' it.

Naturally, the student cannot simply take over the basic idea given by the teacher, nor can the teacher transfer it into the student's mind. The student constructs his or her own images, ideas and explanation models. But the teacher can invite the student to invest adequate meaning of mathematical concepts by didactic means. This can be done by supporting the process of teaching and learning with transposing *basic ideas* (which refer to mathematics) on the one hand, while having purposeful sensitivity for the *individual images* (which are in the student's mind) on the other hand. By doing this, the intention should be to help the student create adequate individual images which represent the epistemological kernel of the concept they refer to, so that they do not lead to misconceptions (Bender 1991).

In this context, 'generation of basic ideas' does *not* mean sampling a collection of mental models which are obtained from the 'perfect' disciplinary structure of mathematical subject matter and therefore are 'perfect' and 'valid' forever. This would be a wrong image both of mathematical subject matter and the individual learning process. Quite the reverse: The generation of basic ideas in the long run is a dynamic process in which there are changes, reinterpretations and substantial modifications to every basic idea. In particular, if the individual is going to be involved with new mathematical subjects, she or he will have to modify and extend his or her system of mental models. Otherwise, the basic ideas, which have been successful for so long, could become misleading 'tacit models' (Fischbein 1989) when dealing with new mathematical subjects (Oehl 1962). Thus the individual scope of mathematical contents and procedures has its roots in the development of a network of many figurative and symbolic basic ideas. Without basic ideas, however, the mathematical operation becomes an inanimate formalism which is cut off from the areas of application and reality.

A didactic usefulness of this model may lie in the simplicity of its explicative categories. Similar analyses seem to be practicable in the field of school and teacher training because they are feasible without immense psychological tools.

Of course there are more thorough and more precise psychological instruments to analyze the student's strategies and mistakes resulting from these strategies, as can be found in the comprehensive research about misconceptions. In contrast to this, the concept of basic ideas and individual images has the advantage, because of its normative categories, that the analysis leads to indications of constructive repairing of the analyzed interferences which can be directly realized in class. How to do this, including discussions with students on the reflections of their images and ideas in class, is shown in another paper (vom Hofe 1992).

The way of analyzing developed in the present paper tries to combine two fundamental didactic ideas which only at a quick glance seem contradictory:

- the idea of general compulsory basic ideas as normative guidelines which come from the traditional mathematics education, and
- the idea of learning in micro worlds and the combined subjective experiences which come from the area of psychology.

These two points of view do not contradict each other but can supplement one another within a holistic view of mathematical thinking and doing. The conception of basic ideas and individual images shown in this paper offers a possibility for such an integrative view. It offers, on the one hand, methodical and theoretical possibilities for combining normative didactic conceptions with descriptive and interpretive working methods. On the other hand, it can offer help for the teacher as to how to develop a purposeful sensitivity for students' actual images and ideas. Thereby it might help the teacher to follow and support the generation of students' basic ideas thus contributing to a greater understanding of the students' ways of thinking and leading to a more meaningful mathematical education.

NOTES

1. Naturally, such concepts are not restricted to the German speaking area, but can be found in other countries as well. In the Anglo-American area they are, for instance, called 'intuitive meaning' (Fischbein 1979), 'use meaning' (Usiskin 1991) or 'inherent meaning' (Noss 1994). Apart from these approaches there are several authors who deal with the generation of mathematical concepts using terms of psychological or epistemological discussion which overlap with the term 'Grundvorstellungen', for instance 'Prototypen' (Dörfler 1990) or 'commonly shared frames' (Davis 1984). However, a discussion of these terms, of their respective theoretical backgrounds and of their interrelations would be far more than can be done in this chapter.

2. In contrast to Pestalozzi and Herbart, J. Wittmann uses the term 'Anschauung' as reflected by Gestalt psychology at the beginning of the twentieth century.

3. This process established the tradition of German 'Stoffdidaktik', which became an important school of thought in German mathematics education.

4. See especially Breidenbach (1957); Oehl (1962); Griesel (1971); Kirsch (1979); Blum (1979). The term 'Stoffdidaktik' is used in an ambiguous way in the scientific community today. According to its advocates, 'Stoffdidaktik' is the most important school of thought in German mathematics education, established since the middle of the twentieth century. 'Stoffdidaktik' analyzes the 'Stoff' – that is the mathematical content or procedure which is to be taught – to make it accessible to the student. This must be done carefully in different respects, especially concerning the disciplinary structure of mathematics, the application in real world situations, and the pedagogical and psychological state of development of the students. The aim of 'Stoffdidaktik' is the development of corresponding lesson units in which the 'Stoff' that is to be taught is related appropriately to mathematics on the one hand, and is in accordance with the state of development of the respective students on the other (Breidenbach calls this 'sach- und kindgerecht'). According to this, the teaching–learning process can be described as an interplay between mathematics, reality and the individual. For further clarification, see, for example, Blum (1985). In contrast to this, there are also some authors who understand 'Stoffdidaktik' simply as an elementarization of the disciplinary mathematical knowledge for teaching and learning purposes. In particular, authors who assume that the above-mentioned advocates of 'Stoffdidaktik' overemphasize the mathematical structure as background theory for mathematics education tend to characterizations like this; for example, Strässer (1994).

5. In the German original, Ingo is speaking of 'Plusgrade' and 'Minusgrade' which is translated here word-for-word with 'plus degrees' and 'minus degrees'.

REFERENCES

Bauersfeld, H.: 1983, 'Subjektive Erfahrungsbereiche als Grundlage einer Interaktionstheorie des Mathematiklernens und-lehrens', in H. Bauersfeld et al. (eds.), *Analysen zum Unterrichtshandeln*, Aulis Deubner, Cologne, 1984, 1–56.

Bender, P.: 1991, 'Ausbildung von Grundvorstellungen und Grundverständnissen – ein tragendes didaktisches Konzept für den Mathematikunterricht – erläutert an Beispielen aus dem Sekundarstufenbereich', in H. Postel, A. Kirsch and W. Blum (eds.), *Mathematik lehren und lernen: Festschrift für Heinz Griesel*, Schroedel, Hannover, 48–60.

Blum, W.: 1985, 'Einige Bemerkungen zur Bedeutung von "stoffdidaktischen" Aspekten am Beispiel der Analyse eines Unterrichtsausschnitts in der Arbeit von J. Voigt', *Journal für Mathematik-Didaktik* **85** (6/1), 71–76.

Blum, W.: 1991, 'Applications and Modeling in Mathematics Teaching – a Review of Arguments and Instructional Aspects', in M. Niss, W. Blum and I. Huntley (eds.), *Mathematical Modeling and Applications*; Horwood, Chichester, 10–29.

Blum, W. & A. Kirsch: 1979, 'Zur Konzeption des Analysisunterrichts in Grundkursen', *Der Mathematikunterricht* **79** (3), 6–24.

Breidenbach, W.: 1957, *Methodik des Mathematikunterrichts in Grund- und Hauptschulen*, Band 1 – Rechnen, Schroedel, Hannover.

Davis, R. B.: 1984, *Learning Mathematics*, Croom Helm, Beckenham, Kent.

Diesterweg, F. A. W.: 1850, 'Wegweiser zur Bildung für deutsche Lehrer', Vierte Auflage (Erstausgabe 1834), in F. A. W. Diesterweg (ed.), *Wegweiser zur Bildung für deutsche Lehrer und andere didaktische Schriften*, Volk und Wissen, 1962, Berlin.

Dörfler, W.: 1990, 'Prototypen und Protokolle als kognitive Mittel gegen Bedeutungslosigkeit und Entfremdung im Mathematikunterricht', in *Die Zukunft des Mathematikunterrichts*, Landesinstitut für Schule und Weiterbildung, Soest, 102–109.

Fischbein, E.: 1979, 'The Intuition of Infinity', *Educational Studies in Mathematics*, **10**, 3–40

Fischbein, E.: 1983, 'Intuition and Analytical Thinking in Mathematics Education', *Zentralblatt für Didaktik der Mathematik*, **2**, 68–74.

Fischbein, E.: 1987, *Intuition in Science and Mathematics: An Educational Approach*. Reidel, Dordrecht.

Fischbein, E.: 1989, 'Tacit Models and Mathematik Reasoning', *For the Learning of Mathematics*, **9**, 9–14.

Fischbein, E.: 1990, 'The Autonomy of Mental Models', *For the Learning of Mathematics*, **10**, 23–30.

Griesel, H.: 1971, *Die Neue Mathematik für Lehrer und Studenten, Band 1: Mengen, Zahlen, Relationen, Topologie*, Schroedel, Hannover.

Griesel, H.: 1973, *Die Neue Mathematik für Lehrer und Studenten, Band 2: Größen, Bruchzahlen, Sachrechnen*, Schroedel, Hannover.

Hentschel, E. J.: 1842, *Lehrbuch des Rechenunterrichts in Volksschulen* (2 Teile, Erstausgabe Weissenfels 1842, achte Auflage 1868), Verlag von Carl Merseburger, 1868, Leipzig.

Herbart, J. F.: 1804, 'Pestalozzis Idee einer ABC der Anschauung als ein Zyklus von Vorübungen im Auffassen der Gestalten' (Zweyte, durch eine allgemeinpädagogische Abhandlung vermehrte, Ausgabe; Erstausgabe: 1802); Göttingen: 1804; in *J. F. Herberts sämtliche Werke*, Hermann Beyer & Söhne, Langensalza, 1887.

Kirsch, A.: 1979, 'Ein Vorschlag zur visuellen Vermittlung einer Grundvorstellung vom Ableitungsbegriff', *Mathematikunterricht*, **3**, 25–41.

Kirsch, A.: 1987, *Mathematik wirklich verstehen*, Aulis Deubner, Cologne.

Kühnel, J.: 1916, *Neubau des Rechenunterrichts*, Erster Band; zweite Auflageauflage Julius Klinkhardt, 1919, Leipzig.

Lawler, R. W.: 1981, 'The Progressive Construction of Mind', *Cognitive Science 1981* **5** (1), 1–30.

Malle, G.: 1988, 'Die Entstehung neuer Denkgegenstände – untersucht am Beispiel der negativen Zahlen', in W. Dörfler (ed.), *Kognitive Aspekte mathematischer Begriffsentwicklung*, Hölder-Pichler-Tempsky, Vienna, 259–319.

Minsky, M.: 1982, *Learning Meaning*, Artificial Intelligence Laboratory, MIT, Cambridge, MA.

Noss, R.: 1994, 'Structure or Ideology in the Mathematics Curriculum', *For the Learning of Mathematics* **14** (1), 2–10.

Oehl, W.: 1962, *Der Rechenunterricht in der Grundschule*, Schroedel, Hannover.

Pestalozzi, J. H.: 1803, 'Anschauungslehre der Zahlenverhältnisse', in *Pestalozzi, Sämtliche Werke*, 16. Band, Walter de Gruyter & Co., 1935, Berlin & Leipzig.

Piaget, J.: 1947, *La Psychologie de l'Intelligence*, Librairie Armand Colin, Paris.

Strässer, R.: 1994, 'À Propos de la Transposition Franco-Allemande en Didactique des Mathématiques', in M. Artigue, R. Gras, C. Laborde and P. Tavignot (eds.), *Vingt Ans de Didactique des Mathématiques en France*, N. Balacheff (ed.), Bibliothèque Recherches en Didactique des Mathématiques, la Pensée Sauvage, Grenoble, 161–176.

Usiskin, Z.: 1991, 'Building Mathematics Curricula with Applications and Modeling', in M. Niss, W. Blum and I. Huntley (eds.), *Mathematical Modeling and Applications*, Horwood, Chichester, 30–45.

Vom Hofe, R.: 1992, 'Grundvorstellungen mathematischer Inhalte als didaktisches Modell', *Journal für Mathematik-Didaktik* **13** (4), 345–364.

Vom Hofe, R.: 1995, *Grundvorstellungen mathematischer Inhalte*, Texte zur Didaktik der Mathematik, N. Knoche & H. Scheid. (eds.), Spektrum, Heidelberg, Berlin, Oxford.

Wittmann, J.: 1929, *Theorie und Praxis eines ganzheitlichen Unterrichts*, 4. Auflage W. Crüwell Verlag, 1967, Dortmund.

Wundt, W.: 1907, Grundtiss der Psychologie, Verlag von Wilhelm Engelmann, Leipzig. Transl: Outlines of Psychology. W. Engelmann, Leipzig and G. E. Stechert, New York.

Rudolf vom Hofe
University of Augsburg, MNF,
Didaktik der Mathematik,
D-86135 Augsburg,
Germany

STEPHEN LERMAN

RESEARCH ON SOCIO-CULTURAL PERSPECTIVES OF MATHEMATICS TEACHING AND LEARNING

A feature of the changing trends in research in mathematics education during recent years has been the growing interest in and focus on the social context of the mathematics classroom (Bishop 1988; Keitel 1989; Lerman 1994; Nickson 1992; Nickson & Lerman 1992). The role played by the social context in the development of the individual or of groups has been theorized, implicitly or explicitly, in many ways. What demarcates current interests is a move away from the identification of social factors as the realm of the affective to a concern with the part that the social and cultural environment plays as a whole in the development of the child. This chapter commences with a discussion of the models offered by different theoretical frameworks in their interpretations of the role of the social context, in order to identify and distinguish what are increasingly being called socio-cultural theories (Cobb 1994a) from more individualistic approaches.

The major orientation of research in education has been towards the individual; indeed it may almost be seen as naturally so. Consequently, psychology, which may be thought of as a language with which to describe individual actions, is a predominant feature of the landscape of mathematics education research, as evidenced by the fact that the leading international society concerned with research in our community is the International Group for the Psychology of Mathematics Education. Psychological research in education ranges from case studies, which endeavor to provide rich data of the interactions of a small number of subjects, to more focused studies of large samples of learners. The aim of such research is to make some statements about how individual cognition develops. That is not to suggest that cognitive psychology has ignored social interactions; on the contrary, they are seen as a major, if not the major, source of material from which the individual builds her or his knowledge and understanding. Even in attempts by the artificial intelligence community to model the human mind and cognition, modern dynamic systems take as central the interactive nature of the process of cognitive growth, whereby the mind and its environment construct and modify each other (Campbell & Dawson 1995).

Fully sociological approaches to mathematics education have not been so prevalent, although that is changing (e.g., Dowling 1995). In this chapter, I am concerned with changes within psychology that can perhaps be described as a move to the sociological, or at least towards the development of cultural

Sierpinska, A. and Kilpatrick, J. Mathematics Education as a Research Domain: A Search for Identity, 333–350.
© *1998 Kluwer Academic Publishers. Printed in Great Britain.*

psychology, and I describe some of the major research directions in mathematics education that have been stimulated by those changes.

1. THEORETICAL FRAMEWORKS AND THE SOCIO-CULTURAL DIMENSION

As I have suggested above, cognitive psychology has always taken account of the nature and function of social interactions in the intellectual development of the child. The distinctions and differences in perspectives arise in the ways that the interaction between the individual and the social are theorized. At one end of the spectrum the social, and for that matter the physical, environment are seen to affect the cognition of the individual in that, by definition, the range of possible constructions that she or he can make are limited. At the other end of the spectrum, the human mind is seen as constituted discursively, through practices, and in particular through language that carries the specificities of social contexts and practices and regulates human functioning. Indeed, in some formulations it makes no sense to speak of an internal mental plane (Gergen 1985). These differences in conceptualizing the relationship between the social and the individual cannot be ignored or inadequately explicated without the danger of confusion.

The dominant psychology at the time that Piaget began writing was behaviorism. Piaget's radical step was to insert the active interpreting subject between the stimulus and response. He argued that the stimulus, rather than being a feature of the physical object, actually comes from the subject in acting on the environment, and the response, rather than being a learned reaction, is an interpretation by the subject of the experience. Thus, for Piaget, learning is a process of continual cognitive reorganization. 'Intelligence organizes the world by organizing itself' (Piaget 1937, quoted in von Glasersfeld 1989, p. 136). This organization and reorganization occurs through the processes of empirical abstraction and reflective abstraction, in extensional and constructive generalizations (Sierpinska & Lerman 1996). These are processes located entirely in the individual, although they are governed, according to Piaget, by structuralist principles (Goldin 1990). 'Observation and experiment show as clearly as can be that logical structures are constructed, and that it takes a good dozen years before they are fully elaborated; further, that this construction is governed by special laws' (Piaget 1971, p. 62). In regard to experience of the world, Piaget sometimes seems to accept constraints imposed by the world as it is, but for the most part he emphasizes the dependence on the interpretation by the individual. 'There is no cognitive elaboration in which subjects appeal to pure experience, since ... observables are always interpreted, and a "fact" necessarily implies an interaction between the subject and the objects in question' (Piaget & Garcia 1989, p. 27). These interactions, or operations, are interiorized actions, they are reversible and they link with other operations into total structures (Piaget 1964, p. 8). Further, they are stage-related, in that the sets of structures are different at the different developmental stages of the child (p. 10). Piaget cites four factors that lead to the development of one set of structures into the other: maturation; experience of the physical world; social transmission; and equilibration, the latter being the fundamental one (p. 13). Regarding social transmission, the central focus of this chapter and an epithet for teaching, Piaget subordinates it to development. If the child is not ready, in terms of having constructed the necessary pre-formed structures, the information cannot be assimilated. Since these structures themselves are formed through the spontaneous process of embryogenesis, teaching cannot do other than exercise the child within her or his particular cognitive state; it certainly cannot transmit knowledge, nor can it effect the formation of the structures of a subsequent stage.

Piaget's ideas form the foundation of the neo-Piagetian theory of radical constructivism (von Glasersfeld 1989, 1990a), which has developed in particular from Piaget's argument that the function of cognition is adaptive in the biological sense of the word (Piaget 1937, in von Glasersfeld 1991, p. 16). Adaptation results from reaction to perturbations to the cognitive system of the individual and leads to the rejection of what is not adapted (von Glasersfeld 1991, p. 16). Thus what becomes significant is the fit of one's conceptual framework with one's experiences, not the goal of the achievement of a match with the world:

To embark on a *radical* constructivist path, thus, means to relinquish the age-old untestable requirement that knowledge must *match* the world as it might 'exist' independently of our experience.... From my perspective, those who merely speak of the construction of knowledge, but do not explicitly give up the notion that our conceptual constructions can or should in some way *represent* an independent 'objective' reality.... Their constructivism is *trivial*. (von Glasersfeld 1991, pp. 16–17, italics in original)

With all forms of social interactions, including language, playing a secondary role, subsidiary to the interpretive lens of the individual's cognition, radical constructivism leads one quite far down the path of solipsism. 'We come to see knowledge and competence as products of the individual's conceptual organization of the individual's experience' (von Glasersfeld 1983, p. 66). 'We construct our understanding through our experiences and the character of our experience is influenced profoundly by our cognitive lenses' (Confrey 1990, p. 108). In particular, the role of the teacher can only be interpreted as offering activities that seem to her or him as appropriate to the developmental stage of the child or children, and that may lead to perturbations for the child or some of the children. The job of the teacher is:

a task of inferring models of the students' conceptual constructs and then generating hypotheses as to how the students could be given the opportunity to modify their structures so that they lead to mathematical actions that might be considered compatible with the instructor's expectations and goals. (von Glasersfeld 1990b, p. 34)

Von Glasersfeld suggests that social interactions have a privileged role in conceptual development only in that 'the most frequent source of perturbations for the developing cognitive subject is interaction with others' (von Glasersfeld

1989, p. 136). Some constructivist researchers are not satisfied with this interpretation of the role of social interactions, however, and draw instead on versions of social constructivism:

Although the primacy of focus of each of conventionalism and radical constructivism is sacrificed in social constructivism, their conjunction in it serves to compensate for their individual weaknesses. (Ernest 1991, p. 86)

We can observe that when we talk of students' constructive activities we are emphasizing the cognitive aspect of mathematical learning. It then becomes apparent that we need to complement the discussion by noting that learning is also a process of acculturation. (Cobb, Yackel & Wood 1992, p. 28)

The fundamental orientation of the work in our own classroom springs from the radical constructivist principle and an integrated and compatible elaboration of the role of the social dimension in these individual processes of constructing as well as the processes of social interaction in the classroom. (Bauersfeld 1992, p. 2)

Piaget's constructivism and radical constructivism are very clear on the nature of the learning process (von Glasersfeld, personal communication, is of the opinion that, were Piaget alive today, he would endorse radical constructivism). The complementarity that is described in these quotes, however, leads to the question of what is the nature of the social learning, or acculturation as Cobb et al. call it, as distinct from constructive learning, given that the writers root their approach in radical constructivism. At the micro-level, Bauersfeld's interactionism (1992) attempts to characterize the other manner in which children learn. 'The core part of school mathematics enculturation comes into effect on the meta-level and is "learned" indirectly' (Bauersfeld, quoted in Cobb 1994a, p. 10). Cobb (1994a) explains this comment as follows:

Bauersfeld's reference to indirect learning clarifies that the occurrence of perturbations is not limited to those occasions when participants in an interaction believe that communication has broken down and *explicitly negotiate* meanings. Instead, for him, communication is a process of often *implicit negotiations* in which subtle shifts and slides of meaning occur outside the participants' awareness. (p. 15)

Elsewhere he says that learning is 'profoundly influenced by' social interactions (Cobb 1994b, p. 9).

Whilst constructivists of one form or another do not challenge Piaget's fundamental claim that the process of learning is that of the individual's reorganization of her or his schemata, the explanation for any other kind of learning appears to remain rather poorly described. The shifts and slides that Cobb refers to could also describe the ways in which these writers try to expand learning beyond the solipsistic constructivism. Elsewhere I have developed a critique of what I see as the confusion of social constructivism (Lerman 1996).

To summarize, for radical constructivists the learner's interactions with the social environment are the major source of perturbations to her or his conceptual structures, and may lead, through accommodation or assimilation, to conceptual development. For 'trivial' constructivists, although writers do not describe themselves thus, the process by which children develop is a process of individual construction. However, since their knowledge will correspond to objective reality if they have learned or understood correctly, there needs to be an appeal to some process of validation or confirmation. For social constructivists, the negotiation of meanings in social situations, perhaps taking place implicitly, on a meta-level, is as important as the individual constructions of the learner. In my view, the process of that learning is not clearly elaborated.

2. THEORETICAL FRAMEWORKS AND THE SOCIO-CULTURAL PERSPECTIVE

In the previous section I have discussed, albeit briefly, various interpretations of the role that the socio-cultural dimension is seen to play when the gaze of the researcher is on the individual and her or his developing cognition. In this section I attempt to characterize the socio-cultural perspective on learning, recognizing, of course, that this perspective is not singular but is as diverse as the individualistic.

Following Vygotsky (1924/1979), the starting point I take is the development of consciousness. As for Piaget, Vygotsky's work began when the dominant view in psychology was that of behaviorism, which took human consciousness as a given and as an *a priori* unanalyzable feature of human behavior. In developing a Marxist psychology, Vygotsky intended to extend the behaviorists' materialist program to deal with the whole human subject, which could not exclude consciousness. On the contrary, if the whole human subject was the product of life, society, culture and history, then the manner in which consciousness comes about – that is to say, the process whereby the animal homo sapiens becomes the social human being – is the proper domain of study of psychology. Thus Vygotsky inserted the mediation of culture, in the form of meaning, between the stimulus and response link. The cognizing subject is already the socialized human being, and the mental activities associated with the active subject such as discriminating, generalizing, voluntary attention, memory and so on are themselves products of communication between the infant and the adults around (Harré & Gillett 1994); they are social acts (Luria 1973, p. 262).

The notion of the mediation of culture, in the form of tools and in particular of language, leads to a key claim in Vygotsky's work, particularly in terms of the relationship between the individual and the social world around her or him; that is, his emphasis on the primary role of the intersubjective:

Any function of the child's development appears twice, or on two planes. First it appears on the social plane and then on the psychological plane. First it appears between people as an interpsychological category and then within the child as an intrapsychological category. (Vygotsky 1978, p. 63)

Consequently, Vygotsky identifies learning as leading development, quite the reverse of Piaget's view, and leads to his important and revolutionary notion of the zone of proximal development (Newman & Holzman 1993). For Vygotsky, learning is a process of pulling the child into her or his tomorrow rather than exercising where she or he is today. Vygotsky described learning in the zone of proximal development (ZPD) as taking place in interaction with a teacher or a more informed peer. What the child can imitate today, she or he will be able to do with assistance tomorrow and alone thereafter. These are key features of research that draws on a Vygotskian socio-cultural perspective, and they make teaching and learning an essentially integrated social activity and focus for research in education.

Critics in the radical constructivist tradition focus on the problem of how knowledge that is in the social plane becomes the individual's; what is the process of internalization?

Vygotsky's notion of internalization is an observer's concept in that what the observer regards as external to the child eventually becomes in some way part of the child's knowledge. But Bickhard (in press) has pointed out that there is no explanatory model of the process. (Steffe 1993, p. 30)

This is an entirely appropriate doubt, as it is perhaps the crucial distinction between the two perspectives, those of Piaget and of Vygotsky. Piaget's point of departure, as a philosopher, from the major traditions of innatism and empiricism was that of the inadequacy of those two traditions in making sense of the individual having knowledge. He proposed instead that one should focus on the process of knowledge construction, on a genetic epistemology. Any perspective which argues from the intersubjective to the intrasubjective appears to fall into either innatism or empiricism. However, Vygotsky's psychology was not driven by philosophical concerns; he was not concerned with the nature of knowledge in an ontological sense. For him, drawing on Marx, the individual is born and matures at a certain time and in a certain place, and whatever that society has to offer – call it culture, customs, world views or knowledge – becomes the milieu of the individual (in fact, each individual is simultaneously positioned in many subcultures, including gender, ethnicity, class etc.; Evans & Tsatsaroni 1994). The difference in perspectives is not a minor one and again relates to Vygotsky's focus on consciousness. Leont'ev (1981), drawing on Vygotsky's writings, describes the process of internalization: 'The process of internalization is not the transferal of an external to a pre-existing, internal "plane of consciousness"; it is the process in which this plane is formed' (p. 57).

The assumption of a pre-existing internal plane does indeed pose the question of how the interpsychological becomes the intrapsychological, but the dissolving of the boundaries between subject and object, between the individual and the culturally-mediated 'world', by viewing the internal plane as being formed in internalization removes the question altogether. Indeed, making the formation of the internal plane a materialistic process further opens the study of consciousness to semiotics and to cultural psychology.

Vygotsky's epistemology is also a genetic one, but one that looks at the function of particular cultural tools in society and the process of their acquisition by the child. Vygotsky's use of the notion of mediation by tools, particularly cultural tools and the semiotic mediation of language, situates people in their time and place. Vygotsky (1986) criticizes Piaget on just this point:

This attempt to derive the logical thinking of a child and his entire development from the pure dialogue of consciousness, which is divorced from practical activity and which disregards social practice, is the central point of Piaget's theory. (p. 52)

In resisting the tendency to study aspects of human activity separately, proposing instead a unit for the analysis of human activity that encompasses all elements as interrelated, the unit of meaning, Vygotsky (1986) drew affect and cognition into one:

When we approach the problem of the interrelation between thought and language and other aspects of mind, the first question that arises is that of intellect and affect. Their separation as subjects of study is a major weakness of traditional psychology. (p. 10)

He argued that all actions are driven by goals, needs and motives (Leont'ev 1978).

I have attempted here to outline the central distinctions that identify a fully socio-cultural psychology from those that draw socio-cultural factors into an individualistic psychology. This is not the place to rehearse all the arguments about the differences between a Vygotskian program and a Piagetian one (see Cobb 1994a; Confrey 1994, 1995a, 1995b; Lerman 1992, 1993). Suffice it to say that a number of researchers take the Vygotskian approach to be the most fruitful framework for their research, and the remaining function of this chapter is to describe some of that body of work.

3. RESEARCH DIRECTIONS

A recent special issue of *Educational Studies in Mathematics* (vol. 31, nos. 1 & 2) is entitled 'Socio-Cultural Approaches to Teaching and Learning Mathematics', and the 1995 meeting of the International Group for the Psychology of Mathematics Education chose the same title as its theme for the meeting. The articles in the former and the plenary and other presentations at the latter illustrate some of the current areas of research within our community. Taken together with the debates in the literature about Piaget and Vygotsky, and the articles by constructivists that seem to be reacting to critiques from socio-cultural theories (e.g., Cobb 1994a; Steffe 1993; Wood 1994), it seems clear that interest in socio-cultural theories is growing rapidly and so too is its influence on research in mathematics education. Below I attempt to describe a number of the main directions of that work; however, the choice of those directions and of references to published research work is intended to be exemplary rather than exhaustive.

Cognition in School Practices and in Out-of-School Practices

In the last decade there have been a number of studies of the mathematical practices of children in various contexts outside of the school that have revealed a
gap between what children are expected to accomplish in school and what they
can accomplish outside school (Carraher 1988; Nunes, Schleimann & Carraher
1993; Saxe 1991; Schliemann & Acioly 1989). The evidence seems to suggest
that, far from being abstract and context free, mathematical cognition is situated
differently in different practices, of which school mathematics is one, albeit
strongly associated with cultural capital. A number of theoretical interpretations
have been offered. Saxe (1991) proposes a theory of emergent goals, in which
the goals of the participants in a social practice develop and use cognitive tools
appropriate to that practice, in an interactive process with the objects. Saxe's
perspective extends the notion of viability of cognitive tools to relate to particular practices in distinct ways, driven by personal and group goals. Lave
describes this phenomenon as 'situation specificity' (Lave & Wenger 1991),
whereby the rationality and efficacy of processes and procedures are regulated
by the situations in which learning, or being apprenticed in, those processes and
procedures took place. Nunes et al. (1993), after initially working in a Piagetian
psychological framework, concluded that it was unsatisfactory as an explanatory
theory for their findings:

Nor can cognitive developmental theories of the structuralist type handle this kind of data. If two
tasks involve the same cognitive operations, there should be no appreciable gap in the solutions....
In the search for some theoretical explanation that could help us understand this phenomenon, we
did find one theory that can accommodate these findings and, what is more important, can lead to
further hypotheses about the differences and similarities between school and street mathematics; the
theoretical framework proposed by Vygotsky and Luria. (pp. 26–27)

Masingila (1993), Masingila, Davidenko & Prus-Wisniowska (1996), Boaler
(1996) and others have drawn on notions of situated cognition to examine the
problem of 'transfer' of mathematical knowledge, the relationship between in-
school and out-of-school mathematical practices, and to try to draw links
between them. Masingila et al. (1996) argue that mathematics in school can
offer generalizability across contexts whilst out-of-school situated practice does
not, and that this distinction can be used to develop connections between the two
sets of practices. They examine problem solving in the classroom on real-world
problems through which students learn mathematical content and in which they
are also able to examine those practices, reflect on them and generalize the
mathematical tools and procedures. Boaler (1996) compared the mathematical
work of students in two different schools in the UK with similar intake in terms
of social class and ability, one where the students learnt techniques and skills
from a strongly textbook-centered system, the other where students learned
through working on problems. In her longitudinal study, Boaler compared the
achievements of the students across a whole range of tests, including numeracy,
standardized national tests at age 16, and a 'real world' problem. In relation to

the last test, the students were also examined on the mathematical content of that task in a pencil-and-paper test some time earlier. She found that 'the [process-focused school] students gained significantly higher grades in all aspects of the applied task; they were also significantly more able to "transfer" their knowledge from test situations to applied situations' (p. 20). On the standardized national tests at age 16, both groups of students performed equally, although they were disadvantaged in different ways. The textbook-centered school students were not able to answer questions that differed from the ones used to teach the techniques, and they did not perform well on those questions. The students in the process-focused school found some of the examination questions dealt with content that they had not met whilst working through problems. The evidence from both groups of students supports the situation specificity both of rationality and of processes and procedures.

Drawing on the same theoretical frameworks, a number of researchers have looked at the classroom as a cultural setting, as a specific situation, within which cognition develops (e.g., Brodie 1995; Matos 1995; Santos 1995). Using naturalistic studies, they identify social practices in the classroom within which meanings are negotiated. The authority of the teacher; the framing offered by the text; the social relationships within and between groups of students; the teacher's choice of whether, when and how to intervene; and the social mores of the classroom, are all crucial factors in the rationality as well as content knowledge gained by the students.

Semiotic Domains of Meaning

Lave's *situation specificity* has a parallel in the study of the metaphors and cultural tools used in the classroom. Boero (1992) uses the notion of *fields of experience* and Lins (1994) *semantic fields* to describe a particular domain of meaning around a cultural tool such as a calendar (Boero, Dapueto, Ferrari, Ferrero, Garuti, Lemut, Parenti & Scali 1995) or around a metaphor (he uses the term *kernel*) such as the 'balance' for solving equations (Lins 1994). Boero's notion is taken to mean 'a sector of human culture which the teacher and student can recognize and consider as unitary and homogeneous' (Boero et al. 1995, p. 153). He makes some important fine distinctions between the kinds of real-world problems teachers can use. For example, some are highly mathematized (money, time, etc.), whereas others are laden with notions that clash with the mathematical (the transmission of hereditary characteristics). An awareness of these distinctions is essential to avoid unnecessary confusion of meanings and can allow real-world situations to be used usefully for mathematical activity where the clashes might lead to further 'scientific' mathematics, as in his use of sun shadows (Boero et al. 1995, p. 158). This latter field is sufficiently close to students' everyday experience to enable them to bring meanings to the problem, yet sufficiently unmathematized in daily life to avoid clashes between scientific and everyday meanings. It cannot, however, avoid the associations that every student brings to contexts used in school mathematics (Evans & Tsatsaroni

1994; Walkerdine 1988), within which students are positioned differently, sometimes as powerful and at other times as powerless.

Lins's (1994) term *semantic fields* emphasizes the linguistic and semiotic significance of the contexts offered in school mathematics. Lins argues that knowledge must be seen as a combination of statements of belief and their justification, which will relate to the kernel in which the student is operating, and the interlocutor; that is, the person to whom the student is justifying those statements (although the other person does not need actually to be present). He suggests that in equations such as $3x + 5 = x + 2$, represented by a balance, meanings will easily be made for taking 2 from each side, but not for taking 100 from each side of the equation $3x + 100 = 90$. Drawing on the work of Davydov (e.g., 1990), Lins offers examples of contexts in which students can learn to operate with symbols as objects, going beyond the apparent meanings available from the real-life possibilities of the kernel. Development can take place within a semantic field or in a shift between semantic fields, the former enriching ways of producing meaning, the latter enriching the potential to produce new knowledge.

The work described in both situated cognition and semiotic domains of meaning recall Vygotsky's emphasis on the mediation of sign systems:

The trend in Vygotsky's thinking near the end of his life is clear. He was searching for a way to relate the psychological functioning of the individual with particular socio-cultural settings, specifically with the setting of formal instruction. The theoretical mechanism that he used to specify this relationship was grounded in his theme of semiotic mediation; his line of reasoning was to identify the forms of speech or discourse characteristic of particular social settings and examine the impact their mastery has on mental functioning (on both the interpsychological and the intrapsychological plane). (Wertsch 1990, p. 116)

Teaching and Learning

An underlying principle of the research of Boero and his colleagues, and others, is the drawing together of teaching and learning. Research problems and methods are couched in terms that reflect that unity. Thus in their abstract Boero et al. (1995) introduce their research as follows: 'The purpose of this contribution is to investigate some cognitive and didactic issues regarding the relationship between "mathematics" and "culture" in teaching-learning mathematics in compulsory school' (p. 151).

Vygotsky's ZPD refers to learning from adults, teachers or more informed peers. Much research has looked at the last, often in short-term studies. This focus may reflect the hesitancy of researchers to focus on teaching after many decades of reaction against transmission models. Boero et al. (1995) are brave enough to suggest that the traditional mode of teaching, prevalent in most parts of the world, should be a legitimate area of study. Its drawbacks are known ('It is greatly selective and supplies knowledge which turns out inert for most students', p. 164), but it is clear that many students learn successfully from this style of classroom. Indeed, 'it is enough to consider that most intellectual classes of our society, including the specialists who have helped sciences reach the levels we know today, have been trained through it in the last centuries' (p. 164).

Bartolini Bussi (1996) has studied the teaching–learning of students in Italian schools from second to fifth grade, using problems and artifacts from the history of mathematics and culture in Italy from the Renaissance to the present. In particular, they have worked on the problem of perspective. Her aim is to analyze 'the functions of semiotic mediation in a long-term teaching experiment on the plane representation of three-dimensional space by means of perspective drawing' (ibid. p. 1). The long-term nature of the study is considered essential in order to research the activity of teaching–learning and consequently the functioning of the ZPD, and the processes of internalization and of semiotic mediation. Short-term studies are more likely to reveal 'the relationships between actions and operations' (p. 7). She emphasizes: the need to consider teaching–learning as a unity; to see the action of the ZPD in relation to the interactions and interventions of the teacher, with cultural tools, and the students, as well as peer interactions; the social construction of knowledge as the history of the culture of scientific progress; and the semiotic significance of the fields of experience.

Cultural Tools

Vygotsky's notion of cultural tools offers a different way of conceiving of the influence of the history of mathematical ideas or the changes in technology in society. Developments in mathematics change forever the meanings available, through the changes in symbols, structures and language that become accessible. The history of negative numbers, for example, shows that there were diverse manifestations in different places in the world, across thousands of years. There is evidence of the appearance of negative numbers in solutions of problems in Babylonian clay tablets that suggest an attitude to the existence of negatives that was no less advanced, and in some ways more accepting, than attitudes of some eminent European mathematicians in the early nineteenth century (Henley 1990). This history would suggest the existence of a significant epistemological obstacle. However, the whole structure of the number system has subsequently been restructured, with new symbols and language available. Introduced to school students in that language, there is no reason to assume that those epistemological obstacles will appear (Lerman 1995a). Crawford (1996) makes the same points in relation to the history of mathematics and widens it to consider technology as cultural artifacts:

Vygotsky worked in a time of enormous change in post revolutionary Russia. That is perhaps one reason that his ideas are particularly relevant to education in the information era – another time of rapid change in the context of human development in many countries. Vygotsky's insistence on non-absolute forms of consciousness seems less radical in an era of computerised 'virtual reality' for auditory, visual and, most recently, tactile sensory domains. (p. 44–45)

The research discussed above (Bartolini Bussi 1996; Boero et al. 1995) draws explicitly on the significance of cultural tools – in particular, problems such as sun shadows and the calendar – and historical artifacts from the history of mathematics and art. Barton (1996), in a review and critique of ethnomathematics, proposes using cultural artifacts such as Maori triple weaving patterns and sports results to build bridges into mathematical culture and at the same time provide material for (perhaps new) mathematical analysis.

Activity Theory

A rich inheritance from the Vygotskian school in terms of educational research has been activity theory (Leont'ev 1978). Although Vygotsky pointed out the centrality of the notion of the acting person, he emphasized 'meaning' as mediating the world for every individual. Through tools, and cultural tools in particular, society and culture are mediated for the child ('It is the world of words which creates the world of things'; Lacan 1966, p. 155), although thought and language are to be seen as a dialectic. Language offers the child inherited historical-cultural meanings, but each participant in the conversation or other activity uses those tools to intersubjectively reshape meanings in communication and action. According to Leont'ev's view, the *activity* orients the participants and provides the initial meaning and motivation. Meanings have a social ontogeny, whereas sense is the individual's perspective. The *actions* of individuals within the activity are always motivated by sense, which incorporates cognition, culture and affect. Finally, there are *operations*, the specific moves, often automatic, that individuals make in response to specific phenomena. To take a well-known example of Leont'ev's, in a group *activity* of hunting, one person, the beater, will have the role of circling around the prey to beat the undergrowth and usher the prey towards the hunters. The *actions* of that person only make sense within the context of the activity. If there is a boulder in the way, the beater will go around it; this is an *operation*. The notion of activity incorporates the cultural artifacts, the social setting and the motives and goals of the participants into a whole, thus embodying in a research perspective the unity of human actions that Vygotsky sought as a unit of psychology.

Teaching–learning, in the work described above by Bartolini Bussi (1996), is seen as an activity in Leont'ev's sense, in which the motive is the social construction of knowledge:

This motive is realised, on the one side, by the format of discussion, that gives value to joint activity in mathematical experience and, on the other side, by the systematic recourse to tools for semiotic mediations, that are explicitly drawn from the existing culture. Theoretically, in a Vygotskian approach the two things cannot be conceived without each other. (p. 37)

In an attempt to engage student teachers, in the final year of their course, with their still unchallenged assumptions about the role of the teacher, Crawford and Deer (1993) devised an activity in which the students had to work in groups to develop a program of mathematics that was centered on the children's environment, rather than a prescribed syllabus. The students found this very hard and experienced 'initial ecstasy, shock of recognition, crisis, realism and commitment' (p. 116). The outcome was at least a recognition by the students of having a wider range of skills upon which to draw and in many cases new-found confidence in their ability to create 'a very different learning environment ... from the one that they had experienced themselves' (p. 118). Elsewhere Crawford (1994) writes:

> The course was designed to create a 'zone of proximal development' for student teachers as a way of expanding their knowledge of the dialectic process of teaching and learning through conscious experience of the process. They were engaged in a learning activity. (p. 6)

The Interface of the Individual and the Social

Socio-cultural theories, in common with social versions of constructivism, enactivism (e.g., Davis 1995) and dynamic systems theories, are concerned with the ways in which individuals and their environments co-construct each other. In the former, the notion of activity captures the sense of the uniqueness of each whole environment in which people act. Thus mathematical learning situations are creations of the participants, with the mathematical culture a major component that is present via the teacher, the texts, and the differentiated knowledge and experience of the students. What socio-cultural theories offer are languages in which to describe the process whereby the environment constructs the individuals, as well as the reverse. Brown (1996) draws on Habermas to develop a theory in which individual learners reconcile their constructions with the framing of the socially determined code of the mathematics teacher.

Evans (1993) argues that Foucault's work on the architecture of knowledge, in which he examines the knowledge/power of discursive practices, captures the way in which individuals are constructed by and within those practices. Evans argues that discursive practices are not clearly bounded; they are continually changing, and one switches from one discursive practice to another. In a series of interviews, he asks mathematical questions set in different social contexts, and attempts to identify the discursive practice that is called up by the question in its context, for a particular person. He criticizes the simplistic notion that giving real-world contexts for mathematical concepts provides 'meaning' for students, a 'meaning' that supposedly exists in some absolute sense and is illustrated by or modeled in that real-world context. He identifies school mathematics practice as one of a range of practices that might be called up for an individual. When that practice is called up, the interviewee will focus on the mathematical calculation required, and can achieve the 'success' we search for in the classroom. Evans applies a range of readings successively to his data, the final one being psychoanalytic.

Pimm (1994) also draws on psychoanalytic theory in examining teacher–student interactions in the classroom. He highlights the overlap between words used in the mathematical context and in everyday life and the potential for slippage between them. He points out that this potential slippage raises questions about the notion of 'meaning' in mathematics, especially given that it is common for mathematics teachers to simply deny any connection, as if that resolves the issue.

There has been considerable interest in equity issues in mathematics education for a number of years, particularly in relation to gender (e.g., Burton 1986) and ethnicity (e.g., Frankenstein 1981), but also to some extent in relation to class (Mellin-Olsen 1987). Socio-cultural theories, with their language of description for the positioning of subjects in discourses, offer new and different resources. Most of the well-known studies focus their attention on changing the mathematics classroom through its being led by an emancipated teacher who can empower her or his students. Ellsworth (1989) argues that the multiple socio-cultural positionings of the actors in the classroom setting are characterized by shifting relationships of power and powerlessness, of voice and its lack, through the many overlapping and separate identities of gender, ethnicity, class, size, age, etc., to say nothing of the 'unknowable' elements of the unconscious. Unless these positionings are given voice, 'emancipatory authority' is as repressive as any other. Post-structuralism, with its emphasis on discourses that are to be seen as partial narratives in which knowledge and interests are intimately tied, offers new readings of the classroom and of research. Activity theory, drawing on Vygotsky's ZPD, which emphasizes a social ontogeny, with cultural tools as well as the needs, motives and goals of the participants constituting the activity, also offers new readings of the classroom and of research. I have suggested elsewhere (Lerman 1996) and rephrase below, that Vygotsky's approach, emphasizing the semiotic, is very close to poststructuralist ideas, despite the fact that he died many decades before those ideas were developed. Recent publications on equity in mathematics education to some extent reflect these different perspectives, which are still relatively new to the mathematics education community (e.g., Lerman 1995b; Secada, Fennema & Adajian 1995).

4. CONCLUSION

Changes in the dominant research paradigms of communities such as ours do not take place through the succession of 'better' ideas over less good ones, as discussions in the substantial literature surrounding the Kuhn–Popper debates of the 1960s and 1970s show. Those changes are associated with other factors: ideas from the disciplines surrounding mathematics education being drawn in and recontextualized; the 'if-you-want-to-get-ahead-get-a-theory' drive in new research; new generations (not age-related) of researchers; research funding initiatives and innovative journals; and many others. Research paradigms or perspectives do not disappear, although they may perhaps become less popular. I believe that one must expect and demand some attempt at coherence between theories and research methods, some attempt by researchers to make clear under which conditions 'this' can be seen as a case of 'that'.

Aspects of cultural studies have, over the last three decades, become established features of educational research in a number of fields including policy studies, sociology of education, second language learning, literature, and others. Their effect on research in mathematics education is only very recent. Whilst we have been concerned about the role of culture for some years, we have only recently begun to draw on that body of literature to inform and support research. The 'discovery' of Vygotsky's work has offered, in my view, an entry into the domain of theory which is radically distinct from the individualistic focus of the Piagetian psychological literature that has so dominated mathematics education outside of the old Eastern Block, and which takes the social, cultural and historical situation of people as the starting point for study. I believe that we can expect to see substantial and important developments in these areas in the future.

REFERENCES

Bartolini Bussi, M. G.: 1996, 'Mathematical Discussion and Perspective Drawing in Primary School', *Educational Studies in Mathematics* **31**, 11–41.

Barton, B.: 1996, 'Making Sense of Ethomathematics: Ethnomathematics Is Making Sense', *Educational Studies in Mathematics* **31**, 201–233.

Bauersfeld, H.: 1992, 'Classroom Cultures from a Social Constructivist's Perspective', *Educational Studies in Mathematics* **23**, 467–481.

Bishop, A.: 1988, *Mathematical Enculturation: A Cultural Perspective on Mathematics Education*, Kluwer Academic Publishers, Dordrecht.

Boaler, J.: 1996, 'Open and Closed Mathematics Approaches and Situated Cognition', in *Proceedings of the Third British Congress on Mathematical Education*, Vol. 1, 152–159.

Boero, P.: 1992, 'The Crucial Role of Semantic Fields in the Development of Problem Solving Skills' in *Mathematical Problem Solving and New Information Technologies*, Springer-Verlag, Berlin, 77–91.

Boero, P., Dapueto, C., Ferrari, P., Ferrero, E., Garuti, R., Lemut, E., Parenti, L., & Scali, E.: 1995, 'Aspects of the Mathematics–Culture Relationship in Mathematics Teaching-Learning in Compulsory School', in *Proceedings of the Nineteenth International Meeting of the Group for the Psychology of Mathematics Education*, Recife, Brazil, Vol. 1, 151–166.

Brodie, K.: 1995, 'Peer Interaction and the Development of Mathematical Knowledge', in *Proceedings of Nineteenth International Meeting of the Group for the Psychology of Mathematics Education*, Recife, Brazil, Vol. 3, 216–223.

Brown, T.: 1996, 'Intention and Significance in the Teaching and Learning of Mathematics', *Journal for Research in Mathematics Education* **27**, 52–66.

Burton, L.: 1986, *Girls into Maths Can Go*, Holt, Rinehart & Winston, London.

Campbell, S. & Dawson, A. J.: 1995, 'Learning as Embodied Action', in R. Sutherland & J. Mason (eds.), *Exploiting Mental Imagery with Computers in Mathematics Education*, Springer-Verlag, Berlin, 233–250.

Carraher, T. (1988). 'Street Mathematics and School Mathematics', in *Proceedings of the Twelfth Annual Conference of the International Group for the Psychology of Mathematics Education*, Veszprem, Hungary, Vol. 1, 1–23.

Cobb, P.: 1994a, 'Where is the Mind? Constructivist and Sociocultural Perspectives on Mathematical Development', *Educational Researcher* **23**(7), 13–20.

Cobb, P.: 1994b, Theories of Mathematical Learning and Constructivism: A Personal View Paper presented at Symposium on Trends and Perspectives in Mathematics Education, Institute for Mathematics, University of Klagenfurt.

Cobb, P., Yackel, E. & Wood, T. (1992). 'A Constructivist Alternative to the Representational View of Mind in Mathematics Education', *Journal for Research in Mathematics Education* **23** (1), 2–33.

Confrey, J.: 1990, 'What Constructivism Implies for Teaching', in R. B. Davis, C. A. Maher & N. Noddings (eds.), *Constructivist Views on the Teaching and Learning of Mathematics* (Journal for Research in Mathematics Education Monograph No. 4), 107–122.

Confrey, J.: 1994, 'A Theory of Intellectual Development (Part 1)', *For the Learning of Mathematics* **14** (3), 2–8.

Confrey, J.: 1995a, 'A Theory of Intellectual Development (Part 2)', *For the Learning of Mathematics* **15** (1), 38–48.

Confrey, J.: 1995b, 'A Theory of Intellectual Development (Part 3)', *For the Learning of Mathematics* **15** (2), 36–46.

Crawford, K.: 1994, Vygotsky in School: The Implications of Vygotskian Approaches to Activity, Learning and Development Paper presented at First International Conference 'L. S. Vygotsky and School', Eureka Free University, Moscow.

Crawford, K.: 1996, 'Vygotskian Approaches to Human Development in the Information Era', *Educational Studies in Mathematics* **31**, 43–62.

Crawford, K. & Deer, E.: 1993, 'Do We Practise What We Preach?: Putting Policy into Practise in Teacher Education', *South Pacific Journal of Teacher Education* **21** (2), 111–121.

Davis, B.: 1995, 'Why Teach Mathematics? Mathematics Education and Enactivist Theory', *For the Learning of Mathematics* **15** (2), 2–9.

Davydov, V. V.: 1990, *Types of Generalization in Instruction* (Soviet Studies in Mathematics Education Vol. 2; J. Kilpatrick (ed.); J. Teller, trans.), National Council of Teachers of Mathematics, Reston, VA.

Dowling, P.: 1995, 'A Language for the Sociological Description of Pedagogic Texts with Particular Reference to the Secondary School Mathematics Scheme SMP 11–16', *Collected Original Resources in Education*, 19.

Ellsworth, E.: 1989, 'Why Doesn't This Feel Empowering? Working Through the Repressive Myths of Critical Pedagogy', *Harvard Educational Review* **59** (3), 297–324.

Ernest, P.: 1991, The Philosophy of Mathematics Education, Falmer, London.

Evans, J.: 1993, Adults and Numeracy, Unpublished Ph.D. thesis, University of London Library.

Evans, J. & Tsatsaroni, A.: 1994, 'Language and "Subjectivity" in the Mathematics Classroom', in S. Lerman (ed.), *Cultural Perspectives on the Mathematics Classroom*, Kluwer Academic Publishers, Dordrecht, 163–182.

Frankenstein, M.: 1981, 'A Different Third R: Radical Math', *Radical Teacher* **20**, 14–18.

Gergen K. J.: 1985, 'The Social Constructionist Movement in Modern Psychology', *American Psychologist* **40** (3), 266–275.

Goldin, J.: 1990, 'Epistemology, Constructivism and Discovery Learning in Mathematics', in R. B. Davis, C. A. Maher & N. Noddings (eds.), *Constructivist Views on the Teaching and Learning of Mathematics* (Journal for Research in Mathematics Education Monograph No. 4), 31–47.

Harré, R. & Gillett, G.: 1994, *The Discursive Mind*, Sage, London.

Henley, A.: 1990, The History of Mathematics as a Pedagogical Aid, with Special Reference to Negative Numbers. Unpublished M.Sc. dissertation, South Bank University.

Keitel, C. (ed.): 1989, *Mathematics Education and Society*, Unesco Document Series No. 35.

Lacan, J.: 1966, *Ecrits I*, Seuil, Paris.

Lave, J. & Wenger, E.: 1991, *Situated Learning: Legitimate Peripheral Participation*, Cambridge University Press, New York.

Leont'ev, A. N.: 1981, 'The Problem of Activity in Psychology', in J. V. Wertsch (ed.), *The Concept of Activity in Soviet Psychology*, Sharpe, Armonk, NY, 37–71.

Leont'ev, A. N.: 1978, *Activity, Consciousness and Personality,* Prentice-Hall, Englewood Cliffs, NJ.

Lerman, S.: 1992, 'The Function of Language in Radical Constructivism: A Vygotskian Perspective', in *Proceedings of Sixteenth Meeting of the International Group for the Psychology of Mathematics Education,* Durham, New Hampshire, Vol. 2, 40–47.

Lerman, S.: 1993, 'The Position of the Individual in Radical Constructivism: In Search of the Subject' in J. Malone & P. Taylor (eds.), *Constructivist Interpretations of Teaching and Learning Mathematics,* Curtin University of Technology, Perth, 105–112.

Lerman, S. (ed.): 1994, *Cultural Perspectives on the Mathematics Classroom,* Kluwer Academic Publishers, Dordrecht.

Lerman, S.: 1995a, Teaching–Learning Negative Numbers: A Classroom Incident. Unpublished paper, Centre for Mathematics Education, South Bank University.

Lerman, S.: 1995b, 'Mathematics Teaching for Empowerment and Equity: An Achievable Goal?', in *Proceedings of the Third British Congress on Mathematical Education,* Vol. 1, 17–30.

Lerman, S.: 1996, 'Intersubjectivity in Mathematics Learning: A Challenge to the Radical Constructivist Paradigm?', *Journal for Research in Mathematics Education* **27**, 133–150.

Lins, R. C.: 1994, 'Eliciting the Meanings for Algebra Produced by Students: Knowledge, Justification and Semantic Fields', in *Proceedings of Eighteenth International Meeting of the Group for the Psychology of Mathematics Education,* Lisbon, Portugal, Vol. 3, 184–191.

Luria, A. R.: 1973, *The Working Brain,* Penguin, Harmondsworth.

Masingila, J.: 1993, 'Learning from Mathematics Practice in Out-of-School Situations', *For the Learning of Mathematics* **13** (2), 18–22.

Masingila, J. O., Davidenko, S. & Prus-Wisniowska, E.: 1996, 'Mathematics Learning and Practice in and out of School: A Framework for Connecting These Experiences', *Educational Studies in Mathematics* **31**, 175–200.

Matos, J. F.: 1995, 'Ethnographic Research Methodology and Mathematical Activity in the Classroom', in *Proceedings of Nineteenth International Meeting of the Group for the Psychology of Mathematics Education,* Recife, Brazil, Vol. 1, 211.

Mellin-Olsen, S.: 1987, *The Politics of Mathematics Education,* Kluwer Academic Publishers, Dordrecht.

Newman, F. & Holzman, L.: 1993, *Lev Vygotsky: Revolutionary Scientist,* Routledge, London.

Nickson, M.: 1992, 'The Culture of the Mathematics Classroom: An Unknown Quantity?', in D. Grouws (ed.), *Handbook of Research on Mathematics Teaching and Learning,* Macmillan, New York, 101–114.

Nickson, M. & Lerman, S. (eds.): 1992, *The Social Context of Mathematics Education: Theory and Practice,* South Bank Press, London.

Nunes, T., Schliemann, A. & Carraher, D.: 1993, *Street Mathematics and School Mathematics,* Cambridge University Press, New York.

Piaget, J.: 1964, 'Development and Learning', in R. E. Ripple & V. N. Rockcastle (eds.), *Piaget Rediscovered,* Cornell University Press, Ithaca, NY, 7–19.

Piaget, J.: 1971, *Structuralism,* Routledge & Kegan Paul, London.

Piaget, J. & Garcia, R.: 1989, *Psychogenesis and the History of Science,* Columbia University Press, New York.

Pimm, D.: 1994, 'Attending to Unconscious Elements', in *Proceedings of the Eighteenth International Meeting of the Group for the Psychology of Mathematics Education,* Lisbon, Portugal, Vol. 4, 41–48.

Santos, M.: 1995, 'Mathematics Learning as Situated Learning', in *Proceedings of the Nineteenth International Meeting of the Group for the Psychology of Mathematics Education,* Recife, Brazil, Vol. 1, 222.

Saxe, J.: 1991, *Culture and Cognitive Development: Studies in Mathematical Understanding,* Lawrence Erlbaum, Hillsdale, NJ.

Schliemann, A. D. & Acioly, N. M.: 1989, 'Mathematical Knowledge Developed at Work: The Contribution of Practice versus the Contribution of Schooling', *Cognition and Instruction* **6** (3), 185–221.

Secada, W. G., Fennema, E. & Adajian, L. B. (eds.): 1995, *New Directions for Equity in Mathematics Education*, Cambridge University Press, Cambridge.

Sierpinska, A. & Lerman, S.: 1996, 'Epistemologies of Mathematics and of Mathematics Education', in A. Bishop (ed.), *International Handbook on Mathematical Education*, Kluwer Academic Publishers, Dordrecht, 827–876.

Steffe, L.P.: 1993, Interaction and Children's Mathematics. Paper presented at American Educational Research Association, Atlanta, GA.

Von Glasersfeld, E.: 1983, 'Learning as a Constructive Activity', in *Proceedings of the Fifth Annual Meeting of the North American Chapter of the International Group for the Psychology of Mathematics Education*, Montreal, Vol. 1, 41–69.

Von Glasersfeld, E.: 1989, 'Cognition, Construction of Knowledge, and Teaching', *Synthese* **80**, 121–140.

Von Glasersfeld, E.: 1990a, 'An Exposition of Constructivism: Why Some Like It Radical', in R. B. Davis, C. A. Maher & N. Noddings (eds.), *Constructivist Views on the Teaching and Learning of Mathematics* (Journal for Research in Mathematics Education Monograph No. 4), 19–29.

Von Glasersfeld, E.: 1990b, 'Environment and Communication', in L. P. Steffe & T. Wood (eds.) *Transforming Children's Mathematics Education*, Lawrence Erlbaum, Hillsdale, NJ, 30–38.

Von Glasersfeld, E.: 1991, 'Knowing Without Metaphysics: Aspects of the Radical Constructivist Position', in F. Steier (ed.), *Research and Reflexivity*, Sage, London, 12–29.

Vygotsky, L.: 1924/1979, 'Consciousness as a Problem in the Psychology of Behaviour', *Soviet Psychology* **17**, 5–35.

Vygotsky, L.: 1978, *Mind in Society*, Harvard University Press, Cambridge, MA.

Vygotsky, L.: 1986, *Thought and Language* (revised edition, A. Kozulin, ed.), MIT Press, Cambridge, MA.

Walkerdine, V.: 1988, *The Mastery of Reason*, Routledge, London.

Wertsch, J. V.: 1990, 'The Voice of Rationality in a Sociocultural Approach to Mind', in L. C. Moll (ed.), *Vygotsky and Education: Instructional Implications and Applications of Sociohistorical Psychology*, Cambridge University Press, New York, 111–126.

Wood, T.: 1994, 'Patterns of Interaction and the Culture of the Mathematics Classroom', in S. Lerman (ed.), *Cultural Perspectives on the Mathematics Classroom*, Kluwer, Dordrecht, 149–168.

Stephen Lerman
South Bank University,
London,
UK

CLAIRE MARGOLINAS

RELATIONS BETWEEN THE THEORETICAL FIELD AND THE PRACTICAL FIELD IN MATHEMATICS EDUCATION

The question of balance between theory and practice in mathematics education research has been a much discussed topic in the present ICMI Study. But the meanings of the words 'theory' and 'practice' seem to vary from one country to another and from one person to another. Therefore I need to clarify my own position. This will be the theme of the first part of the chapter. The second ('Taking facts into account') will be devoted to the problem of the transfer of knowledge from the practical field to the theoretical field, and the third ('Taking phenomena into account') – to the transfer of knowledge in the reverse direction.

1. VOCABULARY DISTINCTIONS

I regard 'didactique des mathématiques' (which can perhaps be rendered by 'mathematics education', in English) as the science of the production and the spreading of mathematical knowledge. I call *opinion* (or didactic opinion) any idea about mathematics education (see Figure 1). I call didactic *statement* a didactic opinion which can be verified, and I call didactic *fact* a didactic statement which can be positively verified.

I can now introduce the word 'theory'.

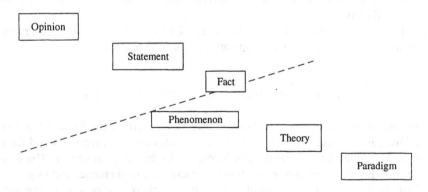

Figure 1 *Taking facts into account*

Sierpinska, A. and Kilpatrick, J. Mathematics Education as a Research Domain: A Search for Identity, 351–356.
© *1998 Kluwer Academic Publishers. Printed in Great Britain.*

A didactic *theory* is a coherent structure which makes possible the production of didactic statements and didactic facts in a specific field of mathematics education. A theory never allows for the production of statements on reality as a whole but only on part of it.

I call a didactic *phenomenon* a didactic fact which can be produced by a didactic theory.

Finally I call didactic *paradigm* a coherent set of didactic theories. The coherence of a paradigm cannot be easily obtained. One has to know the connections between theories. For example, given a particular fact one has to know which theories are capable of producing it or interpreting it.

We now have a set of terms. The line between 'fact' and 'phenomenon' separates in a certain way the field of knowledge resulting from theory and the one resulting from practice.

We shall now consider the relations between these fields.

In every field of knowledge there are isolated statements. These statements can be verified, that is to say, they can be facts. In mathematics education, many of these facts originate from the teachers. For example, 'Pupils have difficulties with the concept of limits'. These statements can be experimentally verified with a statistically analyzed questionnaire. But these facts do not automatically become phenomena if there is no theory to substantiate them. These facts do not therefore have any meaning. They are known, but they are not understood. I feel the production of simple facts is marginal in the scope of research. Let us say facts can be found but we do not look for them. As long as these facts remain isolated, they are questions for new research.

Mathematics education has been studied by many people since antiquity. This is why we have inherited many opinions, statements, and facts. The opinions can inspire us. It is part of the history of the researcher or of the research. But there is no necessity to take the opinions into account. On the other hand, facts must be taken seriously. Researchers should know enough of them to nourish their research. Confrontation with facts should enable researchers to stay close to practical reality.

I would now like to describe the work of two institutes existing in France which foster the knowledge of facts originating from practice.

2. IREM AND COREM: TAKING FACTS INTO ACCOUNT

Founded about twenty years ago, IREMs (*Instituts de Recherche sur l'Enseignement des Mathématiques*) are institutes within universities and their managers must be mathematicians belonging to the same university. The task of an IREM is to develop research in the teaching of mathematics and to participate in the in-service training of mathematics teachers. The persons taking part in the working groups inside an IREM belong to universities or secondary schools, and sometimes to primary schools. The importance of the research activity in these institutes varies from one IREM to another and in duration. The IREMS are not research laboratories. I will return to this point in the last part of the chapter.

In the history of the French *didactique*, the meeting of theoreticians and teachers in these institutes has been one of the levers of research in mathematics education. French papers often have a theoretical character, or a theoretical basis, which often baffle the foreign reader. This theoretical effort is born, however, from encounters with practice and not vice versa. In particular, it is the desire to give meaning to facts, or to produce facts, which motivate, in our country, the production of theory. The interaction between university researchers and teachers around a common project appears to me as necessary for the development of research.

The other institute which I want to introduce is the Center for Observation and Research on the Teaching of Mathematics, or the COREM.

It is a very original institution, which has played and continues to play a fundamental and exemplary part in French research in mathematics education. Created twenty years ago by Guy Brousseau, the COREM is institutionally tied to the IREM of Bordeaux. It is located in a school for 3 to 10 years-old children from an ordinary background and where there is no pre-selection. This school has a special building which comprises a big classroom where video and audio recorders are available. For the needs of research, some mathematics lessons may take place in this room and may be observed. Its purpose is not to be an experimental school but rather a center for observation. The daily interaction between Guy Brousseau, the teachers, and the pupils of this school has enabled him to identify a great number of facts (1978, 1986). This interaction is actually a very stimulating challenge for the researcher since he or she cannot escape reality.

From a naïve point of view, one could think that a center for observation could be just that: it would be sufficient to advise the teachers coming to the school that they might have to give lessons in this room and be observed. However, there is more to this concept, as will be seen from the last part of this chapter where other details of the functioning of the COREM will be described.

The observation of facts is very important: it guides motivation and research questions. But it does not mean that it is good to try and give a meaning to all the facts; that is, to transform all the facts into phenomena. It is the ultimate aim of all science but this aim is not reached by taking all the facts one by one and giving an explanation for each and every one of them. This temptation would lead to ad hoc mini-theories which would be an obstacle to the building of a paradigm. The major concern of the mathematics education research community are connections between theories, not theories in isolation.

There are thus facts which, without being ignored, must be identified and somehow 'put on the back burner' until they can appear as real phenomena. Only then, will these phenomena be research results.

3. TAKING PHENOMENA INTO ACCOUNT

Scientific research takes place in the theoretical field and facilitates the production and interpretation of phenomena. The production of phenomena implies an experimental approach that is not in the practical field but is obviously related to it. This experimental approach, when it takes place in a classroom over several hours, implies a relationship with the 'practitioner': the teacher of the class. It is within this experimental frame that the relationship between the theoretical and the practical field first begins. But before entering into more general considerations, I would like to give some information on the part played by teachers in research classrooms.

Experimental Research and Teachers

In French research, the teachers play various roles but, most of the time, they are integrated into research teams. They often become very active in these teams thereby producing their own autonomous research activity. Within certain structures (notably in the IREMs) these teachers can get teaching time credits (half-time being the maximum), but they have to apply every year and the credits are not easy to obtain. The absence of a real part-time 'researcher status' in primary or secondary schools is a deterrent to the development of more interactive relations between university researchers and teachers of different levels. The need for such a status was outlined in Topic Group 14 'Cooperation between Theory and Practice in Mathematics Education' at the Seventh ICME congress in Québec in 1992.

If we turn now to experimental research in the classroom, what awaits a researcher is the testing of statements originating from theory, which means the production of phenomena. This type of research involves many persons, and therefore has demanding deontological aspects. From the point of view of the pupil, the minimum requirement is that the experiment does not hinder his learning, meaning that he learns at least as well as usual. But the motivation of the teacher who participates in this type of research derives frequently from the need to fight against the aging of teaching situations. The teacher not only expects changes but also improvement: he or she is oriented towards 'innovations'. There are therefore frictions between the motivations of the researcher and those of the teacher or frictions between innovation and research.

If we examine the functioning of IREMs we can see this problem very clearly. In some IREMs, for instance the IREMs of Paris, Marseille or Bordeaux, the term 'experiment' often means academic type of research. But in other IREMs, for instance the IREMs of Poitiers or Lyon, the term 'experiment' refers more likely to innovation. Both these 'experiments' are interesting but they do not have the same meaning, nor the same scope.

In a classroom experiment, there will be an interaction between the opinions and the statements of the teacher and the statements originating from theory.

This interaction is a characteristic feature of the classroom experiment. Under these conditions, we understand that if the teacher's opinions and the researcher's theory are conflicting with each other, it will not be possible for the classroom experiment to take place; at least, not as it had been planned. But it is not only the teacher's opinions which participate in the didactic contract of the classroom, it is also his actions, voluntary or not. If the usual didactic contract is too far removed from the didactic contract necessary for the experiment, the same failure will occur.

Example of an Experimental Centre: The COREM

I now return to the COREM institute to outline the constraints of a real experimental center in mathematics education.

The teachers of the Jules Michelet school where the COREM is located are not directly nominated by the administration but their application is examined. Each teacher of the school benefits from a teaching time credit. He or she meets every week with a team of researchers, and he or she makes some observations in the classrooms. The mathematics lessons taught in the school are discussed even if these lessons are not part of a research project. On the other hand, when an experiment is to take place in the school, it is described and discussed by the whole team. Without going into details, it can be realized that the experimental classroom is much more than a simple observation classroom. What can also be seen is that the Michelet school has become an experimental school for mathematics, even if that is beyond its scope.

4. THE COMPLEXITY OF RELATIONS

The complex interaction which I have described here in the frame of an experiment enables us to understand what happens in the spreading of knowledge derived from a theoretical field. In the practical field, any report of a classroom experiment is understood as an innovation. We can say there is a fundamental misunderstanding in that point. The results of experimental sciences are never the experiences themselves, but the phenomena or facts explained by a theoretical analysis and validated by experiments or observations.

This is why I think it is very important to try and spread the theories and the phenomena, and not the experiments directly, which are part of the history of research. The researcher will perhaps see his or her theoretical constructions being transformed into opinions in the practical field. I feel it is a normal, and probably inevitable, transformation. But if the ideas thus transformed make it possible for researchers to give a coherence and a meaning to a set of facts, and thus make more reasonable decisions, then our work will have borne its fruit.

REFERENCES

Brousseau, G.: 1978, 'L'observation des Activités Didactiques', *Revue Française de Pédagogie* **45**, 130–140.
Brousseau, G.: 1986, 'Fondements et Méthodes de la Didactique des Mathématiques', *Recherches en Didactique des Mathématiques* **7** (2), 33–115.

Claire Margolinas
Institut Universitaire de Formation des Maîtres d'Auvergne,
36 av. Jean Jaurès,
F-63400 Chamalières,
France

JOHN MASON

RESEARCHING FROM THE INSIDE IN MATHEMATICS EDUCATION

1. INTRODUCTION

What are the significant products of research in mathematics education? I propose two simple answers:

1. *The most significant products are the transformations in the being of the researchers.*
2. *The second most significant products are stimuli to other researchers and teachers to test out conjectures for themselves in their own context.*

I said my answers were simple. They are at least short, and perhaps they appear simplistic. But they are actually rather complex. For beneath such answers, and supporting the efficient and effective generation of significant research products, lies an immense tangle of psychological, cultural, and social, forces and actions. The notion of *being* is central to the enterprise, drawn from Eastern roots and from Heidegger's notion of *being-in-the-world* rather than being muddled by inappropriate distinctions between subject and object (Winograd & Flores 1987).

In this introduction, I advance some direct justification for my claim, consider the notion of validity, and make some conjectures about future developments. I have argued elsewhere that:

- researching from the inside can be every bit as systematic and disciplined as traditional (*extra-spective*) research;
- researching from the inside provides a much needed balance to traditional research;

In the second section, by looking at what researchers do and the sorts of products that are currently fashionable, I advance the thesis that:

- the seeds of researching from the inside are present in traditional research;
- the future in educational research will involve even further movement towards and development of researching from the inside.

The third section develops the notion of *working on being*, since it lies at the heart of what research in mathematics education is ultimately about, however

357

Sierpinska, A. and Kilpatrick, J. Mathematics Education as a Research Domain: A Search for Identity, 357–377.
© *1998 Kluwer Academic Publishers. Printed in Great Britain.*

theoretical or practical: providing direction for, and fostering and sustaining professional development, amongst those concerned with mathematics at all levels.

2. JUSTIFICATION

The transformation that takes place in educational researchers while undertaking a study is the most significant product of their research, for it is *their* questions that change, *their* sensitivities that develop, *their* attention that is restructured, *their* awarenesses that are educated, *their* perspectives that alter. In short, it is their *being* that develops. What I choose to encompass within the term *being* will become more clear in the third section. For the moment, I mean the presence that the individual has, her or his manifest wisdom and situational insight, rather than book-knowledge.

Education is about personal development, about transformation of the individual in a social context. Reports of investigations, observations and research inquiries in education cannot be employed directly in the building of an edifice of knowledge about education. Rather, each individual in each generation, supported and confined by current expression, seeks to reconstruct for himself or herself, within social practices, the insight and wisdom attained in previous generations.

Whereas mathematics appears to be able to proceed by building layer upon established layer of definitions and theorems, even in mathematics individuals have to re-examine the statements upon which they build, for there are subtle shifts in interpretation, meaning, and usage. What was once unproblematic may easily become problematic. In mathematics education, everything remains problematic. Readers of educational reports have to interpret what they construe from what they read, within their own context. They decide whether it seems relevant, informative, or productive to pursue. The most researchers can hope for is that their report stimulates others to investigate their own situation. The track record of solutions prepared by some for others is not a happy one.

What makes mathematics education different from other aspects and disciplines in education is the quixotically dual nature of mathematical objects and structures, seen as extant and discovered, or as personally and socially constructed (Davis & Hersh 1981; Sfard 1994). Neither position is sustainable on its own, yet both speak to experience. The interaction of this dual status and the psycho-sociological forces that operate when people encounter particular ideas forms a unique core for our discipline.

My justification proceeds therefore by describing the actions of researchers in terms of self-transformation and other-stimulation, leading to predictions that mathematics teaching will develop most effectively when teacher-practitioners are supported in experiencing similar transformations in how they conceive and carry out their practice. My aim in describing the Discipline of Noticing (Davis 1990; Mason 1984, 1991a, 1991b, 1992b; Mason & Davis 1988, 1989) and Research from the Inside (Mason 1992a, 1994a, 1994b) is to provide teachers and other practitioners with a systematic and epistemologically well-founded methodology to enable them to develop their teaching as an acknowledged form of valid research.

3. DIRECTIONS OF POSSIBLE DEVELOPMENT

I think the most significant development in research in mathematics education in the past decade, apart from the explosion of activity, has been the emergence of teacher/educator as professional practitioner, by which, with Schön (1983), I mean active and continuing exploration of and work on one's own practice. More and more teachers and lecturers are asking questions about their practice, adopting an inquiring, exploratory approach to their teaching, rather than carrying out a definitively specific program in which they 'deliver the curriculum to their students'. There are pockets of presumed certainty, with schemes and programs of training designed to 'fix' or bypass teachers, but it always emerges that trained behavior without educated awareness is not sufficient to equip people to respond to changing and challenging conditions. Uncertainty prevails in the end. The *Grr*'s keep on cycling:

Grumbling Griping Groping Grasping Grappling Gripping

Grumbling about how things are,
Griping about specific frustrations,
Groping for some alternative,
Grasping at some possibility,
Grappling with a possible solution,
Gripping hard to 'something that works', then *Grumbling* and *Griping* as the substance seems to leak out and a new cycle begins.

Initial grumbling often arises while examinations are being marked ('How could they have learned so little?'), but the griping rarely lasts until the next cycle of teaching starts, and so there is little groping for alternatives.

I take the Zen-like view that it is in a continuing atmosphere of not-knowing that knowing can come about. Descartes noted in *Discourse 1* that 'the only profit I appeared to have drawn from trying to become educated, was progressively to have discovered my ignorance' (quoted in Khan 1983, p. 17), a sentiment that has been echoed in every age. Instead of trying to avoid loss of certainty, instead of searching desperately for secure ground on which to stand firm, it is possible to accept tension between knowing and not knowing as a productive and inescapable source of energy and security. In a line from Dewey and

Peirce back to Bacon, 'continuing and persistent doubt' has been promoted as the basis of sensible methodology: 'the Sinews of Wisdom are slowness of belief and distrust' (Bacon 1609, p. 276).

In education, such an approach is all the more necessary as a means for remaining alive to the tensions of teaching. Conflict and stagnation only develop when people claim certainty.

4. VALIDITY

People have always agonized over how to be certain about what they know, and how to find out things with some justifiable confidence in their results. The well-spring of research and the source of interest in epistemology and methodology since recorded time is due to the fundamental problem of whether what you think you have seen or thought is actually true. Philosophers through the ages have discussed the meaning of truth, validity, and 'meaning' in an unending search for somewhere to stand. Archimedes is said to have remarked that if he had a place to stand he could move the earth, and metaphorically the same remark applies: If we can find somewhere certain to stand, we could move mountains.

Physical science is concerned with knowing about the material world. It is possible to isolate factors and conditions and so find somewhere to stand. The Western intellectual tradition derived from Plato and Aristotle has made great advances by seeking knowledge that is independent of the researcher and that is reached with rigorous and unassailable method. Its success is due to the fact that simple objectivity can be contradicted. But education is different. It is about sensitivity to and transformations in others. The only certain place to stand is in the most unlikely place: ourselves.

Instead of looking for the chimera of provable certainty, influenced by simple accounts of the pre-eminent role of mathematics as a source of certainty, it is possible to find a satisfactory position that acknowledges subjectivity, embraces it, and finds objectivity in transcending rather than rejecting apparent subjectivity. Once you let go of the possibility of guaranteed, provable certainty, then numerous opportunities open up:

- Assertions can be permitted to be appropriate to time, place, and people, without pretending to universal validity (Shah 1964), and can be translated into invitationals (McWilliams 1993).
- Assertions can be found to help make sense of past experience, to 'fit' (von Glasersfeld 1984) with current experience, to inform future practice.
- Assertions can be taken as conjectures, which may require modification as conditions change.
- Assertions can be checked out with a wider and wider range of people, thus enlarging the community of those who find the assertions helpful, meaningful, or informative.

● The more precisely one can state the conditions in which an assertion applies, the more robust it is, the more widely recognized and accepted, the more confidence we are likely to have, but the more our sensitivity to subtle changes is blunted.

Once the mechanism of simple laws and cause-and-effect is abandoned, you enter a world of relativism in which you have to keep looking down at the ground because it is constantly shifting. You cannot march ahead confident that what you have found out before will remain true in the future. Instead you have to be constantly re-checking in new circumstances.

Popper made a definition out of falsification, arguing that unless a theory could be falsified, it could not be objective (Popper 1972). Falsification is a much more complex matter outside of science. Just because someone does not recognize an assertion, or better, does not recognize a distinction, it does not follow that they have falsified that distinction. Rather, they have shown that the distinction is not appropriate for them at that time, place, and in their current state. We learn as much about them as we do about the proposed distinction. Falsification is localized rather than global.

Some knowledge has transitory domains of validity, so that an assertion uttered in a moment has validity for those people at that time in that place, but does not stand immutable throughout time. For example, I am suggesting that being, or presence, is something worth working on. But for a novice teacher there are more immediate concerns of control and working practices. This does not mean that working on being is universally invalid, just inappropriate at certain times.

At another time and place, I might want to work on absence, noting that the presence of a teacher may be exactly what diverts student attention and interferes with their own construal. In other words, if I do for people what they can already do for themselves I am not teaching them, but rather entrapping them; if I only do what they really cannot yet do for themselves, I may be more effective in teaching them. But even that depends on my state of awareness of what they can and cannot do at the time, for if there is no awareness, there is only activity. If I am unaware that students can do something for themselves, I may be tempted to do it for them, and so support both local dependency and global conviction that learning consists of submitting to authority and performing as required. Vygotsky's (1978) notion of *zone of proximal development* (ZPD) speaks to this situation, but it has been simplified into teaching slogans that demonstrate what huge transformations can take place when expert awareness (manifested in Vygotsky's ideas) is brought to articulation as advice for others and then turned into mechanical practices. A slogan like ZPD can be and is used to justify a wide variety of conflicting practices. Almost any practice can be construed as 'working in the zone' if meaning is stretched enough.

Failure to get others to recognize a distinction, a perspective, a particular stressing and ignoring, may be due to flaws and imperfections in that distinction, but it may also be due to presentation that the particular examples do not speak to those people at that time, or requisite sensitivities may not yet have emerged. Kant (1781) observed that 'all our knowledge falls within the bounds of possible experience' (p. A146, B185), which expresses much the same sentiment.

How can I guard against fooling myself? How can I guard against the bigot who charismatically leads a group of disciples? Ultimately there is no firm line to distinguish persuasion from perception: a succession of charismatic tyrants, preachers, and entertainers is convincing evidence that it is all too easy to be misled. When testing of proposals in one's own experience is replaced by compulsory acceptance, then it is clear that persuasion is dominating perception. But the converse does not hold. Just because I seem to find some assertion fits with my experience, it does not follow that that assertion is valid. At best it is locally valid, for a time, for me. Instead of demanding proof from outside, as in outer research, I can be convinced by the extent to which it helps me make sense of my past, and informs my future activity, all mediated through confirmation by an ever-growing cadre of colleagues.

5. FORMS OF RESEARCH

Mathematics education has seen a variety of informative research programs:

- long-term studies and surveys of the behavior of many children;
- material development and study of what happens when teachers are inducted into and use that material;
- tracking teacher beliefs and teacher behavior;
- case studies of a few teachers or a few children;
- clinical interviews;
- studies of what children *can* do, and of what children currently *do* do; and
- action research of implementation of changes in practices.

These are all forms of *outer* research. The researchers observe something other than themselves. They attempt to explain, rather than to experience. Researchers observe, collect data, analyze and report on patterns found in that data, but throughout, the purpose is to seek ways of distancing the observer from the phenomenon in order to obtain observer-free knowledge. Action research, informed by developments in qualitative methods from directions such as ethnography and phenomenography, has become a popular paradigm for teacher-as-researcher, but most forms of action research turn into outer research as soon as the participants distinguish between observation, action, and evaluation, or try to objectify their activities by distancing themselves from the action.

Inner Research

Inner research is about developing sensitivity, whether to mathematical ideas, to pedagogical possibilities, or to the thinking of other people. Sensitivity means noticing (making distinctions): the architect notices features of building not seen by novices; the musician does the same in music; the mathematician notices mathematical structure and sees the world through mathematical eyes. There is a continuing tradition of mathematicians expressing their view of mathematics and of the world, displaying their sensitivities, though very often those sensitivities are obscured by, even confused with, mathematical content. Most books that intend to be *about* mathematics end up expounding mathematics, caught in the *transposition didactique* (Chevellard 1985).

Being aware, attending to that awareness, and being explicitly articulate about sensitivities is no mean task. Yet surely I need to be sensitive to the structure, importance, and techniques of a topic if I am to assist others to alter the structure of their attention (Mason & Davis 1988). Attending to my own struggles with mathematics, and seeking examples in my own experience that mirror or parallel the difficulties my students seem to have is an essential component to being able to relate to my students' experience, and is the basis for most of the materials produced by the Centre for Mathematics Education at the Open University. For example, to appreciate difficulties with subscripts, I found an exercise that produces a similar effect on more experienced mathematicians (Mason 1989). Constructing exercises that highlight shifts in the form or structure of attention is a critical component of sharing insights with others.

Inner research also involves a shift from dependency on chains of cause-and-effect reasoning to a perspective in which there is mutual-interaction and co-determination (Varela, Thompson & Rosch 1991). The whole changes not because of the action of particular parts, but through the presence of a multiplicity of factors that all participate in that change. Cause-and-effect reasoning produces simplification, which is attractive and effective when appropriate; mutuality permits transformation while maintaining an in-dwelling complexity.

6. WHAT RESEARCHERS SEEM TO DO

I take research to be about finding out about a world, whether the material world, the worlds of behavior, images and forms, worlds of other people's experience, or my own inner world of experience. The material world is external and can be dealt with by measurement. However, a world of forms is not entered by counting or measurement, but through sensitivity. Even the material world is sometimes most effectively dealt with via a world of forms (as in Aboriginal Dreamtime, the use of real numbers in physics, or images in modeling).

This section considers relationships between what researchers traditionally do and the practices of inner research. The conclusion reached is that aspects of inner research lie at the heart of all research. Inner research is not in competition with outer research, but is available to be brought to the surface so as to enhance all research in education.

Researchers Participate in a Community

Research takes place within a community, which shares not only discipline, epistemology and ontology, but also a value system as to what is worth studying, what is tractable, and what is informative. The community maintains standards of disciplined study and presentation through public repetition and negotiation of practices. Such a community provides a conserving force in tension with struggles to extend, develop and vary practices. As the community grows, differences in practices develop, and unless there is collective acceptance of variation, factions and schisms develop. This process can be seen in religious groups and sectarianism, as well as in mathematics education research.

The need for individuals to achieve recognition and acceptance by colleagues is a driving social force, as well as an epistemological necessity. Only by testing one's articulations of insights and descriptions of practice against others, engaging in trying to persuade them and being persuaded by them, can one validate one's own criteria and avoid solipsism.

Researchers Are Systematic and Disciplined

The essence of any research is that there be a systematic and disciplined approach to withdrawing from full engagement in doing, with attention directed to finding out about that doing. Methodology is supported by a consistent epistemology (including explicit recognition of assumed and consequent ontology) and that norms for justification and validation of conjectures and assertions are maintained and developed. Inner research depends on constantly re-validating distinctions and frameworks with colleagues.

Researchers Seek Invariance within a Specified Domain of Change

One of the struggles for any research methodology is to obtain results which are invariant over some specifiable domain of potential variation, and to develop means for validating that invariance. For example, the results and conclusions may be (thought to be, claimed to be, whether implicitly or explicitly)

- independent of the particular students, classroom, teacher, school, culture, society, and independent of the particular researcher;
- generalizable to other similar situations (what constitutes 'similar' is one aspect of clarification of both the research paradigm in general, and the particular study being undertaken); and
- robust over time.

In education it is often difficult to specify a domain of validity because there are so many factors involved that are not usefully studied from a cause-and-effect perspective. Distinctions and frameworks that arise from inner research have a domain of validity consisting of the perceiver-researcher, and the situations in which the distinction comes to mind and is found to be informative.

Researchers Distance Themselves from Phenomena

Research is a form of distancing from doing, from practice, from what is being studied. Distancing is usually taken to mean drawing back so as to have an overview. But you can also distance yourself by moving inwards, by developing an inner monitor (Mason, Burton & Stacey 1984; Schoenfeld 1985) whose presence in mathematics is evidenced by little thoughts that arise such as 'Is this calculation appropriate?' and by catching yourself about to make little slips. In teaching, the inner monitor's presence is evidenced by the arising of alternative possibilities of how to respond to the situation as currently perceived, through increased sensitivities and triggered by preparation done earlier in imagining acting in fresh ways.

Researchers Observe and Inspect

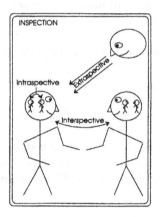

I have found it useful to play with the Latin root *specare* meaning *to look*, which is the basis of words like *spectacles*, *speculate*, *inspect*, and *respect*, to produce words like extraspection (looking from outside), *intraspection* (observing oneself), and *interspection* (sharing and negotiating observations with others). These three terms serve to distinguish research perspectives. Traditional research methodologies are based on extraspection. Intraspection describes the development of awareness and awareness of that awareness, of inner monitors, which forms one dimension of inner research. Interspection describes the interactions between colleagues who go beyond extraspection and begin to form a shared or taken-as-shared (Cobb 1991) world of experience.

Researchers Notice and Mark

Another way of thinking about observation is through the experience of noticing. To notice is to make a distinction, to bring into existence a separation. The distinctions I make in the world that I perceive are reflections of sensitivities to which I have been enculturated or educated. I have goals and intentions that the research is intended to serve (e.g., improving my own teaching, the experience for pupils, other people's teaching, etc.). A researcher seeks new distinctions, or new relationships between distinctions, and introduces these to the community. How you learn to notice, and how you assist others to notice what you notice, is an important question no matter what the overt paradigm. It is an essential ingredient of methodology and epistemology, and also of promulgation of research findings. In this sense, *noticing* lies at the core of any and every research paradigm.

Noticing can be subconscious, in the sense that if someone else reports an observation you can recognize that you too noticed what is being described, but that you could not have initiated that description. The term *marking* refers to noticing that is sufficiently sharp to enable you to initiate a description of what is noticed. An important role for educators is to resonate teacher noticing by re-marking on things that the educator has noticed, thereby bringing awareness into consciousness and enabling it to inform future practice. There are important analogies between the triads of

not-noticing – noticing – marking

and

reacting – responding – initiating.

In both cases, the first is automatic and done without conscious awareness; the second is done from awareness as if by intuition, with actions triggered from outside or metonymically; the third is performed from oneself, from conscious awareness.

Researchers Make Distinctions, Develop Frameworks

Any research paradigm produces an ontology: what is studied, and the distinctions made within that discipline, become objects in and of themselves. They are called into existence by virtue of the research. Classification lies at the heart of perception and is at the heart of all research in education. Classification is how people simplify complexity so that they can think and communicate effectively. It is also the means by which people are induced to overlook details that do not fit with the classification. Outer research seeks classifications that are robust and transferable from researcher to researcher. Inner research recognizes that classifications are evidence of sensitivities, which are in flux and may alter and develop over time.

A classification introduced from outer research may be ignored by others, or it may be taken up and modified; instruments of inquiry such as questionnaires may be validated and employed by generations of workers as part of their own studies taking for granted that the instrument continues to function as described. A distinction, introduced as a result of inner or outer research, will only remain alive as long as people continue to find that it informs their practice. Dawkins (1976) called these *memes*, the cognitive version of genes. A distinction may be integrated into thinking, or it may be played with for a while and then fall to the wayside.

Where several colleagues find that they recognize each other's descriptions of incidents and phenomena, a body of shared experiences accumulates, which serves to enhance noticing in the future. By 'noticing the noticing', or in other words by labeling a collection of experiences with a descriptive term, that term can contribute to a framework that enhances noticing in the future. It is as if something inside generalizes from the particular, triggers the label, and so provides access to the collection of responses and comments on such situations which have been accumulated through collegial sharing.

Researchers Collect and Analyze Data

Application of any methodology produces data, which are collected, analyzed, and synthesized into research products. Data arise from making of observations, and their collection constitutes a first level of abstraction from the phenomenon studied, in much the same way that summarizers such as means and standard deviations are abstractions from numerical data. The data are not the phenomenon, but merely extracted samples, indicators, and précis of the phenomenon. Unfortunately, analysis often turns data into the subject and focus of research rather than using them as a means of searching, reporting and justifying assertions about the original phenomenon. For example, detailed tapes and transcripts of past events frequently become the phenomenon studied, with insightful remarks about the tape transcript that may or may not have anything to do with the event that generated it. And unless some effort is made to re-locate insights in practice, such research is unlikely to have little long term effect on anyone other than the researchers themselves.

The data that are collected in the practice of inner research consist of fragments of experience (Mason 1988), in the form of brief-but-vivid descriptions of striking incidents. The term *brief but vivid* signals that justification and explanation have been minimized and descriptions made in terms that would enable others present to recall the same moment (accounts of rather than attempts to account for situations). Our sense of an event is the story we weave using the salient fragments that are readily recalled, which Bruner (1991) described by saying that we organize our experience and our memories of human events mainly in the form of narrations. Feyerabend (1991) concluded in his analysis of epistemological positions that all you can do, if you really want to be truthful, is to tell a story.

The hallmark of a useful data-fragment is that other people recognize the situation even though they were not present, and are even able to offer similar fragments themselves. Through sharing and negotiating ('that does not strike me as similar to what I experienced ...'), colleagues develop a shared vocabulary that not only facilitates communication in the future, but also serves to awaken sensitivities so that 'similar' situations are noticed in the future. McWilliams (1993) put it nicely when he proposed replacing the indicative mood in assertions and observations with an invitational mood. Instead of ascribing qualities to events, inner research invites undertaking an *as-if* trial, to see if the proposed distinction or quality really does help make sense of the past or inform future practice.

Researchers Have Personal Sensitivities

Few researchers that I know are able to read and keep up with everything written in their field. Even papers written about research in their particular specialized topic are not treated equally: some are pondered and ideas incorporated, while others are seen but ignored. I suggest that the reason is that everyone has certain sensitivities, and that when a paper does not resonate with or evoke those sensitivities, then the contents are literally 'not seen', or go 'unmarked'. If a paper does not either contradict or confirm our own view, if the distinctions it makes do not immediately strike us as potentially informing, then we are likely to ignore what they say. If a researcher claims results that seem obvious, or do not accord with our experience, then we are unlikely to develop them ourselves.

In other words, as members of a research community, we notice what we are attuned to notice, and what strikes us as contradictory or surprising. Our attunement is constantly developing of course, but our response at any time is based on the extent to which what we construe from what we read or hear conforms with and informs experience, and informs practice in the future. I use *inform* here in both its figurative sense of information and its literal sense of providing form. Expanded and made precise, these are the elements of validation for the discipline of noticing. I claim therefore, that despite assertions about criteria for importance and validity in different paradigms, when it comes to practice, most researchers have additional criteria for validation based on their own response to the assertions made in any research report.

The observations of others are of course of great assistance: We cannot expect to locate all or even most potentially useful distinctions entirely by ourselves, nor to discard some without extensive study. We depend therefore on colleagues to alert us to potentially fruitful distinctions which sharpen our sensitivity and inform future actions.

Researchers Seek Resonance with Others

A critical component of inner research is reaching outwards by seeking resonance in the experience of others. The articles (Mason 1980, 1989, 1992a, 1992b) are all products of inner research, presented in a manner consistent with the discipline. They invite the reader to check conjectures against experience (generated by participating in immediate activity, and by recall of past events). Furthermore, the research methods I have used have enabled me to speak to teachers' experience, so that they feel that I have experience of the classroom and know what it is like (despite having only visited classrooms sporadically). I work on my own awareness, and look for analogous situations which cause me to struggle the way I see students struggling mathematically. The characteristic style of Open University Centre for Mathematics Education materials is that they involve the teacher in immediate experience, reflection on that experience, and then search for similar experiences from the past, all in order to sharpen sensitivities so as to inform future practice.

Researchers Learn the Most

It has been noted by almost everyone engaged in research that the person who learns the most is the researcher, no matter how well reports are written, how widely findings are communicated, or what other products produced. Questions that seem clear and focused are refined, changed, even transformed; conjectures are modified, sometimes confirmed, sometimes denied, and sometimes remain indeterminate. Throughout this process, the researchers' perspective develops, their sensitivities increase. They see more, make more and finer distinctions. They experience shifts in what they find fruitful to study, in the distinctions they are able to make, in their appreciation of the researched situation and of similar situations, and what constitutes a *similar* situation. In this sense, the researchers can be said to be learning about themselves as much as about the situation being studied. Instead of thinking of the situation being studied as being acted upon by the researcher, there is a sense in which the studying of the situation performs an action upon the researcher. Usually they emerge feeling they know less than when they started, and find themselves qualifying and hedging when asked apparently straightforward questions by agencies with commitment to universal certainty. To be an expert is to know that you do not know, and that is frustrating for people who want clear simple answers.

Support for this modernist view can be found in many of the natural sciences, as well as in the social sciences, not to say in Plato. No matter what the discipline, astute researchers learn about themselves, and in struggling to be ever more disciplined and systematic, in order to be confident in their findings and compelling to their colleagues, uncover propensities in themselves. Indeed, in the *Thaetetus,* Plato has Socrates describe himself as a midwife for ideas, and he seems more interested in the process of inquiry than in the actual results of those enquiries.

Researchers Research Themselves

My late colleague Joy Davis was working on a thesis that the topics that people choose to study in mathematics education are topics that are problematic for them in their own practice. Even though many research students in education start out wanting to 'study' some aspect that they would like others to 'take on board', their work often reveals that they have some uncertainties, some personal sense of inadequacy in that area, and their view of what they wish to promulgate usually develops and deepens considerably, even to the extent of no longer being in a position to promulgate some single 'thing'.

Theoretical justification for this position lies in the observation that if I am confident in myself about some aspect of mathematics education, I have no need to go preaching about it to others. I am even unlikely to be aware of it *as* an issue, unless it is in fact an issue for me. But if I am aware that I am not doing as much of 'it' as I wish or intend, I may find promulgating the ideas to others a way of trying to goad or pacify myself. As you may imagine, and can soon find out for yourself, this is not a comfortable position to take for someone who writes and preaches! But it is salutary and grounding to bear it in mind when writing, researching, or conducting workshops.

Incidents that remain salient in memory, even for a short time, can be taken as the phenomena under study, or as indicators of it. They can also be taken as a mirror in which to see oneself, to locate one's interests and concerns. Jackson (1992) illustrates mirroring beautifully, through choosing to use things that stood out for him as stimulus to inquire and interrogate his own perceptions and assumptions rather than intrude on the teacher whose lessons he was observing. Italo Calvino (1985) described this process as follows: 'The universe is a mirror in which we can contemplate only what we have learned to know in ourselves' (p. 107). And Hymenehosts Storm (1985) in the American Indian tradition, based his book on the adage that 'the world is a mirror for the people', evoking the ancient notion of augury as a reading of the world.

7. RESEARCH PRODUCTS

The more familiar and overt products of research, namely reports, articles, books, professional development materials, and classroom materials all suffer from what might be called a *research transposition*, following Chevellard's (1985) *transposition didactique*. Genuine questions about what it is like to be a student doing the things students do (and do not do) are quickly turned into researchable questions using preference scales, interview transcripts and so forth, just as attempts to assess what students *can do* are transformed into training behavior that can be mastered without generative awareness.

The insight and effect of having participated in carrying out research is transformed into something quite different when it is presented to others. Research reports tend to be like describing a car journey from a helicopter: major turns and traffic conditions may be reported, but not the views seen and moods experienced by those inside the car. Researchers very rarely report on, much less communicate transformations in themselves. Examples of reports and theses with which I am familiar and which *are* structured around accounts of the research process include Morgan (1977), Yates (1978), Brown (1988), Atkinson (1989), Williams (1989), and Chatley (1992).

In order for research activity to affect anyone other than the researchers themselves, it is necessary that an action take place, in which the practitioner's experience is spoken to directly, and acted upon by the practitioner:

It's never enough to just *tell* people about some new insight. Rather, you have to get them to experience it in a way that evokes its power and possibility. Instead of pouring knowledge into people's heads, you need to help them grind a new set of eyeglasses so they can see the world in a new way. (Seeley Brown 1991, p. 109)

You cannot persuade people to act differently through logic and appeal to rational argument and evidence. You have to create a social practice and speak directly to people's experience. Speaking *to* rather than *about* experience means resonating awareness, whether previously articulated or not. For example, when someone says 'You said just what I have been thinking, but I wasn't aware of it until you said it so clearly', you know that experience has been touched and an action is beginning. Providing experiences that remind people of their past and that highlight some sensitivity is another way to speak to experience. Factual accounts of data collection and analysis tend not to.

The test of effectiveness of research is whether future action is informed; whether the individual is able to choose to respond rather than react automatically to a situation. And this is what I mean by developing one's being. The processes of teaching and learning mathematics form an organic whole. You cannot make significant changes just by poking them in one or two places. Rather, you have to support the organism itself in evolution. Piaget (quoted in Evans 1973) put it as follows:

Knowledge does not begin in the I, and it does not begin in the object; it begins in the interactions.... There is a reciprocal and simultaneous construction of the subject on the one hand and the object on the other. (p. 20)

Insight is a personal construct that is produced through participation in an action, supported and influenced by a collegial conjecturing environment. It is co-produced, not instigated. It evolves. It is best supported by an experiential approach in which exercises are developed and honed that tend to highlight significant noticing that can then be developed into action-informing frameworks.

From this perspective, to be alive and awake as a teacher means to be working on, or researching one's teaching (and more generally, one's practice), to be working on one's being. I predict that disciplined *inner* research by teachers of their own experience (including their practice, their awareness, and their emotional-energy, as well as the stories they construct to make sense of that experience) will form the core of the significant developments in mathematics teaching and learning in the future, augmented and informed by sensitivities located through other more traditional *outer* research styles.

8. RESEARCH AS WORKING ON BEING

Habits and Freedom

Most of the time, people react to stimuli. In roles as students, teachers, educators and researchers, we develop characteristic reactions through enculturation and personal construction in order to reduce energy drain. Our reactions constitute our role. In order to cope with the variety of situations that develop, it is essential to automate certain responses so that they become rehearsed reactions that can flow automatically without agonizing all the time over making choices. Actions taken are the working out or unfolding of decisions made hours, months or years earlier. But the trouble with habits is that they sink below the level of awareness, and then are not available for inspection: habit forming can be habit forming.

One aspect of inner research is to seek out such habits and make them available for re-questioning from time to time, to maintain a conjecturing 'as-if' stance towards the whole process of teaching and learning. Learning diaries and personal journals can serve to awaken inquirers to issues that are of significance to them, simply by looking back over the incidents that were considered sufficiently significant to record, and locating patterns and commonalities in these.

Occasionally, there is a moment of awakening, a moment when an opportunity presents itself, or more accurately, when there is fleeting awareness of alternative possible behavior. Such a moment is a *moment of noticing*. Study of teachers noticing in the moment – that is, teacher decision making – provides a case study in transition from outer to inner research. Early studies (McNair 1978a, 1978b; Peterson & Clark 1978; Peterson, Marx & Clark 1978) focused on interviews and observations by extraspective researchers. More recently, teachers are finding that such decisions are available to self study, and that it is precisely in such moments that the taste of freedom, of being alive to the situation, is felt (Baird & Northfield 1992; Billington 1992; Chatley 1992; Rosen 1991; Williams 1989).

For an opportunity to exist in the moment, there has to be a convergence of recognition (of some typical situation), with awareness (of alternative action or behavior). Thus there is *no* opportunity when you become starkly aware of the unstoppable flow of events in which you are embroiled (awareness alone is not enough). There is *no* opportunity if you have prepared a dozen different strategies, but none of them occur to you in the midst of an event (alternative actions are not sufficient). The two have to come together.

The Discipline of Noticing is a form of personal and collective inquiry into how to sharpen moments of noticing so that they shift

- from the retrospective 'I could have....' or 'I should have....',
- to the present 'spective' 'I could...',
- by means of the descriptive but non-judgmental post-spective review, 'I did ...', and the pro-spective preparation of 'I will...',

by imagining oneself in a similar situation in the future, entering that moment as vividly as possible, and mentally carrying out each of the actions amongst which one wishes to choose in the moment. It is one direction in which inner research can be conducted.

Being

Heidegger (1927) is famous for use of the term ~~being~~, with the word crossed out to indicate that what is referred to is not some thing that is, but a domain of awareness that can be neither deduced nor represented and through which one contacts (spiritual) reality. We experience the being of others as a presence, as (metaphorical) stature, as the atmosphere that they generate in a room. It is bit like a magnetic force that attracts, not in the manner of charisma which comes at you, but in the manner of seduction, drawing you to trust, to be with, to bear with, and to be yourself. People with real being have a presence that is almost palpable, through the extent and scope of their awareness, their present moment (Bennett 1966). They enable others to excel themselves. I recall vividly a musician in whose presence I was able to improvise publicly in a manner not experienced before or since.

Being encompasses energies and state. It is experienced as consciousness; not ordinary consciousness but full and sensitive contact with reality in an awakened (rather than dreamy, identified, or automatic) state. In a sense, being is the ability to bear reality without resorting to displacement and story telling.

In discussing roles of a therapist, Patrick Casement (1985) observes:

As far as possible, the therapist's presence therefore has to remain a transitional or potential presence (like that of a mother who is non-intrusively present with her playing child). The therapist can then be invoked by the patient as a presence, or can be used by the patient as representing an absence. (p. 29)

And there are direct parallels with teaching. An intrusive teacher has a presence that is all too manifest and that blocks students from doing what they can; a teacher with being may at times be present as an absence.

Ability to Bear

When a teacher asks a question, and then is able to hold silence and attract students to want to express themselves, she or he displays being. It takes considerable self-confidence and considerable effort (often in holding back or withdrawing, in creating absence) in order to generate conditions in which students are willing to participate in a conjecturing atmosphere. When a teacher invites students to talk in pairs about some issue, it takes considerable being to bear the loss of immediate control, in the expectation that the action generated will produce individual and collective construal.

The focus and locus of a teacher's attention delimits what actions and interactions she or he can initiate. If attention is on answers, and on getting students to

get correct answers, then possible actions are more severely limited than if the
teacher is aware of a variety of mathematical thinking processes, themes, and
connections. Willingness to forgo immediate facility, simplicity, and directness
for greater gains in awareness and understanding requires pedagogic and
mathematical being.

Developing Mathematical Being

A teacher's mathematical being is critical, because how teachers deal with mathematical questions, how they reveal their enthusiasms and interests, how they
conduct themselves, can have a major influence on their pupils. Working on
mathematics at one's own level (individually and with colleagues), not simply to
learn more mathematics, but to increase one's sensitivities to processes of learning, to frustrations and exhilarations, to mathematical themes and processes, is
an important part of developing one's being. Learning to work collaboratively in
a conjecturing atmosphere with colleagues and in-dwelling recent experiences
analogous to those experienced by students is a particularly useful focus (see
e.g., Griffin & Gates 1988; Mason 1989). Working on one's mathematical being
enables one to be mathematical in front of and with one's students by drawing
on enhanced awareness of what it means to think mathematically. Lebesgue
(quoted in Schmalz 1993) put it strongly: 'The only instruction which a professor can give, in my opinion, is to think in front of his students' (p. 252).

Working on Pedagogic Being

Working on specific incidents from the classroom – describing them briefly but
vividly, seeking resonances with analogies in colleagues' experience, and locating alternative gambits that could be employed in similar situations – contributes
to one's being, because it provides the ground for greater choice and opportunity
in the future. Wheeler (1986) advocated description of classroom incidents as
the currency of exchange between teachers, and this approach has been developed by many groups including the Centre for Mathematics Education at the
Open University (for extended descriptions, see Open University 1988, 1992;
and Mathematical Association 1991). The roots lie very deep. Plato advocated
use of the particular (as in 'particular incident') in the way he set out his
dialogues, while commenting through Socrates on the difficulty that discussants
have in dealing with abstraction without using the particular. Effective writers
are those who are able to resonate reader experience through the clever use of
brief but vivid evocations.

Pedagogic being requires a split awareness that simultaneously participates in
the flow of student activity and yet observes the process. Without this split, students are not in the presence of desirable higher psychological functioning that
comprises mathematical thinking. At the same time, pedagogic being requires
access to alternative pedagogic practices to invoke, as well as to mathematical

processes, heuristics, and connections to other topics. So just as a mathematics emerges from explicit awareness of the awarenesses implicit in number and geometry (dynamics of relationships and dynamics of mental imagery) as suggested by Gattegno (1971), mathematical pedagogic being emerges from explicit awareness of awareness of those awarenesses. No wonder teaching is such a complex task, so rarely manifested fully, and so difficult to research.

9. SUMMARY

I am convinced that failure is the only possible outcome for any approach to research in mathematics education in which researchers hand their results to curriculum developers and teachers, who are then expected to apply them in their practices. Education simply does not work that way, either in how other people can be influenced to try out 'ideas', or in how new ideas are located and developed in the first place. In order to maximize the effectiveness of research effort in mathematics education it behoves us to reflect on what we go through when we undertake research, and to try to offer and support practitioners in undertaking similar or analogous personal reconstruction of research activities, in order that they may develop their practice in a manner that is self-consistent, theoretically sound, and practically well-founded. Research from the inside is an essential component of any effective educational research. Outer and inner research do not displace or replace each other, but rather are complementary.

REFERENCES

Atkinson, D.: 1989, Theory and Practice in Art Education. Unpublished Ph.D. thesis, University of Southampton.

Bacon, F.: 1609, reprinted 1851, *Of the Proficience and Advancement of Learning*, William Pickering, London.

Baird, J. & Northfield, F.: 1992, *Learning from the Peel Experience*, Monash University, Melbourne.

Bennett, J.: 1966, *The Dramatic Universe*, Vol. 3, Routledge, London.

Billington, E.: 1992, Beyond Mere Experience: The Role of Reflection in The Learning of Mathematics. M.Phil. thesis, Open University, Milton Keynes.

Brown, A.: 1988, Language Interaction Patterns in Lessons Featuring Mathematical Investigations. Unpublished Ph.D. thesis, University of Southampton.

Bruner, J.: 1991, 'The Narrative Construction of Reality', in Ammanti & Stern (eds), *Rappresentazioni e Narrazioni*, Laterze, Roma-Bari, 17–42.

Calvino, I.: 1985, 'Mr. Palomar', Harcourt Brace & Jovanovitch, London.

Casement, P. :1985, *On Learning from the Patient*, Tavistock, London.

Chatley, J.: 1992, The Use of Experiential Learning Techniques in Mathematics Education: A Personal Evaluation. M.Phil. thesis, Open University, Milton Keynes.

Chevellard, Y.: 1985, *La Transposition Didactique*, La Pensée Sauvage, Grenoble.

Cobb, P.: 1991, 'Some Thoughts About Individual Learning, Group Development, and Social Interaction', in F. Furinghetti (ed.), *Proceedings of PME XV*, Vol. I, Assisi, 231–238.

Davis, J.: 1990, 'The Role of the Participant Observer in the Discipline of Noticing', in
    H. Steinbring & F. Seeger (eds.), *The Dialogue between Theory and Practice in Mathematics
    Education: Overcoming the Broadcast Metaphor,* 4th SCTP Conference, Institut für Didaktik der
    Mathematik, Bielefeld, 167–176.
Davis, J. & Mason, J.: 1989, 'Notes on a Radical Constructivist Epistemethodology Applied to
    Didactic Situations', *Journal of Structural Learning* **10**, 157–176.
Davis, P. & Hersh, R.: 1981, *The Mathematical Experience,* Harvester, Brighton.
Dawkins, R.: 1976, *The Selfish Gene,* Oxford University Press, Oxford.
Evans, R.: 1973, *Jean Piaget: The Man and His Ideas,* E. Duckworth (trans.), Dutton, New York.
Feyerabend, P.: 1991, *Three Dialogues on Knowledge,* Blackwell, Oxford.
Gattegno, C. 1971, *What We Owe Children: The Subordination of Teaching to Learning,* Routledge,
    London.
Griffin P. & Gates, P.: 1988, *PM753: Preparing To Teach a Topic,* Open University, Milton Keynes.
Heiddegger, M.: 1927, *Existence & Being,* W. Brock (trans.) 1949, 'Introduction', Vision Press,
    London.
Jackson, P.: 1992, 'Untaught Lessons', Teachers College Press. New York, 37–56.
Kahn, M.: 1983, Hidden Selves: Between Theory and Practice in Psychoanalysis, Moresfield
    Library, London.
Kant, I.: 1929, *Critique of Pure Reason,* N. Smith (trans.), MacMillan, London (originally published
    in 1781).
McNair, Kathleen,: 1978a, 'Capturing Inflight Decisions: Thoughts while Teaching', *Educational
    Research Quarterly* **3** (4), 26–42.
McNair, Kathleen,: 1978b, 'Thought and Action: A Frozen Section', *Educational Research
    Quarterly* **3** (4), 16–25.
McWilliams, S.: 1993, 'I Accept, with Pleasure, the Invitation(al)', *Proceedings of the Tenth
    International Conference of Personal Construct Theory,* Townsville, Australia (preprint).
Mason, J.: 1980, 'When is a Symbol Symbolic?', *For the Learning of Mathematics* **1** (2) 8–12.
Mason, J.: 1988, 'Fragments: The Implications for Teachers, Learners, and Media Users/Researchers
    of Personal Construal and Fragmentary Recollection of Aural and Visual Messages',
    Instructional Science **17**, 195–218.
Mason, J.: 1984, 'Towards One Possible Discipline of Mathematics Education', presentation to
    ICME 5 Adelaide, in *Theory of Mathematics Education,* Institut für Didaktik der Mathematik,
    Bielefeld, Paper 54, 42–55.
Mason, J.: 1989, 'Mathematical Abstraction Seen as a Delicate Shift of Attention', *For the Learning
    of Mathematics* **9** (2) 2–8.
Mason, J. & Davis, J.: 1989, 'Notes on a Radical Constructivist Epistemethodology Applied to
    Didactic Situations, Proceedings of PME XIII (G. Vergnaud, J. Rogalski, M. Artigue, eds.) Paris,
    Vol. 2, p. 274–281.
Mason, J.: 1991a, *Awakening the (Re)Searcher Within,* Open University Inaugural Lecture, Centre
    for Mathematics Education, Open University, Milton Keynes.
Mason, J.: 1991b, 'Epistemological Foundations for Frameworks Which Stimulate Noticing',
    Proceedings of PME-NA **13** (2), 36–42.
Mason, J.: 1992a, 'Researching Problem Solving From the Inside', in J.P. Ponte, J.F. Matos,
    J.M. Matos & D. Fernandes (eds.), *Mathematical Problem Solving and New Information
    Technology: Research in Contexts of Practice,* Series F, No. 89, Springer-Verlag, Berlin, 17–36.
Mason, J.: 1992b, *The Discipline of Noticing,* Sunrise Research Laboratory, RMIT, Melbourne.
Mason, J.: 1994a, 'Researching from the Inside in Mathematics Education: Locating an I–You
    Relationship', in J. Ponte & J. Matos (eds), *Proceedings of PME XVIII,* Lisbon, University of
    Lisbon, 176–194.
Mason, J.: 1994b, *Researching from the Inside in Mathematics Education: Locating an I–You
    Relationship,* extended version, Informal Publication 5, Centre for Mathematics Education, Open
    University, Milton Keynes.
Mason, J. & Davis J.: 1988, 'Cognitive and Metacognitive Shifts', in M. Borbás (ed.), *Proceedings
    of PME XII* **2**, 487–494.

Mason, J., Burton L. & Stacey K.: 1984, *Thinking Mathematically*, Addison Wesley, London.

Mathematical Association,: 1991, *Developing Your Teaching*, Stanley Thornes, Cheltenham.

Morgan, J.: 1977, *Affective Consequences for the Learning and Teaching of Mathematics in an Individualised Learning Programme*, Dime Report, University of Stirling.

Open University: 1988, *ME234: Using Mathematical Thinking*, Open University Course, Open University, Milton Keynes.

Open University: 1992, *EM235: Learning and Teaching Mathematics*, Open University Course, Open University, Milton Keynes.

Peterson, P. & Clark, C.: 1978, 'Teacher's Reports of Their Cognitive Processes during Teaching', *American Educational Research Journal* **15** (4), 555–565.

Peterson, P., Marx, R. & Clark, C.: 1978, 'Teacher Planning, Teacher Behavior, and Student Achievement', *American Educational Research Journal* **15** (3), 417–432.

Popper, K.: 1972, *Objective Knowledge: An Evolutionary Approach*, Oxford University Press, Oxford.

Rosen, J.: 1991, 'Moment-by-Moment Decisions a Teacher Makes in the Classroom: A New Approach to Teaching Style', *SET Research Information for Teachers* 2, Item 3, ACER, Melbourne.

Schmalz, R.: 1993, *Out of the Mouths of Mathematicians*, Mathematical Association of America, Washington, DC.

Schoenfeld, A.: 1985, *Mathematical Problem Solving*, Academic Press, New York.

Schön, D.: 1983, *The Reflective Practitioner: How Professionals Think in Action*, Temple Smith, London.

Seeley Brown, J.: 1991 (January–February), 'Research that Re-invents the Corporation', *Harvard Business Review*, 102–117.

Sfard, A.: 1994, 'Reification as the Birth of Metaphor', *For the Learning of Mathematics* **14** (1), 44–55.

Shah, I.: 1964, *The Sufis*, Jonathan Cape, London.

Storm, H.: 1985, *Seven Arrows*, Ballantine, New York.

Varela, F., Thompson, E. & Rosch, E.: 1991, *The Embodied Mind: Cognitive Science and Human Experience*, MIT Press, Cambridge, MA.

Von Glasersfeld, E.: 1984, 'An Introduction to Radical Constructivism', in P. Watzlawick (ed.), *The Invented Reality: How Do We Know what We Believe We Know – Contributions to Constructivism*, Norton, London.

Vygotsky L.: 1978, *Mind in Society: The Development of the Higher Psychological Processes*, Harvard University Press, London.

Wheeler, D.: 1986, 'Epistemological Approaches to the Study of Mathematics Education II', Proceedings of PME IX, London.

Williams, H.: 1989, Tuning-in to Young Children: An Exploration of Contexts for Learning Mathematics. Unpublished M.Phil. thesis, Open University, Milton Keynes.

Winograd, T. & Flores, F.: 1987, *Understanding Computers and Cognition: A New Foundation for Design*, Addison Wesley, New York.

Yates, J.: 1978, *Four Mathematical Classroom: An Enquiry into Teaching Method*, Report for the Faculty of Mathematical Studies, University of Southampton.

John Mason
Centre for Mathematics Education,
Open University,
Milton Keynes
MK7 6AA,
UK

THOMAS A. ROMBERG

THE SOCIAL ORGANIZATION OF RESEARCH PROGRAMS IN MATHEMATICAL SCIENCES EDUCATION

1. INTRODUCTION

The goal for research in mathematics education should be to produce new knowledge about the teaching and learning of mathematics. Because students learn most of their mathematics in school classrooms, I believe the primary mission of a program of research should be to identify the major components of classrooms that promote mathematical understanding and to clarify some of the organizational features that contribute to or hinder the operation of such classrooms.

Today in the United States three aspects of research programs in mathematics education warrant attention. First, the agenda should focus on the current international reform efforts for school mathematics. The teaching and learning of mathematics in American classrooms need to change as a consequence of several related factors: developing new technologies; changes in mathematics itself; changes in the use and application of mathematics; new knowledge about learning, teaching, and schools as sociopolitical institutions; and renewed calls for equity in learning mathematics, regardless of race, class, gender, or ethnicity. Thus, a contemporary research program should support the reform efforts by providing reliable knowledge about important aspects of reform. Current school mathematics operates within a coherent system; reform will occur only if an equally coherent system replaces it. Research can provide information needed to help design such a system.

Second, because mathematics should be considered a plural noun, different groups of scholars should be investigating teaching and learning within a particular mathematical domain (e.g., whole numbers, fractions, decimals, integers, algebra, geometry, statistics). Each domain has its own independent characteristics (terms, signs and symbols, rules for use). In this sense, as the noted mathematician William Thurston (1990) has stated:

Mathematics isn't a palm tree, with a single long straight trunk covered with scratchy formulas. It's a banyan tree with many interconnected trunks and branches – a banyan tree that has grown to the size of a forest, inviting us to climb and explore. (p. 7)

Thus, each group should be studying aspects of different important trunks to the tree of mathematics. Also, other groups could be investigating factors that influence the learning and teaching of mathematics for understanding in classrooms (e.g., assessment practices, staff development).

379

Sierpinska, A. and Kilpatrick, J. Mathematics Education as a Research Domain: A Search for Identity, 379–389.
© *1998 Kluwer Academic Publishers. Printed in Great Britain.*

Third, because researchers in mathematics education are faced with limited resources to study important issues related to the reform movement, it seems both logical and prudent to utilize existing human and financial resources to create networks of scholars working on common issues. One strategy that could be followed to accomplish this goal is based on studies about the sociology of science. In particular, Diana Crane's (1972) examination of the growth of scientific knowledge in many fields suggested that the social institution responsible for the rapid growth of knowledge in a field is the *invisible college*. She used this term to describe a small group of scholars in the same field of study who set priorities for research, recruited and trained students, communicated with one another, and thus monitored the changing structure of knowledge in their field.

Note that this way of organizing a research program is at odds with the predominant pattern for conducting research in most fields. For example, McGrath and Altman (1966) found that in many fields typical studies are usually done in relative isolation. Their image is of an isolated scholar working with one or two assistants on small problems, publishing results in journal articles, with the value of that work indicated by the number of such articles produced. This image has led to emphasis upon methodological rigor rather than on more fundamental substance and theory. One way to publish rapidly is to apply the same procedures over and over, using new variables or slight modifications of old variables.

The invisible-college strategy involves the consideration of five related ingredients deemed necessary for rapid, yet substantial, progress to be made in producing reliable knowledge about the teaching and learning of mathematics. The first ingredient is the existence of a small group of scholars with common interests. The second ingredient involves the desire of the group to produce reliable knowledge about phenomena of common interest. The third ingredient is the group's general agreement on an ideology and associated methods of inquiry for generating reliable knowledge. The group's commitment to communicate regularly, using an agreed-upon vocabulary and syntax, is the fourth ingredient. The final ingredient is the continual monitoring of the progress of knowledge by the group.[1]

2. SCHOLARLY GROUPS

It is only when a group of scholars have agreed on an area of study; have chosen a language to describe it; have accepted an approach to test conjectures about it; have agreed to its utility; and shared their ideas, data, findings, and reports that rapid progress in a scientific field can be made. It should be noted that the members of the group need not all have the same background or experience with respect to the phenomena chosen. For a focus on the teaching and learning of mathematics, one would expect mathematics education researchers, mathematicians, classroom teachers, educational psychologists, educational sociologists, teacher educators, and others to be involved. Furthermore, the investigations the group conducts should be seen as embodying some of the elements of a craft. This notion directs attention both to the personal autonomy and the collective responsibility of the group in creating and testing conjectures. Researchers must be creative, but at the same time they are not free to do anything they please. As with other crafts, there must be agreement in a broad sense about what areas for investigation are significant, what procedures are acceptable, and what work is likely to be declared acceptable. Such agreements arise from discourse and debate within the group and are not simply impositions of external standards.

Invisible colleges are commonly formed under the leadership of one or two senior scientists and include their graduate students and a few others who are socialized into the group. The senior scientists surround themselves with collaborators and students in order to produce a greater volume of work and promote a particular line of inquiry, which in turn defines the important problems for research in that area. Although there is considerable variability in the manner in which such groups evolve, Crane (1972) presents evidence that many groups begin with the publication of seminal, innovative, theoretical work, which attracts other scholars. Then the diffusion of these ideas occurs via a process she calls *contagion* (p. 70). Initially, the group is a loose federation of scholars 'infected' by ideas. Over time the group grows in size, begins to develop a social structure (holds conferences, publishes collaborative papers, and so forth) and defines both the nature of problems to be addressed and how they are to be studied. Following 'infection', three things appear to be important if a group is to become productive. First, the implications of the conceptual, or theoretical work have to be amenable to scholarly exploration. It must be possible to raise conjectures and study them. In fact, a major aspect of the group's work is to set guidelines for how research in the area is to be conducted. Second, the group leaders need to be aggressive in their recruitment of new members. Without such leadership, groups are unlikely to endure. Third, and most important, a pattern of regular and reciprocal communication among members must be established. Crawford (1970) found that communication among scholars appears to take at least two forms. The most common is nonreciprocal communication, which occurs when one scholar just asks another for advice. Ongoing reciprocal communications characterize well-functioning invisible colleges.

3. PHENOMENA OF COMMON INTEREST

To support the vision of the reform movement about mathematics classrooms and the fact that mathematics is composed of a set of domains, four related features of the teaching and learning of mathematics are critical.

1. To clarify the important mathematical ideas in each domain.
Mathematics is not a static discipline comprising a large collection of concepts and skills to be mastered in some order, but a dynamic discipline comprising interrelated sets of signs and symbols, rules for the use of those representations,

problem situations that have given rise to the invention of those signs and symbols, and strategies for investigating and solving problems that can be represented by those signs and symbols. Furthermore, the availability and uses of new technologies are changing, in fundamental ways, the problem situations, methods of representation, and strategies used.

2. To investigate student learning in each domain.
The learning process for every student involves the cognitive construction of ideas via the social interaction that occurs both in and out of school. This cognitive perspective should be used to describe and clarify the processes of learning when students formulate, represent, reason about, and solve problems in each content domain. This perspective about learning centers on the notion that to learn mathematics involves 'doing' mathematics. One does mathematics by abstracting, inventing, proving, and applying. Students then construct their mathematical knowledge from these purposeful activities. For example, the eminent mathematician George Pólya (attributed by Kilpatrick 1987) felt that we understand mathematics best when we see it being born, either by following in the steps of historical discoveries or by engaging in discoveries ourselves.

3. To investigate the content, pedagogical knowledge, and the beliefs of teachers about the nature of mathematics and student learning in each domain.
Instruction should focus on making classrooms discourse communities. Students need to share their ideas, reasons, and strategies with other students in an open, supportive environment. Teachers must not simply listen but also hear what students say. Teachers and students alike must recognize that not all mathematics is learned in a classroom setting. Students must be encouraged to make links between real-world mathematics and classroom mathematics: They should be encouraged to seek out applications of mathematics in the world around them and to continue their sense-making and mathematical construction outside the classroom.

4. To assure that teachers who provide an environment that promotes deep thinking have a deep understanding of the mathematics appropriate to the instructional level at which they teach.
This understanding must include knowledge of the mathematics they are teaching and how it is related to the mathematics that is prior to and beyond the instructional level of students, knowledge of how students reason about and come to understand the mathematics at their instructional level, and knowledge of the nature of mathematical activity. It is this knowledge that allows teachers to competently guide their students through the twists and turns of rich mathematical discourse.

Other groups of scholars should investigate other crosscutting issues such as assessment or staff development. For example, a group could study procedures that are potential models for assessing the full range of mathematical performances for various purposes. Furthermore, their work should be based on the assumption that any model of assessment should be aligned with the curricular and instructional goals of the reform movement.

4. RELIABLE KNOWLEDGE

By reliable knowledge, we mean a claim 'to know' that is substantiated as trustworthy for some given purpose. The gathering of evidence and the construction of an argument are the means by which researchers substantiate conjecture. This task is an arduous, ongoing one that requires a substantial amount of training and effort; in the more complex cases, it taxes the patience and ingenuity of the most gifted thinkers. Nor does the task, once achieved, stay finished or complete; it has to be continually re-achieved because what constitutes both a reasonable argument and given purposes continually change.

It also must be noted that most of the body of past, as well as contemporary, research addresses a past vision of schooling and mathematical literacy. For example, much research on effective teaching has been based on predicting residualized-mean-gain scores on standardized tests. Such research does not support the pursuit of reliable knowledge about the new vision of mathematical literacy. New questions are now being posed with expectations of different outcomes. New research needs to address questions using methods of gathering information in light of the reform. That is not easy, for as Shulman (1988) has argued, education is not a discipline: Rather,

education is a field of study containing phenomena, events, institutions, problems, persons, and processes, which themselves constitute the raw material for inquiries of many kinds. The perspectives and procedures of many disciplines can be brought to bear on the questions arising from and inherent in education as a field of study. (p. 5)

One problem is that each perspective brings with it its own set of terms, methods, and procedures. Thus, one of the initial tasks for a group as it is forming is to understand each other's vocabulary and research methods. Gradually over time a group develops its own technical terminology to explicate its statements. Finally, because theories are designed to interpret experience, they must be verified in experience. Theoretical statements are informally tested by assessing their fit with commonplace observations and reflection. Then, within disciplines, rationalized schemes for verifying statements exist in custom and tradition. Cronbach and Suppes (1969) referred to such schemes as *disciplined inquiry,* which they defined as 'inquiry conducted and reported in such a way that the argument can be painstakingly examined' (p. 15). Hence, other tasks that the group must decide on include how such statements can be validated by evidence that either supports the causal inference or falsifies the assertion, how evidence is to be aggregated and synthesized, how reports are to be written, and so forth.

5. COMMON IDEOLOGY

Merton (1957) has argued that there is a strong relationship between the particular structure and goals of research groups, their general world-view, and the productivity of the group. For such groups in mathematics education, the goal is to create information about some aspect of classrooms that promote understanding of mathematics by all students. Thus, studies need to be about the rules that underlie and govern the social actions between teachers and students in classrooms with respect to mathematics. The structure underlying such studies must be based on the belief that knowledge is situated and personal, that pupils learn by construction as a consequence of experiences, that the job of teaching is to create instructional experiences for students and negotiate with them intersubjective understandings gained from those experiences, and that the organization and technology of the classroom and school are arranged so that all of the experiences of the students can be regarded as rich and meaningful.

The world-view of research that is consistent with these beliefs is what philosophers call the *Symbolic Paradigm*. The goal of this approach to research is to understand how humans relate to the social world they have created. This world-view is based on a belief that social life is created and sustained through symbolic interactions and patterns of conduct and that the unique quality of being human is found in the symbols people invent to communicate meaning or to interpret the events of daily life. It assumes that language and its use within a culture – the interactions and negotiations in social situations – are assumed to define the possibilities of human existence. In fact, it is through the interactions of people that rules for governing social life are made and sustained. Note that this is a dynamic view of the world. It often stands in sharp contrast to the conservative, static view of the world held in the natural sciences: 'The idea of "rules" shifts attention of researchers from the invariant nature of behavior to the field of human action, intent and communication' (Popkewitz 1984, p. 40). Also note that agreement on a world-view is necessary for productive work. Continual disagreement prevents contemporaries from building on each other's work.

From this perspective, the methods used by researchers to gather, synthesize, and report evidence about the teaching and learning of mathematics in classrooms involve a variety of field-based methods borrowed from sociology, anthropology and political science. The most common procedure is to carry out case studies. This method involves organizing and reporting information about the actions, perceptions, and beliefs of an individual or group under specific conditions or circumstances. The researcher is interested in telling an in-depth story about a particular case. Evidence is gathered through observations, interviews, and artifacts. The synthesis of this evidence for a case involves building an argument about patterns that define the range of findings that can be expected.

6. COMMUNICATION

To be productive, a research group needs to agree upon a language to express its ideas; it also needs to meet and share ideas regularly. The vocabulary used must contain, along with a set of logical words common to all science, descriptive words or concepts used by the group to characterize the key ideas about the teaching and learning of mathematics in classrooms. As Scheffler (1974) argues, there are two types of language available to communicate by scholarly groups. The *language of ordinary experience* includes both literal and figurative languages. Literal language identifies facets of reality in conventional, consensual terms that retain their primary meanings. Figurative language is a discourse of linguistic devices such as metaphor, simile, or personification that is used suggestively to portray facets of reality that resist formulation in literal terms. *Technical terminology* can be viewed broadly as terms embedded within networks of characterization, explanation, or definition that acquire conventional meanings within those networks. Such terms can be borrowed from ordinary language and redefined within a network of definition, or they can be invented.

Initially, group members communicate in the language of everyday experience. Highly developed groups tend to express their ideas in technical terminology because such groups are characteristically proposed by scholars who require a language suitable for precise communication. Hempel (1965) argues that ordinary language tends to be inadequate in meeting these requirements because it often lacks precision and uniformity in usage. Thus considerable effort is expended toward developing languages with which to communicate.

To share ideas about teaching and learning in classrooms within specific mathematical domains, scholars need to communicate with each other on a regular basis. In fact, Barton and Wilder (1964) found that lack of interpersonal communication seriously weakened intellectual development in applied reading research. Productive reciprocal sharing usually occurs via a variety of informal means: sending a letter to friends for comments about ideas, asking colleagues to react to drafts of papers or reports, and attending meetings and engaging in conversations at professional conferences.

7. MONITORING

The final aspect of an invisible college is that its members monitor the progress of work in the field. The argument here is that although individual studies may be important, it is the chain of studies in an area that is critical. Two aspects of monitoring have been identified. The first is motivational: It involves leaders in the group encouraging others to carry out specific investigations. Leaders make it apparent that rapid recognition within the group will be a consequence of contributing in this manner.

The second aspect of monitoring is more formal. It involves members of the group supervising the formal publication channels for research. Usually this involvement means becoming an editor of a research journal, editing a series of books in the field, and even serving on review panels for funding agencies. In this manner, the group defines and reifies the kinds of investigations that are valued. For emerging fields of research, however, penetrating the formal publication system is often daunting. The interaction between a complex and volatile research front and a stable and much less flexible publication system is often problematic. Thus, it is not uncommon for a group to start its own journal or book series. The control of formal publication procedures is critical for the growth of the group. Although operating groups may have an effective informal communication system, it is only through formal publications that others (e.g., scientists, policy makers, teachers) come in contact with the ideas of the group.

8. AN EXAMPLE: NCRMSE

From its organization in 1987, the goal of the U.S. Department of Education's National Center for Research in Mathematical Sciences Education (NCRMSE) housed at the University of Wisconsin–Madison was to identify the major components of classrooms that promote mathematical understanding for all students and to clarify some of the organizational features that contribute to or hinder the operation of such classrooms. Its research program was carried out by seven organizationally independent, but related, working groups.

Although the productivity of established invisible colleges is clear, when we decided to try this strategy, it was not obvious that such groups could actually be formed in each of the seven areas we proposed to study. We were not starting with an innovative theory. Rather, we were starting with reform movement rhetoric about the learning and teaching of mathematics. The movement was attracting a number of persons, but powerful rhetoric could not guarantee that those involved would produce good research. In actuality, the formation of four of the seven working groups of NCRMSE grew out of the existence of loosely associated groups of scholars who already had a common interest. For example, a group studying whole number arithmetic had been sharing ideas since 1979 following the Wingspread Conference on Addition and Subtraction (reported in Carpenter, Moser & Romberg 1982). By 1987, this group had begun to coalesce. At that time, three other groups (Quantities, Algebra, and Mathematics Assessment) were in the early stages of formation as a consequence of initial conferences on those topics held by the Research Agenda Project of the National Council of Teachers of Mathematics in 1987–1988 (reported in Charles & Silver 1989; Hiebert & Behr 1989; Wagner & Kieran 1989) and by NCRMSE in 1988–1989 (reported in Carpenter, Fennema & Romberg 1993; Romberg 1992b; Romberg, Fennema & Carpenter 1993). Some of the participants in these groups were invited to be affiliated members of working groups of the Center. When the Center was re-funded, a new group was organized in each of the first three years

(1991, Implementation of Reform; 1992, Geometry; 1993, Statistics). Although there were several scholars working in each area, the groups had no history of collaborative work. Thus, they had to be organized from scratch.

In summary, seven working groups were formed to carry out the research program of the Center. These groups were *Whole Numbers*, directed by Tom Carpenter and Elizabeth Fennema; *Quantities*, directed by Judy Sowder; *Algebra*, directed by James Kaput; *Geometry*, directed by Richard Lehrer; *Statistics*, directed by Susanne Lajoie; *Assessment*, directed by Tom Romberg; and *Implementation of Reform*, directed by Walter Secada.

Overall, the Center's long-term objective was to develop a new 'theory' about the teaching and learning in classrooms that would promote student understanding of mathematics consistent with the assumptions and philosophy of the reform movement. By *theory*, we mean a set of statements about the causal relationships between and among a number of variables used to describe features of classroom instruction in mathematics. Theoretical statements have two features that make them different from most other causal statements (e.g., speculations, conjectures, hypotheses). First, there is a group of persons who finds the theoretical statements useful; and second, they have been put to the test several times and survived. Note that 'causal' statements are of the form 'if A, then B', which implies that if A occurs, B follows it, and when B occurs, A has preceded it. If this is the case, we can assert that 'A is a possible cause of B'. Note that this is not a claim that 'A is *the* cause of B'. Such causal claims must rest on other persuasive evidence about the relationships between and among variables. Nevertheless, causal statements allow researchers to develop predictions supported by data and logic.

So that there would be some overall coherence to the work of the Center, the five groups studying mathematical domains (Whole Numbers, Quantities, Algebra, Geometry, and Statistics) were charged with the four tasks discussed earlier:

● To clarify the important mathematical ideas in their domains.
● To investigate student learning in their domains.
● To investigate the content, pedagogical knowledge, and the beliefs of teachers about the nature of mathematics and student learning in their domains.
● To assure that teachers who provide an environment that promotes deep thinking have a deep understanding of the mathematics appropriate to the instructional level at which they teach.

From the studies carried out in response to these tasks by members of each of the working groups, we assumed that general principles could be derived and verified. Such principles provided the Center with the building blocks for a theory of mathematics instruction.

The other two working groups were formed to study other factors that influence the learning and teaching of mathematics in classrooms. The purpose of the Assessment Working Group was to investigate procedures that are potential models for assessing the full range of mathematical performances for various purposes. The work of this group was based on the assumption that any model of assessment should be aligned with the curricular and instructional goals of the reform movement. The Implementation of Reform Working Group studied reform as it was happening. Beyond developing new knowledge about how schools and school personnel react to the current reforms in mathematics education, the group's intent was to use that evolving knowledge to inform practitioners and policy makers about the options and choices they face, about the decisions (both conscious and by default) that others have made when facing similar choices, and about the consequences of those decisions. We assumed that information derived from the work of these two groups would clarify important aspects of the reform process.

9. SUMMARY

It is premature to judge whether or not the strategy of forming working groups based on sociological notions about invisible colleges was effective. The criteria for effectiveness includes both whether the group was formed and did function and whether the products of its investigations were numerous and useful. Each of the seven groups was established, but they functioned at different levels, and the degree of productivity varied. Nevertheless, the scholars associated with the Center's program identified and substantiated many salient features in classrooms that promote mathematical understanding; they began to uncover other phenomena we suspect will be substantiated; they found evidence that challenges many myths about the scope and sequence of current mathematics instruction, what students are capable of learning, what instructional contexts motivate different students, and so forth.

NOTE

1. For background information on the reform movement, see Mathematical Sciences Education Board 1989; National Council of Teachers of Mathematics 1989; Romberg 1992a.

REFERENCES

Barton, A. H. & Wilder, E. D.: 1964, 'Research and Practice in the Teaching of Reading: A Progress Report', in M. B. Miles (ed.), *Innovation in Education*, Bureau of Publications, Teachers College, Columbia University, New York, 361–398.

Carpenter, T. P., Fennema, E. F. & Romberg, T. A.: 1993, *Rational Numbers: An Integration of Research*, Lawrence Erlbaum, Mahwah, NJ.

Carpenter, T. P., Moser, J. M. & Romberg, T. A.: 1982, *Addition and Subtraction: A Cognitive Perspective*, Lawrence Erlbaum, Mahwah, NJ.

Charles, R. I. & Silver, E. A. (eds.): 1989, *Research Agenda in Mathematics Education: The Teaching and Assessing of Mathematical Problem Solving*, National Council of Teachers of Mathematics, Reston, VA.

Crane, D.: 1972, *Invisible Colleges: Diffusion of Knowledge in Scientific Communities*, University of Chicago Press, Chicago.

Crawford, S.: 1970, *Informal Communication among Scientists in Sleep and Dream Research*, unpublished doctoral dissertation, University of Chicago.

Cronbach, L. J. & Suppes, P. (eds.): 1969, *Research for Tomorrow's Schools: Disciplined Inquiry for Education*, Macmillan, New York.

Hempel, C. G.: 1965, *Aspects of Scientific Explanation*, Free Press, Glencoe, IL.

Hiebert, J. & Behr, M. (eds.): 1989, *Research Agenda in Mathematics Education: Number Concepts and Operations in Middle Grades*, National Council of Teachers of Mathematics, Reston, VA.

Kilpatrick, J.: 1987, 'George Polya's Influence on Mathematics Education', *Mathematics Magazine* **60** (5), 299–300.

McGrath, J. E. & Altman, I.: 1966, *Small Group Research: A Synthesis and Critique of the Field*, Holt, Rinehart and Winston, New York.

Mathematical Sciences Education Board: 1989, *Everybody Counts: A Report to the Nation on the Future of Mathematics Education*, National Academy Press, Washington, DC.

Merton, R. K.: 1957, *Social Theory and Social Structure*, Free Press, Glencoe, IL.

National Council of Teachers of Mathematics: 1989, *Curriculum and Evaluation Standards for School Mathematics*, National Council of Teachers of Mathematics, Reston, VA.

Popkewitz, T.: 1984, *Paradigm and Ideology in Educational Research: The Social Functions of the Intellectual*, Falmer, London.

Romberg, T. A.: 1992a, 'Problematic Features of the School Mathematics Program', in P. W. Jackson (ed.), *Handbook of Research on Curriculum*, Macmillan, New York, 749–788.

Romberg, T. A. (ed.): 1992b, *Mathematics Assessment and Evaluation: Imperatives for Mathematics Educators*, State University of New York Press, Albany, NY.

Romberg, T. A., Fennema, E. F. & Carpenter, T. P.: 1993, *Integrating Research on the Graphical Representation of Functions*, Lawrence Erlbaum, Mahwah, NJ.

Scheffler, I.: 1974, *The Language of Education*, Thomas, Springfield, IL.

Shulman, L. S.: 1988, 'Disciplines of Inquiry in Education: An Overview', in R. M. Jaeger (ed.), *Complementary Methods for Research in Education*, American Educational Research Association, Washington, DC, 3–17.

Thurston, W. P.: 1990, 'Letter from the Editor', *Quantum* **1** (1), 6–7.

Wagner, S. & Kieran, C. (eds.): 1989, *Research Agenda in Mathematics Education: Research Issues in the Learning and Teaching of Algebra*, National Council of Teachers of Mathematics, Reston, VA.

Thomas A. Romberg
University of Wisconsin–Madison,
1025 West Johnson Street,
Madison, WI 53706,
U.S.A.

YASUHIRO SEKIGUCHI

MATHEMATICS EDUCATION RESEARCH AS SOCIALLY AND CULTURALLY SITUATED

1. INTRODUCTION

The theme of the International Commission on Mathematical Instruction study 'What Is Research in Mathematics Education, and What Are Its Results?' has also been of increasing concern among Japanese mathematics educators. One reason is a gap in interests between researchers and mathematics teachers. Mathematics teachers tend to expect direct and specific prescriptions to solve the problems they experience in their classrooms, whereas researchers tend to address problems raised in research communities that are not always directly related to teachers' concerns.

Another reason is the emergence of new research areas and new methodologies in mathematics education. For example, we are increasingly seeing sociological and anthropological studies in mathematics education journals and conferences. Some researchers wonder whether those studies really belong to mathematics education research.

In addition, as the opportunity to participate in international conferences has increased, we have come to experience problems in communicating our country's research interests and perspectives and in appreciating other countries' research. Language has come to cause very few troubles because English is commonly used in international conferences. Increased specialization of research topics does not cause a problem either, insofar as studies are frequently published internationally. A much more serious problem is that educational practice and research are socially and culturally situated. Our international research community is not very much like that of pure mathematics or the natural sciences. In the latter, research is internationally shared, and international interests are overriding. It does not make much sense to talk about 'domestic' or 'regional' mathematics. But in mathematics education, each country has its domestic educational problems and interests, and they shape its research practice significantly. 'In each country of the world, education is strongly influenced by unique national cultural factors that shape the goals, expectations, and instructional patterns in each school subject' (Fey 1989, p. 4). This condition also holds true in educational *research*. In this chapter, I explore the social and cultural aspects of mathematics education research and discuss some problems they caused in the international research community.

Sierpinska, A. and Kilpatrick, J. Mathematics Education as a Research Domain: A Search for Identity, 391–396.
© *1998 Kluwer Academic Publishers. Printed in Great Britain.*

2. ON RESEARCH PROBLEMS

There are many journals of mathematics education in Japan. Many studies are published there every year. Also, there are many meetings of mathematics educators around the country, and many papers presented. The majority of these studies are done by mathematics teachers, and their research questions are directly related to classroom practice in Japan. They usually relate their goals to those of the official course of study, which is determined by the Ministry of Education. Compared with these studies, the numbers of journals and meetings for academic research are very limited. Most research studies presented at international conferences like the International Congresses of Mathematics Education have been conducted by researchers in universities or research institutions, and they present papers that although expected to interest international audiences nevertheless often reflect domestic issues.

There may be significant differences in domestic issues among different countries. For example, mathematics education researchers in the United States have recently appealed to the need for attention to problem solving, classroom communication, and cooperative learning (e.g., National Council of Teachers of Mathematics 1989). It seems a little difficult for Japanese educators to appreciate such goals, however, partially because they think that they have already been striving to meet them for a long time, or already been casually engaged in such practices in schools (especially in elementary schools). On the other hand, Japanese educators are much more interested in current issues raised by the introduction of a new course of study by the Ministry of Education: for example, how to facilitate self-reliant learning, or how to cultivate computer literacy.

3. ON RESEARCH FRAMEWORKS, THEORIES, AND METHODS

The philosopher of science Mary Hesse (1980) has pointed out that any theory is value-laden, and that an awareness of the value behind a theory is crucial, especially in the social sciences. In natural sciences, also, criteria for theory choice contain value judgments (e.g., simplicity); however, 'these have tended to be filtered out as theories developed' (p. 188):

We can observe by hindsight that in the early stages of a science, value judgments (such as the centrality of man in the universe) provide some of the reasons for choice among competing underdetermined theories. As systematic theory and pragmatic success [i.e., increasingly successful prediction and control of the environment] accumulate, however, such judgments may be overridden, and their proponents retire defeated from the scientific debate. (pp. 190–191)

In the social sciences, on the other hand, the pragmatic criterion does not work as a filtering-out mechanism because researchers may have overriding value goals other than pragmatic success (Hesse, 1980, p. 193). We have to be conscious that educational theories are value-laden. Without appreciating the values behind it, people other than a theory's proponents would find it difficult to understand. When I came to the United States as a graduate student in 1986, I was struck by the growing interests in constructivism among North American mathematics educators. For a while, it was difficult for me to appreciate the significance of the constructivism debates. I wondered why researchers were so busy criticizing straw men like naive realism, formalist mathematical philosophy, behaviorist psychology, process–product research, knowledge-transmission pedagogy, and so on. One difficulty was that I did not imagine that there were still many people who consciously or unconsciously espoused these straw-men views in the United States.

Another difficulty was when I tried to understand the relation between the constructivist epistemological claims and many nice educational implications the constructivists proposed. Well-known principles of constructivism may be formulated as follows:

1. Knowledge is not passively received either through the senses or by way of communication. Knowledge is actively built up by the cognizing subject.
2. a. The function of cognition is adaptive, in the biological sense of the term, tending towards fit or viability.
   b. Cognition serves the subject's organization of the experiential world, not the discovery of an objective ontological reality. (von Glasersfeld 1990, pp. 22–23)

These principles certainly help us to deconstruct traditional views of research and practice in mathematics education. But to deduce any educational implication from them, it is necessary to discuss the value goals of the education toward which the constructivists strive. There was little explicit discussion of them in the debates on constructivism, however. Allen (1992) points out some value goals underlying constructivist arguments that fit 'America's tradition' well:

The debate over constructivism (and the post-modern crisis which it reflects) is not merely about cognition but about morality. Multiculturalists argue, for instance, that the systems which assume (and *teach*) one determinable, consistent, and objective set of standards for representing reality have long been used to justify economic oppression, cultural hegemonism, and ecological exploitation. The issue can be expressed more forcefully this way: Insisting that our view is in fact a 'God's Eye view' is not merely a matter of hubris. It is a potentially immoral claim because it can be so easily lead to actions against 'unbelievers' and 'aliens', because it can be used to rationalize our dominion over nature,... and because it impedes the development of human consciousness ... and therefore of spiritual liberation. (Allen 1992, p. 201)

We can also find similar value or moral positions underlying constructivist arguments in mathematics education; for example, in their strong respect for children's own construction of mathematics, and interactive communication between a teacher and children, though the positions have usually been suggested implicitly (cf., Noddings 1990).

When constructivism was publicized among U.S. mathematics educators, researchers in other countries began to conduct research claiming to be 'based on' constructivism. But, it is necessary to consider what value goals they espouse, whether they shared the same value goals as the constructivist educators in the United States, and whether they were facing similar educational problems as the United States.

4. ON RESEARCH RESULTS

When we read a research report of a study conducted in a foreign country, we usually wonder whether the research results would be valid in our country. This thinking is reasonable because there may be significant differences in the educational situations of the two countries that may affect the results. The research results consist of part of a complete research practice that is socially and culturally bounded. Research participants, settings, unit of data analysis, interpretation, educational implications are all socially and culturally constrained. The reliability and validity of research results are, therefore, also socially and culturally bounded.

For reporting a research study, who the audience of the paper is going to be significantly changes how the study is described and how extensive the details of description will be. I conducted a classroom study on mathematical proof using ethnographic methods in the United States (Sekiguchi 1991). When I was writing a paper on the study, I tried to write for mathematics educators in the United States, assuming their knowledge of their own educational context. I recently conducted another ethnographic study of a mathematics classroom in my own country and wrote papers about it in English and later in Japanese. When writing the English paper, I wondered how much it should assume about the background of Japanese mathematics education. In addition, I tried to play down a certain 'romanticism' (Silverman 1993) that foreigners tend to hold towards Japanese mathematics education because of the well-known high scores of Japanese students in international mathematics surveys (e.g., McKnight et al. 1987), and provide a more realistic description of Japanese classrooms. On the other hand, when writing the Japanese paper, I was not sure how interested Japanese mathematics educators would be in reading a description of ordinary teaching practice in their country: most of the time they are more interested in 'good', 'successful', or 'innovative' teaching practices.

International conferences are usually expected to provide a good opportunity to learn about other countries' educational practices and to find cultural factors affecting the research results (Fey 1989, p. 4). But the opportunity would be lost if we did not coordinate a conference deliberately. We are not interested in every aspect, but those aspects of social and cultural situations that are critical to the problem at issue. What aspects? If a researcher does not know other countries' education very well, he or she cannot figure out what cultural aspects are crucial, and may fail to include relevant data in the paper. Also, the reporting styles of experimental research or survey research, which are still often used in research reports, often make it difficult for the audience to appreciate the cultural aspects.

Regarding the evaluation of research reports in mathematics education, there can be no mechanical way to identify or apply criteria. It is possible to draw up general guidelines for educational research (e.g., Eisenhart & Howe 1992). How to apply them in the actual process of reviewing research, however, requires communication and negotiation between the reviewers and the reviewees on social and cultural contexts of research, and on the values of the research in consideration of trade-off relations among its strengths and weaknesses. The issue here is no longer epistemological, as positivists have believed, but practical – a matter of what value goals our research community intends to share:

Criteria [for distinguishing knowledge from belief and good from bad research] are best thought of as characterizing traits that are worked out as we go along as part of the social practice of inquiry. They are not abstract standards that determine decisions about good versus bad research, they are expression of our values, of what we expect from research at any given time under any given conditions.... The attempt to distinguish knowledge from opinion and good research from bad research is a practical and moral task, not an epistemological one. (Smith 1993, p. 160)

Because our value goals in the international community are diverse, we have to admit here a pluralism on the acceptability of research, as Hesse (1980) says about theory choice in the social sciences:

Whether or not we wish to use epistemologically loaded terms like 'cognition,' 'knowledge,' 'objectivity,' and 'truth' for acceptable theories in social science, a consequence of my arguments is that criteria of acceptability are *pluralist* – as pluralist as our choices of value goals. (Hesse 1980, p. 202)

5. CONCLUDING REMARKS

Educational research is essentially local practice, the major part of which consists of practical studies in socially and culturally bounded places and communities. The international research community of mathematics education should be conscious that research practice is socially and culturally situated. Understanding and enhancing the quality of research require cultivation of this consciousness among the research community, and constant dialogue between researchers from different societies and cultures to appreciate research practice as a whole.

REFERENCES

Allen, B. S.: 1992, 'Constructive Criticisms', in T. M. Duffy & D. H. Jonassen (eds.), *Constructivism and the Technology of Instruction: A Conversation*, Lawrence Erlbaum, Hillsdale, NJ, 183–204.

Eisenhart, M. A. & Howe, K. R.: 1992, 'Validity in Educational Research', in M. D. LeCompte, W. L. Millroy & J. Preissle (eds.), *The Handbook of Qualitative Research in Education*, Academic Press, San Diego, CA, 643–680.

Fey, J. T.: 1989, 'The Emergence of Mathematics Education as an International Problem Field', in T. J. Cooney (ed.), *American Perspectives on the Sixth International Congress on Mathematical Education*, National Council of Teachers of Mathematics, Reston, VA, 4–5.

Hesse, M.: 1980, *Revolutions and Reconstructions in the Philosophy of Science*, Indiana University Press, Bloomington, IN.

McKnight, C. C., Crosswhite, F. J., Dossey, J. A., Kifer, E., Swafford, J. O., Travers, K. J. & Cooney, T. J.: 1987, *The Underachieving Curriculum: Assessing U.S. School Mathematics from an International Perspective*, Stipes, Champaign, IL.

National Council of Teachers of Mathematics: 1989, *Curriculum and Evaluation Standards for School Mathematics*, National Council of Teachers of Mathematics, Reston, VA.

Noddings, N.: 1990, Constructivism in Mathematics Education', in R. B. Davis, C. A. Maher & N. Noddings (eds.), *Constructivist Views on the Teaching and Learning of Mathematics* (JRME Monograph No. 4), National Council of Teachers of Mathematics, Reston, VA, 7–18.

Sekiguchi, Y.: 1991, 'An Investigation on Proofs and Refutations in the Mathematics Classroom', *Dissertation Abstracts International* 52, 835A-836A. (University Microfilms No. 9124336)

Silverman, D.: 1993, *Interpreting Qualitative Data: Methods for Analysing Talk, Text and Interaction*, Sage, London.

Smith, J. K.: 1993, *After the Demise of Empiricism: The Problem of Judging Social and Education Inquiry*, Ablex, Norwood, NJ.

Von Glasersfeld, E.: 1990, 'An Exposition of Constructivism: Why Some Like It Radical', in R. B. Davis, C. A. Maher & N. Noddings (eds.), *Constructivist Views on the Teaching and Learning of Mathematics* (JRME Monograph No. 4), National Council of Teachers of Mathematics, Reston, VA, 19–29.

Yasuhiro Sekiguchi
Yamaguchi University,
Japan

PART V

EVALUATION OF RESEARCH IN MATHEMATICS EDUCATION

GILA HANNA

EVALUATING RESEARCH PAPERS IN MATHEMATICS EDUCATION

I would like to address, from two different perspectives, the question of the criteria used in evaluating research papers in mathematics education. The first perspective, which I will call *theoretical,* encompasses the principles on which an evaluation is said to rest. A scholarly journal would normally spell out such principles in a policy statement, which would take the form of a set of guidelines and a list of criteria, supported by a rationale. In most cases such a policy statement would represent the best efforts of the board of editors in setting out how they believe the merits of a research paper should be judged. The editors would expect that reviewers (themselves included) would have a good understanding of these criteria and would apply them consistently to research papers submitted to the journal.

When it comes to the application of such criteria, however, there is actually a great variation from case to case in the weight that is given each of them. Thus the second perspective I want to present is the practical one of how criteria for judging the merits of a research paper are interpreted and applied in actual cases. I base my presentation of this perspective on my experience as co-editor of *Educational Studies in Mathematics (ESM)* and as a reviewer for *ESM* and a number of other journals spanning the fields of curriculum studies, mathematics education, and research methods. I have sifted through referees' reports to try to understand why reviewers differ so widely in their judgments of the same paper, and in particular to examine how reviewers use and interpret the criteria and whether they seem to be using criteria other than those stated in the guidelines of the journal. Although we would all like to believe that there is a direct and straightforward association between settling on criteria and having them carried out, that is certainly not the case. Indeed, there is a big gap between theory and practice.

1. THE THEORETICAL PERSPECTIVE: GUIDELINES AND OFFICIAL POLICIES

In setting out their policy for reviewers, most journals include a general statement such as that found on the inside front cover of the *Journal for Research in Mathematics Education (JRME)*:

The journal ... is devoted to the interests of teachers of mathematics and mathematics education at all levels. [It] is a forum for disciplined inquiry into the teaching and learning of mathematics. It encourages the submission of a variety of types of high-quality manuscripts: reports of research, including experiments, case studies, surveys, philosophical studies, and historical studies; articles

399

Sierpinska, A. and Kilpatrick, J. Mathematics Education as a Research Domain: A Search for Identity, 397–407.
© *1998 Kluwer Academic Publishers. Printed in Great Britain.*

about research, including literature reviews and theoretical analyses; brief reports of research; critiques of articles and books; and brief commentaries on issues pertaining to research.

In addition to a general statement of this sort, the policy usually articulates specific expectations from research papers, though in most cases these are also broad statements that focus on quality and significance. Quality is of great concern to a scholarly journal, of course, since its reputation rests upon the quality of the papers it publishes:

Articles accepted for publication *must* be of high quality and *must* make significant contributions to the field. In particular, the work presented should be well conceptualized, should be theoretically grounded, and should move the field forward in clearly identifiable ways. (*JRME* 1996)

In addition, many journals ask reviewers to address specific questions. In particular, they would like to see discussions of all or most of the following questions:

1. What is the theoretical framework within which the research was conceived?
2. How is this research paper related to others? In what ways is it different? In what ways is it similar?
3. What does it add to what is already known?
4. What kind of data was used?
5. What systematic data-gathering methods were adopted, and why?
6. What mode of analysis was chosen, and why? Would a different paradigm have been more appropriate?
7. Do the conclusions follow from the data?
8. Are the specific questions addressed in the research important and likely to be of interest to readers of this journal?

In the case of papers that are of a strictly theoretical nature, that is, papers that do not rely on data collection, reviewers are usually asked to comment on the plausibility of the assertions made, on the flow of the argument, on the correctness of the reasoning used, on the kind of evidence presented, and again on whether the issue under discussion is of relevance to mathematics education and of sufficient interest to the readers of the journal.

In addition, most scientific journals demand that potential articles display the attributes of originality, significance of substance and content, usefulness, readability, succinctness, and accuracy, while also insisting upon adequate literature review, appropriate attributions and citations, and a clear demonstration of current knowledge in the field.

2. THE PRACTICAL PERSPECTIVE

It is no secret that there is often a large gap between the expectations of a journal as set out in its editorial guidelines and the manner in which research projects, even valuable ones, are actually conceived, carried out and documented. To a degree, of course, such guidelines represent an ideal that few researchers live up to fully. That is the background of reality against which editors and reviewers must make their assessments, a reality that complicates the already rather un-structured process by which potential papers are submitted for peer review in the first place. There are three points to keep in mind: (1) most papers fall short of the ideal, (2) the process of selecting candidate papers is subject to the vagaries of what gets submitted and what the editors know is being worked on, and (3) the assessments are subject to the varying opinions and thoroughness of the reviewers. The rest of the chapter expands on the assessment process.

As mentioned earlier, very different interpretations of the guidelines will be brought to bear upon papers selected for review, no matter how well the criteria have been spelled out and no matter how detailed the instructions to the review-ers. In the first place, reviewers will always bring their own biases. These biases are part and parcel of the process, and in fact not necessarily undesirable. In many cases, the editor selects reviewers because they are known to have specific knowledge in certain areas of mathematics education and specific biases in their approach to research in mathematics education. In the second place, there is also room for valid differences in applying any set of criteria. Not only will review-ers usually have differing interpretations of each of the criteria, they may also assign quite different weights to them. Originality may rank very high in the mind of one reviewer, while another may see it as no more than a bonus to be taken into consideration after significance and accuracy have been demonstrated. And in may cases reviewers, consciously or not, will add criteria of their own.

In what follows, I describe those factors that in my experience have often played a crucial role in the acceptance or rejection of research papers in *Educational Studies in Mathematics.*

Personal Concept of Quality

A journal relies heavily on the opinions of its reviewers and on their personal concept of quality. Editors will concede that these very personal views are idio-syncratic and not amenable to easy formulation, but they do not consider this to be a weakness in any way. On the contrary, they take pride in the wisdom and strong opinions of their reviewers. This view has been eloquently put by the editor of *Science:*

There will be no lobbying or phoning of review board members to influence their decisions. These individuals were chosen for their Solomonic wisdom, their mercurial response times, and their ency-clopedic knowledge, qualities achieved by being exceedingly busy scientists. (Koshland 1985, p. 4684)

In any case one must rely primarily on personal concepts of quality, because a generally accepted definition of quality in mathematics education research has remained elusive. At a symposium on criteria for scientific quality and relevance in the didactics of mathematics held in Gilleleje, Denmark, in 1992, for

example, three of the seven speakers, Dörfler (1993), Kilpatrick (1993) and
Sierpinska (1993), addressed the question of criteria for scientific quality in the
didactics of mathematics, but did so with admittedly limited success. As Jeremy
Kilpatrick (1993) said there, 'Even if ultimate criteria for evaluating the quality
of that research cannot be specified, mathematics educators need to continue to
hold to standards for their research' (p. 32).

Assessment of Originality

Many reviewers rank originality very high in their personal list of criteria. They
would like an article to present fresh ideas in the field or, if the ideas are not new,
at least a new way of looking at them. The innovation might be an interesting new
method, a perspective not previously used in this particular context, or an innovative approach to old research questions. Typical comments made by reviewers are:

'This analysis of assessment data, whether a classroom, state, national or international assessment,
can hardly be called a new line of research. The techniques used are traditional.'

'The issue treated is a timely one. The research has been carried out competently. However, there is
nothing new and interesting in this article. We learn very little from it.'

'Although the analysis has been thoroughly executed, the paper does not establish authoritatively
that the approach adopted to analysis was adequate to give the best answers to the cited questions
from the available data. Nor does it yet show that the more general findings and policy implications
are sufficiently novel and important to merit publication.'

Although reviewers look for original ideas, they are sometimes reluctant to
accept papers that are too daring in their use of new methods. That may be
because it is difficult to assess the usefulness of new methods, let alone their
scholarly status, before they become accepted in several fields of educational research. I might cite papers in the field of mathematics education based on research methods such as action research, personal knowledge, classroom diaries,
and autobiographical writing. These are part of qualitative methods for which
criteria such as reliability, validity, and generalizability cannot be applied. They
would be classified as ethnographic research, but are not as well known to reviewers as the more classical ethnographic methods such as classroom interactions or observational research. The moral of the story seems to be that it is good
to be original, but not too much so.

Presence of Certain Elements

A reviewer will often seem to insist on the presence of a certain element, such as
the use of a fashionable research paradigm or of a current theory, whether accepted or controversial. In the absence of such use, the reviewer will certainly
want to see a good indication that the author is well versed in the particular
paradigm or theory and has not avoided it out of ignorance.

The recent popularity of constructivist approaches to mathematics education, for example, has caused many reviewers to search for their presence in research papers and to interpret their absence as a deficiency. The current interest in alternate ways of assessing mathematics learning, ways still insufficiently articulated, has also clearly given some reviewers grounds for rejecting papers that use a more traditional mode of assessment, even when these papers treat interesting and important topics.

A further example that comes to mind has a European flavor. In papers dealing with the teaching situation, many European reviewers seem to expect to find the concept of *contrat didactique,* as if this concept were an indispensable tool of analysis rather than one of several perhaps equally useful ones. Yet another example is that of the reviewers who interpret all educational phenomena from the stance of critical pedagogy. These reviewers are inclined to expect other researchers to analyze problems of mathematical education in terms of oppression, resistance, empowerment and political positioning. Their judgments of research papers tend to be colored by the presence or absence of such tools of analysis.

It may be quite legitimate, of course, to attach significance to such a specific element when reviewing a paper, in that its presence does demonstrate that the author is aware of an existing body of research. There is justifiably an expectation that each new piece of research be well situated in the existing corpus, and that it use effectively the ideas and techniques of analysis appropriate to the topic. Nevertheless, the absence of a particular element can lead to a negative assessment of an otherwise valuable paper. Authors not making use of fashionable paradigms, techniques or theories would clearly be well advised to mention them and justify their avoidance.

Usefulness

Reviewers will also scrutinize a paper for the extent to which the research reported is potentially valuable for informing practice or for adding some new dimension of understanding. Although research and practice might seem to be at opposite poles and to have conflicting aims, reviewers nevertheless look to research to inform educational practices in the immediate future or to demonstrate that it can do so at a more distant time. The questions most often asked are: For what reasons was the research undertaken? Why do we need to know this? Who needs to know? What are the benefits stemming from this research? Given the inevitable tension that exists between research and practice, these questions often seem unfair, but they are repeatedly asked.

The following reviewer comments are quite typical:

'The authors have done an excellent job in offering a model for ...'

'A drawback of this paper is that it deals with an important property of a new model without giving us any information of the usefulness of the model.'

'The goals of this paper are clearly stated. In general the paper is successful at illustrating the effect of ... but is less successful arguing for the use of the Y model for teaching mathematics. What the authors have here is potentially interesting and useful, but I think that the paper has not connected these goals to interesting curricular or methodological questions, and this severely weakens the paper for acceptance in our journal.'

Readability and Length

Authors of research papers often overlook these very crucial aspects. Reviewers value conciseness and clarity. They do not want to plow through dense prose to find out what the author is saying, nor do they want to read lengthy arguments where shorter ones would suffice. This point can be illustrated by some comments made by reviewers of a paper that was eventually rejected:

'This was a very frustrating paper to read, and the ongoing temptation was to set the paper aside with the conclusion that it is unreadable. But two factors kept me going: first, my respect for the brilliance of the author, and second, the flashes of clarity and even poetry that occurred on occasion.

Unfortunately, as it stands, the paper should not be published – unfortunately because the author has much of value to say. He needs to find a way to say it.'

Another reviewer added:

'I do think the paper raises important issues, but I think it is unnecessarily difficult to read.'

Similar very critical comments were made on a paper that was eventually accepted after a number of revisions. The reviewers emphasized, however, that the work itself was important and potentially very useful, and the author accepted the editor's suggestion that the paper be recast in clear and readable language: 'You need a major rewrite of this important work if you are going to reach the wide audience it deserves.'

Wordiness is most often the enemy. I must point out in passing that in this respect there seems to be a cultural difference between North American and European authors, the latter often submitting papers that are much too long for what they have to say. The following is representative of comments made when such unnecessarily long papers are submitted:

'Concerning the style of the paper I think it is excellent from a didactical point of view (helping the reader understand...), but it is taking too much space of a journal like ours. Many passages can be shortened, as I indicated in the margins.... I think the paper should be accepted, but, the ratio of important results to the required space should be increased by shortening the present contents.'

Unfortunately, papers in mathematics education are not getting any shorter. On the contrary, the trend in the past 20 years has been towards longer papers, according to Lester and Lambdin (this volume). The average length of papers published in the *JRME*, for example, went from nine pages in 1973 to 20 in 1994. Indeed, it is now not uncommon for journals to receive manuscripts of 40 to 50 pages (about 20,000 words). This development is alarming. If a piece of research work cannot be summarized in no more than 15 pages (about 8,000 words), then a journal may not be the appropriate medium for its dissemination, and the author should consider publishing a monograph or a book. The average issue of a mathematics education journal has about 100 pages and usually includes four to six papers and one or two book reviews. This format is dictated, as it should be, primarily by the desire of the readership for the early publication of a broad range of research results in summary form. By necessity, then, papers should be in the range of 4,000 to 8,000 words.

Thoroughness of the Review

Reviewers are expected to review papers carefully, but unfortunately they do not always do so. Reviews are sometimes hastily and perfunctorily done; a reviewer may even miss the point of a paper altogether. It would seem that in many cases, reviewers, having noticed the use of prevalent paradigms, assume that the research has been carried out competently, when even a slightly more careful scrutiny would have revealed errors, misuse of methods, or improper conclusions.

That is why it is very important to have papers examined by a number of reviewers. In my view, the best number is four. One is certainly not enough, while two do not give the editor enough to go on, especially when the two reviewers make contradictory judgments. Three is better than two, but it is still not good enough, given that some reviewers are much less thorough than others. With four reviewers there is a better chance that the editor will obtain a good overall idea of the strengths and weaknesses of the paper. The editor will then be in a good position to justify rejection, acceptance or conditional acceptance of a paper:

'While all reviewers agree that there is material for a publication in this paper, they also concur that much work remains to be done before the paper can actually be published....

In my opinion the review process has worked very well for this paper and has produced many insightful and constructive comments. All in all, my recommendation is to reject the paper and to encourage submission of a revised version which would then go through the same review process a second time.'

Timing

There is certainly an element of chance in determining which articles are accepted for publication. Manuscripts are usually evaluated in relation to others, not solely against an absolute standard. Many manuscripts worthy of publication must be rejected because there is a limit to the number of papers a journal can publish. Since there are fluctuations in the number of papers submitted, even a paper of very high quality, unfortunately, may be rejected if it happens to be submitted at a time when there is a backlog of several others of similar quality. The opposite and similarly unfortunate event may occur when articles of borderline value are accepted simply because they happen to arrive at a slack time in the life of a journal.

Biases against Certain Modes of Research

It is a bit odd that many reviewers who are themselves researchers in mathematics education consider statistical methods as unsuitable research tools. They may dismiss out of hand a paper that uses an experimental design derived from research in psychology, or that relies in whole or in part upon a particular type of statistical analysis, even though the statistical method used may be appropriate and even essential to the research.

Presumably these reviewers dismiss certain statistical methods because at some point in the past these valuable research tools have been misused by well-meaning but incompetent researchers. Yet the dissatisfaction in the field over the years with the results of quantitative methods is certainly matched by the dissatisfaction with the results of some of the qualitative ones – and in any case, abuse of a research method is not a reason to abandon it.

Because they are reluctant to see statistical analyses in research papers, most journals in mathematics education are not likely to publish manuscripts that use quantitative research methods. Much excellent research about mathematics education is routinely published in journals such as *Journal of Educational Measurement, Applied Measurement in Education, Educational Measurement: Issues and Practice* and certain others, simply because the more general mathematical education journals seem to have turned their back on quantitative methods.

I should mention, however, that the Australian *Mathematics Education Research Journal* does not seem to share this aversion to quantitative research. In the September 1993 issue, it published five articles (Bourke 1993; Clements 1993; Menon 1993a, 1993b; Rowley 1993) devoted to the strengths and weaknesses of statistical significance testing, and the various alternatives to it. It is significant that a general journal on mathematics education found it appropriate to foster such a discussion and in so doing to encourage researchers to learn productive research techniques, statistical and non-statistical.

3. CONCLUSION

Defining precise criteria for the assessment of research in mathematics education would probably be a fruitless task, if not an impossible one. The best one can hope for is a broad set of guidelines that recognizes mathematics education research as a complex and difficult endeavor and that accordingly allows for experimentation with new research paradigms, for the use of tried and true approaches, and for original combinations of new and old.

For this very reason, it is all the more important that competent practitioners make themselves available as reviewers. It is also important that all reviewers consider manuscripts thoughtfully and responsibly, remaining open to new ideas and keeping in mind the broad spectrum of approaches, even as they bring to bear their own specific research experience.

The evaluation of research manuscripts is not a simple process. Reviewers will often disagree, and they will justify their judgments with arguments that may or may not be compelling. They will almost always bring their own biases to their evaluations, and sometimes they will even fail to understand the point of a manuscript. This reality underscores the importance and the responsibility of the editor, with whom rests the ultimate decision on whether a manuscript is to be accepted, rejected outright, or returned for modification. For the editor, the advice of the reviewers is extremely valuable, however, even when it is conflicting. Weighing the arguments of the reviewers as carefully as possible, and consulting with other members of the editorial board where necessary, the editor must then seek to make a decision that best reflects the scientific value of the manuscript and its potential contribution to the continued good reputation of the journal.

REFERENCES

Bourke, S.: 1993, 'Babies, Bathwater and Straw Persons: A Response to Menon', *Mathematics Education Research Journal* **5** (1), 19–22.

Clements, M. A.: 1993, 'Statistical Significance Testing: Providing Historical Perspectives for Menon's Paper', *Mathematics Education Research Journal* **5** (1), 23–27.

Dörfler, W.: 1993, 'Quality Criteria in the Field of Didactics of Mathematics', in G. Nissen & M. Blomhøj (eds.), *Criteria for Scientific Quality and Relevance in the Didactics of Mathematics,* Danish Research Council for the Humanities, Roskilde, 75–87.

Journal for Research in Mathematics Education: 1996, Information for Reviewers, unpublished manuscript.

Kilpatrick, J.: 1993, 'Beyond Face Value: Assessing Research in Mathematics Education', in G. Nissen & M. Blomhøj (eds.) *Criteria for Scientific Quality and Relevance in the Didactics of Mathematics,* Danish Research Council for the Humanities, Roskilde, 15–34.

Koshland, D. E.: 1985, 'An Editor's Quest', *Science* **227**, 4684.

Menon, R.: 1993a, 'Statistical Significance Testing Should Be Discontinued in Mathematics Education Research', *Mathematics Education Research Journal* **5** (1), 4–18.

Menon, R.: 1993b, 'Take Off Those Blinkers, Mate! Response to Bourke, Clements and Rowley'. *Mathematics Education Research Journal* **5** (1), 30–33.

Rowley, G.: 1993, 'Response to Menon', *Mathematics Education Research Journal* **5** (1), 38–29.

Sierpinska, A.: 1993, 'Criteria for Scientific Quality and Relevance in the Didactics of Mathematics', in G. Nissen & M. Blomhøj (eds.), *Criteria for Scientific Quality and Relevance in the Didactics of Mathematics,* Danish Research Council for the Humanities, Roskilde, 35–74.

Gila Hanna
Ontario Institute for Studies in Education of the University of Toronto,
252 Bloor Street West,
Toronto, Ontario,
Canada M5S 1V6

KATHLEEN M. HART

BASIC CRITERIA FOR RESEARCH IN MATHEMATICS EDUCATION

Those who are employed in education in the 1990s feel, quite rightly, that they are blamed for the world's ills. The shortcomings of the education system of a country are often high on the political agenda, and they are given considerable media space. In the United Kingdom, there have been 14 major educational 'reforms' in the last 14 years. Central to these changes is the belief of the Government that education is too important to leave to the professionals or those who have qualifications in education, and that decisions should be made by the representative of the person in the street. Governors of schools now have considerable power, and lessons in those schools are by law inspected every four years. Both governors and inspectors do not need to have educational qualifications – by government design, not because of a grave shortage of qualified personnel. In such a climate when the education workforce is de-professionalized, it is important that those doing research in mathematics education state the criteria by which they work and take decisions worthy of a profession.

1. RESEARCH: THE WORD

The word *research* has a perfectly honorable meaning different from *knowledge, thought* or *argument*. Mathematics education research is pursued (a) as part of postgraduate studies by a master's degree or doctoral student, (b) as part of the expected work of an academic and (c) within the profession of researcher. Those engaged in mathematics education as an academic field are usually employed in universities or colleges of education, and their main role is teaching. They are very often trainers of teachers as well as working with postgraduate students. Applicants for entry into such an academic career are judged on experience and research record. Entry to the profession is decided on the quality of the candidate's research by those already embarked on the career, so some consensus on the meaning of *research* is assumed, although perhaps not stated.

Students are accepted by an academic department to pursue a research degree. The institution (university usually) judges that the supervision will be adequate and that the student will proceed to success. Usually others from the same institution (and sometimes also from another university) are involved when the research thesis is examined. It is assumed that there is some degree of agreement amongst all concerned as to what is meant by *research,* since the student's

Sierpinska, A. and Kilpatrick, J. Mathematics Education as a Research Domain: A Search for Identity, 409–413.

examination is not seen as a forum for argument on the meaning of the activity. It may be that this assumption of agreement among mathematics education researchers is incorrect. We will not know unless the issue is discussed.

2. METHODOLOGY

Mathematics education as a separate field of study has been developed over the last 30 years; it tended initially to employ research methods taken from pure science. Twenty years ago pretests and posttests and statistical tests with degrees of significance predominated. More recently, researchers have adopted, and often adapted, the methodology of social science and psychology. These fields have their own experts who dictate the expected degree of rigor in the research carried out. Mathematics education has not necessarily accepted the same rigor.

3. CONSUMERS

Teachers and curriculum developers should be major consumers of research in mathematics education. Those engaged in research need to bear in mind the teacher's need for valid information and so design accordingly the methods by which evidence is collected. It should no longer be sufficient for advice given to teachers to be based on a 'good idea'. Those giving advice need to provide evidence of the effectiveness of that good idea. A recurrent theme in mathematics education discussions is the lack of impact of research on mathematics teaching. Whilst we fail to distinguish between what is mere suggestion and points for which evidence is available, teachers are wise to regard the advice we give with skepticism. Many mathematics teachers have been urged to employ teaching methods that they find unworkable, or ineffective when *they*, rather than a clinical experimenter, put them into operation. Researchers have not attempted (or possibly not been funded to attempt) to isolate the fundamental ingredients of successful teaching that are independent of personality and robust even with a class of 30 children.

4. REVIEWERS

Reviews of articles for publication and proposals for research are often carried out by others in the same field. There is no accepted basis on which proposals are reviewed, and they may be condemned simply on a difference of opinion. Until there is a stated consensus of the requirements by which work can be labeled 'research', a professional approach to reviewing is likely to be absent. There is a grave danger that mathematics education research will not be funded because peers reviewing proposals condemn methodology different from that in which their own work is set. In an atmosphere when education amateurs are preferred to professionals, that is particularly destructive.

The ICMI-affiliated International Group for the Psychology of Mathematics Education (PME) has an annual conference for which research reports are submitted and accepted after review. In 1989, a discussion group at the Paris meeting considered a list of very basic criteria to be met by work purporting to be research. The organizers of the group (Figueras, Hart, Lester & Nunes) suggested that mathematics education research could be seen as historical, philosophical, ethnographic, experimental, case study, survey, and so forth (as suggested by Jaeger 1988). Within all these methods and approaches, one could insist that research be disciplined inquiry in response to a problem or question. The list of essential and minimum criteria for this disciplined inquiry was:

1. There is a problem.
2. There is evidence/data.
3. The work can be replicated.
4. The work is reported.
5. There is a theory.

5. THERE IS A PROBLEM

Within the discipline of education, the word *research* has status, and work thus titled is meant to inform. Teacher recipients of such information seldom question its authenticity. It is assumed that others closer to the topic of inquiry have checked it for validity. Research needs to be focused, and so it seems eminently suitable that the problem or hypothesis, which is the object of focus, be stated. The question might be changed and reformulated during the pilot study, but the main work would focus upon it and attempt to answer it.

6. THERE IS EVIDENCE/DATA

Data are usually paramount in research of experimental or quasi-experimental design, and these specific pieces of information are often the most interesting aspect of research for the teacher who will read the resulting reports and articles. The researcher herself might find the subsequent analysis and theory building more interesting, but what the teacher recognizes as of immediate value is likely to be a statement of performance with which her own pupils can be matched – not necessarily simply to see whether they do as well, but also to see if the same errors are present, the same methods used, the same amount of time elapsed, even the same style of teaching employed. In historical research, documents, archives, libraries form the data from which conclusions can be drawn. Case studies have to be written after observation and interview; the raw data may be

from logs, notes, or tapes and need replicating or supporting from other sources
to validate key observations.

7. THE WORK CAN BE REPLICATED

Researchers seek to produce results that are generalizable to a wider population
than that under study. It is arguable that results that apply only to the immediate
situation under study are of little value in education, when so much information
is needed if teaching is to improve. If one wishes to generalize, then it is imperative that the study can be replicated. Even if the study is particular to one instance, it is necessary that all definitions and descriptions be sufficiently precise
to make it possible for a replication to be carried out. The study need not be
actually replicated, though that is desirable, but it should be possible to do so.
The samples will not be the same when made up of human beings but will be so
when the subject of study is a set of documents or archives. The results may not
be the same; indeed, in a repeated survey taken some years later, the purpose
may be to show the difference in response to the same instruments. The intention of this criterion is that the work is clearly defined and so becomes more
precise, understandable and applicable.

8. THE WORK IS REPORTED

The need for research to be reported stems from the desire that it be available for
comment and criticism as well as be made of use to the education community.
Most researchers and academics need to publish to survive, so their work is
likely to be disseminated at conferences and in papers. The current popular idea
that teachers do research in their own classrooms is valid only if they report
what happened in an environment that will judge the utterance as research.
Teachers who reflect on their teaching and the experiences they provide for their
classes and make subsequent adjustments are being good teachers and not necessarily researchers. The two activities are different. By equating reflection on
practice with 'research', we encourage teachers to think that this is a special activity and not one expected of *every* classroom practitioner. Authors should be in
a position to answer criticism of their research, so ideally the criticism should be
addressed initially to the individual who carried out the research.

9. THERE IS A THEORY

Some disciplines from which methods of research have been borrowed demand
that the inquiry be firmly placed within a theory and have as its focus the finding
of evidence to support or refute that theoretical standpoint. Mathematics education research of the 1960s and 1970s tended to seek for evidence to support or to

undermine the theories of Piaget, Ausubel and Gagné. More recently, these theories of learning have been seen as part of a framework in which research is set, without accepting all of the theoretical stance. Much of the work reported recently has, however, been seen as the start of theory formation, with research necessarily leading to theory rather than being carried out to provide information for practice. The PME participants were inclined to expect the research to be set within a theoretical framework.

10. CONCLUSION

The above minimal criteria could form the basis for those who fund, those who judge and those who do research in order that decisions with some common attributes can be made. Important decisions are made now using *personal* criteria for judging research but not producing them for scrutiny. Research evidence is vital to move what is now speculation onto a firmer base in the educational field. In order to be a profession, we need to show a measure of unity to the users of our research.

REFERENCE

Jaeger, R. M. (ed.): 1988, *Complementary Methods for Research in Education*, American Educational Research Association, Washington, DC.

Kathleen M. Hart
Shell Centre for Mathematical Education,
University of Nottingham,
Nottingham NG72RD,
UK

FRANK K. LESTER, JR. & DIANA V. LAMBDIN

THE SHIP OF THESEUS AND OTHER METAPHORS FOR THINKING ABOUT WHAT WE VALUE IN MATHEMATICS EDUCATION RESEARCH

1. THE SHIP OF THESEUS: A METAPHOR FOR OUR EVOLVING FIELD

Recently we repaired a very old, wooden rowboat, patching leaks in the hull, replacing the seat, applying a fresh coat of paint, and wondering all the while how many times these tasks had been repeated over the decades that the boat has been in use. Our work also reminded us of the ancient Greek story of the ship of Theseus – a story of personal identity that has challenged philosophers throughout the ages (Nozick 1981).

One of the planks of the ship of Theseus needed repair, so it was replaced. Time passed, and a second plank was replaced as well. And so the story continues. The planks of the ship were removed – one by one, over time – and as each plank was removed, it was replaced by a new plank. Was it still the ship of Theseus after one plank had been replaced? Most of us would not question its identity. The replacement of one plank by another does not make it a different ship; it is just the same ship with one plank different. Was it still the same ship after ten planks had been replaced? Although its hull was now somewhat newer, we probably would not consider it a different ship. In fact, over time, each and every plank in the ship might be removed and replaced, but if that occurred quite gradually, the ship would still be considered the ship of Theseus. The identity of something over time does not require it to keep all the very same parts.

But the story continues. It turns out that each time a plank was removed from the ship of Theseus it was stored in a warehouse, and much later the collection of old planks was assembled into a ship again. Two ships floated now in the harbor, side by side. Which one, wondered the Greeks, was the ship of Theseus?

The story of the ship of Theseus reminds us of how the young and evolving field of mathematics education research has been gradually redefining itself over time, although few have taken note of the numerous but subtle changes as they have occurred. Which 'planks' in the mathematics education research 'ship' have been replaced? A few telling indicators of the changing nature of our field are changes we have noted in the nature of manuscripts submitted for publication in the *Journal for Research in Mathematics (JRME)*.[1] During the early days of the journal (i.e., 1970–1980), the predominant methodology in the field was statistical in nature (mostly hypothesis testing and regression analysis designs); by 1983, about one-third of all research reports in the journal were

Sierpinska, A. and Kilpatrick, J. Mathematics Education as a Research Domain: A Search for Identity, 415–425.
© *1998 Kluwer Academic Publishers. Printed in Great Britain.*

non-statistical. The trend away from reliance upon statistical methods has continued: in 1993, five-eighths of all manuscripts submitted utilized some methodology other than statistics. (One-half used various non-statistical methods exclusively, and the remaining one-eighth used some combination of quantitative and qualitative methods.) Mathematics education seems to have outgrown its singular reliance on statistical techniques in favor of methodologies developed in such disparate research disciplines as anthropology, psychology, history, philosophy, and sociology, as well as the various 'natural' sciences. Use of non-statistical methodologies has led to more narrative reporting, and a concomitant trend toward longer articles. In 1973, the average length of an article published in the journal was slightly more than 9 pages. By 1993, the average length had grown to 20 pages – more than twice the length of 20 years earlier.

If the changes we have described seem minor, remember the ship of Theseus. When one or two 'planks' are replaced, a 'ship' does not seem significantly changed. Yet, looking back after some time, we might well wonder whether we are still dealing with the same entity. Indeed, at some point we might argue that the new ship is no longer the same ship as the original. Furthermore, if we agree that the new ship is, in fact, different, then we will probably want to consider whether the new ship represents an improvement or a decline in quality from the original. This chapter is about what we, as researchers, should consider in determining whether the changes that are taking place in our field represent improvements or declines in quality.

2. JUDGING THE QUALITY OF RESEARCH: SIMPLY A MATTER OF TASTE?

Judging the identity and the value of a change is, in many respects, a matter of taste. For example, a cook may be able to substitute margarine for butter in a brownie recipe with no perceptible change in the taste of the final product. (Brownies, a popular American treat, are a dense, chewy chocolate cake baked in a rectangular pan and served in squares as finger food.) Similarly, honey might be used as a healthy replacement for white sugar, or egg substitute as a low-cholesterol replacement for whole eggs. However, if an overly health-conscious baker attempts to make all three substitutions to the recipe at once, as well as using carob instead of real chocolate and whole wheat flour instead of white, the end result will probably no longer be considered worthy of being called a brownie. The treat will have changed in form so much as to have lost both its identity as a brownie and its value to those who crave the traditional taste of a brownie. How many recipe substitutions are too many? That question is not easily answered, although brownie connoisseurs probably use some set of implicit criteria (such as sweetness, chewiness, and chocolate flavor) in judging the tastiness of their favorite treats.

It is important to note that connoisseurs, whatever their field, usually are able to come to agreement when asked to critique an object in their domain of expertise. Perhaps even more important, whether they agree or disagree they are able to engage in debate and discussion because they have developed shared notions of the dimensions of quality in their field. For example, Charles Montgomery, expert on American decorative arts, spent decades looking at objects; comparing their proportion, ornament, and textures; and eventually developing a 14-point system for evaluating antiques that demystified the process by which connoisseurs in his field make decisions about the authenticity and quality of early American artifacts (Montgomery 1961).

In our own field – in response to changes in the types of manuscripts received by the *JRME* – a few years ago the Editorial Board began to consider criteria for judging the quality of research manuscripts. In particular, the Board was interested in identifying criteria that can be applied to manuscripts based on a wide variety of ideological and methodological traditions. And, since various educational researchers in North America had recently engaged in the development of standards to guide judgments about the quality of research (e.g., Howe & Eisenhart 1990; Shulman 1988), the Editorial Board's deliberations were guided by these previous efforts. Yet, the Board felt that it is important to consider criteria that are specific to the field of mathematics education – as it is has evolved today – since academic disciplines are in many ways unique. As explained in a recent *JRME* editorial, academic disciplines

differ with regard to the nature of the questions asked, the manner in which questions are formulated, the process by which the content of the discipline is defined, the principles of discovery and verification allowed for creating new knowledge within the discipline, and the criteria used to judge the quality of research within the discipline. (Lester 1993, p. 2)

Especially when a field is undergoing change, or when one wishes to be able to initiate newcomers to the field, it is useful to be as explicit as possible about criteria so that they can be openly shared, discussed, and debated.

3. THINKING ABOUT WHAT WE VALUE IN MATHEMATICS EDUCATION RESEARCH

Although the current concern about criteria used to judge research in mathematics education may seem like a new development, interest among North American mathematics educators in the development of criteria for evaluating research in the field can be traced back nearly 30 years. From the late 1960s to mid-1970s, various checklists and sets of criteria were created to assist mathematics education researchers in evaluating research reports and proposals (e.g., Coburn 1978; Romberg 1969; Suydam 1972). Although these efforts tended to be better suited for evaluating experimental and quasi-experimental studies, the claim was made by the developers of at least one of them that the 'criteria [are] general enough to reach across several categories of studies' (Coburn 1978, p. 75). The 'several categories' referred to by Coburn included experimental studies, clinical studies (both observational and teaching experiments), and organizational studies.

At the research presession held in conjunction with the April 1994 annual meeting of the North American National Council of Teachers of Mathematics (NCTM), a discussion was held about the sorts of criteria the *JRME* Editorial Board had been considering. About a month later, at the International Commission on Mathematics Instruction study conference held at the University of Maryland in connection with the present book, these criteria as well as criteria developed by others were discussed. (In particular, criteria that were considered included those proposed by Dörfler 1993; Hanna, this volume; Hart, this volume; Kilpatrick 1993; and Sierpinska 1993.) Then, in November 1994 the criteria were discussed once again at the annual meeting of the North American section of the Psychology of Mathematics Education in Baton Rouge, Louisiana. A fourth forum was held in April 1995 at the research presession prior to the annual NCTM meeting in Boston, Massachusetts. The discussions at these four meetings generated numerous suggestions about the value of having criteria and how our existing criteria might be changed. The various sets of criteria generated widespread interest among those who took part in the discussions at the four meetings, but it would be inaccurate to claim that they were generally accepted. In fact, there were some who were opposed to attempts to establish criteria and others who simply saw no need to establish any sort of criteria for assessing the quality of research in the field. What resulted from these discussions were a number of questions; among them, the following were particularly prominent:

1. Is it reasonable to expect one set of criteria to be suitable for every type of research? An ancillary question might be: Suppose you are asked to judge the quality of a report of an ethnographic study (or a case study, or a quasi-experimental study, etc.). Does the list of criteria apply to this type of study?
2. Are some criteria more important than others? Which ones and why?
3. Are there criteria that *are not* on the list, but *should be*? Similarly, are there criteria that *are* on the list, but *should not be*? If so, which criteria should be added or removed and why?
4. What, if anything, makes these criteria especially relevant for research in mathematics education? That is, are these criteria just as appropriate for judging the quality of research in, for example, science or language education or is there something special about some or all of them that makes them particularly appropriate for mathematics education research?

These and similar questions caused us to rethink the criteria presented at those meetings. The result of our deliberations is a set of seven rather general criteria that can be applied to designing, conducting, and reporting of a wide range of types of research in mathematics education. Before presenting these criteria, we briefly discuss some of the assumptions upon which they are based.

Assumptions Underlying Our Criteria

An explication of the assumptions (principles) about research in our field and how this research might be evaluated is of utmost importance if a set of criteria is to be broadly applicable. Four basic assumptions have guided our thinking.

First, we are assuming that there is a need to propose possible criteria. This assumption may seem trivially obvious, but we feel it deserves to be made public. By stating that there is a need to discuss criteria for judging the quality of research in our field, we are insisting that open criticism is a natural and important aspect of research. We are also saying that criticism cannot occur at the whim of the individual, but must take place in view of the public standards (viz., criteria for quality) to which members of the community feel themselves bound (Longino 1990). Kilpatrick (1993) insists: 'We should maintain a healthy skepticism toward our own and others' work plus an openness, even an eagerness, to subjecting that work to critical tests' (p. 32). For us, that means that criticism should be a vital part of any research program and that, in order to keep it from being superficial and empty, criticism should be guided by clearly articulated principles.

Second, any set of criteria for judging research in mathematics education must have *mathematics education* as its primary focus. That is, in order for the criteria to be useful in evaluating research in mathematics education, they must resonate with issues and questions about the teaching and learning of mathematics. Moreover, the single most important consideration in evaluating research in mathematics education must be the extent to which the research is situated in issues that the mathematics education research community regards as important and worthwhile.

A third assumption is that the criteria must necessarily be general, and rather abstract, in order for them to be applicable to research involving methods and techniques borrowed from various disciplines. Furthermore, since it is unrealistic to expect any mathematics education researcher to master all of these disciplines, judgments about methodological issues and principles of evidence particular to a discipline must be the responsibility of scholars of those disciplines. (See also Howe & Eisenhart's 1990 discussion of the necessity of having abstract criteria.)

Fourth, we do not regard the criteria we are proposing as immutable principles, nor do we expect them to be used uncritically or in a mechanical way. As the nature of mathematics education research continues to change, these criteria will probably also change. Moreover, we shudder to think that someone might suggest turning our set of criteria into a checklist to be used to assign a numerical score to a research proposal or report.[2]

Criteria for Judging the Quality of Mathematics Education Research

We intend the seven criteria discussed in this section to be useful for evaluating all aspects of the research process (conceptualization and design, question formulation, conduct of the study, data analysis and interpretation, preparation of reports, etc.) and all types of research activities.[3] Also, with the exception of the first criterion – *worthwhileness*, which seems, without doubt, the most important criterion in the set – no ordering of the criteria with respect to importance is intended.

Worthwhileness

Worthwhileness for mathematics education is the *sine qua non* of any research endeavor. The relevance for mathematics education of a research study is of paramount importance in any attempt to judge its quality. Worthwhileness has to do with the potential of a research study for adding to and deepening our understanding of issues associated with mathematics teaching and learning. Also, research that leads the field in new directions is generally more worthwhile than closed-ended research.

The worth of a research study is a function of, among other factors, when the study is proposed or conducted. Thus, a study that is in the forefront of research during one period of time may be considered of little value at some later date. Put another way, in order for a research study to satisfy the worthwhileness criterion, it must resonate with the issues and questions that are regarded as interesting and important to mathematics educators at a given point in time.

We recognize that, as Howe and Eisenhart (1990) point out, evaluations of worthwhileness can be quite difficult to do and are subject to extreme bias on the part of the evaluator. Nevertheless, rather than allowing tacit judgments of worthwhileness to establish directions for mathematics education research, it is preferable to make judgments of worthwhileness open to public scrutiny.

Some key indicators of worthwhileness might include: the study generates good research questions, the study contributes to the development of rich theories of mathematics teaching and learning, the study is clearly situated in the existing body of research on the question under investigation, and the study informs or improves mathematics education practice.

Coherence

Among the weaknesses most often cited by reviewers of manuscripts submitted for publication in the *JRME* is the mismatch between the research question and the research methods and analysis techniques employed to answer the question. Thoughtful researchers first give serious attention to identifying interesting and worthwhile research questions and then to selecting the research methods and techniques that best fit the nature of those questions.

Related to the match between question and methods is the issue of whether the research design will generate evidence that is appropriate for the question being asked. What is considered evidence in one discipline (e.g., anthropology) may not be the same as what is accepted as evidence in another discipline (e.g., physics). It is important that the research remain true to the principles of evidence deemed appropriate for the research methods employed.

Competence

It is not enough that a research study involve relevant, interesting questions and be carefully conceptualized, designed and reported. In addition, the conduct of the study itself must include the effective application of appropriate data collection, analysis and interpretation techniques. Principles (some tacit, others explicit) for conducting interviews, designing instruments, reducing data, selecting samples and so on have been developed within various disciplines to guide researchers in carrying out their studies. Although these principles should not be followed slavishly, competent researchers always consider them in order to ensure that every aspect of their studies is appropriately and carefully carried out.[4]

More and more frequently, it seems, mathematics education researchers are employing methods and techniques in their research for which they have received little or no direct training. Typically, a prospective researcher trained in the United States will have had at most one graduate seminar related to conducting a particular type of research – hardly enough to ensure competence. It seems essential that mathematics education graduate programs offer students courses or experiences providing direct and substantial attention to the research traditions of several disciplines (e.g., anthropology, psychology and sociology). However, since it is unreasonable to expect graduate programs to provide adequate preparation in conducting research based on so many different traditions, and since researchers cannot always anticipate the types of expertise they will need, it may often be the case that mathematics educators need to collaborate with researchers in other fields when the questions they want to study require use of research methods outside their domain of expertise.

Openness

Openness involves two qualities. First, in planning and conducting their investigations, good researchers are cognizant of the personal biases and assumptions that underlie their inquiry and, to the extent that it is possible to do so, they make these biases and assumptions public. Peshkin (1988) insists that researchers should strive to 'disclose to their readers where self and subject became joined' (p. 17). By so doing researchers are better able to clarify for their readers the bases for the conclusions they draw.

Second, the research methods and techniques used should be described completely enough to allow members of the research community to scrutinize them. In reporting research results, the researcher should provide the reader with a clear sense of how the data were collected, what data were used to make interpretations, and how the data were analyzed.

Both qualities – acknowledgment of personal assumptions and openness to scrutiny by the research community – are essential ingredients of any research that is likely to move the field forward.

Ethics

Considerations of ethics have been all but ignored in previous efforts to establish criteria for judging research in mathematics education. These considerations have to do with two concerns: (a) the manner in which the research has been conducted in relation to the research subjects (often students or teachers), and (b) acknowledgment of the contributions of others. The first concern involves matters such as informed consent, confidentiality, and accurate (to every extent possible) portrayal of situations and persons involved in the research. The second concern includes acknowledgment of the contributions of all persons who contributed to the research project, as well as open recognition of individuals whose research has influenced the present research.

Although ethical considerations sometimes necessitate compromising the quality of the data that are collected, they are fundamental to any determination of the quality of a research study.

Credibility

This criterion has to do with the extent to which sensible, thoughtful and open-minded readers find the claims and conclusions made in a research report believable. That is, the claims made and conclusions drawn should be justified in some acceptable way. Research findings should be grounded in data or evidence of some sort, not merely in rhetoric. That is, the credibility of research should not rely solely on surface plausibility, nor should it depend upon researcher eloquence.

Moreover, the arguments and interpretations provided in a research report should be presented in a manner that makes it possible to verify or refute the conclusions drawn. Here the openness and credibility criteria intersect and support each other.

Other qualities of good research reports

A research report that is lucid, clear, and well organized is likely to be more valuable and useful than one that does not possess one or more of these qualities. Similarly, the research community usually values conciseness over verbosity, and directness over obscurity. To some extent the aforementioned qualities are intangibles – difficult to identify and to evaluate, but frequently the qualities that distinguish good from mediocre research.

Originality, one of Kilpatrick's (1993) and Sierpinska's (1993) criteria for evaluating research, is another example of an intangible that sets good research apart from other types. An original study is not necessarily one that has never been done before. Rather, originality can also result from looking at an old question in a novel manner: using a new technique of analysis, synthesizing evidence in a different way, or providing a new interpretation for old data. However, Kilpatrick (1993) cautions that originality does not mean a lack of connection to the existing body of relevant research. Rather, he notes, 'it refers to the way in which evidence is marshaled and portrayed so as to cause the reader to think again' (p. 25).

Consideration of intangibles such as those discussed above can go a long way toward improving the design, conduct and reporting of a research study. Similarly, mathematics education researchers may be able to move the mathematics education research community forward by taking time to identify and debate the intangibles that we value.

4. AN EVOLVING FIELD REQUIRES EVOLVING CRITERIA FOR QUALITY

Certain metaphors have aided us in thinking about how we can identify and evaluate quality research in the ever-changing field of mathematics education. The dilemma posed by the Ship of Theseus helps us realize that it is no simple matter to define mathematics education once and for all – changes may be imperceptible, but important. Comparing two brownie recipes helps us see that without standards, we may risk loss of quality if too many changes take place too quickly. A third metaphor may bring us full circle in thinking about how standards are established and used. At the Henley Royal Regatta, an international rowing contest that takes place near London in July each year, 'women cannot wear dresses above the knee or trousers or culottes; men cannot remove their jackets and ties unless it reaches 80 degrees in the shade' (Dowd 1994, p. 72). Michael Sweeney, chairman of the event, compares the regatta's 154-year-old dress code with that of several other tony British sporting events: 'I've never been to Wimbledon. I'm told their standards are not our standards. How to put it delicately? There are less elegant sights to be seen at Ascot and Wimbledon' (Dowd 1994, p. 73). Most who attend the regatta value the fact that the dress code has never changed, although there are some rowing enthusiasts who would prefer to see the event's standards change with the times.

However we might feel about preserving societal traditions and standards, we would nevertheless argue that what we regard as good research must be responsive to changes in the field. In fact, that was our fourth assumption in posing the criteria discussed in this chapter. The primary purpose for identifying these criteria was to attempt to make explicit the tacit principles mathematics educators use *today* in making judgments about the quality of inquiry. Having proposed criteria, the mathematics education community can then engage in a much-needed, and on-going, dialogue about what researchers value.

NOTES

1. At the time this chapter was written, we were, respectively, editor and review editor of the *JRME*, the only North American journal devoted entirely to the publication of mathematics education research reports.

2. Thomas Schwandt (1996) argues forcefully against the 'necessity of regulative norms for removing doubt and settling disputes about what is correct or incorrect, true or false'.

3. Throughout this paper, the word *data* should be interpreted broadly to refer to any type of information used to justify claims.

4. Unfortunately, over the past few years referees of manuscripts submitted for publication in the *JRME* have been paying too little attention to the statistical procedures used in quantitative studies. There seems to be a trend toward taking for granted that such procedures have been carried out in a competent manner.

REFERENCES

Coburn, T. G.: 1978, 'Criteria for Judging Research Reports and Proposals', *Journal for Research in Mathematics Education* 9, 75–78.
Dörfler, W.: 1993, 'Quality Criteria for Journals in the Field of Didactics of Mathematics', in G. Nissen & M. Blomhøj (eds.), *Criteria for Scientific Quality and Relevance in the Didactics of Mathematics*, Danish Research Council for the Humanities, Roskilde, Denmark, 75–87.
Dowd, M.: 1994 (June), 'Always an England', *US Air Magazine*, 70–73, 104, 106.
Howe, K. R. & Eisenhart, M. A.: 1990, 'Standards for Qualitative (and Quantitative) Research: A Prolegomenon', *Educational Researcher* 19 (4), 2–9.
Kilpatrick, J.: 1993, 'Beyond Face Value: Assessing Research in Mathematics Education', in G. Nissen & M. Blomhøj (eds.), *Criteria for Scientific Quality and Relevance in the Didactics of Mathematics*, Danish Research Council for the Humanities, Roskilde, Denmark, 15–34.
Lester, F. K.: 1993, Editorial. *Journal for Research in Mathematics Education* 24, 2.
Longino, H. E.: 1990, *Science as Social Knowledge*, Princeton University Press, Princeton, NJ.
Montgomery, C. F.; 1961, 'Some Remarks on the Science and Principles of Connoisseurship', *The Walpole Society Newsletter* [as reported in a brochure of the Winterthur Library, Winterthur, Delaware].
Nozick, R.: 1981, *Philosophical Explanations*, Belknap Press, Cambridge, MA.
Peshkin, A.: 1988, 'In Search of Subjectivity – One's Own', *Educational Researcher* 12 (5), 17–22.
Romberg, T. A.: 1969, 'Criteria for Evaluating Educational Research', *Investigations in Mathematics Education* 2, x–xi.
Schwandt, T. A.: 1996, 'Farewell to Criteriology', *Qualitative Inquiry* 2(1), 58–72.
Shulman, L. S.: 1988, 'Disciplines of Inquiry in Education: An Overview', in R. M. Jaeger (ed.), *Complementary Methods for Research in Education*, American Educational Research Association, Washington, DC, 3–17.
Sierpinska, A.: 1993, 'Criteria for Scientific Quality and Relevance in the Didactics of Mathematics', in G. Nissen & M. Blomhøj (eds.), *Criteria for Scientific Quality and Relevance in the Didactics of Mathematics*, Danish Research Council for the Humanities, Roskilde, Denmark, 35–74.
Suydam, M. N.: 1972 (February), 'Instrument for Evaluating Experimental Research Reports', *Introduction, Compilation of Articles – Annotated Compilation of Research on Secondary School Mathematics, 1930–1970* (Vol. 1), (Final Report, USOE Project No. 1-C-004), The Pennsylvania State University, University Park.

Frank K. Lester, Jr.
School of Education, Room 3056,
Indiana University,
Bloomington, Indiana 47405 1006,
U.S.A.

Diana V. Lambdin
School of Education, Room 3266,
Indiana University,
Bloomington, Indiana 47405 1006,
U.S.A.

JUDITH T. SOWDER

ETHICS IN MATHEMATICS EDUCATION RESEARCH

Scenario 1: You are conducting a case study of a mathematics organization in your country. You have interviewed many members, including several within the organization's hierarchy, and have promised confidentiality. Your funding agency asks you to share your interviews in a forthcoming meeting. What should you do?

Scenario 2: You have been investigating the effect of teachers' mathematical understanding of rational numbers and proportional relationships and the manner in which this understanding affects their teaching. You have worked closely with a small group of teachers for over two years, providing them with rich opportunities to develop their understanding, and observing what changes in their teaching. You have formed friendships with these teachers. Now you are writing about what changes you observed. How do you deal with issues of confidentiality and anonymity, particularly since some of what you want to say is not something some of these teachers would like to read about themselves?

Scenario 3: You have undertaken a study to determine what happens in California schools where the faculty and administration take seriously the mandate to implement the State's *Mathematics Framework.* Your team selects twelve schools for study, and visits each school for several days. You conclude that in many of these schools, the teachers and administration are not implementing the *Framework*; rather, the changes are superficial and the underlying view of mathematics as an accumulation of facts, rules, and skills to be learned continues to guide the instruction in the school. You were granted funding for the study based on the understanding that you would profile these schools in a series of articles that would help other teachers and administrators better understand what 'implementing the *Framework*' means. You know that profiling schools where the implementation is superficial will help readers understand the difference between deep change and superficial change. But within the districts where the schools are located, it is well known that the schools have been studied, so anonymity is simply not possible. Publication of your report could be extremely embarrassing to the teachers and principals at some of the schools, could cause those schools to give up on implementation efforts, and could alienate these districts as future sites for research projects. What should you do?

Sierpinska, A. and Kilpatrick, J. Mathematics Education as a Research Domain: A Search for Identity, 427–442.
© *1998 Kluwer Academic Publishers. Printed in Great Britain.*

All of these situations are similar to real questions investigators have faced; the one dealing with teachers' understanding and teaching of rational numbers and proportional relationships is one with which I have been struggling. As I grappled with these issues and explored some of the relevant literature, I realized that what I was learning might be helpful to many of my fellow researchers in mathematics education who have increasingly selected qualitative research methodologies to investigate questions of importance, and who have faced ethical problems not encountered in their previous quantitative research.

1. WHAT ARE WE TALKING ABOUT? AND WHY?

Ethics is not a new concern to any of us. It permeates everything we do. We make many ethical decisions in our professional lives – the way we treat our colleagues and our students, the manner in which we treat and report data, the care with which we plan our classes, review papers for journals, use our time – all reflect the ethical life we lead. Smith (1990a) has described ethics as referring to 'that complex of ideals showing how individuals should relate to one another in particular situations, to principles of conduct guiding those relationships, and to the kind of reasoning one engages in when thinking about such ideals and principles' (p. 141). Within the realm of research, Eisner and Peshkin (1990) remind us that ethical conduct

is not just the simple matter of avoiding placing at risk those whom in our research projects we variously call the researched 'others', 'subjects', and 'respondents'. It is the infinitely more complex challenge of doing good, a consideration that places researchers at odds with one another as they raise entirely different questions about the location of good in the conduct of research. (p. 243)

There are always alternatives to be considered, choices to be made, and many times these choices involve other people. When these choices are value-related, we have an ethical problem.

Typically, problems arise when an investigator has selected a method to investigate a scientifically important question but realizes that the methodology selected can bring harm to the participants (Sieber 1982). For example, an interview situation might be overly stressful for the mathematics-anxious students participating. New teachers might lose their jobs if anonymity cannot be preserved and the research reveals that their backgrounds in mathematics are inadequate to teach. Or, as in a case we have faced, a teacher may become seriously demoralized when coming to believe that her teaching has been totally ineffective and that she is unable to become the type of teacher idealized by the reform movement in mathematics education. Most of the ethical problems that arise in research situations follow from the ways in which the researcher relates to 'the researched' (Eisner & Peshkin 1990). The manner in which these relationships influence the research process, together with related issues of informed consent, confidentiality, and anonymity, and finally issues reflecting the validity of the researchers' interpretations, the control over data, and debts to funding agencies will be discussed here. But first we need to situate our discussion within the historical boundaries of research ethics, the philosophical theories that guide interpretation, and the newer concerns brought about by qualitative methods.

2. HISTORICAL BACKGROUND

Moral considerations, as they apply to research, have their origins in the field of medicine. World War II and the Holocaust served as catalysts for attention to moral issues (Beauchamp, Faden, Wallace & Walters 1982). Basic principles for conducting experiments that involve human subjects were proposed in the Nuremberg Code, which has served as a model for subsequent governmental regulations. In the United States, the National Commission for the Protection of Human Subjects of Biomedical and Behavioral Research issued a report in 1979 (usually referred to as the Belmont Report) that set forth three basic underlying principles to guide research involving human subjects (Office for the Protection of Research Risks 1993). These principles, still used as guidelines by University Institutional Review Boards, are respect for persons, beneficence, and justice. Respect for persons recognizes the individual's personal dignity and autonomy; beneficence calls for protection from harm and the maximization of good outcomes; justice requires an equitable distribution of the benefits and costs of research. According to the report, these principles are applied through consideration of informed consent, risk–benefit assessment, and fair procedures for selecting subjects. The Department of Health and Human Services responded to the report in 1981 with significant revisions of its human subjects regulations. The most recent revision of those regulations in 1991 involved adoption of the Federal Policy for the Protection of Human Subjects (called the Common Rule) by all government agencies that conduct or support human subjects research (Office for the Protection of Research Risks 1993).

In universities in the United States, these regulations are carried out by a local Institutional Review Board, which reviews and approves research plans based on information provided concerning consent, the instruments used, and the competence of the investigators. Similar regulations apply in many countries, according to conversations I have had with researchers in those countries. In this country, most mathematics education research has been exempted from review because the purpose of the research has been to improve instructional practice, much like what happens in schools as part of normal efforts to improve and evaluate instruction, and the research was therefore considered low risk (Howe & Dougherty 1993). Qualitative research, however, is more open-ended with regard to participant selection, interactions with participants, and modes of analysis, and is not as likely to be considered risk-free and therefore exempt from review. These differences cause additional difficulties for the researcher because the review process, both here and elsewhere, has usually been set up to accommodate quantitative methodology, making the demands more difficult for the qualitative researcher to meet. Some researchers argue forcefully that social and behavioral sciences need different regulations than those instituted for bio-medical research, for example: 'Since fieldwork involves such a different methodology, with unique and often unpredictable benefits and risks (for both researcher and researched), this regulatory system is inappropriate in many ways, particularly on the issue of consent, to the ethical dilemmas of fieldwork' (Wax 1977, p. 29, cited in Macklin 1982, p. 193).

Professional organizations have also been paying increasing attention to ethical concerns. Within the social and behavioral sciences, several controversial studies (e.g., the Milgram obedience experiments of 1963, and the Tearoom Trade study of male homosexual behavior by Humphreys in 1970, both de-scribed in Reynolds 1979) led professional organizations such as the American Psychological Association, the American Sociological Association, and the American Anthropological Association to adopt guidelines for assuring that re-search participants are treated in an ethical manner. The American Educational Research Association, to which many mathematics education researchers belong, published its own code of ethics in 1992.

3. THEORIES OF MORALITY

While codes of ethical behavior might offer guidelines, they do not tell the re-searcher what to do in specific cases. Interpretation is needed, but any individ-ual's interpretation is highly dependent on the personal ethical position taken by that person. Several authors (e.g., Bunda 1985; Deyhle, Hess & LeCompte 1992; Flinders 1992; May 1980; Reynolds 1979) have discussed the manner in which different ethical positions influence decision making when researchers are faced with an ethical dilemma. Understanding these positions will help anchor later discussions in this chapter. Below is my summary of these authors' categoriza-tions of ethical stances.

Several positions can be summarized under two general categories of theories. The first, called a *teleological* ethic (from the Greek word for an archery target), is oriented towards a particular goal or outcome. Research should be inde-pendent, that is, not controlled by the priorities of outside funding agencies; it should be fundamental because knowledge is a fundamental good; it should not be considered an absolute good because although knowledge is intrinsically good, one is not allowed to use any and all means in pursuing it (May 1980). The best known variation of this ethic is utilitarianism, which falls within this category because its proponents are 'result-oriented, consequentialist thinkers' (May 1980, p. 360). Their method for resolving ethical problems is to consider all possible outcomes and choose the one that maximizes the resultant net good or 'happiness'. This 'principle of utility' (Bunda 1985) is the basis upon which actions are judged. Its major attraction, that it offers an exact way to predict and weigh harms and benefits, is also its major difficulty, since the precision de-manded is hardly ever possible. Furthermore, who is to benefit? The individual or the group? How is the net good, 'happiness', to be defined? May points out that this cost–benefit analysis can lead one to either side of ethical debates over such issues as whether covert research can be justified.

This form of utilitarianism, referred to in the literature as *act-utilitarianism,* is to be distinguished from *rule-utilitarianism,* in which a general set of principles or standards is established that maximizes the chances of choosing an action that will lead to a greater good and thus resolve ethical problems. There are obvious difficulties with this form of utilitarianism. Who sets the guidelines? How specific can they be? What if they lead to contradictions – how is a choice to be made? Yet both forms of utilitarianism have had considerable influence in social and political arenas (Reynolds 1979).

An example of this consequential-oriented position can be found in the influential work of Brody (1981) on making ethical decisions in medicine. Once a problem is acknowledged to exist, Brody advocates making a list of alternative actions, making a tentative choice, determining the consequences, comparing the consequences with one's personal values, and looping back to the next action if the consequences are unacceptable. Smith (1990a) finds this decision-making process attractive for educational fieldwork.

In the second major position, referred to as *deontological* (from the Greek word for obligation or duty), rules of behavior based on duties and rights play a larger role in ethical decision-making than does the consideration of consequences. This duty-based position stems primarily from the work of Kant (1785, reprinted 1959). Kant's moral laws, which he called categorical imperatives, are considered to be universal in nature. They define appropriate action in all situations and carry obligations or duties on the part of individuals. These principles demand that I act as I would have others act and that I treat others as ends, never as means. Covert research or actions, for example, that would undermine confidentiality would not be undertaken by a researcher following these principles. May (1980) identifies three major problems in applying this ethic to research. The first is that there is no room for exceptions, such as extraordinary situations that require some form of covert research but do no harm to the individual. The second problem is that the ethic is too individualistic. And the third is that it focuses only on general obligations, whereas there are often specific obligations that also deserve consideration. Yet this ethical stance is attractive to many. Lincoln (1990) is an example of a researcher who advocates Kant's principles as guides for qualitative research:

The two Kantian principles are also appropriate in that they force examination of personal ethical principles with virtually every act 'committed' in the name of sciences. Searching self-examination is an excellent path to united consciousness as well as to the value clarification and explication demanded by a new generation of qualitatively oriented researchers. (p. 293)

Rules of behavior can also be based on a consideration of the rights of others rather than our duties towards them. Moral rights (as distinguished from legal rights) become criteria for action: dignity, liberty, equality, autonomy. There are, of course, duties here also; the rights of one imply obligations on the part of another. But the principles of behavior focus on individual rights rather than obligations. Rawls's *A Theory of Justice* (1971) has perhaps the clearest explication of this position.

There are three additional positions or theories on which ethical decisions might be based. The first is critical theory and its application for advocacy research. Critical theory holds that 'knowledge must be situated historically and cannot be a matter of universal and timeless abstract and abstractly related principles' (Smith 1990a). Critical theorists believe that knowledge is not the exclusive right of the policy-maker; it should be used for the purposes of emancipation. The researcher writes not about but rather on behalf of the participant group (May 1980). This view naturally leads to advocacy as a research goal. Soltis claims that in critical research 'there seems to be the presumption that the very purpose of all human research is to raise our consciousness regarding ethically suspect arrangements embedded in the structure of our sociocultural world' (1990, p. 255). Now there are serious ethical considerations to be made concerning the *purposes* of research in addition to those concerning the *processes* of research.

An ethical position attractive to many qualitative researchers derives moral behavior from regard for others rather than from obligations to others, and for this reason is called *relational ethics* (Flinders 1992):

A proponent of relational ethics would readily accept that moral behavior often upholds utilitarian standards by leading to good consequences [and] might also conform to deontological canons such as honesty and justice, [but] should be informed above all else by a caring attitude toward others. (p. 106)

This ethical position is closely aligned with the *covenantal ethics* advocated by May (1980):

At the heart of a covenant is an exchange of promises, an agreement that shapes the future between two parties.... It acknowledges the indebtedness of one to another.... It emphasizes gratitude, fidelity, even devotion, and care. (p. 367)

This position is the basis of some feminist research, and is expressed well in Noddings's (1986) work on caring:

In educational research, fidelity to persons counsels us to choose our problems in such a way that the knowledge gained will promote individual growth and maintain the caring community. (p. 506)

Here again, purpose as well as processes must be given careful consideration.

The fifth position discussed here has also found renewed interest within educational research. Flinders (1992) calls this *ecological ethics* and warns that its relevance is difficult to understand without first coming to a better understanding of ecological thought. He notes that ecology does not refer necessarily to the environment but rather to a set of interdependent relationships. The classroom can be seen as an environment 'populated not by foliage, insects, and fungi, but by language, relationships, and ideas. These cultural dimensions of the classroom ... are what connect it meaningfully with the world beyond its four walls' (p. 108). In viewing the classroom as an ecological system, we are forced to understand that no part of the system can gain control over the entire system; that attempting such control would threaten the ecology of the system. This ethical framework takes one beyond cultural sensitivity; individuals must be recognized as part of a larger system, so that rather than avoiding harm to the individual, we focus on avoiding harm to the system, that is, to the environment.

Researchers do not always choose one of these positions but rather follow principles reflecting an amalgam of positions. The Protection of Human Subjects regulations are an example of a blend of the utilitarian and Kantian perspectives. The utilitarian position is expressed in the beneficence standard, which calls for an assessment of risks versus benefits, while the Kantian principle of treating people as ends rather than means can be found in the respect-for-persons standard.

4. WHAT IS DIFFERENT ABOUT QUALITATIVE RESEARCH?

Much of what has been said thus far relates to all educational research. The point I am trying to make here, however, is that various ethical issues need to be considered differently in the qualitative and the quantitative domains. This point becomes clear when we begin to compare and contrast the two types of research. Lincoln (1990) describes qualitative research as participative, cooperative, or experiential, and contrasts it with quantitative research by saying:

Qualitative research is increasingly seen not simply as a set of findings that reflect nonnumerical or quantitative data but, rather, as a set of social processes characterized by fragile and temporary bonds between persons who are attempting to share their lives and create from that sharing a larger and wider understanding of the world. (p. 287)

Researchers do not usually alternate between quantitative and qualitative research depending upon the question to be investigated; as Lincoln indicates, the choice goes beyond that of methodology. Her description is echoed by Bogdan and Biklen (1982) when they claim that doing qualitative research is more like establishing friendships than forming contracts. Wax (1982) has argued that the analysis of ethical problems generated by qualitative research has been impoverished by the experimenter–subject model that dominates review boards. (Of course, there are ethical considerations related to quantitative research; see, e.g., Krombrey 1993, for a discussion of ethics applied to quantitative data analysis.)

Soltis (1990) delineates four purposes associated with qualitative research. The first is to describe: 'The classic and pervasive purpose of qualitative research has been to adopt, create, and use a variety of non-quantitative research methods to describe the rich interpersonal, social, and cultural contexts of education more fully than can quantitative research' (p. 249). A second purpose is evaluation, that is, to augment quantitative judgments with the richer and more wide-ranging assessments that qualitative research can offer. The third purpose is intervention – to bring about change and to understand its effects. The fourth is political, cultural, or social critique. Each of these purposes brings with it a range of different ethical issues. For example, descriptive studies might invade participants' privacy; evaluative studies often provide negative information harmful to others; intervention studies can be manipulative of other people; critique raises questions of what is good and what is evil and how one is to know either.

The remainder of this chapter is devoted to a discussion of particular problems associated with qualitative research. These problems are categorized in terms of those facing the researcher during the research planning stage, the implementation stage, and the analysis and reporting stage.

5. PROBLEMS ARISING DURING THE PLANNING STAGE

The purpose for undertaking the research, the risk to participants, and questions of obtaining informed consent need to be examined before beginning a research project. There are ethical considerations within each of these arenas. Possible harm to participants is discussed in the section on conducting research, but is mentioned here because of the need to consider risks and benefits during the planning stage of a research study.

Purpose of Research

The first criterion is that the research must be worth doing – surely it is unethical to waste people's time undertaking research that one does not believe to be worthwhile. But others may believe it is worthwhile, and 'the availability of funds for some activities and the lure of fame and fortune for others can be very seductive tempters into projects that one cannot defend on grounds of importance or significance' (Smith 1990b, p. 273). In other cases, an investigator might find a question to be both interesting and worthwhile, but impossible to investigate without bringing harm to others. For example, it might be impossible to preserve anonymity in reporting on a case study contrasting the teaching behaviors of an effective teacher and an ineffective teacher where both teachers have been teaching for several years. One might have to compromise by selecting a novice teacher who would not yet be expected to be effective. Related to the issue of worthwhileness is the obligation to advance knowledge through research, and to conduct research in a competent manner. For those of us new to qualitative research, there are obligations to gain competence in a field where the methodology is unfamiliar, where issues such as validity take on new meaning, and where the manner in which data are analyzed and the style in which they are reported are unpracticed.

Informed Consent

The requirement of informed consent stems from the principle of respect for persons, that is, recognizing the individual's personal dignity and autonomy. The requirement is not simply a pragmatic one designed to reduce the responsibility of the researcher; investigators continue to be responsible for any negative effects suffered by participants (Reynolds 1979).

Each of the four elements of informed consent – that the participant is fully informed, is competent to give consent, fully comprehends the conditions of consent, and gives consent voluntarily – raises particular concerns for qualitative research. Full information usually includes such aspects as the research goals, the participant's role, why and how the participant was selected, the risks and benefits of participation, the ways in which the data will be used, and the possibility of withdrawing participation at any time. This requirement provokes investigators to examine the risks and benefits during the planning stage and might lead them to discover previously unknown risks to participants (Capron 1982).

One of the characteristics of qualitative research, however, is that initial questions are often found to be inadequate and new research questions are frequently uncovered during the process of the investigation, with consequent adjustments in the goals of the investigator and in the role of the participant (Deyhle et al. 1992; Howe & Dougherty 1993). 'It is therefore legitimate to ask if qualitative researchers can inform others when they themselves do not know what twists and turns their work is likely to take' (Flinders 1992, p. 103). Competence and comprehension can be concerns, particularly when students are the participants. Legally, we must obtain consent from parents, but ethically, children too should be allowed to exercise autonomy. Do children (or their parents) fully understand what consent means? Usually, in our research, this question may not be a concern, but there are instances when it is – as when a student is placed in an experimental mathematics program and scores lower on external examinations because of lack of adequate preparation. We may argue that the external examination is not testing what should be tested, but there might nonetheless be harm of some sort forthcoming if the score is low. Even when competence for giving consent is not a concern, such as with a teacher participant, comprehension of subtle risks associated with the research may still be problematic. Finally, voluntariness is of special concern in long-term studies. For example, if close associations are formed between the investigator and participant, the participant may feel some obligation to continue even though participation has become overly burdensome or stressful.

Diener and Crandall (1978) argue that while informed consent is desirable in most research and essential when there is substantial risk, it should not be absolutely required. 'When field studies do not significantly affect subjects' lives, informed consent becomes irksome and time-consuming for all parties and may be both ethically and methodologically undesirable' (p. 39). Observing and recording the behavior of a group of people in a natural environment is a particular instance in which informed consent is impossible, since the information provided could change the setting and behaviors. Unobtrusive observation of classroom group problem-solving behavior might be such an instance. Foregoing informed consent (or requesting this of the IRB), however, should not be taken lightly – there are ways of making a considered decision. Diener and Crandall suggest, for example, obtaining ethical opinions from others, including some members of the subject group.

This is not to argue that qualitative researchers should be relieved of their obligation of obtaining consent in most cases. Howe and Dougherty (1993) point out that the intimacy and open-endedness of qualitative research 'significantly muddy the ethical waters' (p. 18) to the extent that most qualitative studies conducted in schools today should no longer qualify for exemption from review by an Institutional Review Board. They concede that the issue of informed consent is 'especially tangled and contested' (p. 19) for qualitative studies, but conclude that requirements for informed consent should be perhaps even more demanding. For example, periodic reaffirmations could be required when participation is long-term.

Informed consent will continue to be a sticky issue, however, as long as review boards continue to judge consent questions in qualitative research with the same guidelines they use for biomedical research, which is where the requirement originated. Wax (1982) has argued that 'the analysis of the ethical problems generated by scientific investigations has been impoverished because of the insistence on regarding all research as if structured about the model of experimenter-subject' (p. 33), and that informed consent is often irrelevant to ethical assessments of fieldwork. He suggests that a different type of relationship underlies good fieldwork – that of mutual benefit and support, or what he calls reciprocity.

Flinders (1992) adds clarification to this change of focus by relating the informed consent principle to the utilitarian framework, and reciprocity to the deontological perspective. When discussing relational ethics, he chooses to use the term *collaboration* rather than *informed consent* or *reciprocity* since the term *collaboration* better describes the researcher's access to information: 'Collaborative relations denote shared affinity as well as interdependence' (p. 107). And within the ecological framework, he believed this notion is more appropriately termed *cultural sensitivity,* as a reminder that 'neither we nor our participants are able to exercise absolute control over such processes as informing, reciprocating, or collaborating with others' (p. 109). As discussed earlier, this consideration of an ethical issue from within different moral perspectives helps us understand where others are coming from, and why resolution is so difficult.

6. PROBLEMS THAT ARISE WHILE CONDUCTING QUALITATIVE RESEARCH

'Much of what is most controversial with regard to ethical practice follows from the researchers' conception of how they should relate to those who join them in their research projects as the researched' (Eisner & Peshkin 1990, pp. 244–245). Qualitative researchers conceive the research process and consequently their role within that process quite differently from the positivist-oriented researchers (Deyhle et al. 1992). The relationships they form with participants can often be best described as friendships, with all of the shared interdependencies a friendship connotes. It becomes difficult to separate the personal from the professional relationship, and ethical problems can arise that are either or both personal and professional in nature (Soltis 1990). Two problem areas are considered here: the avoidance of harm and maximization of benefits to participants, and the effects of the researcher's presence on the research setting.

The Avoidance of Harm and Maximization of Benefits

The principle of beneficence demands that we protect subjects from harm and increase good outcomes for them whenever possible. As Cassell (1982) points out, 'the potential harms of fieldwork typically are less immediate, measurable, serious, and predictable than those associated with other research modes' (p. 17). Adopting a Kantian perspective, she suggests that participants are more likely to be wronged than harmed; that is, they might be treated merely as means to the researcher's ends, thus discounting or violating their own essential humanity. She points out that while the notion of wronging people may have been in the minds of those formulating federal requirements, obtaining informed consent does not solve the problems of wronging participants.

Just as the notion of avoidance of harm stems from the utilitarian or teleological perspective, and the notion of avoidance of wrong from the Kantian or deontological perspective, other ethical positions present other ways of conceptualizing this issue. In a relational framework, the researcher may be more concerned about avoidance of imposition, while within the ecological framework the researcher is concerned about avoidance of detachment, that is, the failure to recognize an individual as part of a larger system to be protected (Flinders 1992).

While consideration of benefits may be weighed against possible harms within the utilitarian perspective, a person acting within a deontological framework would find, as Cassell (1982) expresses it, that this is like weighing apples against fish. While people can be compensated for harms, it makes little sense to talk about compensation for wrongs. Yet we do owe some benefits to those who assist us as participants (Deyhle et al. 1992). There may be material compensations to be made. Cassell (1982) reminds us that one benefit we have to offer is knowledge, and very often this knowledge can be used to help those researched.

Effect of the Researcher's Presence on the Research Setting

It is naïve to assume that our presence has no effect on the outcomes of our research. We are morally obliged to attempt to recognize and examine these effects. In-depth interviewing may have unanticipated long-term effects. During the interview process, for example, our questions and nonverbal responses may push the interviewee in one or another direction (Ayers & Schubert 1992). Our presence may be changing not only what is observed, but changes may also be taking place in the manner in which we as investigators interpret what is happening (Merriam 1988).

7. PROBLEMS ASSOCIATED WITH THE REPORTING AND PUBLICATION STAGE

The publication stage is perhaps the most problematic in terms of the ethical issues involved when undertaking qualitative research. During the planning stage, publication seems distant and the issues are often not foreseen and evaluated. The problem areas examined here are related to interpreting data; confidentiality, anonymity, and the use of data; and responsibilities to outside agencies and research sites.

Interpreting Data

Merriam (1988) reminds us that data collected by an investigator have been filtered through the particular theoretical positions and biases of that researcher. Investigators are sometimes not even aware of these biases, thus making it difficult to guard against excluding data that may contradict them. Diener & Crandall (1978) offer valuable advice for dealing with these issues:

There is simply no ethical alternative to being as nonbiased, accurate, and honest as is humanly possible in all phases of research. In planning, conducting, analyzing, and reporting his work the scientist should strive for accuracy, and whenever possible, methodological controls should be built in to help experimenters and assistants remain honest. Biases that cannot be controlled should be discussed in the written report. Where the data only partly support the predictions, the report should contain enough data to let readers draw their own conclusions. (p. 162)

Confidentiality, Anonymity, and the Use of Data

The basis for preserving anonymity (or confidentiality) stems from the principle of respect for persons. In qualitative research, preserving anonymity during data collection is usually not possible, so the ethical issue becomes one of confidentiality – not disclosing the participants' identities in any written reports. Even protecting confidentiality is a difficult process. As Soltis (1990) points out:

It is not hard to imagine scenarios in which the identity of those studied and reported on in articles and books can only be thinly disguised. The ethically sensitive researcher will worry over his or her personal invasion of the privacy of others. (p. 251)

During the planning stage it is important that the investigator acknowledge to participants how the data will be used, and the difficulties in preserving privacy.

A strict demand for confidentiality also prevents the researcher from giving public credit to the assistance given by participants (Flinders 1992).

A closely related issue is that of ownership of the data. I was recently at a meeting of the directors of a research center where we were told that the funding agency wanted all of the data from all research centers to be stored at some central place. There was great consternation and discussion of ownership of data, and of the need to 'cleanse' data in order to insure confidentiality before submitting it to a sponsoring agency since an investigator can lose control over data and its subsequent use once reports have been submitted.

Johnson (1982) has provided helpful guidelines for the *ethical proofreading* of fieldwork manuscripts that I summarize here:

1. Decide whether what you want to say is important enough to warrant the consequences of discovery of identities of participants.
2. Read your manuscript carefully: where words are judgmental change them to be only descriptive. If you keep judgmental statements, be clear about your purpose for doing so.
3. If you need to describe unflattering characteristics, try to generalize first and then give specifics so that a participant will be less singled out.
4. Examine your perspectives and biases, both positive and negative, and the ways in which they affected the way you viewed and wrote about the participants. How can you increase objectivity?
5. Caution the participants that it is never easy to read about oneself as described by someone else.
6. If the project is large, make certain that all members of the investigative team understand the ways in which the data can be used. Who owns the data? Who has access to it? Who has final say about publication?
7. Have several people do an 'ethical proofreading' of what you write. You might want to include one or more participants as readers.

Responsibilities to Outside Agencies and Research Sites

Much of the research we undertake is supported by grants and contracts, primarily funded by government agencies. As Smith (1990b) points out, 'tax dollars have their own kind of ethical imperative' (p. 259). One of the obligations incurred is the preparation of a final report, usually within a set time frame. This obligation can sometimes be very difficult to fulfill if an investigator does not yet feel that the data have been accurately and fairly analyzed and portrayed at the time the report is due. Projects and funding cycles seem to have their own, often conflicting, time frames over which we have little control.

There are also often obligations to report to participant groups, to school districts, to administrators, or to other groups associated with the research site. Since we are ethically bound to make any reports promised, we should carefully consider, during the planning stages, what information we will be prepared to share. But there are times when we also owe information to participants that may not have been promised ahead of time but that will be of some benefit to them. The most serious ethical dilemmas occur when we feel some obligation to share information but realize that it may be unwelcome or even harmful to the participants in the study.

8. CONCLUSION

Ethical decision making is difficult. There are times when we face conflicting ethical demands, and a decision must be made – a decision with which we want to be satisfied in the long run. Making such a decision demands both diligence and thoroughness. 'When none of the possible solutions is without serious pitfalls, one must seek the least unsatisfactory alternative, and anticipate any likely shocks, disadvantages, or discomforts, so that it is possible to live with that best of various unpleasant alternatives' (Sieber 1982, p. 7).

Researchers have responsibilities to several groups: to participants, to funding agencies or sponsors, to the scientific community, to society at large, and last but not least, to themselves. While it might seem that responsibilities to participants are in conflict with responsibilities to others, Johnson (1982) points out that if research and publication affect any one group, then all are affected in some manner.

Diener and Crandall (1978) have nicely summarized issues of sensitivity and responsibility in ethical decision-making in a statement about the ethical researcher. I conclude with their summary:

Ethical decisions are made by concerned and knowledgeable people who realize the value implications of their choices. The ethical researcher is concerned about the well-being of research participants and about the future uses of the knowledge, and he accepts personal responsibility for decisions bearing on them. The basic ethical imperatives are that the scientist be concerned about the welfare of subjects, be knowledgeable about issues of ethics and values, take these into account when making research decisions, and accept responsibility for his decisions and actions. (p. 215)

REFERENCES

Ayers, W. & Schubert, W: 1992, Winter, 'Do the Right Thing: Ethical Issues and Problems in the Conduct of Qualitative Research in Classrooms', *Teaching and Learning* 6 (2), 19–24.

Beauchamp, T. L., Faden, R. R., Wallace, R. J., Jr. & Walters, L. (eds.): 1982, *Ethical Issues in Social Science Research*, Johns Hopkins University Press, Baltimore, MD.

Bogdan, R. C. & Biklen, S. K.: 1982, *Qualitative Research for Education: An Introduction to Theory and Methods*, Allyn & Bacon, Boston, MA.

Brody, H.: 1981, *Ethical Dimensions in Medicine* (2nd ed.), Little Brown, Boston, MA.

Bunda, M. A.: 1985, 'Alternative Systems of Ethics and Their Application to Education and Evaluation', *Evaluation and Program Planning* 8, 25–36.

Capron, A. M.: 1982, 'Is Consent Always Necessary in Social Science Research?', in T. L. Beauchamp, R. R. Faden, R. J. Wallace, Jr. & L. Walters (eds.), *Ethical Issues in Social Science Research*, Johns Hopkins University Press, Baltimore, MD, 215–231.

Cassell, J.: 1982, 'Harms, Benefits, Wrongs, and Rights in Fieldwork', in J. E. Sieber (ed.), *The Ethics of Social Research: Fieldwork, Regulations, and Publication Regulations*, Springer-Verlag, New York, 7–31.

Deyhle, D. L., Hess, G. A., Jr. & LeCompte, M. D.: 1992, 'Approaching Ethical Issues for Qualitative Researchers in Education', in M. D. LeCompte, W. L. Millroy & J. Preissle (eds.), *The Handbook of Qualitative Research in Education*, Academic Press, San Diego, CA, 597–641.

Diener, E. & Crandall, R.: 1978, *Ethics in Social and Behavioral Research*, University of Chicago Press, Chicago.

Eisner, E. W. & Peshkin, A. (eds.): 1990, *Qualitative Inquiry in Education: The Continuing Debate*, Teachers College Press, New York.

Flinders, D. J.: 1992, 'In Search of Ethical Guidance: Constructing a Basis for Dialogue', *Qualitative Studies in Education* 5 (2), 101–115.

Howe, K. R. & Dougherty, K. C.: 1993, 'Ethics, Institutional Review Boards, and the Changing Face of Educational Research', *Educational Researcher* 22 (9), 16–21.

Johnson, C. G.: 1982, 'Risk in the Publication of Fieldwork', in J. E. Sieber (ed.), *The Ethics of Social Research: Fieldwork, Regulations, and Publication Regulations*, Springer-Verlag, New York, 71–91.

Kant, I.: 1959, *Foundations of the Metaphysics of Morals*, Bobbs-Merrill, Indianapolis, IN. (originally published 1785).

Krombrey, J. D.: 1993, 'Ethics and Data Analysis', *Educational Researcher* 22 (4), 24–27.

Lincoln, Y. S.: 1990, 'Toward a Categorical Imperative for Qualitative Research', in E. W. Eisner & A. Peshkin (eds.), *Qualitative Inquiry in Education: The Continuing Debate*, Teachers College Press, New York, 277–295.

Macklin, R.: 1982, 'The Problem of Adequate Disclosure in Social Science Research', in T. L. Beauchamp, R. R. Faden, R. J. Wallace, Jr. & L. Walters (eds.), *Ethical Issues in Social Science Research*, Johns Hopkins University Press, Baltimore, MD, 193–214.

May, W. F.: 1980, 'Doing Ethics: The Bearing of Ethical Theories on Fieldwork', *Social Problems* 27 (3), February, 358–370.

Merriam, S. B.: 1988, *Case Study Research in Education: A Qualitative Approach*, Jossey-Bass, San Francisco, CA.

Noddings, N.: 1986, 'Fidelity in Teaching, Teacher Education, and Research for Teaching', *Harvard Educational Review* 56, 496–510.

Office for the Protection of Research Risks: 1993, *Protecting Human Research Subjects: Institute of Review Board Guidelines*, Department of Health and Human Services, Public Health Service, NIH, Washington, DC.

Rawls, J.: 1971, *A Theory of Justice*, Harvard University Press, Cambridge, MA.

Reynolds, P. D.: 1979, *Ethical Dilemmas and Social Science Work*, Jossey-Bass, San Francisco, CA.

Sieber, J. E.: 1982, 'Ethical Dilemmas in Social Research', in J. E. Sieber (ed.), *The Ethics of Social Research*, Springer-Verlag, New York, 1–30.

Smith, L. M.: 1990a, 'Ethics, Field Studies, and the Paradigm Crisis', in E. G. Guba (ed.), *The Paradigm Dialogue*, Sage, Newbury, CA, 139–157.

Smith, L. M.: 1990b, 'Ethics in Qualitative Field Research: An Individual Perspective', in E. W. Eisner & A. Peshkin (eds.), *Qualitative Inquiry in Education: The Continuing Debate*, Teachers College Press, New York, 258–276.

Soltis, J. F.: 1990, 'The Ethics of Qualitative Research', in E. W. Eisner & A. Peshkin (eds.), *Qualitative Inquiry in Education: The Continuing Debate*, Teachers College Press, New York, 247–257.

Wax, M. L.: August 1977, *Fieldwork and Research Subjects: Who Needs Protection?*, Hastings Center Report 7.

Wax, M. L.: 1982, 'Research Reciprocity Rather than Informed Consent in Fieldwork', in
    J. E. Sieber (ed.), *The Ethics of Social Research: Fieldwork, Regulations, and Publication
    Regulations*, Springer-Verlag, New York, 33–48.

Judith T. Sowder
CRMSE, 6475 Alvarado Rd., Suite 206,
San Diego State University,
San Diego, CA 92120,
U.S.A.

PART VI

MATHEMATICS EDUCATION AND MATHEMATICS

ANNA SFARD

A MATHEMATICIAN'S VIEW OF RESEARCH IN MATHEMATICS EDUCATION: AN INTERVIEW WITH SHIMSHON A. AMITSUR

The late Shimshon Abraham Amitsur, the eminent Israeli mathematician, is known worldwide for his contributions to algebra. His outstanding achievements were rewarded with many international prizes and distinctions. Retired since 1990, he was a creative researcher until the last days of his life.

From the late 1950s, Professor Amitsur was active in the field of mathematics education. With a group of colleagues and students who volunteered to follow his lead, Amitsur devoted much of his time and energy to repeated attempts to revamp mathematics curriculum for the Israeli secondary schools. His position with regard to research in mathematics education, which is well known to the interviewer thanks to her more than a decade-long collaboration with the interviewee, seems to be typical of the community of research mathematicians. For today's researchers in mathematics education, Amitsur's opinions may sometimes sound somewhat provocative.

The interview was held in April 1994, just before the ICMI study conference and only four months before Professor Amitsur's untimely death. In the interview, Professor Amitsur commented on some of the issues raised in the Discussion Document and expressed his views on related matters.

ON AIMS AND METHODS OF MATHEMATICS EDUCATION

Interviewer: What are, in your opinion, the main goals we should strive to attain in the field of mathematics education?

Shimshon A. Amitsur: Research should focus on finding ways to improve the teaching of mathematics. In school, we try to teach two things: mathematical facts and logical thinking. The question is in what ways one may try to attain these targets. I'm not sure the two abilities can be developed in the same manner. Facts should be taught in one way, and logical thinking should be developed in another way. Moreover, different populations require different methods: Some students may learn facts easily even if they have a distinct difficulty with logical thinking; with other students, it may be the other way round.

I: Is it possible to separate these two abilities?

A: Of course not. They are not independent. I would rather say that we're talking here about two partially coinciding normal curves, or rather about one

Sierpinska, A. and Kilpatrick, J. Mathematics Education as a Research Domain: A Search for Identity, 443–458.
© *1998 Kluwer Academic Publishers. Printed in Great Britain.*

curve with two peaks. A person may be quite proficient in calculations but at the same time unable to cope with a problem requiring a little non-routine thinking.

There is also another educational objective. In a sense, it is only ancillary. It's the ability to translate real-life problems into mathematical problems, and vice versa. We should be careful not to mix up this particular ability with the capacity for logical thinking. The latter plays a central role in proving, whereas the issue of translation belongs, in fact, to linguistics. We should remember that people may differ quite substantially in their linguistic skills.

Let me add a remark on what people call 'understanding mathematics'. A number of different things go into this term. Quite often, we tend to confuse knowledge with translation skills and with the capacity for logical thinking. A student may understand mathematical facts, and at the same time be weak in proving and drawing conclusions. Such a student may be very proficient in, say, differential calculus while being unable to translate from the real world to mathematics. Without this special ability, he or she won't do too well in statistics. If, in addition, a person cannot visualize, he or she may be weak in geometry as well. We know about quite a few well-known mathematicians who were far from strong in geometry, and particularly in proving geometrical theorems.

I: Could you be a little more specific about what you have in mind when you say 'logical thinking'? Do you mean simply a mastery of deductive processes?

A: Yes, this is exactly what I mean: the mastery of deductive processes. One should keep in mind one additional factor: intuition, the ability to transfer ideas from one place to another. Be the logical skills as well developed as they might, if a person doesn't have intuition, he or she is not very likely to succeed in mathematics. When facing more than one possible path, a person without an intuition has no tools to decide where to go and what to choose.

I: What is the source of mathematical intuition?

A: I wish I knew!

I: Well, it's too easy an answer. I assume you wouldn't be a mathematician if you didn't have mathematical intuition. Thus, you should be able to say a word or two about it.

A: I think that intuition depends, to a great extent, on experience. As years go by, experience helps to develop intuition. But when you think about young people, people who have very little experience.... Well, they often do have intuition, but it is difficult to say where it comes from.

My own mathematical intuition is not general. It is limited to certain topics. I specialize in algebra, and this is where my intuition works best. I can give you an example showing how the power of my intuition goes down when it comes to other kinds of mathematics. I once wrote a work in analysis. Some time later, a fine mathematician who specialized in number theory took the same formulae I used and without a visible effort solved problems on which I had been working unsuccessfully for a long time. He had an intuition that led him in a completely different direction. Questions about intuition – where its roots are, in what variety it may come, how it develops – all these questions are among the topics which should be explored by those doing research in mathematics education. It would be very interesting to know if, indeed, intuition comes in variety of colors and shapes.

ON THE GOALS OF MATHEMATICS EDUCATION RESEARCH

I: Speaking about research in mathematics education, when were you exposed for the first time to the subject?

A: In the beginning of the sixties.

I: What was the source of your attraction to the field?

A: I participated in the project organized by Professor Frenkel. The university faculty was having a series of meetings with mathematics teachers in secondary schools. On this occasion, I realized that there was a serious problem with our curriculum. It was obsolete. I could see no connection between mathematics (as I understood it) and what was taught in schools. I decided to try to introduce some changes to the program. I succeeded in getting the support of the mathematics inspector in the Ministry of Education and together with him started meeting with teachers. And when you meet teachers, you suddenly grasp that there are different ways of thinking and different ways of seeing mathematical concepts. It was then that I realized the potential importance of research in mathematics education. I felt it could be useful to investigate these differences.

I: Are the ways teachers think about mathematics substantially different from those of mathematicians?

A: No. But there are many differences between the teachers themselves. You will find a wide variety of thinking styles and of conceptions – and this is true about mathematicians as well. I wish I could know more about it in order to be able to find the right way to 'talk mathematics' to everybody. In fact, we may assume that the student is a mirror of the 'upper system' – of his or her teachers. Thus, the more we know about teachers, the more we know about learners.

During this period, I used to go to all kinds of lectures on education. More often than not, it was a bitter disappointment. Many sophisticated terms were thrown on the audience when the lecturer tried to analyze different types of conceptions and ways of reasoning. Nobody explained, however, what one is supposed to do with all this, how all this was to inform practice.

The claim that there are many different types of mathematical thinking is, of course, a matter of guess and of rough assessment. But mathematicians feel it is true: there are different kinds of mathematical minds. We know that there are 'geometricians' – people who turn out particularly clever when it comes to geometry. There are 'analysts', and there are 'logicians'. The different types do not always listen to each other, and more often than not they are unable to communicate. In fact, every good mathematician is endowed with a mixture of the different abilities, but the proportions vary. In those days, in addition to our worries about the curriculum, we thought it would be useful to find principles which could help us to distinguish between the different types of mathematical minds and give us tools to diagnose them. Our aim was to find a kind of formula with which we could come to the teachers and say: 'Listen, this may help you understand the situations which are likely to arise in your classroom. This is going to help you in coping with these situations.' We wanted them to realize that there are different types of students, and that their different needs should be taken into account in classroom teaching.

I: You said that your interest in research stemmed from your interest in curriculum. Were you actually involved in any kind of research?

A: No, I wasn't. All I did was an analysis of teachers' reactions to our new curriculum. I used to talk to the participants of the experiment. These people volunteered to our project. They were really good teachers. They opened our eyes to the problems which are likely to emerge in the classroom.

I: Did you intend to turn all this into a systematic research?

A: No, I had no such intention. My only aim was to bring about positive attitudes toward the curriculum. Typically, mathematicians believe that all one has to do to improve teaching is to introduce a change in contents or in the way things are presented. They seem to ignore the question of the teachers' attitudes towards these changes. They also tend to forget that teachers work under much pressure and many social constraints.

We tried to put our ideas into practice, and we concluded that it was possible. First, we approached the good students. I promised myself that no harm would be done to the students even if the experiments failed. With the support of the Ministry of Education, we organized special matriculation examinations for those who participated in the project. We were careful to build the exams so that they wouldn't include anything that wasn't a part of our program and that their level of difficulty would be what the students were used to. We held weekly meetings with the teachers. Their reactions were very interesting. There were teachers who had a very good sense of the whole system and were able to appreciate its weaknesses and strengths. But the final decisions were ours. Only we could decide which subjects had to be learned and what contents should be included. Some of the teachers objected to our approach, and we made it possible for them to voice their disagreement. We had some very tough, fierce discussions. Sometimes, people would lose their temper and start shouting. In the end, we usually found a compromise, and if not, then we would just impose our decisions on the teachers – we had no choice.

I: Did you try to document all this in any way?

A: No, we didn't. Today I regret it. But we just didn't think about keeping any kind of records.

I: Are you acquainted with recent literature on research in mathematics education? Are you used to reading on the topic?

A: Only if I am forced to. In the beginning, I used to read quite a lot, but I found most of the papers non-significant and uninteresting. They tackled such questions as how one can prove a mathematical fact in a simple way, or whether it is possible to teach a certain subject in school. I wasn't interested in this kind of work. It dealt with marginal details, whereas I was looking for an analysis of the whole system. This I didn't find. But in those days I read quite a lot.

I: The words 'those days' refer to the seventies, right?

A: Yes, more or less.

I: A radical change has taken place in educational research since then. If you read some more recent works, you would probably have difficulty recognizing them as belonging to the same genre. I'm not sure you would like it more; still, it's something different.

A: Last year I participated in an international conference on research in science and mathematics education in Jerusalem. I listened to quite a few talks. Many times I could feel that the presenter somehow missed the point. For example, some people reported on having discovered certain phenomena. And then, instead of trying to pinpoint the sources of these phenomena, they would discuss techniques for better teaching. But it may well be that the phenomena they were talking about were natural, and it would be a mistake to try to fight them. Rather, one should try to use them.

I have also some questions about the method. I'm not sure about the soundness of the techniques of interview and statistical analysis. And there is one other point here. After all, the students are not just guinea pigs. One cannot use them and then throw them away. The mathematics they learn while participating in teaching experiments is the only mathematics they will ever get. One must remember that. Of course, this puts quite tight constraints on educational research.

I: It's like experiments in medicine.

A: Exactly. Except that in medicine one can still use animals before going to people. Here, in the research in mathematics education, people are the only 'material'. It's the researcher's duty to develop such methods of study that will have no effect on their lives.

I: You mean, they should not have an *adverse* impact, right?

A: Yes. But one shouldn't assume a positive influence either. This is a basic principle. Indeed, if you assume that there is a positive influence, your whole study becomes biased.

I: Is there anything like a 'clean' research, one which is free from biases and pre-judgments (except, perhaps, for the research in mathematics)?

A: Of course there *are* studies which are completely unbiased.

I: But people always start with certain assumptions, prejudgments...

A: This is not what *biased* means. Good research is not biased. I could bring quite a few samples of biased research. Such research is used by politicians all the time. When we wrote our textbooks, we had certain goals in mind. Thus, somebody may say that while we were writing our texts we were guided by a certain intention, and therefore we were biased. But the real question is not about the contents you choose but about the way you analyze and evaluate the outcomes. It is here that you can be really biased. Indeed, in this sense we are usually biased, at least to some extent. If a teacher didn't succeed in teaching new material, we often say that it was his fault: he wasn't prepared to introduce the changes, he was biased himself. These are natural difficulties that a researcher should try to overcome somehow.

I: And how about the aims of the research in mathematics education? What is the role of this research?

A: First, there are theoretical aims. For instance, we may be interested in individual differences in the ways people think about mathematics. We may ask, for example, which of two paths is usually more effective in learning mathematics: the one that leads from concrete and detailed to abstract and general, or the one that goes the other way round. We may wish to know how the different cognitive preferences are distributed in the population. These are the theoretical goals of research: We try to find rules of mathematical learning and thinking. But there are also more practical aims. The research should provide answers to questions which arise in connection with the preparation of teachers to the introduction of new programs. Empirical studies should provide information about the possible difficulties. On the basis of this information, one is in a position to warn the teachers that they may come across different kinds of difficulties. The teacher must be aware that the source of these problems may be not so much in students' lack of understanding as in the fact that different people think differently.

I: I conclude from this that you see educational research as important and necessary.

A: I don't believe in research which tries to answer the question how to teach a certain subject, or whether the students can or cannot learn this concept or another.

I: But you did say that one does need research, sometimes. What kinds of studies do you regard as worthwhile?

A: The ones that can really help teachers and curriculum designers. The results of research should help in planning teaching and building curricula. The outcomes of theoretical and empirical studies should never be the only, not even the principal, basis for decisions about instruction. Curriculum planners are guided by goals which must be considered as primary to any other consideration. Research only deals with constraints of which one should be aware when one comes to realize these goals. I am quite extreme in this view. Research can only provide us with some basic information about prospective difficulties, and it can guide me in the choice of tools. However, it cannot dictate us what to teach. I won't give up important topics of study only because research tells me that somebody tried to teach it and the students haven't grasped it.

I: Do you recall any kind of finding in mathematics education that you would be prepared to call 'a result'? Have you ever felt that something that has been said by mathematics education researchers opened your eyes, taught you a new thing?

A: I can recall almost nothing of this sort. I say 'almost' because something like that might have happened, once or twice. I don't recall right now, but I guess that I did learn from this research about a phenomenon or two of which I wouldn't be aware otherwise. These things might even have had a certain impact upon my thinking in the longer run. The point is, most of the studies I am familiar with aim at showing that certain things are unlearnable, and this is something I cannot put up with.

In my eyes, the main reason for research in mathematics education – and in this respect the case of mathematics education is quite different from that of science education – is that mathematical thinking is something unique, peculiar and unlike all other types of thinking. It is substantially different even from scientific thinking, although people view these two as almost the same. Mathematical thinking is unique in that it is completely abstract, non-empirical. You never know where you're going to when you do mathematics. Investigating this very special type of thinking is certainly a worthwhile endeavor.

ON THE RELATIONSHIP BETWEEN RESEARCH IN MATHEMATICS EDUCATION AND CURRICULUM DEVELOPMENT

I: This brings us to the question of the ways in which new mathematics curricula are built. What are your criteria for deciding whether a certain mathematical topic should be included in the school program or not?

A: This is a very complex problem, and there are two sides to the story. On the one hand, there is a question of the factual knowledge which the students should acquire; and on the other hand, there are the abilities and skills that the learners have to develop. The issue of factual knowledge is relatively straightforward. I have no difficulty with deciding and saying explicitly what should and what shouldn't be included. The second component is more elusive. When we talk about it, it is very important to keep in mind the basic fact that has been acknowledged for a long time now: For certain pupils, learning mathematics provides the only opportunity to exercise and develop their thinking and reasoning skills. In this sense, mathematics classes are similar to sport lessons: The latter are for many the only occasion for physical exercise and for the improvement of their physical skills. The students won't get the necessary intellectual practice just from reading history books or even from studying physics. The question, of course, remains open whether there is a transfer from one domain to another. It seems plausible that transfer does occur, at least in some of the students. How can one bring it into open? This is, once again, one of the most intricate questions.

I: When you say 'transfer', do you mean a transition to domains other than mathematics itself, or are you talking about transferring skills from one topic to another within mathematics itself?

A: Both in mathematics and beyond. For example, one can expect mathematical skills to be transferable to physics and to chemistry – two fields in which one must use mathematical systems.

I: I guess you refer mainly to logical-deductive reasoning.

A: Yes. Logical thinking, formal thinking.

I: When you decide about contents, about subjects to be taught, what means do you use to estimate their possible impact on students' thinking and reasoning?

A: I must say that in the past, when we developed our curricula, we always appealed to our intuition rather than to any systematic data or explicit arguments.

We had no background knowledge on which to base this kind of estimation – thus our recourse to intuition. For instance, it is a well-known fact that geometry has always been used by mathematicians themselves and that it has been attracting many students. On the other hand, we knew that geometry has the highest failure rate. We decided, therefore, to divide the students into two groups: those who will be learning geometry and those who won't.

I: Why teach geometry at all? Only because of its presumable impact on students' thinking?

A: If I could be sure that certain students cannot learn the subject, I would probably give up trying to teach it to them. However, nobody has managed to prove, as yet, that there are students who gain nothing out of geometry.

I: Would it be right to say that the reason for your insistence on teaching geometry is your belief in its potential contribution to students' reasoning skills rather than your concern about students' geometrical knowledge?

A: Indeed, I am not so much concerned about students' knowledge of geometrical facts. One doesn't need a systematic course in Euclidean geometry to learn them. One gets acquainted with geometrical facts through visual heuristic means and through problems in planimetry, stereometry, and trigonometry. The facts may be learned without solving problems that require more than technical skills.

Regarding the ways in which the contents of curricula are decided upon, everything depends on the purpose. If you believe that the students need a language and techniques that will help them in science, then you know this is what they should be taught. If research in mathematics education could give us a precise estimation of the percentages of people who have a preference for different ways of thinking, than we might well have changed some of curricular components. We would change them in such a way that, while the objectives remain the same, the route toward them would be different.

I: This leads me to a question I wanted to ask anyway. You said at a certain point that research cannot alter the contents of a curriculum. On the other hand, if you aim at an improvement of reasoning skills, then only research can tell you what topics are the best means toward this goal.

A: Let me help myself with a metaphor. Assume a group of people has to climb a mountain. First, they undergo a physical examination. As a result, it is possible to know in advance that there are people in the group who will be able to make it to the top very easily, almost running. Some others will have to be helped and led at a slower pace. Still others will only make it if they rest every half an hour. Finally, some people will need a very long time to climb the whole mountain. Medical research may provide us with information about people's ability to make the effort, but the goal remains the same all along: to reach the top.

I: All this is in force provided you believe that reaching the top is, indeed, your only purpose. However, somebody may aim at a different target. For example, he or she may wish to give the people in the group a good exercise in order to increase their fitness and improve their physical skills. Mountain climbing is chosen as a kind of activity which is likely to help them with that. In this case, it seems that the mountain itself should be selected according to the information provided by medical research, don't you agree?

A: There may be, indeed, a difference in our views on the aims: You may aim at one goal while I am aiming at another. Anyway, there are accepted contents of mathematical curricula, at least at the elementary and secondary levels, if not in the universities. It is our duty to prepare the students for future studies and for the life in the modern society with all this advanced technology around. These are the considerations which determine my goals.

I: Geometry is perhaps the most salient example of a subject that was given up by many mathematics educators as a result of research findings. It was a general agreement that the main goal in teaching geometry is the development of students' thinking, and the research – first that by van Hiele and then by many others – has shown that the teaching and learning of the subject doesn't bring the desired results. This is why there is a tendency today in the world to turn away from geometry in school curricula.

A: Initially, we also thought about such possibility, but then we arrived at the conclusion that it would be a great loss. I got well acquainted with the van Hiele model, I once knew it in all its details. Today I'm no longer sure that their theory is true. I'm not convinced that the kids who haven't developed geometrical intuition would be able to do any better in other subjects. The van Hieles remain silent on this point.

MATHEMATICS FOR ALL: WHY AND HOW

I: It's my impression that you're convinced beyond any doubt about the necessity of teaching mathematics to everybody. Why?

A: Indeed, I have no doubt about that. I believe that learning mathematics is equally as essential as learning reading and writing. My argument is simple: Today, information is conveyed to us through the language of mathematics. It's our duty to prepare young people for life in modern society, and mathematics is the language in which the modern society speaks. The problem is that through mathematics we also wish to teach logical thinking – no better tool for that has been found so far. As I said already, I have no means yet to distinguish between the students for whom this way to teach thinking may be successful and those for whom it wouldn't work.

I: Who should play the leading role in planning mathematics curriculum? Mathematicians? Teachers? Researchers?

A: Mathematicians, there can be no doubt about that. My experience taught me that teachers and researchers are too cautious, too susceptible to fear and doubt. They know that nothing will bring satisfactory results, that anything we try is bound to evoke negative reactions and social pressures. This awareness makes them unable to force the system to implement changes. Thus, they would give up trying, and later they would bring you all kinds of excuses for the fact that the things didn't go the way they should. Our experience shows, however, that when one works with good teachers, then anything is possible, and even the most difficult of changes may be successfully implemented.

I: You point out the problem of resistance to change on the part of the teachers. Do you believe that the frequent failure with curricular reforms results from teachers' refusal to take part in the endeavor?

A: Yes, but researchers should also be blamed. They resist changes no less than the teachers. In our case, they didn't do the job they were expected to do: They should have investigated the reasons behind our failures. They should have looked for the obstacles that prevented us from reaching our target. But they have not done it, and this is how we missed the opportunity to learn about the ways in which the seemingly unfeasible could have been turned into the workable.

I: In spite of all this, don't you think that there is still room for researchers in the process of building mathematics curricula?

A: Yes, there is room for researchers, provided they serve only as consultants. If I wish to build my own business, I'll take a consultant to advise me on different matters; however, the consultant wouldn't tell me what to do, and all the decisions will be mine.

I: The New Math movement that began in late fifties was initiated and led by mathematicians. Everybody agrees today that all this ended in one of the most spectacular failures in the history of educational reform.

A: From the very beginning, we objected to the excessive formalism typical of the new math programs. True, we tried to introduce the concept of set very early, in order to make it the basis for everything else. But soon we (or at least I) started to have doubts about this as well. Definitions of *set* and *function* require so much time.... Intuition, no doubt, works much better than rigorous formal thinking.

I: During all these years we have been working together, I noticed time and again that when you examined teaching materials, you used to label them in a peculiar way. Your verdicts are often something like 'this is mathematics', 'this is not mathematics'. I always asked myself about your criteria for this kind of judgment.

A: Let me answer with an example. Somebody teaches inequalities and draws a graph. This is an exercise in finding a visual expression for a mathematical phenomenon. However, if I wish to solve an inequality or to prove a theorem, I cannot use these tools. My final goal is that students are able to do all this by mathematical means only. When I say 'this is mathematics', I know exactly, even if not explicitly, what kind of thinking is involved in the process at hand, what are the aims of whatever is being done and what kind of facts may be necessary. When we teach mathematics, we often have recourse to phenomena that might be interesting, but are not truly mathematical.

I: Like, say, the relationships between algebraic and graphical aspects?

A: Yes. We have to teach it, but mainly in order to give the student an opportunity to get acquainted with the everyday mathematical language.

I: Don't you agree that graphical representation may improve students' understanding quite substantially? Research seems to show that, indeed, the ability to translate from one representation to another is one of the central factors in mathematical understanding.

A: Well, I don't know.... It may be true, ... but when I emphasize this ability at the expense of all others, I am in danger of inadvertently skipping some important stages in a pupil's mathematical development. It is definitely not enough to be able to discover mathematical facts with the help of graphs. One has to be able to prove the facts in truly mathematical ways. He must also be able to draw conclusions.

I: I see. You demand that in addition to 'laboratory observations' the students are taught to implement more or less complex deductions, like those performed by 'grown up' mathematicians.

A: Yes. The observations may certainly help the student, exactly like seeing shapes and bodies helps in geometry. It's definitely true. I am all in favor of using this additional sense – the sense of seeing – which in the past didn't play almost any role in mathematics. But in mathematics proper, in mathematics itself, what I really have to know is something different: it is how to draw conclusions from things I know about things I don't know yet. This cannot be done with pictures and other visual representations. In mathematics, I teach not only facts, but also special ways of thinking. The student must know how to construct proofs – proofs that won't depend on the specific drawings produced by the particular computer I use.

ON THE POSSIBILITY OF MUTUAL UNDERSTANDING BETWEEN MATHEMATICIANS AND MATHEMATICS EDUCATION RESEARCHERS

I: As we both know only too well, research in mathematics education is still regarded at some universities as but a distant relative of the 'true' science, as something that only aspires to the title of academic discipline. This is certainly a view held by many mathematicians. Is this opinion justified? Can anything be done about it?

A: To strengthen the status of the research in mathematics education, one has to prove its usefulness. The onus of proof is on the researchers themselves. They have to show that they have a theory of mathematical thinking which convincingly explains observed phenomena. Only when they can provide such a theory will mathematics education turn into a true academic discipline. Not even one day earlier.

I: You say that mathematics education researchers have yet to convince you that their efforts are worthwhile. I wonder, however, whether you, mathematicians, are really open to the possibility of being convinced. What kind of proof would be acceptable to you?

A: I am acquainted with quite a few types of research in mathematics education. None of them convinces me, and let me explain why. Some of the work I've seen is the same kind of work each teacher does in her classroom. She would teach a subject for one year, and, since there always are students who have difficulties, she would decide to teach it differently the next year. Then there would be another student with a different kind of difficulty. These are the stories of individual students, while what I need is information about the distribution of different ways of mathematical thinking in the population. The studies of errors and misconceptions also seem problematic. I am not always convinced that what the researchers call *error* or *misconception* should indeed be called this name. To me it seems more plausible that the student is just unable to relate a given concept to a new context, a context to which the things were never applied before.

I: In spite of the difficulty, I would like you to try to clarify what you have in mind when you require 'proof' of usefulness. After all, a mathematician and mathematics education researcher may well have quite different things in mind when they claim that they have proved something. I can imagine a researcher coming to you and saying, 'I think I have shown this and that, here is the evidence', and you would say that what he calls evidence doesn't convince you.

A: I won't accept it if – as it happens only too often – some important part is missing. This is supposed to be evidence, but for what? For the fact that a certain phenomenon exists? For me it is not enough. I want to know what one can do to change the phenomenon. Just saying 'These are the facts; the students do not understand this topic' is not what I need. I don't want to be told by the researchers that some topics are unlearnable.

I: And how about the methods? What are your requirements regarding the way in which research is done and reported?

A: If a study in mathematics education is to be called scientific, it must bring more than a description of facts and analysis of results. It is very important that it is done in such a way that there is a possibility of control and of verifying the results.

I: You seem to be talking about something we used to call a 'scientific method'. Is such method possible in mathematics education?

A: If it is possible in psychology, I can't see why it shouldn't be possible in mathematics education.

I: How do you define scientific method?

A: As I said, an experiment, an empirical study, in order to be scientific must be controllable. True, it is not too easy to say what *controllable* means. In medicine, in chemistry, you just require that if other people repeat the experiment, they get the same result. In mathematics education, it is somewhat more complex because there are so many parameters, and some of these parameters cannot be the same each time. Thus, in mathematics education there is a need for a special definition of *controllability*.

I: Indeed, we know only too well that no experiment in mathematics education is fully replicable.

A: Fine, so it is another thing mathematics education researchers are obliged to find an answer to: What does it mean that an experiment is controlled? This means a certain type of replicability.... . This means that all the relevant parameters are explicitly dealt with so that another person may come and suggest how these parameters might be changed or reproduced.

I: You seem to be convinced that all these classical attributes of what we call 'scientific method' are attainable in mathematics education research, and it is only a matter of finding appropriate definitions and tools. Is that right?

A: If it isn't possible, then there is no room for this discipline. But I am sure it is possible, or at least one may find a satisfactory approximation.

I: So, if the researchers want their field to be recognized as a fully-fledged academic discipline, they have to come up with a special kind of 'scientific methods'. Incidentally, does it mean that you expect this kind of research to be practiced within mathematics and science departments?

A: No, not necessarily. In the Discussion Document, Balacheff is quoted as saying: 'Most of us want to develop this research field within the academic community of mathematicians.' I think it is a great mistake. The subject matter of research in mathematics education and the subject matter of mathematical research are two completely different things. Mathematics involves processes of discovery that use a very special kind of rules and techniques. All these have nothing in common with the rules and techniques of mathematics education that deals with human beings. I once had a discussion with a lawyer. I accused him of trying to be too strict and precise in his definitions. He was surprised. 'Why, isn't it what you mathematicians like and expect?' To which I said , 'Oh no, not in this case! We mathematicians do not kill people with our formulas'. One must always remember that there is this important difference also between mathematics and mathematics education.

I: I don't think Balacheff would disagree with that. The only thing he was trying to say is that since research in mathematics education has much to do with mathematics itself, it is important that the communities of mathematicians and mathematics education researchers coexist peacefully and collaborate with each other. This can only be achieved through recognizing and respecting the differences.

A: As far as recognition is concerned, I already told you: The research in mathematics education has yet to prove itself, and the onus of proof is on mathematics education researchers.

As to Balacheff's suggestion, keeping mathematics education within mathematics departments may have a certain advantage, after all. Indeed, mathematics education cannot be separated from the world of mathematics. The question is, who is the father of the subject, who should claim priority in decisions and responsibility. The case of mathematics education reminds me of that of computer science a few decades ago. The latter was placed within mathematics departments, or in departments of statistics (and never in the social science departments) only because it has a strong mathematical and statistical component. But once a research subject comes of age, it may be treated as an academic discipline in its own right, and there is no need anymore to tie it to the existing departments. Mathematics education may still be too young for this.

Anna Sfard
The University of Haifa,
Haifa,
Israel

RONALD BROWN

WHAT SHOULD BE THE OUTPUT OF MATHEMATICAL EDUCATION?

The question of 'output' is a good place to start in considering an activity. The quality of mathematics education is important for the future social and economic strengths of our countries. We are also all interested, for example as taxpayers, in the output of activities which are publicly funded. How should we judge this?

The Washington conference was meant to discuss the value of research in mathematical education. But can we give reasonable expectations for research in mathematical education if we do not know what we can reasonably want from mathematical education itself? Otherwise, it might even be suggested that an aim of a conference on research in mathematical education is to improve the status of the research and of the researchers.

To write a full and scholarly view of output as suggested by the title would be a very ambitious undertaking. I am also conscious that a good deal of the readership for this article will be people who have had experience of teaching in schools, of teaching the teachers over many years, and who are well aware of the work that has been done in mathematical education. It seems advisable therefore to talk about the areas in which I have had experience, and leave the readers to debate the possible analogies and relevance to wider issues.

The use of the word 'should' in the title is normative, and might suggest that I think I know all the answers. To the contrary, what I hope this article will do is direct attention to some interesting questions.

In discussing 'output' I am going to start at the 'top' end, namely that of research in mathematics. This may at first seem foolish, since any comprehension of what goes on in research in mathematics is generally thought to be way beyond the needs and comprehension of the vast majority of those who go through the educational process.

Against this I would set the argument that it would be desirable for there to be a clearer view in mathematical education as the nature of the mathematical beast, in order to know how to deal with it. If it is unclear as to what is mathematics, what are its main achievements, and what constitutes performance in it, then what hope is there of teaching it in a clear way, or of coming up with new practical hints as to how it should be taught more effectively?

I wondered after attending the Washington conference whether this itself was not the crucial unasked question, which required debate. One speaker referred to the paucity of practical hints which were available. It is more difficult to give practical hints for teaching an activity not itself well defined. Not all the

Sierpinska, A. and Kilpatrick, J. Mathematics Education as a Research Domain: A Search for Identity, 459–476.
© *1998 Kluwer Academic Publishers. Printed in Great Britain.*

resources of psychology, linguistics, or philosophy, will help education and training in the sport of, let us invent one, say, Yarangoo, if the rules of Yarangoo, and the method by which one wins a game of Yarangoo, are not understood in the first place. For this, it does not matter whether the level of teaching is that of a beginner or a first division player.

I find some support for this view from articles I recently came across in *Mathematics Teaching*, December, 1986, 'Special Issue: The roots of mathematical activity'. Rafaella Borasi writes that: 'Several recent research studies investigating the difficulties that many students encounter in learning mathematics have suggested that what students believe to be the nature of mathematics may influence considerably their possibilities for success in the subject' (Borasi 1986). Phil Boorman explains how his outlook on teaching mathematics is shaped by his view of mathematics as the study of pattern and structure: 'Now, if I am right in defining mathematics in this way, everyone is very, very good at "doing" mathematics – if they were not they would be dead – for all thought is mathematical' (Boorman 1986).

Here is another example of how a view of mathematics can affect teaching and assessment. A colleague in London, say Fred, tells the story of how his son, say Bob, was given as an investigation the problem of determining the number of diagonals of a regular polygon. Bob and two friends came in one day when Fred was watching television and started discussing the problem. They tried a few low dimensional cases, came up with a general formula, tested it out for pentagons, saw they had got it wrong and needed to divide by two, and so arrived at the correct formula. Fred was delighted with their progress. The teacher was not so delighted. One of the boys just wrote down the answer and got 0. It appeared they were supposed to write a nine page project, testing special cases, drawing graphs, and so on. As the teacher explained: 'They had to learn: There is more to mathematics than thinking.' None of this had been explained before they started the project.

Since the Washington conference was broad in the areas of study which were used to try and illuminate the question of what should be research in mathematical education, it is right that mathematics itself, how it is done, what is its value, what it advances towards, and the way it advances, should have some pride of place. Otherwise, we might have Hamlet without the prince.

A further point is the psychological truism that the behavior of people is more similar if they are compared at the limits of their ability. The struggles that we have to understand and master advanced new ideas could give some sympathy for those who are also struggling at a much lower level of performance, and perhaps suggest some ways of helping their struggles.

An advantage of the research viewpoint is that it counters the assumption that there is always a 'right' mathematics, and that the problem is necessarily to get people to behave in the way assumed by this mathematics. To the contrary, from the research point of view, we are especially interested in those parts of mathematics where we feel uncomfortable, since this feeling could be a pointer to a different approach being required. It is easy to forget the way in which that which was once research becomes, maybe over centuries, part of everyday mathematics, and so a part of mathematics teaching. Sometimes mathematics teaching is forced into the latest view of 'modern' mathematics, to its detriment. Views even on basic mathematics are changing, not only partly in response to research needs and the developing language to describe new kinds of structures and the ways in which structures interact, but also in response to new understandings of what mathematics is about and how it works.

David Tall in the abstract of his ICMI lecture at the International Congress of Mathematicians in Zürich in 1994, writes: 'There is thus a significant difference between the flexible thinking of the mathematician and the form in which the product of that thought is eventually communicated.' One of the ways mathematics progresses is in the closing of this gap, by finding forms, structures and languages in which the way we think can be more properly, more accurately, and more understandably expressed. These new forms then allow for more elaborate patterns of thought, intuition, calculation, and deduction. For example, mathematics was held up for centuries for lack of the 'trivial' concept of zero. (I owe this point in this context to A. Grothendieck, the twentieth century master of the development of concepts.) Were children once beaten for not being able to add properly in Roman numerals?

As an example, a striking change in basic thinking in mathematics is the move away from set theory as a 'foundation' for mathematics, and a realization that a more flexible and intuitive approach is needed (cf. Brown & Porter 1994). The tools for this move are provided by category theory. In this method, emphasis is placed not on set theory, and not on the 'elements' of a set, since these are somewhat counterintuitive when one comes to large sets such as the real numbers. Instead, emphasis is on the functions, and so on the relations between structures, through the homomorphisms between them. The constructions that can be made on objects are defined by the relations of these constructions to all other objects. In this view, functions are removed from the passive role as a set of ordered pairs, as a logical device used as part of a thrust to give mathematics a 'safe' foundation, and instead functions return to their key and intuitive role in mathematics; the expression of motion, of change.

Thus the difficulties which pupils, students, and professors have, and will always have, in learning concepts, procedures, and skills in mathematics is one aspect of a process with two variables: the learner and the mathematics. The processes of teaching, and so of trying to understand in order to explain to others, have often been a stimulus to the development of mathematics. It would be fascinating if philosophy and the social sciences of education, linguistics, and psychology, which figured so much in the discussions at the Washington conference, can help in this development.

1. THE OUTPUT OF RESEARCH IN MATHEMATICS

It could be surprising to you that there might be any question about this. However the public and Government are, in general, unclear as to what mathematicians produce and how valuable what they produce is. Mathematics undergraduates, even very able ones, are often unclear that any research goes on in mathematics. This is not surprising since they are usually not required to look at original papers, nor to acquire any understanding of current work in progress.

Whereas astronomers, physicists, biologists, engineers, chemists, and so on, have worked hard to convince the public and Government as to what they are doing, what is its value, and what constitutes a significant advance, it is less clear that this has happened in mathematics. The biggest splash recently has been for the solution of Fermat's last theorem. Thus the solution of famous problems is often advertised as the main success of mathematics, and this view is encouraged by mathematicians. Certainly the solution of such problems will bring fame within mathematics to the solver. What, though, will be the effect on science, technology and the general public?

The history of mathematics shows that the contribution of mathematics to science and technology has been, not so much in the solution of its own classical problems, but instead to provide a precise and developing language for the invention, representation, and discussion of certain concepts and relationships, together with a mode of deduction, verification, calculation with and exploration of these.

In places where mathematics can be applied, it has been able to say: This is true, and that is false. This has enabled mathematics to reveal astonishing elaboration of patterns and structure, to provide tools for applying these structures, and to show new problems. The fact that mathematics is, if you like, the science, craft, and art of pattern and structure explains why it underlies so many other scientific and technical areas, which are themselves seeking to understand the patterns and structures in nature.

We should avoid, though, the idea that the interest of mathematics rests on these applications, rather than the joy of the investigation itself. Indeed, the structures which have been forced on mathematicians by the logic of their arguments have sometimes seemed weird and strange to others, and an argument for the weirdness of mathematicians. Yet some of these structures have later, many years later, found their true place in applications. A good example is the theory of fractals, and of chaos. The notion of fractal compression is being developed to compress data in hard disks of personal computers, and so to become in world wide use.

Within mathematics there has long been a need to decide what is good mathematics. This need occurs most crucially at the sharp end of publication. Authors have to decide for which journal their work is good enough. Referees and editors of journals have to decide whether or not a submitted paper should be accepted, and while competition for space in the top journals continues to increase, this decision has to be made on the basis of which are the 'best' papers. Yet the question of what is good mathematics is little debated at the professional level, and an understanding of this is not explicitly part of a qualification for an undergraduate degree or even postgraduate. By contrast, students of, say, design or musical performance are introduced to this question as a basic object of their studies.

To give some focus to these questions, I would like to give some account of what I have been doing mathematically for the last thirty years. In any case, I like explaining this background and the way this research has gone has influenced my views on mathematics as a whole. There has been a 'reach for the stars' aspect of this research, and it has been satisfying that some basic foundations of this route have now been firmly laid down.

After my Ph.D. in topology, not quite sure of the direction I wished to go, I embarked, as a kind of displacement activity, on writing a topology text. Its principal aim was to explain the main basic results in algebraic topology, particularly the notion of a cell complex as a way of representing a space as constructed out of 'nice' and comprehensible bits, namely the 'cells' or 'balls', and the use of the so-called fundamental group as a topological invariant.

A crucial part of this theory is the calculation of the fundamental group, in which a main tool is what is known as the Van Kampen theorem, first found in the mid-1930s. It shows how the fundamental group of a big space can be obtained if the space is the union of two 'nice' parts, such that the fundamental groups of these parts, and of their intersection, is known. However, to use this theorem, the intersection of the parts has to be connected, to have only one 'piece', and this prevented the theorem from being used to calculate the fundamental group of a basic example of a space, the circle, where the result is the additive group of integers. This example had to be determined by another method, the use of the exponential function from the real line to the circle, which wraps the line around and around the circle, like a rope around a bollard. Of course, this wrapping method is a nice, intuitive, and important method, but still I found the diversion unaesthetic.

In 1965 I came across a 1964 paper of Philip Higgins on the applications of groupoids to group theory. The notion of groupoid was introduced by Brandt in 1926, as a tool for extending important work of Gauss on the composition of quadratic forms, from the case of two variables to that of four variables. A groupoid should be thought of as a group in which the multiplication is not everywhere defined, so that many identities are allowed. Intuitively, the notion of groupoid corresponds to that of traveling between many points, so that a journey from London to New York can be composed with one from New York to Tokyo, but not with one from Washington to Montreal. By contrast, with a group, you always return to the starting point. (The word groupoid is, unfortunately, also used for the quite different notion of a set with a binary operation.)

The notion of groupoid was found in topology as the fundamental groupoid. It was recognized in the 1950s that groupoids gave a nice account of one aspect of the theory, that of change of base point. Perhaps this aesthetic feature should earlier have been taken as a clue that something potentially important was going on. However, groupoids were regarded as something of a curiosity, since the real interest was felt to be in the widely used notion of abstract group, recognized as one of the central concepts of mathematics.

To my surprise, I found in 1965 that the notion of groupoid solved my expository problem, since the Van Kampen theorem extended neatly to the fundamental groupoid, and so yielded an elegant determination of the fundamental group of the circle. Indeed, one obtained a simpler proof of a more powerful theorem. Later, in 1967, a famous analyst, George W. Mackey, told me how he had been using groupoids for years in ergodic theory. All this convinced me that the thrust of my book should be on groupoids, and that this area of basic theory was most naturally expressed using that term and that language. This became the pattern of the book (Brown 1968, 1988).

The excitement of this extension from groups to groupoids was the very wide and important uses of groups in mathematics and science, particularly in applications of symmetry. The obvious question raised by this was: To what extent can parts of mathematics and science which use groups be better served by using groupoids?

I was brought up in homotopy theory, which studies higher dimensional versions of the fundamental group, called higher homotopy groups, basically by replacing what might be thought of as loops of strings, by their higher dimensional versions, the n-spheres, which for $n = 2$ are just the surfaces of balls. Could there be higher homotopy groupoids and a higher dimensional version of the Van Kampen theorem? By now I was very familiar with the proof of this theorem, since I had written it out say five times in various versions. It seemed quite clear that if one had the right language of higher homotopy groupoids, then the proof of the Van Kampen theorem would generalize, at least to dimension two rather than one, and probably even further. So I had an outline proof in search of a theorem. Unfortunately, the theorem itself could not be formulated because the statement required several concepts which had not been defined, and which indeed were unclear.

This started me out on a long road. Every so often I would say 'Tonight's the night!', and start to write the basic paper in the area, giving the basic definitions and propositions, only to find it drifting into the sands, and getting thoroughly stuck. The questions, though, always seemed to recur and the drawing of a few diagrams kept on strengthening the conviction that they must represent some real mathematics. There was a bit of the old adage: 'If a fool will but persist in his folly, he will become wise!'

Gradually, with clues and methods from here and there, through collaborations with Chris Spencer in 1971 and 1972, with Philip Higgins for fifteen years from 1974, with Jean-Louis Loday in 1981 to 1987, with research students who inputted key ideas and results, and contributions from others, a large theory and method took shape. It really is true that a higher dimensional group(oid) theory exists, which is not available for groups alone, and that a number of phenomena in topology, and even in group theory itself, can be better understood, and new results found and proved, from this viewpoint. The higher dimensional Van Kampen theorems also express, and have their roots in, some long standing traditions in topology and group theory, such as the notion of a cycle, namely 'something' with no boundary, like the surface of a sphere, and also with methods of gluing pieces together to make larger pieces. I have used these vague words deliberately, since the early literature is unclear as to what is a cycle, and how one 'adds' pieces. The later clarifications, in terms of chains and homology, use a trick of working with formal sums which has a considerable success, but does not fully represent the intuitive idea. The intuitive idea is that of gluing bits a and b together, rather than writing a 'formal sum', $a + b$. Finding some mathematics which represents this idea ends up by yielding a statement and proof of the above theorem, and so allowing for new understanding and new calculations in homotopy theory not currently possible by other methods. These calculations are not the main aim of the theory, but they are satisfactory as a test that the ideas are working, and that they do something new.

This theory has not so far solved any really famous problem. Rather, it has solved problems not previously formulated, and suggested a new set of problems and areas of investigation. Also, there is a strong intuitive pull. Draw a square, then divide it into smaller squares, and you can easily convince yourself that there has to be a theory which expresses the way the big square is built up. The algebra which expresses this should be of general importance, since the method of building a complicated object from small standard pieces is quite widespread. It is interesting to see the conceptual and technical advances that are required to express these ideas, even in the simple format used so far.

History shows that, in the long run, new methods win out over new theorems, and over the solution of famous problems. The latter can often be more in the nature of a test of a new method rather than an indication of what the method will eventually achieve.

These experiences have led me to emphasize the conceptual mode of progress in mathematics. This allows for the view that good mathematics can be easy. On the other hand, it should also be said that the technical requirements to set up these concepts of 'higher dimensional algebra', and make sure they work, are considerable and not what I had expected to happen. One aim of mathematics is to set up machinery of which you do not need to know or test all the parts before you use it, just as you can drive a car without knowing the workings of the internal combustion engine. This is the function of lemmas and theorems. Thus it is quite in accord with the practice of mathematics to teach pupils the 'lemma' or 'proposition' that $7 * 8 = 56$, without them having to understand exactly why this is so.

These points are relevant to a recent debate, sparked off by a *Scientific American* article on 'The death of proof' (Horgan 1993; Thurston 1994). A point that I think has not so far been made in this debate is that mathematics increases certainty by the development of new concepts, and the formalization of ways of thinking, so that the framework of an assertion and its proof can become so well structured, so natural, and each part so well tried, that the whole carries conviction. What one calls a good proof is one which is not like finding a route through a maze, but like following a walk through a natural seeming landscape, to a surprising viewpoint.

2. THE OUTPUT OF POSTGRADUATE EDUCATION IN MATHEMATICS

Postgraduate education represents perhaps an extreme of individual involvement in mathematics training. I have had the privilege of taking nineteen people through to a successful doctorate degree. From all these students I have learned a lot, since their problems and approaches to mathematics have all differed considerably. But this number does contrast with the hundreds of undergraduates whom I have taught over the years.

I still find training postgraduates to work for a doctorate a risky business. My overall method has been to involve students in the problems in which I happen to be interested at the time and to discuss frankly how one would reasonably assess progress in this particular field. In this process, some students have made quite crucial contributions to the overall research programs, in ways which I would not have foreseen and quite possibly would not have worked out for myself.

I think that in all cases, work has been done which otherwise would not have been done. It was partly forced by the necessity, particularly from the student's point of view, of some kind of progress, and so for an analysis of how we should proceed. If you have a lot of available questions, and there is no real lack of them in this area, then each one has rather a lesser priority and can perhaps be replaced by another. Some kind of ranking, of value judgment, is crucial for the researcher in deciding what to tackle next. For the student, though, the most important problem in the world is one that he or she is tackling, and there is a special urgency about getting somewhere.

The process of training and discussion, of making students aware of available strategies of work and study, is of course individually intensive, involving, in some cases, many hours of discussion and of reading student's work. In some cases, the opposite has taken place, and I have been instructed how the problem should be tackled! What I have provided then is the context and the problem, the reason for wanting this problem solved, and also help in expressing the mathematical ideas in a way which makes it clear to the reader.

The majority of the problems given to students have tended to be my problems, that is, the problems thrown up in studying this area. This has the possible danger then of leading students into a byway of mathematical progress. Fortunately, these sets of ideas have come more and more to link with, to require and to illuminate known areas. This fact was comforting in judging the progress and prospects for the overall area.

A severe problem in postgraduate training is that of background. Some knowledge is necessary to understand the problem and its context. An even greater background is necessary to understand and master the tools which should be relevant for the study of the problem. Many of these tools are learned 'on the job' and there is a judgment required as to a 'need to know'. There is no easy answer to this. Most problems require for their solution a degree of skill in certain specific areas, from say, group theory to programming in C, and without these skills at a professional level no progress which can be judged worthy will be made.

There are two main procedures which have over the years evolved as important for postgraduate training: writing mathematics and analyzing the aims of a research work.

The Benefits of the Process of Writing Mathematics

The main idea is that writing mathematics to a high standard of exposition is a crucial element of doing mathematics. It took a long time for me to realize for myself that this was crucial in my own mathematical work. As explained earlier, the writing of a book on topology set a course of many years of research work.

So our postgraduate students are set, as part of their work, the task of writing up a piece of mathematics, not just by copying from a text or from various papers, but to give an exposition of one area from a different viewpoint.

The growth in the use of mathematical wordprocessors has been a great help in this process. The tutorial process of instruction can work on a readable typed text, which can be improved, and the process of making mathematics is seen as an iterative method for the production of a finished, accurate, clear and readable work. This allows for an emphasis on the craft of mathematics, and so, on its nature as a process and on exposition. The art comes in the analysis of the qualities in the finished product for which one is looking and the decisions on how to achieve them. It is very helpful to students to have these matters discussed.

For these reasons, I do believe that pupils should be trained in good exposition. The emphasis on examinations as the major means of assessment can militate against the development of good exposition, reflecting understanding and clarity of expression. For example, we should have not only Olympiad type competitions in problem solving, but also competitions in mathematical expositions in exhibitions, in which the aim is to make clear the ideas and points of mathematics, and in which a variety of skills from mathematics, arts, writing and craft are brought to bear.

The Analysis of Aims

The idea here is to explain to a student what it is that might be done, and then to discuss the question why have I, the professor, not done it before? Answers might be: 'Just thought of it', 'Forgot about it', 'No time', 'Considered it too hard', 'Thought it not worth while', 'More study of the background literature needed', 'Not clever enough', 'No time to learn the required skills'.

How should the answer, or answers, to the previous question be used to influence the immediate tactics?

For example, it might be necessary to do a serious study of a particular part of the literature, to learn some skills from a given area, or to evaluate new evidence that the problem might be more important than previously thought.

What Would Be the Expected Results of Achievement of the Immediate Aims?

Some judgment as to the value of the proposed achievement must be made if a student is to spend some time on it.

There needs to be a 'fall forward' position: what do we do if the problem is far easier than had been thought? Where do we go on?

There needs to be 'fall back' position: what do we do if the problem as stated is far harder than anticipated, or, even worse, not as sensible as originally thought?

Is the Advice of the Supervisor Sensible?

It is often hard for students to realize that while it may be the job of the supervisor to have ideas, and to suggest ways forward, it is for the student to evaluate them.

It is useful to work on the following analysis. If 3% of your ideas are good, and you have 100 ideas, then you have 3 good ones. Result, happiness. If 3% of your ideas are good, and you have 10 ideas, then you have problems. Result: misery.

For this reason, it is useful to see how the supervisor copes with failure. Indeed, a part of the success in research has to be the successful management of failure.

I learned a tremendous amount from my supervisor Michael Barratt. I remember thinking after a long session with Michael: 'Well, if Michael Barratt can try one damn fool thing after another, why can't I?' I have followed this method ever since!

Is the Problem a Natural One for the Student?

A colleague, José Montesinos of Madrid, said that his advice to students was to continue with those aspects they found easy! It is difficult to describe exactly what it means to understand mathematics, and even von Neumann commented: 'You don't understand mathematics, laddy, you just do it!' (Ulam 1986). Each person's mental equipment, and natural mode of thought, differs from those of others, by reason of both genetics and of experience. Only trial and observation can find which problems are the most appropriate.

Skill Learning

Of course, if new skills have to be learned to carry out the work, then the usual methodology of skill learning applies, namely:

- Task analysis
- Practice of basic skills

- Moving from the very easy in gradual stages as skill level increases
- Putting together basic skills
- Observation and analysis of performance
- Notions of style and quality
- Conceptualization
- Internalization.

Motivation for skill learning is important, and that is why I have put task analysis at the beginning. A problem of mathematical teaching is to give motivation for the skills which are learnt.

Psychologists have observed that transfer of skills is increased with the complexity of the task (e.g. matching complex objects rather than making fine distinctions). Advice from my first supervisor J. H. C. Whitehead was to learn one small area better than anyone else in the world, and then gradually expand this level of expertise.

Conclusion Concerning the Output of Postgraduate Education

The output of postgraduate education is, in theory, a trained independent worker, with a proven battery of skills and knowledge, and with some idea of how to make judgments in seeking out and inventing problems, reading and evaluating the literature, making progress with problems, writing up the results, and evaluating the results achieved.

The problems my students have been asked to tackle have sometimes turned out remarkably hard, so that the discussion of what might be partial progress has been important. One student said that a good aspect of this program has been the combination of grand prospect, strong intuitive base, and the technical problems that need to be overcome to make this intuition work. Also, I cared strongly about the results and was delighted that on many occasions I was shown how to do things.

The supervisor has a great advantage over the research student in the knowledge of background, context and notion of value. On the other hand, each student is an individual, and each is likely to respond differently to different kinds of mathematics.

I do not think there is any final answer to the methods of postgraduate education. As with many activities, it is possible to point to some avoidable mistakes, such as that of assuming something is 'obvious' without writing down all the details, and also to give some kind of framework to the doctoral process. Part of the problem seems to be to combine a sense of direction, with the ability to take note of promising lines if and when they appear. A considerable part of the difficulty is to acquire the necessary background and skills to understand, evaluate and tackle the problems. Since a lifetime can be spent in acquiring knowledge that might be useful, a pragmatic attitude has to be taken of learning what seems to be necessary to get on with the job.

On the other hand, when a skill is necessary, then it has to be learned. You cannot make an analogy with, or use a method from, an area of which you know nothing.

There is no advantage in reinventing the wheel, except as a learning method. The spirit of Polya's *How to Solve It* (Polya 1957) is relevant, but we need a further trick, against the spirit of his book, namely 'Look it up in the literature'. A student would feel aggrieved if he had been slogging away for six months at a problem the experts know has been already solved, or would fall easily to a range of standard techniques. So there has to be a search for those methods which are in the literature and which might be relevant. We are not playing a party game, and doing research is difficult enough without artificial restrictions. A paper reproving something already known will get short shrift from a referee, unless a new viewpoint or simplification is apparent.

The acquiring of necessary skills is no easy task. The teacher and the student have to allow for time, practice, repetition, and thought. Persistence is important, since it may take a long time of seemingly little progress before it is apparent that real improvement takes place. The speed with which people pick up a new skill is one of the important variations between them. The factor may be more than fifteen times. Some people need more repetition and practice of a particular skill, more, dare I say it, rote learning, than others.

There is also the question of what level of skill is required for the problem at hand. I like the comment of a magician who explained: 'I practice till the difficult becomes easy; the easy becomes habit; and the habit becomes beautiful.'

Even if practice does not necessarily make perfect, it is clear that practice is an essential element of perfection. There is no way to become a good swimmer, or a good musician, without putting in the hours to get the feel of the activity on the basis of sensible coaching and teaching. It is surprising nowadays that the extensive practice of basic skills is accepted as an obvious necessity in sport, but seems not to be so accepted in mathematical education. But how good do we need to be for the purposes at hand? What do we have time for?

Is there a danger that the baby of practice has been thrown out with the bath water of rote learning? You tend to get funny looks at educational conferences when you bring up the topic of rote learning, but have the psychologists analyzed for us the differences between rote learning and practice? I have met education students who have learned by rote that rote learning is a bad thing! Has anyone shown us a royal road to mathematics?

It may be that if we ourselves cannot understand the processes involved in carrying out an activity, then we are left with the notion of practice, and of being shown how to do it, until we get the feel of it ourselves. How many activities can we really understand in a way which helps with the teaching of them? This is the advantage of concentrating on the notion of output, and on the quality of output.

3. THE POPULARIZATION OF MATHEMATICS

The aim overall in popular lectures and exhibitions has been to convey something of the methodology and nature of mathematics to a broad audience. Part of the problem is thus to make abstract ideas concrete.

Among all the subject areas, mathematics has a special difficulty in popularization since the general public have little ideas on what are even some of the most basic objects in mathematics, such as that of a group. Part of the theme of this chapter is that this lack of knowledge is in part traceable to a lack of clarity in the mathematical community, by both researchers and teachers, as to the nature of mathematics itself, and so to a lack of clarity about conveying this to the general public.

The Nature of Mathematics

My popular lectures aimed to convey some related aspects of mathematics. These were exemplified with two demonstrations in the Washington talk which it is worth explaining here, and which were necessary to give on that occasion, since I had brought the things 2000 miles!

Mathematicization

One of these aspects was the aim in mathematics of mathematicization, that is, of expressing an intuitive idea in a format which is sufficiently precise to enable deduction, calculation and proof. There is usually a prior stage to this, that of conceptualization, which also requires its own analysis.

Rules and Laws

Mathematicization usually involves the notion of a rule or law. One example of this which I used in the lectures was the theory of groups, and the calculation with relations on symmetry operations.

The mathematicization of symmetry can be demonstrated through the particular example of the symmetries of a square; which can easily be done for a large audience with a large cut out square with labeled corners. It is easy to show the rules

$$x^4 = y^2 = xyxy = 1$$

where x denotes rotation of the square through 90 degrees clockwise, y denotes reflection in a bisector of two edges, 1 means the operation which leaves the square alone, and $xyxy$ means do first x, then y, then x, then y. The point here is that the representation of an action by symbols, and of consecutive actions by concatenation of symbols, and the use of the symbolic method generally, is one of the greatest difficulties the general public finds in getting a glimpse of what mathematics does. So it is necessary to show this feature in a concrete situation, and then to go further and show the value of the symbolic representation through the use of rules on combinations of these symbols for explicit calculation, and to show how this calculation models real operations.

Another nice feature of this example is the importance of pedantry; in this case, the importance of taking note of the operation 1 of leaving the square alone. Without this, the rules for the symmetry could not be properly expressed, just as our counting system would not work without the use of the number zero.

There is a nice trap here about this representation, which shows the importance of precision. When you rotate the square, the labels on the corners change their position. But one wants to iterate the operation x. So x has to be an operation which works on positions; that is, it moves all the elements in their various positions one place around clockwise.

One has also to be careful, because the notion of clockwise differs between the audience and the demonstrator. This gives an opportunity for another remark on symmetry.

Idealization

This is a standard procedure. The real square is not completely symmetrical. For the mathematicization, we think instead of an ideal square.

Abstraction

This symmetry example also shows the importance of abstraction; namely, representing the real operation of rotation on an ideal square by a symbol, and representing the combinations of operations by combinations of symbols.

This abstraction is usually a stumbling block for the general public. However, it is crucial to the progress of the subject since it allows the notion of analogy, an aspect of mathematics not commonly stressed.

For example, when we write $x + y = y + x$, and $xy = yx$, we are illustrating the commutative law and so illustrating an analogy between addition and multiplication. This process of analyzing and using laws at various levels in the subject is very important. For example, algebraic structures are an important part of mathematics, but there is also a mathematics of algebraic structures, in which such systems are looked at as a whole. Thus analogies work between levels of abstractions as well as at one given level. The symbolic method is a crucial part of the process of abstraction since symbols can represent a variety of things.

Deduction and Calculation

A further part of mathematics is the deduction from and calculation with rules. In the case of the symmetries of a square, we want to deduce from the rules given above that, for example, $xy = yx^3$, $yx = x^3y$. This shows that whenever x is taken past y it changes to x^3. The carrying out of these kinds of deductions and calculations is an important part of mathematics.

The Learning of Algorithms

In some circles, the ability to learn and carry out algorithms is rather decried. To the contrary, the development of algorithms to carry out tasks is one of the goals of mathematics. For the same reason, learning algorithms, learning procedures, is a basic part of the acquisition of mathematical skills just as hitting a ball consistently is a basic part of tennis which can be acquired only by a considerable amount of regular practice. Mathematics involves, of course, far more; namely, the ability to solve problems and develop theories by the planned use of basic algorithms and methods, and also by the effort to see their scope and limitations. These again are skills which can be analyzed, and which can be learnt and acquired.

Surprise

Some of the best mathematics has about it the element of surprise, the revealing of a fact one would not have thought possible. Such a surprise makes one want to explain why this happens. Here is an example.

Dirac String Trick

The Dirac string trick illustrates a surprising feature of space, and also the notion of a rule or law. The apparatus for this is as follows:

Take two squares of card or board, say 18″ square (0.5 m), and on one you draw an arrow, to indicate direction, or place a picture. The corners of the top square are then connected to those of the bottom square by string, or, better still, different colored ribbon. It is a good idea to clip the ribbon to the board by bulldog clips, so that the apparatus can be untangled easily.

Hold the bottom square on the floor by your foot, and rotate the top square through 360 degrees, keeping it horizontal. The ribbons become tangled.

Now rotate the top square in the same direction through another 360 degrees. It appears that the ribbons become more tangled. However, it is possible to untangle them completely, moving the top square up and down to allow room for the ribbons to be manipulated, but without altering the direction of the top or of the bottom square (see Figure 1).

This illustrates the law $x^2 = 1$, where x is now the rotation of the square through 360 degrees.

There is a more subtle point which accords well with my research interests. The proof that the rule holds is obtained by untangling the ribbons. However,

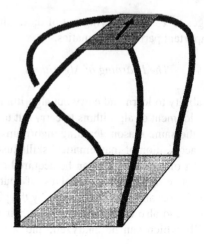

Figure 1

the finding of such an untangling, and the classification of these 'untanglings', is a 'higher dimensional problem' of a much greater difficulty.

It is amusing to try variations on the above, such as rotating the square about a different axis than a vertical one.

Dirac's interest in this trick was the argument that it was a model of the spin of an electron. This analogy is quite a good one, once the exact mathematics of the situations has been spelled out. For more information, see Kauffmann 1987, p. 93.

Knot Theory

Here is another 'trick' which also illustrates some mathematical methodology:

> Make out of copper tubing a pentoil knot as shown in thick lines in Figure 2. (This needs professional help, such as an engineering workshop, to make it look good, but you can make it yourself with thick wire.) Now tie string on according to the following rule:
>
> $$xyxyxy^{-1}\, x^{-1}\, y^{-1}\, x^{-1}\, y^{-1}$$
>
> as shown in Figure 2, and tie the ends of the string together. It is then possible to take the loop of string off the knot without cutting or untying it. That is, in this situation the above complicated formula represents, or equals, 1. It is even possible to give the proof of the formula for bright youngsters familiar with some algebra, such as cancellation.

4. CONCLUSION

There is no space or time here to discuss the wide topics of output at undergraduate and at school level. Each of these represents a very big problem. In the UK,

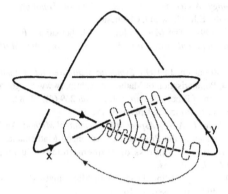

Figure 2

teachers of mathematics at secondary level (11–18 years) come largely from those who have taken a degree in mathematics, so that attitudes acquired at universities have a large influence on teaching. There is, though, a question as to whether in fact any attitudes are deliberately fostered. The debate on the output of undergraduate education has a long way to go!

Here is the elephant analogy for the teaching of mathematics: We can teach the structure of the elephant's trunk, tusks, skin, feet, stomach, and so on. But can the elephant (here the mathematics) really be understood without a global approach, an ecological and evolutionary approach? This is the argument for bringing a wide variety of subjects to bear on the question of the nature of mathematics, and on the problems of teaching mathematics.

On the other hand, there is a danger of all these fun subjects being brought in, and then the actual relevance to the elephant being minimized. Thus, a test of the results of research in mathematical education would seem to be the help they give in understanding the nature of mathematics, and the way in which learners can be helped to be better mathematicians, at the level at which they are currently functioning. It is right here to bring in at this point an approach from a professional mathematician, just as music and sports teaching bring in professionals in those areas.

There is also a worry that students of mathematics are starved of any attempt at a global viewpoint, and a sense of value and of context, an understanding of the place of mathematics as a human endeavor. I am sure that those who came to the Washington conference are aware of these concerns, and that the work of the present study will help to develop not only teaching in mathematics but also a general awareness of these issues.

REFERENCES

Boorman, P.: 1986, 'Mats: Theory and Practice 2', *Mathematics Teaching* **117**, 18–22.
Borasi, R.: 1986, 'Behind the Scenes', *Mathematics Teaching* **117**, 38–39.
Brown, R.: 1968, *Elements of Modern Topology*, McGraw-Hill, Maidenhead.

Brown, R.: 1988, *Topology: A Geometric Account of General Topology, Homotopy Types and the Fundamental Groupoid*, Ellis Horwood, Chichester.

Brown, R. & Porter, T.: 1994, 'The Methodology of Mathematics', *Bulletin of the International Commission on Mathematical Instruction* **37**, 23–37; also, 1995, *Mathematical Gazette* **79**, 321–334.

Brown, R., Quinton, C., Robinson, J.: 1996, *Symbolic Sculpture and Mathematics*, Edition Limitée and the Centre for the Popularisation of Mathematics. http://www.bangor.ac.uk/~mar007/

Horgan, J.: 1993, 'The Death of Proof', *Scientific American* **269** (4), 74–82.

Kauffman, L.: 1987, *On Knots*, Princeton University Press.

Polya, G.: 1957, *How to Solve It: A New Aspect of Mathematical Method*, Anchor Books.

Tall, D. O.: 1994, Understanding the Processes of Advanced Mathematical Thinking, paper presented at the International Congress of Mathematicians, Zürich, 1994, to appear in *L'Enseignement des Mathématiques*.

Thurston, W. P.: 1994, 'On Proof and Progress in Mathematics', *Bulletin of the American Mathematical Society* **30** (2) (new series), 161–177.

Ulam, S. M.: 1986, *Science, Computers, and People: From the Tree of Mathematics*, Cambridge, MA, Birkhäuser-Boston.

Ronald Brown
School of Mathematics,
University of Wales,
Dean St, Bangor,
Gwynedd LL57 1UT,
UK

MICHÈLE ARTIGUE

RESEARCH IN MATHEMATICS EDUCATION THROUGH
THE EYES OF MATHEMATICIANS

In this article, I would like to address the theme of the relationships between mathematicians and research in mathematics education from two different points of view:

1. Firstly, as a researcher in the didactics of mathematics – mainly by using my personal experience as a researcher in the didactics of analysis both at high school and university level,
2. Secondly, as a representative, in some sense, of the French didacticians inside the mathematics community – mainly by using my experience as a member of the national committee in charge of the appointment and promotion of mathematicians.

This article, therefore, belongs to the category of case studies, but even if the reflections presented here are highly dependent on the French situation, they certainly have a more general value.

First, I would like to specify the meaning given here to the word 'mathematician' since different meanings can be given to it. Indeed, one can restrict it to mean somebody producing new mathematics, or one can enlarge it so as to include anyone having a strong professional relation to mathematics; for instance, by teaching or using mathematics intensively. In this text, in order to clarify the discussion, I adopt the first restrictive meaning which seems to be implied by the title of the theme. But it does not mean, in my opinion, that didacticians, since they do not produce new mathematics stricto-sensu, do not have their right place inside the mathematics community (see section 2 below).

Reflecting about the relationships between mathematicians and research in mathematics education means attending not only to the present relationships but also researching what made them as they presently are, and what would be necessary in order to improve them if it appears necessary. For this purpose, what we have to analyze and understand is the dynamics of a process rather than a static state.

In order to analyze this dynamics, a sound analysis of the relationships between mathematicians and the whole world of mathematics education, not only the mathematics education research world, as well as their historical evolution would be necessary. I cannot attempt such an analysis here; nevertheless, by evoking the history of French secondary teaching of Analysis (a subject more or

477

Sierpinska, A. and Kilpatrick, J. Mathematics Education as a Research Domain: A Search for Identity, 477–489.
© *1998 Kluwer Academic Publishers. Printed in Great Britain.*

less corresponding to the Anglo-Saxon Calculus) during this century, I would like to situate the reflection about the present situation in some historical perspective.

1. AN HISTORICAL PERSPECTIVE

The history of the teaching of mathematics, during this century, clearly shows the investment of mathematicians in teaching issues and the fundamental role they have played as a driving power for the evolution and modernization of secondary mathematics curricula.

The 1902 Reform : A Strong and Rewarding Investment

In France, for instance, the beginning of this century was marked by an important reform of the high school curriculum. One objective of this reform was to abolish the supremacy of the classical literary culture in the education of the upper middle class elites. Within this culture, the role of science was put down and science appeared in the curriculum mainly as technical knowledge. The reform wanted to introduce the idea of 'scientific humanities' and give them the same high status as 'classical humanities' (Belhoste 1990). As far as mathematics was concerned, the reform was strongly supported by famous mathematicians such as Borel, Poincaré, Hadamard, Darboux, Tannery, Appell, Bourlet and others who took charge of the elaboration and political negotiation of the new curriculum, explained its spirit in public lectures, and wrote fundamental and basic books[1] for teachers which covered the whole content of the new secondary mathematics.

Clearly, and it seems natural for mathematicians (Fey 1994; Tietze 1994), they conceived their task as consisting of reorganizing the ancient and obsolete mathematics curriculum by adapting it both to the evolution of mathematics and to the needs resulting from the scientific and technical evolution. This modernization process led to the introduction of analysis at the high school level – in their first year for scientific students and in their third year for the others. But I want to stress that the student continued to be considered throughout the reflections and decisions. There was a clear will to adapt the teaching contents and processes to the student's cognitive potentials.

A famous lecture by Poincaré on mathematical definitions (Poincaré 1904) clearly illustrates this fact. He evokes the notions of continuity and derivability of functions, and shows how intuition has been misleading for mathematicians in this domain. He points out that these erroneous intuitions were only overcome, and reasonings made completely rigorous, when rigor entered the definitions themselves. He also points out, however, that this perfect rigor could only be obtained by according a primary role to logic, but that it would be catastrophic to impose this kind of rigor on the students from the beginning. He wrote:

Here we are obliged to go back; no doubt, it is difficult for a teacher to teach something which does not satisfy him entirely; but the satisfaction of the teacher is not the unique goal of teaching; one has at first to take care of what is the mind of the student and what one wants it to become. (Poincaré 1904, p. 265).

Later on in his lecture, we can see how these ideas are applied in a special case: the introduction of the notion of integral in the curriculum:

In order to define an integral, we take all sorts of precautions; we distinguish continuous and discontinuous functions, functions with and without derivatives. All of this is normal at university level. All of this would be execrable at high school level. Whatever the definition you give him, the student will never know what an integral is if he has not been shown integrals before. All your subtleties will leave him indifferent. He thinks he knows what an integral is and he will understand that he doesn't know only when he has gained a strong mastery of integral calculus. The moment you introduce the integral is not the appropriate moment for making all these distinctions. So, what one has to do is very simple: Define the integral as the area included between the x-axis, two ordinates and the curve, show that when one of the ordinates moves, the derivative of the area is precisely the ordinate itself. This was the reasoning used by Newton: Integral calculus was born in that way and, whether one likes it or not, we have to pass where our fathers passed before (ibid., p. 279).

But, we cannot conclude from these quotations that rigor is considered as something unimportant or that these mathematicians would be satisfied by teaching analysis purely in an intuitive way. The importance and the place given to rigor clearly appear in the debates which take place in the framework of the young CIEM (Commission Internationale sur l'Enseignement des Mathématiques, later ICMI), widely echoed in the journal *L'Enseignement Mathématique*, and in the articles published by this journal concerning the inquiry conducted by the CIEM on the introduction of the first notions of differential and integral calculus at high school level all around the world. As Beke writes in the final report of this inquiry,

Our main duty is to introduce the notions of differential and integral calculus in an intuitive way, by starting from geometrical and mechanical considerations, and to gradually rise to the necessary abstractions. All our affirmations have to be true, but we do not have to target the whole truth (Beke 1914, p. 269).

In the same report, he stresses that there does not exist a country where the notion of limit only functions at an intuitive level. Furthermore, he notices that generally Taylor's approximations are excluded from the curriculum as they cannot be introduced properly at high school level.

It is also necessary to stress the role played by the positivist philosophy which was, at the time, the dominant philosophical framework. Positivist ideas support the development of inductive methods and an experimental conception of mathematics which, without neglecting its deductive specificity, has to be tightly linked to the real world and has to be useful in other scientific areas.

This leads to an introduction of analysis with limited ambitions: with an accent put on applications, both mathematical and physical. This analysis is strongly algebraic and algorithmic. Seen from this angle, it is obviously more calculus than analysis (Artigue 1994b).

The young CIEM supported innovations, discussed the reforms carried out in the different affiliated countries and tried to evaluate their effects. The published reports on these discussions attest to the great and international investment of mathematicians, as we have said before. They also attest to the positive feedback these mathematicians received. Ten years after the beginning of the movement, they could see themselves as pioneers of a useful and successful renovation. The following excerpt from a lecture by Bourlet illustrates this fact:

When, 15 years ago – after trying it with my students – I asserted that the candidates to the *baccalauréat* would learn easily the calculus of derivatives. When I asked for the suppression of non-useful speculations and for the introduction of what is really helpful for applications, many 'experts' were taken aback. Today, these candidates learn the differential notation and yet calculate some integrals (do some quadratures); and our first and second year scientific students (in high school) juggle with derivatives (Bourlet 1910, p. 382).

Thus, the investment made by mathematicians was substantial and for this they were well rewarded.

The New Math Reform : A Strong but Not Rewarding Investment

When one looks back at the New Math reform of the 60s, the picture is quite different. In the 50s, once more, curriculum appeared as obsolete and inadequate, especially taking into account the scientific and technical evolution. Once more, famous mathematicians in France, for instance, committed themselves to its reform: Choquet, Dieudonné, Lichnerowicz, Revuz.... Once more, we had a renovation of the subject matter – the content being the main entrance gate for the mathematicians – and once again, attention was paid to the pupils and students and their cognitive functioning.

Psychologists, such as Piaget, appeared as guarantors for this new renovation enterprise, widely supported by the theory of structuralism, whose influence was at the time scientifically predominant.

But this time, difficulties and distortions accumulated in spite of the precautions taken by the noosphere (Chevallard 1985) – loosely speaking, the movers and shakers – for instance, in the formulation of the curriculum. Here it was stressed that mathematics teaching had to avoid the temptation of becoming a pure abstract game, that experimental activities linked to the real world had to come before any introduction of abstract notions, that applications of mathematics to different contexts – everyday life, physical, economical or social contexts – had to be systematically incorporated into the internal mathematical work itself.

Soon, the pioneers were disappointed. Choquet, for instance, denounced the fact that generations of pupils would be mathematically prepared for nothing. Dieudonné denounced a new scholasticism, all the more aggressive and foolish since it stood behind the banner of modernism.

The teaching of analysis was not the main object of the New Math reform whose implementation began in 1970. This had been widely modernized by two successive reforms in 1960 and 1965. One can find an indirect proof of it in the following fact: In the instructions accompanying the presentation of the new national curriculum a very restricted place was devoted to analysis – if one exempts the part concerning the theory of the Riemann integral which appeared as a new object in Terminale (the last year of secondary mathematics courses). But the teaching of analysis was globally shaped by the reform spirit and suffered its formalist influence.

The New Math reform wanted to promote mathematics for all – not only for the upper middle class elite, culturally adapted as had previously been the case. For the same reason, it concerned the whole community of teachers – and not only the elite of teachers, as also had previously been the case. It put to the fore problems of foundations, of axiomatics and the notion of structure, in particular, algebraic structures, which had been for mathematicians a divine revelation. But, confident in their own experience and in the predominant ideas of structuralism, mathematicians were neglecting a fundamental issue: What kind of mathematical culture was necessary in order to accept such an approach to mathematics and to benefit from it?

The failure of the New Math reform was the evident proof that mathematics expertise, complemented by some general psychological and pedagogical principles, was not enough for promoting an effective organization and management of the complex reality of teaching mathematics for all. In order to really understand the failure, it was necessary to accept the idea that other competences, other forms of knowledge, had to be developed. It was also necessary to accept the fact that the observed distortions were not the mere effect of some dysfunction of the educational system, but rather normal phenomena induced by an inadequate and insufficient understanding of the constraints and laws governing this educational system. One had also to consider that if school was to thrive mathematically, specific research in order to better understand these constraints and laws was required. Inside our French didactic community, Y. Chevallard and his colleagues developed extensive research on that theme, within the framework of the theory of didactic transposition (Chevallard 1985, 1992).

Some Consequences of the New Math Reform

The failure of the New Math reform tended to induce an attitude of reserve among the French mathematicians with respect to secondary school mathematics questions. Nevertheless, one positive outcome of this reform was the birth of the IREMs[2]-university-based institutions where mathematics teachers, either from university or secondary schools, continue to be engaged in cooperative work.

Thanks to these institutes, relationships could be maintained between the mathematics community and the mathematical educational world, even if these relationships were restricted by the small number of mathematicians institutionally attached to the IREMs, and the small rate of turnover.

For instance, the reflection on analysis, developed within the framework of the national commission of the IREMs, served as the main basis to a counter-reform which began in the early 80s. This commission denounced the lack of problematics of the analysis being taught, the precocity and the excess of formalization, the domination of qualitative aspects over quantitative ones, the exaggerated attention paid to pathologies and a too predominant role given to the discourse of the teacher with respect to the students' activity. Relying on a conception of mathematics as a human and social activity, as well as a historical one, and situated within a constructivist conception of learning, this commission argued for a more intuitive approach [to analysis] in terms of approximations centered around the problems at the core of the field: study of functions and curves, optimization problems, numerical methods for solving equations, and approximation of numbers and functions (Commission InterIREM Analyse 1981). These proposals directly inspired the 1982s programs (Artigue 1994b).

Another indirect consequence of the New Math reform was the progressive development and institutionalization of research into the didactics of mathematics and the birth of a community of didacticians, mainly from a mathematical origin and attached to the IREMs: A community willing at the same time to preserve privileged links with the mathematics community and to constitute didactic research as a specific and autonomous scientific field.

This progressive development of the didactics community was likely to have some influence on the relations between the mathematics community and the mathematical educational world. There is no doubt that mathematicians have continued to be regularly involved during the last ten years in the national committees successively in charge of the secondary mathematics curricula (COPREM, GREM, CNP[3]) and to have leadership positions in these. There is no doubt that they have been strongly involved in the organization of some important conferences such as 'Mathématiques à Venir' (1987), and 'Les Objectifs de la Formation Scientifique'(1990).

Nevertheless, it seems that the progressive institutionalization of the didactics community tends to withdraw the pressures of the educational world from the mathematicians. In fact, the existence of a community of specialists in the teaching and learning of mathematics inside the mathematics community, or at its border, in some sense allows mathematicians to stand back in a reflexive and critical position. It may also incite them to carry out some responsibility in the domain or to limit their involvement to some activities more directly rewarding; for example, at the border of school mathematics, such as math competitions, math clubs and workshops[4] (I do not deny, of course, the mathematicians' interest for mathematical education!).

So we, mathematicians as well as didacticians, have to be very attentive to the risks inherent in this new situation and have to act energetically in order to create the positive synergy between our respective competences which is necessary for a real improvement of mathematical education, both at secondary and at tertiary levels. Obviously such a positive synergy is not easy to create and is strongly dependent on the quality of the relationships between mathematicians and didacticians. It is the reason why I shall focus on these relations in the second part of this text.

2. MATHEMATICIANS AND DIDACTICIANS

Ten years ago, French mathematicians, globally, did not see a difference between research into mathematics education and a strong pedagogical involvement. Now, it is no longer the case. The functioning of the National Committee of Mathematicians, mentioned at the beginning of the text, obviously attests to this evolution, helped by the progressive development of institutional means of validation for didactic research. Didactics is now considered as a legitimate specialty of research in applied mathematics and didacticians employed at tertiary level in mathematics departments are nationally evaluated with the same criteria as other applied mathematicians. Nevertheless, these privileged links with the mathematics community remain fragile.

As French didacticians are very sensitive to this problem, I would like to elaborate further on this point. Indeed, a great majority of us are convinced that didactic research has to preserve strong and privileged links with the mathematical world; that mathematics as a discipline has to remain the fundamental root of didactics; that we need didacticians with a strong mathematical background; that didacticians from a mathematical origin have to try to preserve their present place within the world of mathematics production and mathematics education at the tertiary level. Moreover, we are convinced that the only way to guarantee that this strong connection with the mathematical world remains effective is to foster it institutionally. We have fought for it during the last ten years, and reasonably succeeded – thanks to the support and the effective help of pure and applied mathematicians who thought that mathematicians had some responsibility to assume with respect to didactic research and agreed, as a consequence, to consider didactic research as a legitimate specialty. But the present equilibrium remains fragile.

For instance, faced with the exponential increase of didactic productions, and faced with the strong development of more theoretical research, mathematicians can feel themselves more and more didactically incompetent: How to give advice on the pertinence of such and such a theoretical frame? How to judge what is really original research and what is not? How to evaluate the quality of didactic research? No doubt, they have been tempted in the past, and they continue to be tempted, to get rid of these embarrassing didacticians and to encourage them to join the community of educational science researchers. I have to confess, too, that a fairly large number of them live with a restrictive vision of science and are still wondering how the study of systems such as educational systems, which are so complex, so dependent on human beings, and so far from their mathematical world, could be the objects of a real scientific work. But, in the past, at the same time that some mathematicians were tempted to get rid of all this business, they more or less consciously feared the development, anarchic and autonomous, of a caste of didacticians which, possibly, could some day take charge of the whole educational business. So, with the help of mathematicians, for both reasonable and unreasonable reasons, we succeeded up to now in maintaining the privileged links with the mathematics community which we think necessary to the development of a didactic research of quality.

These problems are real problems but, in my opinion, we have to look further if we want to reach a clear understanding of the difficulties we have to face to create the positive synergy I previously evoked. Firstly, I would stress that, even if they accept the emergence of didactic research as a new scientific field, French mathematicians, most often, are not convinced of the legitimacy of those who do didactic research, and, as a consequence, have some a priori doubts on the pertinence of the results they obtain. More or less consciously, they tend to see the didactician as a kind of sub-mathematician who finds, in didactic research, a diversion from his or her lack of mathematics productivity. The introduction of a didactic component in the pre-service teacher training, in the framework of the creation of the IUFM[5] and the ensuing polemics and declarations up to the highest sphere of the National Academy of Sciences, clearly illustrate this fact.

Secondly, but certainly more important, we have to face the following fact: For a very long time, mathematicians have been protected from the problems induced by the democratization of teaching. They are no longer spared. They are more and more faced with students, less culturally-adapted, who need, in some sense, to learn what thinking mathematically is all about. Mathematicians are conscious of the increasing discrepancy between the lectures they give and the public they address. However, they do not find in the results of didactic research the means to remedy the problems that, in their opinion, this research should provide. As aptly noticed by G. Brousseau, didactic research is both an object of mistrust and an object bearing excessive expectations:

It ought to show its pertinence by conclusions related to what worries the teachers, the noosphere, the parents, the society; its conclusions ought to be directly communicable to everybody, without any strange conception, any specific vocabulary; these conclusions ought to be sure and proved by classical methods. At the same time, they should be original, more than what can be obtained from mere common knowledge, yet they should be compatible with it. Didactics ought to also show its effectiveness by proposing educational actions or materials which would provoke significant, general and quick improvement of teaching. Knowledge produced by didactic means ought to be visibly applicable to teaching and without any drastic modification of the professors' usual conceptions and means (Brousseau 1994, p. 58).

Didactics cannot meet all these requirements. Moreover, at a first contact, most often, it disturbs and destabilizes: It shows the failures of our usual teaching methods and tends to deprive us, as teachers, of the delusions which help us in our professional life. It shows us how, as actors of the didactic system, we are involved with its misfunctioning.

For instance, didactic research has clearly shown the existing discrepancy between, on the one hand, the usual content of university courses in analysis, which present a highly conceptualized and formalized analysis and, on the other hand, what is at play in the usual assessments – the assessment of elementary and strictly algorithmic competences. (This latter assessment exists in order to guarantee the stability of the didactic system and, as a consequence, is considered as the core of analysis by students.) Research has clearly shown the disastrous effect of such a functioning, even for those who pass. A fairly large number of these students may complete their university courses without really understanding what are for us, as mathematicians, the essential ideas of analysis. For instance, it has been shown that they can succeed without a clear perception of the following essential fact: Two real numbers whose distance is less than ε, for every strictly positive ε, are necessarily equal (Cornu 1981), and, of course, without having understood that this property, with its natural generalizations, is an essential tool which we use in analysis in order to prove that two objects are identical (Legrand 1993). It has been shown that they may have been using differential and integral processes for many years without being able to explain why such processes are required, or not, for the solving of such and such a problem and why, while based on approximation processes, they produce exact results (Alibert et al. 1989).

Even if all of the above is intuitively known to be true, demonstrating it is not pleasant. Accepting the didactic view means accepting to face this destabilization and looking beyond for useful information and production. If I refer once more to analysis, results obtained for more than ten years have certainly produced coherent and effective tools in order to help us understand the cognitive functioning of students in this area and the difficulties they have to face and overcome when learning basic objects and notions of analysis such as real numbers, functions, limits.... (cf. for instance Tall 1991; Dubinsky & Harel 1992). Let us mention that these difficulties were strongly under-estimated by mathematicians at the beginning of this century, as the following excerpt from the Beke's report shows:

The notion of limit is so present in secondary teaching and even at beginning levels (unlimited decimal fractions, area of the circle, logarithm, geometric series...) that its general definition would not be likely to occasion any difficulty (Beke 1914, p. 270).

Research has also identified some key steps and processes in the learning of analysis. It finally proposes some educational material, more or less local, which has been proved, under experimental conditions, to be reasonably effective.

But research also shows that understanding analysis and the gain of a reasonable mastery of the elements of this field, at both conceptual and technical levels, cannot be obtained quickly and without pain: There does not exist a path which would avoid the major difficulties since these have been proved to be real epistemological obstacles, in Bachelard's sense (Bachelard 1938; Artigue 1992). We would like to think of the learning process as a continuous spiral process. It is more like something partly chaotic, a mixture of continuities and ruptures, interweaving different concepts in complex cognitive networks. And teachers have to pilot such a learning process, by elaborating long term strategies and trying to

avoid two opposite risks: too precocious formalization and structural presentations, on the one hand; mere reduction to an algebraic and algorithmic analysis,
on the other. They also have to keep in mind that the intuitive and experimental
approaches, which seem now in some sense unavoidable, will have successful
results only if:

- experimental activities do not become, for the students, mere pottering about
  without any precise finality,
- the necessary ruptures with erroneous or partly erroneous intuitions, the
  necessary adjustments of provisory definitions, statements and formulations
  are explicitly and carefully managed,
- one succeeds in finding an adequate equilibrium between conceptual and
  technical concerns.

Some didactic engineering productions tend to incorporate all these requirements and look, at least under experimental conditions, reasonably effective. But
we have to confess the high didactic cost of these productions and the difficulty
of their survival in standard environments with standard teachers: They are so
far removed from the traditional ways of teaching!

The lack of success, for a large number of our students, of a logic of teaching
based only on the logic of present mathematics has been clearly proved. Now we
want to combine two coherencies: the mathematical coherence which is epistemologically fundamental and the cognitive coherence – as far as we can know it
– of the students.

There is no doubt that it is a difficult and challenging task. Nevertheless, I
would stress that, as a researcher, both from my personal work in this domain
and from the numerous contacts and discussions I have had with other researchers, I have learned a great deal. I am convinced that this didactic knowledge, without making the teaching of analysis miraculously easy, would likely
be helpful for teachers.

But I have to confess that we are far from this point – even with the forms of
didactic knowledge which certainly could be taken into account, at least partly –
without requiring costly and perhaps uncertain changes in the organization of
curricula: forms of didactic knowledge allowing diagnosis, analysis of
difficulties encountered by students, and interpretation of answers and errors.

Without any doubt, didacticians bear a lot of responsibility as they tend to
under-estimate the necessary work required by communication outside the strict
community of researchers. Without any doubt also, the didactic community encounters some difficulty in setting up the problem of adapting and diffusing
products initially elaborated for research needs and experimental work and it
does not involve itself seriously enough in the difficult engineering work
(Artigue 1994a) required by such adaptation and diffusion.

In order to address this issue correctly, didacticians have to recognize that
didactics is only one among the various significant perspectives in mathematics
educational research. Its role is not to take charge of all the issues linked to the

real management of mathematics at school. Due to the privileged link of the didactic approach with the mathematical discipline itself, didactic work relies more or less implicitly on a sound priority being given to the mathematical content and the relationships of teachers and students with this content. Even if we seriously take into account the social character of mathematics learning, the institutional constraints and the way they shape the didactic relations, nevertheless this privileged link with mathematics tends to partly occlude what, in the classroom mathematics life and, as a consequence, in the teaching and learning processes, plays a determining role – without appearing specific to mathematics teaching and learning.

Working on the necessary adaptations of experimental products and eventually specific training for possible users, considering the costs of introducing such products, the possible resistance of the institution and its actors and questioning their compatibility with present teaching – all this is necessary if one wants to have a positive action on the educational system. But this is a specific work, different from the usual research work and has not to be left to the sole responsibility of didacticians. It obviously has to be carried out in a collaborative way, not only by didacticians, but also by mathematicians, by teachers at secondary and tertiary level, and also, at least at certain times, by educational researchers from other origins. The input from secondary school teachers is necessary, not only because they are the potential users of didactic research, but also because they are professional experts: we know well that, in this kind of engineering work, one always has to fill the holes of the scientific work by using empirical and professional knowledge. The educational researchers enable us to take into account the limitations of the didactic perspective.

Didactics of mathematics, as other scientific fields, is both a fundamental field of research and an applied one. There is no doubt that the kind of work discussed above is crucial for developing reasonable and productive relationships between these two facets of the field, and, beyond that, between the didactic community, the mathematical community and, more widely, the whole mathematical educational world.

Moreover, in order to create the synergy I am arguing for in this chapter, didacticians have also to pay more attention to the demands of the educational system whenever they consider that these are compatible with a scientific work.

Faced with the challenge mathematics teaching has now to take up, in promoting mathematics for all and not only for future mathematicians and scientists, I hope to see this kind of synergy of competences develop, with mutual respect, and to keep alive our desire, even if it appears crazy, of being able to move mountains.

ACKNOWLEDGMENT

I acknowledge Mary Winter for her help in the English translation.

NOTES

1. For instance, G. Darboux was the editor of a 'Cours Complet de Mathématiques Elémentaires' (Colin Publisher) with the collaboration of J. Tannery ('Leçons d'Arithmétique Théorique et Pratique'), J. Hadamard ('Leçons de Géométrie Elémentaire'), F. Tisserand & H. Andoyer ('Leçons de Cosmographie'), C. Bourlet ('Leçons d'Algèbre Elémentaire'). We can also mention the famous book: *Notions de Mathématiques*, written by J. and P. Tannery.

2. IREMs: Instituts de Recherche sur l'Enseignement des Mathématiques. The first IREMs were created in 1969. There are now 25 and their activities are nationally coordinated.

3. COPREM: Commission Permanente pour la Réflexion sur l'Enseignement des Mathématiques; GREM: Groupe de Réflexion sur l'Enseignement des Mathématiques; CNP: Comité National des Programmes.

4. The operation '50 Lycées' first and then the association 'Math en Jeans', involving mathematicians in research activities with secondary students, in some sense, are typical of this tendency. We can mention also competitions such as 'Le Kangourou des Mathématiques' and note that 'Math en Jeans', in 1992 and 'Le Kangourou' in 1994, have received the prize 'D'Alembert' awarded by the SMF (Société Mathématique de France) for exemplary actions in the diffusion of mathematics.

5. IUFM: University Institutes for Teacher Training. They were created in 1991.

REFERENCES

Alibert, D. et al.: 1989, *Procédures Différentielles dans les Enseignements de Mathématiques et de Physique au Niveau du Premier Cycle Universitaire*, IREM Paris 7, Paris.
Artigue, M.: 1992, 'The Importance and Limits of Epistemological Work in Didactics', in *Proceedings of PME XVI* (Durham) 3, 195–216.
Artigue, M.: 1994a, 'Didactical Engineering as a Framework for the Conception of Teaching Products', in R. Biehler, R. W. Scholz, R. Sträßer and B. Winkelmann (eds.), *Didactics of Mathematics as a Scientific Discipline*, Kluwer Academic Publishers, 27–40.
Artigue, M.: 1994b, 'Réformes et Contre-Réformes dans l'Enseignement de l'Analyse au Lycée au XXème Siècle en France', in B. Belhoste, N. Hulin and H. Gispert (eds.), *Réformer l'Enseignement Scientifique: Histoire et Problèmes Actuels*, INRP, Paris.
Bachelard, G.: 1938, *La Formation de l'Esprit Scientifique*, Librairie J. Vrin, Paris.
Beke, E.: 1914, 'Rapport Général sur les Résultats Obtenus dans l'Introduction du Calcul Différentiel et Intégral dans les Classes Supérieures des Etablissements Secondaires', *L'Enseignement Mathématique* (16), 246–284.
Belhoste, B.: 1990, 'L'Enseignement Secondaire Français et les Sciences au Début du XXᵉ Siècle', *Revue d'Histoire des Sciences* **XLIII**, 4, 371–399.
Bourlet, C.: 1910, 'La Pénétration Réciproque des Enseignements de Mathématiques Pures et de Mathématiques Appliquées dans l'Enseignement Secondaire', *L'Enseignement Mathématique* (12), 372–387.
Brousseau, G.: 1994, 'Perspectives pour la Didactique des Mathématiques', in M. Artigue, R. Gras, C. Laborde and P. Tavignot (eds.), *Vingt Ans de Didactique des Mathématiques en France*, La Pensée Sauvage éditions, Grenoble, 51–66.
Chevallard, Y.: 1985, *La Transposition Didactique* (2nd ed. 1991), La Pensée Sauvage éditions, Grenoble.
Chevallard, Y.: 1992, 'A theoretical approach to curricula', *Journal für Mathematikdidaktik*, **13** (2–3), 215–230.

Commission interIREM Analyse (ed.): 1981, *L'Enseignement de l'Analyse*, IREM de Lyon, Lyon.

Cornu, B.: 1981, 'Apprentissage de la Notion de Limite: Modèles Spontanés et Modèles Propres', *Proceedings of the Fifth Conference of the International Group for the Psychology of Mathematics Education*, Grenoble, 322–326.

Dubinsky, E. & Harel, G. (eds.): 1992, *The Concept of Function: Some Aspects of Epistemology and Pedagogy*, MAA Notes, 25.

Fey, J.: 1994, 'Eclectic Approaches to Elementarization: Cases of Curriculum Construction in the United States', in R. Biehler, R. W. Scholz, R. Sträßer and B. Winkelmann (eds.), *Didactics of Mathematics as a Scientific Discipline*, Kluwer Academic Publishers, Dordrecht, 1994, 15–26.

Legrand, M.: 1993, 'Débat Scientifique en Cours de Mathématiques et Spécificité de l'Analyse', *Repères IREM* (10), 123–159.

Poincaré, H.: 1904, 'Les Définitions en Mathématiques', *L'Enseignement des Mathématiques* 6, 255–283.

Tall, D. (ed.): 1991, *Advanced Mathematical Thinking*, Kluwer Academic Publishers, Dordrecht..

Tietze, U.: 1994, 'Mathematical Curricula and the Underlying Goals', in R. Biehler, R. W. Scholz, R. Sträßer and B. Winkelmann (eds.), *Didactics of Mathematics as a Scientific Discipline*, Kluwer Academic Publishers, Dordrecht, 41–53.

Michèle Artigue
IUFM de Reims,
32 Bd Ledru Rollin,
51100,
Université Paris 7
France

ANNA SFARD

THE MANY FACES OF MATHEMATICS: DO MATHEMATICIANS AND RESEARCHERS IN MATHEMATICS EDUCATION SPEAK ABOUT THE SAME THING?

If a mathematics educator studies mathematics, is it the same object for him or her as it is for a mathematician who studies mathematics?, ask the authors of the Discussion Document. The very fact that the problem was raised is the first indication of a difference between the mathematician's and the mathematics education researcher's approaches to mathematics. Indeed, while the question imposes itself on the latter, it is not very likely to be asked by the former. Mathematicians seem to have little doubts as to the nature and the identity of the subject they deal with on a daily basis. Most of them would claim that there is only one object that can be called Mathematics and anything that differs from this unique exemplar cannot be given the same name. More often than not, the working mathematician is a Platonist, even if only tacitly: his or her mathematics has an appearance of an ideal, well-defined body of knowledge, faithfully mirroring a certain mind-independent reality of abstract ideas.

Even though the view held by mathematics education researchers might have been not much different just thirty years ago – when mathematics education started to emerge as a discipline in its own right – today a serious conceptual gap seems to exist between the two communities. During these last three decades the epistemological position of the researcher in mathematics education underwent a long chain of metamorphoses and dramatically shifted away from the classical stance of the mathematicians. Following the lead of philosophers of science, and feeling that they were now members of the community of psychologists and sociologists, rather than of scientists and mathematicians, mathematics education researchers arrived at a vision of mathematics that was consonant with the general conception of human knowledge developed by such contemporary thinkers as Kuhn, Feyerabend, Lakatos, Rorty, and Foucault, to name but a few. As a result, they are now torn between two incompatible paradigms – to both of which they have a certain commitment – either willingly or just because of the nature of the 'subject matter' they have chosen to deal with. On the one hand, there is the paradigm of mathematics itself where there are simple, unquestionable criteria for distinguishing right from wrong and correct from false. On the other hand, there is the paradigm of social sciences where there is no absolute truth any longer; where the idea of objectivity is replaced with the concept of intersubjectivity, and where the question about correctness is replaced by the concern for usefulness.

Sierpinska, A. and Kilpatrick, J. Mathematics Education as a Research Domain: A Search for Identity, 491–511.

Thus, if mathematics education researchers are to make a sensible (and generally acceptable) contribution to the theory and practice of learning and teaching mathematics, they have to find their way out of the conflicting allegiances. It seems that the first necessary step towards this goal is to clarify what the word 'mathematics' means to them *in relation to what it means to mathematicians*. At the ICMI Study Conference in Maryland some participants expressed their doubts as to whether or not mathematics education researchers should ask themselves this question: some of them went so far as to wonder whether they should care about the mathematicians' standpoint at all. In this paper, after I tell the story of the mathematics education researcher's drifting away from mathematicians, I shall explain why the resulting gap between the two communities should be viewed by the former (if not by the latter) as a problem requiring attention and thought, and why mathematics education researchers cannot afford the luxury of ignoring the question where they stand with respect to the 'expert practitioners'.

1. THE PROBLEM: WHAT IS MATHEMATICS?

In a paper which, in a sense, summarizes his work, so far, as a mathematics education researcher, Dreyfus (1994) makes 'an argument for ... visual reasoning'. Like many of his colleagues (see e.g. Presmeg 1986; Eisenberg & Dreyfus 1989; Vinner 1989), Dreyfus believes that reasoning based on information provided by graphical representations of mathematical concepts should be regarded as *par excellence* mathematical and, as such, should receive attention of both researchers and curriculum developers. Nowadays, not many educators would contest this view and this fact is clearly reflected in the nature of curricular changes suggested by leading professionals. The spirit of the reform envisioned by mathematics educators is well rendered in the following statement by Schwartz and Yerushalmy (1992):

Solving equations and inequalities using graphical techniques is presented in some algebra curricula as a procedure to be invoked [only] when no analytic techniques are available.... [However,] when the technology permits the easy plotting of functions, this becomes an attractive as well as efficient technique both for solving equations and inequalities and for understanding the essential nature of these constructs (Schwartz & Yerushalmy 1992, p. 285).

As a consequence of this and similar recommendations, visually-oriented activities are being introduced into mathematics classrooms and their place in the curriculum is steadily growing. A representative example of a problem designed according to the guidelines of the 'visual' approach is presented in Figure 1. The majority of researchers would qualify this kind of activity as genuinely mathematical and would therefore embark eagerly on investigating learners' responses, believing that this may be a good opportunity to get a better understanding of typical cognitive processes involved in mathematical thinking.

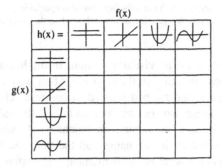

Figure 1 *An example of 'visually-oriented' activity (Yerushalmy 1993)*

The goal: To diagnose different types of rational functions and describe their behavior

The task: Use the software *Algebraic Supposer* to find relationships between 'ratios' of the different graphs

As natural and obvious as this position may seem to educators, it is far from fully admissible in the eyes of some mathematicians. Indeed, the latter would often object to calling activities like the one in Figure 1 'typically mathematical' and they would protest against giving them prominence in curricula. I heard such objections time and again in the course of my nearly two-decades-long collaboration with mathematicians who participated in designing new mathematics curricula for the Israeli high schools. This point was made particularly clear by one of them, a prominent algebraist, Shimshon Amitsur, in a conversation I had with him before the ICMI Study conference (the whole interview is included in this volume):

Question: During all the years we have been working together, I have noticed that when you examine teaching materials you label them in a very peculiar way. Your verdicts are often something like 'This is not mathematics'. I have always asked myself what were your criteria for this kind of judgment.

Answer: Let me answer with an example. Somebody teaches inequalities and draws a graph. This is an exercise in finding a visual expression for a mathematical phenomenon. However, if I wish to solve an inequality or to prove a theorem, I cannot use these tools. My final goal is that the students become able to do all this by mathematical means only.

When I say 'this is mathematics', I know exactly, even if not explicitly, what kind of thinking is involved in the process at hand, what are the aims of whatever is being done and what kind of facts may be necessary. When we teach mathematics we often show to students phenomena that might be interesting, but are not truly mathematical.... Being able to discover mathematical facts by the help of graphs is definitely not enough. One has to be able to prove the facts in truly mathematical ways. He must also be able to draw conclusions....

The observations may certainly help the student; exactly like seeing shapes and bodies helps in geometry. This is definitely true. I am all in favor of using this additional sense – the sense of seeing – which in the past didn't play almost any role in mathematics. But in mathematics proper, in mathematics itself, what I really have to know is something different: It is how to draw conclusions from things I know, about things I don't yet know. This cannot be done with pictures and other visual representations. In mathematics, I teach not only facts, but also special ways of thinking. The student must know how to construct proofs – proofs that won't depend on the specific drawings produced by the particular computer I use.

The debate on the status of visual argument in mathematics is but one of many issues that keep surfacing in the ongoing, often only implicit, discussion between mathematicians and mathematics education researchers. The visions of mathematics held by these two communities seem to differ substantially in more than one respect. To be sure, many mathematicians would admit today, however reluctantly, that mathematics itself changes all the time and the criteria according to which one decides what is 'mathematical' are different now from what they were a few centuries, or even a few decades, ago (Lakatos 1976; Davis & Hersh 1981; Kitcher 1984; Tymoczko 1986; Ernest 1994). However, the changes currently taking place within the context of research in mathematics education do not necessarily parallel those that alter the face of 'professional' mathematics. In fact, the disparity between the two communities would often go much farther than may be understood from the example I just gave. In its most extreme form, the discussion is more about the basic epistemological beliefs and general conception of human knowledge than just about mathematics. Although, on the face of it, mathematicians and mathematics education researchers deal with the same 'subject matter', the fact that they come from completely different paradigms is likely to make their views on mathematics incommensurable rather than merely differing in some points. The difference may go so deep as to render the question, asked by the authors of the Discussion Document, meaningless in the eyes of some mathematics education researchers. Indeed, as I will try to show in the brief review in the next section, the very idea of referring to mathematics as a well-defined self-contained 'object' is incompatible with new conceptual frameworks gaining a growing recognition in certain circles.

2. MATHEMATICIAN'S MATHEMATICS VS. MATHEMATICS EDUCATION RESEARCHER'S MATHEMATICS: A BRIEF HISTORY OF THE COMPLEX RELATIONSHIP

The diversity in people's thinking about mathematics and in their mathematical practices seems to entail some obvious questions: To what meaning of the word 'mathematics' do people subscribe when they identify themselves as researchers in the field of mathematics education? Which is 'the' mathematics that lies at the heart of their professional interest and provides criteria for evaluation of research results? The blatantly naïve question, *What is this elusive object called mathematics and where are we supposed to look for it?* – a question often ridiculed by 'the insiders' (see Mura 1993), and carefully avoided by many cautious 'outsiders' – may nevertheless be of importance for those mathematics educators who feel they have yet to resolve the problem of their professional identity. In order to deal with the issue, however, one has to dive into the general quandaries of human knowledge and tackle such questions as where this knowledge is to be found or how it can be delineated and defined as an entity in its own right – if at all. Considering the weight of these problems, the amount of paper that has been filled through centuries with attempts to solve them, and the stature of the thinkers who voiced their stance on these matters, such a project must seem too ambitious for this brief exposition – to say the least. I will tackle it, nevertheless, aiming at refining the questions rather than at formulating answers.

As a starting point for a future discussion I will survey, in historical order, a number of generic outlooks which – each one in its turn – constituted a general epistemic paradigm for research in mathematics education at one time or another. Inevitably, while doing this, I will make certain extrapolations, ascribing certain views to communities which are certainly neither homogeneous in their thinking nor fully unified in their practices. Such generalizations may have their shortcomings but one can hardly avoid them while trying to contribute to the task of establishing an identity of a professional group.

Along the way I will be contrasting the outlooks adopted by mathematics education researchers with views of the Typical Mathematician – a synthetic character whose convictions and opinions are believed to be representative of the community of today's mathematicians. In a sense, the Typical Mathematician is the very same person who was satirically portrayed by Davis & Hersh (1981) as an Ideal Mathematician.[1] My prototypical character will be ascribed views that are rarely explicitly presented, but can often be read between the lines of the professional mathematical writings. In general, not much systematic work has been done to explore today's mathematician's vision of mathematics. A rare exception is the study by Mura (1993) who investigated the views held by university teachers of mathematics in Canada (and one may assume that this group includes research mathematicians). The other sources, from which I will be drawing to synthesize the portrait of the Typical Mathematician, are my own studies (Sfard 1994), as well as my non-formal knowledge coming from a long collaboration with a number of research mathematicians.

In spite of the scarcity of systematic research, it is not impossible to create a reliable picture of the ontological and epistemological beliefs that shape today's mathematical practices. The existing findings match the opinions held by the many writers (e.g. Davis & Hersh 1981; Rotman 1994; Ernest 1995) who claim that the majority of mathematicians still adhere to the Platonist tradition, whether knowingly and willingly (like Goedel or Thom, and like the majority of my interviewees), or implicitly, as if against their will (like Davis & Hersh's Ideal Mathematician who declares himself a Formalist but acts and feels like a Platonist; or like Mura's respondents who do not necessarily formulate a truly Platonic definition of mathematics, but who nevertheless subscribe to views which usually go hand in hand with this position). As Rotman (1994) stated, even today, 'logicist/Platonist insistence on formalizable exact mathematics and its accompanying programme of foundations and rigor' should be viewed as a mainstream approach to mathematics. For instance, Amitsur's refusal to recognize visually-oriented activities as genuinely mathematical may be interpreted as an indication of his adherence to this traditional school of thought. Indeed, it is within this tradition that 'diagrams, as so-called "merely explanatory", motivational and heuristic devices, have been excluded from what officially counts as mathematical sense at the same time as linear symbol strings ... have been elevated to mathematics proper' (ibid., p. 80).

To sum up, a logicist/Platonist outlook is alive and well in spite of the recent revolutionary developments in the philosophy of mathematics. This is certainly true, even if many of today's mathematicians are crypto-Platonist, rather than declared followers of this old school. No wonder, then, that the story of evolving epistemic frameworks for mathematics education research is a story of mathematics education researchers moving farther and farther away from the place many of them originally came from. While the Typical Mathematician, known for his lack of interest in philosophy (Mura 1993), seems entrenched in his old logicist/Platonist positions, mathematics education researchers follow the lead of a quickly developing philosophy of mathematics (which goes hand in hand with the contemporary philosophy of science, and more generally, with modern non-objectivist, non-Cartesian epistemology) thus steadily increasing the gap between themselves and the mathematicians. This 'drifting away' will now be briefly surveyed in the remainder of this section.

Platonism

In the late sixties and seventies, when a quantum leap in the amount of studies in mathematics education signaled that the hitherto marginal topic began to turn into a fully-fledged domain of research (Kilpatrick 1992), the question 'What is mathematics?' did not seem to bother anybody except a handful of philosophers. The answer was obvious: Mathematicians' mathematics – the one that can be found in professional mathematical texts – was the only true mathematics, the Mathematics-with-the-capital-M. This ostensibly well-defined body of statements was taken as a touchstone of anything a 'mathematizing person' – notably, the student – would do. There was a consensus that mathematicians' mathematics was the 'real thing' while other people's versions were but imperfect imitations. What gave mathematicians' mathematics this unchallengeable status was the tacit belief that mathematicians do not invent mathematics but rather serve as media through which the mind-independent mathematical truth is communicated to the world. In contrast, what the students did and thought could hardly be regarded as mathematics: Rather, it was seen as a mere mixture of correct, partially correct, and mainly incorrect notions; the latter being known in a professional language as *misconceptions*. The non-standard ideas that popped up in the classroom were regarded as 'road accidents' that could be avoided if only the teaching had been better. The idiosyncratic students' conceptions were viewed as the result of failed attempts to convey to them what true mathematics was all about.

The early research in mathematics education leaned heavily on such a Platonic view of the nature of mathematics, even though almost nobody bothered to explicitly spell it out. This Platonic stance was deeply rooted in the long tradition of ontological Realism and epistemological Objectivism (see e.g. Lakoff 1987; Johnson 1987; Rorty 1991). To put it in Ernest's (1985) words,

> By locating the source of mathematics in a pre-existing static structure, Platonism results in a static body-of-knowledge view of mathematics. Platonism discounts both man as a creator of mathematics and the importance of dynamic processes in mathematics. In educational terms this corresponds with the view of mathematics as an inert body of knowledge which instruction transmits to the student (p. 607).

Research reports of this period bring clear-cut distinctions between right and wrong and are usually highly judgmental. The general tone is that of despair over the deplorable condition of mathematics education and of annoyance with teachers who fail to inculcate the correct conceptions. The following exclamation from an unpublished research report, although written already in the 80s, may nevertheless be considered as representative of this style:

> How can we explain and justify the astonishing fact that a new concept in basic mathematics [function, as presented by the Dirichlet-Bourbaki definition] did not diffuse sufficiently into the common practice of teaching and into the minds of professional teachers after a century...

All this is hardly surprising when one recalls that at that time, many of the researchers were former mathematicians brought up in the tradition of logical positivism. Imbued with the culture of mathematics departments, they could not help bringing the mathematicians' attitudes into the field of education.

Mathematicians were also the first reformers of mathematics curricula. Quite understandably, the rigor-minded conceivers of what came to be known as the New Math movement tried to put the 'true' mathematics into school textbooks. Here, again, is a telling example of the mathematician's exhortation to teach the 'correct' mathematics: The author first states that 'some degree of formalization is necessary and unavoidable ... [because in] this form the ideas can be transferred without personal contact into another's mind', and then offers some pedagogical advice:

> It would appear that a lot of trouble can be avoided if we prepare our students carefully, introducing a sound idea of functional dependence within the framework of set theory.... (ibid.).

It may be the mathematicians' unquestionable authority which transferred the certainty about mathematics to the sphere of schooling and which contributed to the 'overwhelming temptation to view the subject matter as given, inevitable, natural' (Noss 1994). The absolute reign of mathematicians in the field of education collapsed only when the great disillusionment came – when there was no doubt left that the school is not a proper place for practicing the 'true' mathematics. Following a spectacular failure of the pedagogical project, inspired by

Bourbaki's philosophy, curriculum developers came to realize that the so-called 'subject matter' cannot be directly imported to the classroom from the mathematician's desk. Before it happens, 'the material' is bound to undergo a didactic transposition (Chevallard 1985). This transposition would often change the face of mathematics beyond recognition. Although more and more people were prepared to acknowledge the inevitability of the compromise, most of them would nevertheless deplore its results.

Constructivism

First doubts about the Platonic framework in mathematics came from the constructivist school. Constructivism has many faces and branches, and the name encompasses philosophies aiming at different issues and having little in common with each other. Some conceive of it as an alternative to Realism; some others as an answer to Objectivism; still others accept both these interpretations. The name has been given to schools of thought in such different fields of inquiry as philosophy of science, philosophy of mathematics, cognitive psychology, and education. In spite of the confusing diversity, some overriding themes, common to the broad variety of seemingly unrelated trends, make it possible to talk about a Constructivist framework – a more or less coherent whole that offers itself as an alternative to the Cartesian paradigm. Let me sketch the basic tenets of Constructivism in just a few sentences.

First, the term may be used to refer to philosophy of mathematics suggested by such mathematicians as Kronecker and Brouwer as a replacement of Platonism. The Constructivist regards, as genuine mathematics, only what can be obtained by a finite construction. Intimately connected with this view is the intuitionist stance according to which 'mathematics is a self-sustained intellectual activity that deals with mental constructions governed by self-evident laws' (*Encyclopaedia Britannica*).

This position, although limited to mathematics, has much in common with the post-Piagetian psychological version of constructivism. Constructivist psychologists challenge the simplistic realist vision of learning as a passive absorption of information and claim instead that 'knowledge cannot simply be transferred ready-made from parent to child or from teacher to student but has to be actively built by each learner in his or her own mind' (von Glasersfeld 1991).

Constructivism brings an air of tolerance and openness towards students' thinking: If knowledge is a personal construction, one should no longer refer to this knowledge as a mere reflection of some ideal exemplar. There simply is no such thing as an ideal exemplar. As a result, the term 'misconception' is replaced by some authors with such notions as *alternative framework* or *novice (naïve) conception* (Driver and Easley 1978; Smith, diSessa & Roschelle 1993). Students' thinking, even if visibly different from that of the experts, is no longer dismissed as devoid of any value. The learners stop being accused of 'thoughtlessness' and 'meaningless behavior'. The researchers began to realize that 'students usually *do* deal with meanings, and when instructional programs fail to develop appropriate meanings, *students create their own meanings'* (Davis 1988). The change of language marks the change of the researcher's attitude. Students' conceptions are now to be taken seriously and never again regarded as a mere result of failed attempts at imitating mathematicians' thinking. With the advent of the constructivist approach to learning, the students are transformed from mere recipients of the God's-eye-view, transmitted by textbooks, into active builders of their knowledge. As an immediate side effect of this change comes the recognition that the learner's mathematics might be in fact much closer to the mathematician's mathematics than the adherents of simplistic dichotomy between right and wrong would admit. When students' learning is watched with an unprejudiced eye, one notices time and again that certain ways of thinking about mathematical ideas, far from being mere 'mistaken conceptions', are natural, inevitable stages in concept development. To put it differently, there are learning invariants that can hardly be changed by the way the concepts are taught and, more often than not, display a striking similarity to early notions suggested by mathematicians themselves through history.

Although constructivism had a sweeping impact on the research community, it soon ran into difficulties because of the unconstrained relativism implied by its early version. It was attacked for its inability to give a thorough explanation of the apparent objectivity of knowledge. Some critics went so far as to accuse it of being a distant cousin of solipsism. The most outspoken protagonists of constructivism did not conceal their own doubts. Recalls Paul Cobb: 'Since 1983 I have been troubled by sociological and cultural phenomena which seemed to fall outside my admittedly immature grasp of constructivism' (in von Glasersfeld (ed.) 1991, p. vii).

The early version of constructivism conspicuously lacked a social dimension. This all-important element was readily provided by the rediscovered writings of the members of the Vygotskian group. The cross-breeding between the old version of constructivism and the new focus on social practices proved successful and resulted in a new brand of the old framework, known as social constructivism (Ernest 1991) or social interactionism (Cobb 1995b).

Social Constructivism

Vygotsky, like Piaget, tells us that knowledge is a human construction and that the source of this knowledge is in human actions; but while Piaget deals with a cognizing person in almost total isolation from anything except his or her physical surroundings, Vygotsky stresses the crucial importance of social interaction. This difference is epitomized in the widely-used, and often abused, Vygotskian concept of zone of proximal development. The ZPD, as it is called for short, may be viewed as a margin left by Vygotsky (and forgotten by Piaget) for all these elements of learning which cannot be accounted for by a reference to the learner's own intellectual activity and must therefore be grounded in interpersonal exchange (many Vygotskian scholars would claim that what I called 'a margin' plays the most central role in learning).

Vygotsky's version of constructivism combines well with the constructivist philosophy of science promoted by Kuhn and his followers. The hallmark of the scientific constructivism is 'the view that subject matter of scientific research is wholly or partly constructed by the background theoretical assumptions of the scientific community and thus is not, as realists claim, largely independent of our thoughts and theoretical commitments' (Boyd, Gasper & Trout 1991). Translated to mathematics, social constructivism becomes a rich and useful doctrine. It brings a new vision of mathematics:

Social constructivism views mathematics as a social construction. It draws on conventionalism, in accepting that human language, rules and agreement play a key role in establishing and justifying the truth of mathematics. It takes from quasi-empiricism its fallibilist epistemology, including the view that mathematical knowledge and concepts develop and change. It also adopts Lakatos' philosophical thesis that mathematical knowledge grows through conjectures and refutations, utilizing a logic of mathematical discovery (Ernest 1991, p. 42).

Like the earlier radical constructivism, this vision of mathematics is dramatically different from the one proposed by Platonists. Unlike Piagetian constructivism, however, social constructivism tries to offer an answer to relativism. Although it does not leave much room for the traditionally conceived objectivity, it makes up for the loss of certainty with the help of the idea of *intersubjectivity*. Today's mathematics education researchers must still tame their need for 'objectively correct conceptions', but they have some substitutes to adhere to: they can speak about conceptions that are *taken-as-shared*, which are built through *negotiation of meaning*, and which are accepted by the *community of practitioners* (Cobb 1995b). Mathematics education researchers working within this framework focus their attention on the process of learning and on classroom interactions rather than on the solitary activity of a lone learner. Their pedagogical advice is to promote the mode of instruction which is based on group work (cooperative learning is one variety of this type).

Mathematics emerges from all this with a changed identity, but with an identity, nevertheless. Constructivist mathematics education researchers do not speak any more about the Mathematics-with-the-capital-M, but they still view mathematical knowledge as a clearly delineated aspect of social activity which can be dealt with as a self-contained whole. As time goes by, and social constructivism gains strength and recognition, however, this position becomes increasingly difficult to sustain. At a closer inspection mathematics turns out to be so deeply embedded in ongoing ever-changing social practices that it seems practically impossible to refer to it as an object in its own right.

In fact, the question whether knowledge can be objectified and seen as a self-sustained entity is not new:

The fundamental philosophical question of the disobjectivation of *knowledge* has been with us for centuries, starting with Plato. Essentially, disobjectivation is the problem of grasping thought and life in their dynamic and 'open' form before they are 'finished', before they become *things*.... [To Hegel and Neo-Kantians] Science ... appeared not as a cumulative account of incontrovertible facts

but as a continuous logical effort to construct the object of inquiry. These early attempts at disobjectivation of the history of science are echoed in the more recent and more radical notions of scientific
revolutions and anarchistic theory of knowledge (Kozulin 1990, pp. 22–23).

With the advent of social constructivism, the urge for 'disobjectivation' can
no longer be suppressed. Recently, two schools of thought came up with proposals of solutions. First, there is the theory of situated knowledge which, in its
most radical form, fully complies with the request of disobjectivation; second,
there are the postmodern and poststructural movements which bypass the
quandary of knowledge, and its status, by focusing on externally observable
attributes: the scientific text and, more generally, scientific discourse.

Situated Knowledge

Futile efforts to find satisfying interpretations of the concept of knowledge at
large, and of 'correct' knowledge in particular, resulted first in historicism: in
the claim that human ideas are so deeply and insolubly immersed in historical
context that reifying them as self-sustained objects of study, which can be dealt
with independently of this context, comes to be viewed as a mistake. Another
step in the same direction is now being made by those who promote the idea of
situated cognition and situated knowledge (Brown, Collins & Duguid 1989;
Lave & Wenger 1991). The doctrine of situated learning is a special edition of
the theory of social practice (and social constructivism) which

emphasizes the relational interdependency of agent and world, activity, meaning, cognition, learning,
and knowing. It emphasizes the inherently socially negotiated character of meaning and the interested, concerned character of the thought and action of persons-in-activity. This view also claims
that learning, thinking, and knowing are relations among people in activity in, with, and arising from
the socially and culturally structured world. This world is socially constituted; objective forms and
systems of activity, on the one hand, and agents' subjective and intersubjective understanding of
them, on the other, mutually constitute both the world and its experienced forms. Knowledge of the
socially constituted world [of which mathematics is a part] is socially mediated and open-ended. Its
meaning to given actors, its furnishings, and the relations of humans with/in it, are produced, reproduced, and changed in the course of activity. In a theory of practice, cognition and communication
in, and with, the social world are situated in the historical development of ongoing activity (Lave &
Wenger 1991, pp. 50–51).

To sum up, adherents of this school deny the possibility of dealing with any
aspect of human practice in isolation from all the other constituents of this
practice. This statement obviously includes knowledge and learning which are but
emergent phenomena in a complex network of social interactions. Consequently,
science or mathematics cannot be considered as self-contained entities any more;
rather they have to be regarded as aspects of ongoing social activities. The
researchers must no longer insist on isolating knowledge from the totality of social
interactions. 'There is no activity that is not situated', say Lave and Wenger
(p. 53); therefore, there is no knowledge that is not situated. 'Even so-called general knowledge only has power in specific circumstances' (p. 32). Consequently,

the authors claim that it would be of no use to 'objectify' either learning or knowledge as objects of study in their own right: '[C]oming to see that the theory of situated activity challenges the very meaning of abstraction and/or generalization has led us to reject conventional readings of the generalizibility and/or abstraction of "knowledge"' (ibid., p. 37). For Lave and Wenger, the refusal to view knowledge as a self-sustained entity means also disobjectivation of learning; namely, '... shift away from a theory of situated activity in which learning is reified as one kind of activity, and toward a theory of social practice in which learning is viewed as an aspect of activity' (ibid., pp. 37–38).

The new mathematics education researcher, therefore, is interested in all the aspects of the situation within which the activity of mathematizing takes place. In his or her eyes, mathematics cannot be separated from a mathematizing person and a mathematizing community. To put it differently, the researcher has to investigate the activity of the community of practitioners in its totality, even if he or she is mainly interested in a particular aspect of learning.

The idea of situated knowledge was readily picked up by mathematics education researchers. First, the voices could be heard against objectivation of knowledge:

The treatment of issues as given objects, as knowledge, that is making-an-object-of-it (*Verdinglichung*), obscures the nature of related processes as currently performed and as highly situation specific 'accomplishments'.... The key difference is with the concept of knowledge as an object and the alternative interpretation of knowing as a current performing.... It is the difference between the use of product in a process versus flexible fixation of meanings in the current flow of social interaction. We therefore avoid the notion of 'knowledge', the notion as weak as 'information', and prefer to speak of 'knowing' or 'ways of knowing' (Bauersfeld 1995).[2]

Then, attempts have been made to apply the idea of situated knowledge and learning to research (see e.g. Cognition and Technology Group at Vanderbilt 1990; Greeno 1991). For all its special appeal, however, the theory does not translate easily into the domain of mathematics. Indeed, because of peculiarities of mathematics, the project is hindered by quite a few dilemmas. In fact, the Typical Mathematician is likely to say that the idea of situated learning is inherently inapplicable to mathematics.

First, in the case of mathematics, it is not all that obvious how the concept of contextualization should be interpreted, what is the meaning of the terms 'cognitive apprenticeship' and 'peripheral participation' (Lave & Wenger 1991), and who are the 'expert practitioners' whose task it is to guide the novices. If the 'context' is interpreted as a real-life situation which is rich enough in mathematical content to become for mathematics students what the craftsman's workshop is for the apprentice, then such a situation is obviously very difficult (if not impossible) to find. No wonder then, that the few pioneering projects in mathematics education which tried to apply the situated learning framework (e.g. the anchored instruction project of the Vanderbilt Group) proved somehow artificial: rather than providing 'real-life context' they created synthetic environments, often no less abstract than the mathematical content they wanted to convey. These proposals 'are confined to school-structured situations or simulated apprenticeships, rather than actual situations that relate directly to the student. They often fail to challenge the "culture of mathematics instruction" that exists in our schools' (Heckman & Weissglass 1994). The critics' credo is that 'mathematics will be learned by more students (1) if taught with other subjects in a real-world context, (2) if practical learning apprenticeships are developed'. The mathematicians, however, with whom I discussed the topics claimed that the critics' (Heckman and Weissglass) own project, conducted in the participants' natural surrounding and aimed at solving an authentic problem faced by the local community, seems to be too poor in mathematics contents to be seen as a true mathematical apprenticeship.

The mathematician's other concern regarding situated mathematics projects is of an even more serious nature, as it has to do with a possible inherent inadequacy of the idea of contextualization to the particular case of mathematics. The exhortation to contextualize, namely to eschew from dealing with 'distilled' mathematical content, may contradict the very essence of mathematization. After all, mathematizing is almost synonymous with abstracting. One may say that, in the most fundamental way, mathematics is about 'flying high' above the concrete, classifying things according to features that cut across contexts, and then synthesizing these detached features into new context-independent wholes. Mathematicians would claim, no doubt, that the ability to strip the bones of abstract structures of the flesh of concrete embodiments is the main source of mathematics' unique beauty and strength. When we restrict ourselves to 'contextualized mathematics', we are tying mathematics back to the concrete and particular. This is where the typical mathematician must begin wondering whether the thing we would then be teaching could still be called mathematics.[3]

Postmodern/Poststructural Approach

The denial of the possibility of Truth-with-the-capital-T and the resulting relativism seem to be at odds with the basic human need for permanence and stability. The concepts of 'objective knowledge' and 'absolute truth' are the result of human eternal pursuit of certainty and security (Rorty 1991). When these are taken away people lose their sense of direction. The once unshakeable belief in logic and reason is what guided humanity for ages. Most mathematicians still enjoy the luxury of certainty, but mathematics education researchers urgently need a replacement.

When everything becomes questionable, temporal, and dependent on human consent, people cling to those aspects of reality which seem least susceptible to relativization. Postmodernism offers the text and speech acts as a more secure, better defined object of systematic investigation than the inherently problematic notion of knowledge. This turn toward language as a primary focus of study was inspired mainly by the late work of Wittgenstein (1958) and was developed into a fully-fledged school of thought when *discourse* acquired the role of the fundamental notion and *discourse analysis* became the basic activity of the researcher.

According to Foucault, one of the most outspoken protagonists of the new para-digm, discourse is an assembly of 'serious speech acts' on which a special method of justification and refutation confers the right to be called knowledge and makes of them objects to be studied, repeated, and passed on to others (Dreyfus & Rabinow 1982, p. 48). Foucault presents himself as a person inter-ested in 'those familiar yet enigmatic groups of statements that are known as medicine, political economy, and biology' – and mathematics, of course:

I would like to show that these units form a number of autonomous, but not independent, domains, governed by rules, but in perpetual transformation, anonymous and without a subject, but imbuing a great many individual works.... I would like to reveal, in its specificity, the level of 'things said', the conditions of their emergence, the forms of their accumulation and connection, the rules of their transformation, the discontinuities that articulate them (Foucault 1972).

The first attempts to apply this way of looking at knowledge to mathematics have been made recently by some thinkers, notably Walkerdine (1988, 1994) and Rotman (1994). Rotman, for instance, chooses to view mathematics through the prism of mathematical writing. He briefly summarizes his 'discursive posi-tion made available by, and quite inseparable from, language' in a sentence in-spired by Peirce: 'There is nothing here but signs whose explication will always rest upon a call for more signs: "Man himself is a sign"' (Rotman 1994, p. 79). This claim may perhaps read as a statement made by a formalist; however, Rotman's further attempt to understand how mathematical objects are brought into being through discursive practices makes it clear that the similarity between his standpoint and that of Hilbert and his followers is deceptive, as the two outlooks come from completely different epistemological frameworks.

Postmodern writers take the social constructivist project a few steps further when they refuse to render any point of view the privileged position once held by 'mind-independent truth' and promote instead the pragmatic approach to knowledge, where socially grounded criteria of intersubjectivity and usefulness replace the idea of objectivity. In this pragmatist vein, Rorty (1991) underlines the importance of consensus within a community and its central role in deciding what should be regarded as 'true' and what dismissed as 'false':

I urge that whatever good the ideas of 'objectivity' and 'transcendence' have done for our culture can be attained equally well by the idea of a community which strives after both intersubjective agreement and novelty.... If one reinterprets objectivity as intersubjectivity, or as solidarity ... then one will drop the question of how to get in touch with 'mind-independent and language-independent reality' (Rorty 1991, p. 13).

The aspect of usefulness is explained by Lyotard (1992):

Scientific reason is not examined according to the (cognitive) criterion of truth or falsity, on the message/referent axis, but according to the performativity of its utterances, on the (pragmatic) axis of addressor/addressee. What I say has more truth than what you say, since I can 'do more' (gain more time, go further) with what I say than you can with what you say (Lyotard 1992, p. 63).

When translated to mathematics, this approach must once again evoke an indignation and protest from the Typical Mathematician. Like the idea of situated knowing, this relativistic vision of knowledge is clearly at odds with his Platonic or as-if-Platonic attitude. There is an inherent incompatibility between the postmodern doctrine and the epistemological foundations on which the whole edifice of his mathematics seems to rest. Thus, the postmodern outlook will have to sink well into the general awareness before its echoes can be heard in mathematics departments – this well-guarded bastion of rationality and objectivity. Indeed, it is highly unlikely – perhaps impossible – that the Typical Mathematician is swept away by the postmodern Zeitgeist. The Mathematician knows only too well that the postmodern relativism is a major threat to his whole intellectual existence. In the postmodern world he, with his vital dependence on the objectivist epistemology, may well turn into a relic of the past. Clearly, this does not apply to mathematics education researchers whose brief intellectual history made the non-objectivist turn almost inevitable. Together with other social scientists, they have been going postmodern for some time now (see e.g. Walkerdine 1988; Williams 1993; Dowling 1993; Brown 1994; Tsatsaroni and Evans 1994; Ernest 1995).

3. LIVING WITH THE DIFFERENCE: POWER GAME OR MUTUAL GAIN?

That the Typical-Mathematician's-mathematics and the mathematics-education-researcher's-mathematics came to be worlds apart seems undeniable. (It is to be understood that in order to avoid the fallacy of objectivation, I use the term 'mathematics' in the way Foucault used the terms 'medicine', 'economics', and, more generally, 'knowledge': by 'Typical-Mathematician's-mathematics' and by 'mathematics-education-researcher's-mathematics' I mean two different kinds of discursive practices, specific to these two communities). It is clear that the difference is too fundamental to be just dismissed or glossed over. Moreover, it seems that trying to fill in the gap in an attempt to make the two mathematics into one would be pointless. Indeed, we are faced here with systems of beliefs as distinct as those which separate incommensurable scientific paradigms, rivaling socio-economical doctrines or different religions. As it happens, the difference often yields animosity. From time to time, a member of one community will voice his or her objections to the conception of mathematics promoted by the members of the other community. The attempts to discredit the alternative outlook may sometimes be viewed as an outcry of an endangered species struggling for survival; more often than not, however, it is but a power game maneuver. The participants of this game delude themselves that they can settle the controversy by a rational argumentation. The two communities – that of Typical Mathematicians and that of new mathematics education researchers – live within paradigms which Kuhn and his followers would call incommensurable; therefore, judging one from within the other is equally absurd as, say, trying to prove or disprove a claim of Lobachevskian geometry using arguments based on the Euclidean axiomatic system.

The only question, therefore, that can and should be asked is whether this disparity matters, and if it does, what can be done in order to make it possible for both sides to live with it while keeping a fruitful conversation going. To put it differently, once we recognize and accept the fact that the worlds of mathematicians and of mathematics education researchers differ in some fundamental ways, the question is what kind of collaboration these two communities could create and what contributions each one of them might make in order to promote what seems to be their mutual goal: finding ways of enhancement of human learning and creativity. The problem has many faces and the answer – if any – depends on the particular aspect one chooses to focus on.

First, there is the issue of *mathematics curricula* and their contents. The Typical Mathematician has little doubt that, being the 'lawful owner' of the discipline, he is also 'lawfully entitled' to having a decisive voice on the kind of mathematics that should be learned in schools. To quote Amitsur once again, mathematics is mathematicians' business, and exclusively so:

Mathematicians [are the ones who should play the leading role in planning mathematics curriculum], there can be no doubt about that.... There is room for researchers, provided they serve only as consultants. If I wish to build my own business, I'll take a consultant to advise me on different matters; however, the consultant won't tell me what to do and all the decisions will be mine (Interview with S. A. Amitsur, this volume).

Be mathematics education researchers' view on mathematics as different from that of the Typical Mathematician as it may, many of them wouldn't deny mathematicians their right to have a major influence on the contents of mathematical curricula (see e.g. Sierpinska's observations on French didacticians (Sierpinska 1995a)). Many, but not all. Moreover, those who do choose to listen to mathematicians may nevertheless view mathematicians' role as more restricted than the latter would wish. For example, the proponents of visually-oriented activities seem undeterred by mathematicians' protests. Mathematics education researchers agreed a long time ago that mathematics comes to schools reincarnated (didactically transposed) as a 'subject matter' and they take it as a matter of fact that while '*mathematics* is mathematician's business, schools have *math*' (Noss 1994). It seems, however, that it would be wise to warn educators against going to the extreme and interpreting the inevitability of the didactic transposition as a sanction to ignore mathematicians' voice. On the contrary, in a somewhat paradoxical way, those who adhere to the social constructivist view of knowledge may be compelled to listen carefully to what 'expert practitioners' have to say. After all, when mathematics is viewed as a social construct, there may be no better definition of mathematics than the one which says that 'mathematics is what mathematicians do'.[4]

Neither curricular considerations, however, nor mathematics education research are restricted to the matters of mathematical contents. In deciding about mathematics programs, but even more so in the research on mathematical learning and teaching, mathematics must be seen not only as a ready-made product, but also as a process. According to what has been said above, many mathematics education researchers would go so far as to say that they view mathematics solely as a form of ongoing human activity. Consequently, they would often prefer to talk about mathematizing or mathematical knowing rather than about mathematics or mathematical knowledge. As Al Cuoco put it in a recent exchange in the Algebra Working Group network:[5]

> It makes no sense to me to think of 'standard mathematics' as being a body of knowledge. We've all seen people talk about trigonometry in ways that have nothing to do with mathematics, and we've also seen people talk about, say, cooking in decidedly mathematical ways. It's been more productive for me (in my teaching) to think of mathematics as a way of thinking, as a collection of habits of mind (Al Cuoco, unpublished).

Similar preference was declared by John Mason in a somewhat different way: 'I find it more useful to think in terms of what sort of [intellectual] *activity* I want to call mathematical' (personal exchange, my emphasis).

Shifting the focus of attention from the ready-made mathematical contents to the activity of mathematizing does not, however, resolve the old dilemma. Once again, the question may be asked, 'Who is to say what kinds of intellectual activity may be regarded as mathematical?'. Mathematics education researchers, while still looking for a satisfactory definition of the notion, may be reluctant to turn to mathematicians for an answer even though, on the face of it, mathematicians should be regarded as an authority also on this matter.

On a closer inspection, mathematicians and their written works may not be the best source of inspiration. As Gauss remarked, mathematicians like to 'make the scaffolding invisible' once the product of their efforts is rendered in its final form. Mathematicians do not choose to reveal the ways their mathematical work is being done. To be sure, there are notable exceptions, like Poincaré (1952), Polyà (1957) or Thurston (1994), who tried to capture the nature of mathematizing by reflecting on their own work as mathematicians. However, to have a better grasp of the nature of processes that can be viewed as mathematical, more input from mathematicians would certainly be desirable. Needless to say, mathematization by mathematicians and mathematization at a school level cannot be regarded as the same thing. Nevertheless, the study of mathematicians' ways of thinking (see e.g. Sfard 1994), is likely to bring some universal messages on the unique nature of mathematical processes and thus contribute also to our decisions as to what should count as children's mathematization.

4. CONCLUDING REMARKS

One more observation would be in place before I finish this chapter. Implicit in most of my remarks was the assumption that mathematicians' vision of mathematics is relatively stable and that the main question is whether this unchanging outlook may or should affect the way mathematics education researchers think about their 'subject matter'. In such a presentation an important aspect remained overlooked: the possibility of an influence going in the opposite direction – from mathematics education research to the mathematics of the experts. A closer look at the developments in today's mathematics will reveal that the likelihood of this 'reverse' influence is not purely theoretical and that whether it really takes place does not depend on anybody's will or deliberate decision. Indeed, there are some clear signs that mathematicians' vision of mathematics cannot remain totally immune to what is happening around, in general, and to what is going on in the field of mathematics education research, in particular. Like Amitsur, more and more mathematicians would admit, for example, that they are 'all in favor of using this additional sense – the sense of seeing – which in the past didn't play almost any role in mathematics' (Interview with Amitsur, this volume). Similarly, Davis (1993) tells us that 'almost two decades ago [he] wrote an article which argued that our view of mathematics ought to be widened so as to allow the inclusion of what [he] called "visual theorems"'. Davis feels now that his 'opinions have found a measure of approbation', and this development may probably be ascribed not only to the advent of computers and the subsequent growth in the availability of graphical representations, but also – and perhaps mainly – to the new insights into the nature of mathematical thinking brought about by research in mathematics education. Since mathematics education research is a young discipline, this is surely only the beginning of what may become, in the future, a 'beautiful friendship' – a truly reflexive relationship in which the difference is respected and the diversity becomes the source of mutual gain.

NOTES

1. For the sake of simplicity, the Typical Mathematician will be referred to as 'he' rather than 'he or she'. The preference for 'he' over 'she' is justified by the actual situation in mathematics departments all over the world and obvious statistical considerations.

2. Compare with the debate on this topic between Cobb (1995a) and Smith (1995).

3. Compare discussion of situated learning from the point of view of the mathematics education researcher in Sierpinska 1995(b).

4. Readers interested in more opinions on the ways of shaping mathematics curricula are referred to the special issue of *Journal of Mathematical Behavior* (Davis 1994) devoted to this issue.

5. The Algebra Working Group is an association of mathematics educators who communicate regularly via Internet on matters pertaining to learning and teaching algebraic reasoning at all grade levels. The AWG is managed by James Kaput, JKAPUT@UMASSD.EDU, under the auspices of the University of Wisconsin National Center for Research in Mathematics Sciences Education.

6. The Algebra Working Group is an affiliation of mathematics educators who communicate regularly via Internet on matters pertaining to learning and teaching algebraic reasoning at all grade levels. The AWG is managed by Dr. James Kaput, JKAPUT@UMASSD.EDU, under the auspices of the University of Wisconsin National Center for Research in Mathematics Sciences Education.

REFERENCES

Bauersfeld, H.: 1995, '"Language Games" in the Mathematics Classroom: Their Function and their Effects', in P. Cobb & H. Bauersfeld (eds.), *Emergence of Mathematical Meaning: Interaction in Classroom Cultures*, Lawrence Erlbaum Associates, Hillsdale, NJ, 271–291.

Boyd, R., Gasper, P. & Trout, J. D. (eds.): 1991. *The Philosophy of Science*, The MIT Press, Cambridge, MA.

Brown, J. S., Collins, A. & Duguid, P.: 1989, 'Situated Cognition and the Culture of Learning', *Educational Researcher* **18** (1), 32–42.

Brown, T.: 1994, 'Describing Mathematics You Are Part of: A Post-Structuralist Account of Mathematical Learning', in P. Ernest, (ed.), *Mathematics, Education, and Philosophy: An International Perspective*, The Falmer Press, London, 154–161.

Chevallard, Y.: 1985, *La Transposition Didactique*, La Pensée Sauvage éditions, Grenoble, France.

Cobb, P.: 1995a, 'Continuing the Conversation: A Response to Smith', *Educational Researcher* **24** (7), 25–27.

Cobb, P. & Yackel, E.: 1996, 'Constructionist, emergent, and sociocultural perspectives in the context of developmental research. *Educational Psychologist*, **31**, 175–190.

Cognition and Technology Group at Vanderbilt: 1990, 'Anchored Instruction and its Relationship to Situated Cognition', *Educational Researcher* **19** (6), 2–10.

Davis, R. B. (ed.): 1994, 'What Mathematics Should Students Learn?', *Journal of Mathematical Behavior* **13** (1), Special Issue.

Davis, P.: 1993, 'Visual Theorems', *Educational Studies in Mathematics* **24**, 333–344.

Davis, P. & Hersh, R.: 1981, *The Mathematical Experience*, Penguin Books, Harmondsworth.

Davis, R.: 1988, 'The Interplay of Algebra, Geometry, and Logic', *Journal of Mathematical Behavior* **7**, 9–28.

Dowling, P. C.: 1993, 'Theoretical "Totems": A Sociological Language for Educational Practice', in C. Julie, D. Angelis, and Z. Davis (eds.), *Political Dimensions of Mathematics Education 2: Curriculum Reconstruction for Society in Transition*, Maskew Miller Longman, Cape Town, 124–142.

Dreyfus, H. & Rabinow, P.: 1982, *Michel Foucault: Beyond Structuralism and Hermeneutics*, Harvester Wheatsheaf, New York.

Dreyfus, T.: 1994, 'Imagery and Reasoning in Mathematics and Mathematics Education', in D. Robitaille, D. Wheeler & C. Kieran, (eds.), *Selected Lectures from the 7th International Congress on Mathematics Education*, Les Presses de l'Université Laval, Sainte-Foy, Québec, 107–122.

Driver, R. & Easley, J.: 1978, 'Pupils and Paradigms: A Review of Literature Related to Concept Development in Adolescent Science Students', *Studies in Science Education* **5**, 61–84.

Eisenberg, T. & Dreyfus, T.: 1989, 'Spatial Visualization in the Mathematics Curriculum', *Focus on Learning Problems in Mathematics* **11** (1), 1–6.

Ernest, P.: 1985, 'The Philosophy of Mathematics and Mathematics Education', *International Journal for Mathematics Education in Science and Technology* **16** (5), 603–612.

Ernest, P.: 1991, *The Philosophy of Mathematics Education*, The Falmer Press, London.

Ernest, P. (ed.): 1994, *Mathematics, Education and Philosophy: An International Perspective*, The Falmer Press, London.

Ernest, P.: 1995, 'Editorial: Criticism and the Growth of Knowledge', *Philosophy of Mathematics Education Newsletter* **8**, 2–4.

Foucault, M.: 1972, *The Archaeology of Knowledge*, Harper Colophon, New York.

Greeno, J.: 1991, 'Number Sense as Situated Knowing in a Conceptual Domain', *Journal for Research on Mathematics Education* **22** (3), 170–218.

Heckman, P. & Weissglass, J.: 1994, 'Contextualized Mathematics Instruction: Moving beyond Recent Proposals', *For the Learning of Mathematics* **14** (1), 29–33.

Johnson, M.: 1987, *The Body in the Mind: The Bodily Basis of Meaning, Imagination, and Reason*, The University of Chicago Press, Chicago.

Kilpatrick, J.: 1992, 'A History of Research in Mathematics Education', in D. Grouws (ed.), *Handbook of Research on Mathematics Teaching and Learning*, Macmillan, New York, 3–38.

Kitcher, P.: 1984, *The Nature of Mathematical Knowledge*, Oxford University Press, Oxford.

Kozulin, A.: 1990, *Vygotsky's Psychology: A Biography of Ideas*, Harvester Wheatsheaf, New York.

Lakatos, I.: 1976, *Proofs and Refutations*, Cambridge University Press, Cambridge.

Lakoff, G.: 1987, *Women, Fire and Dangerous Things: What Categories Reveal about the Mind*, The University of Chicago Press, Chicago.

Lave, J. & Wenger, E.: 1991, *Situated Learning: Legitimate Peripheral Participation*, Cambridge University Press, Cambridge.

Lyotard, J.-F.: 1992, *The Postmodern Explained*, University of Minnesota Press, Minneapolis, MN.

Mura, R.: 1993, 'Images of Mathematics Held by University Teachers of Mathematical Sciences', *Educational Studies in Mathematics* **25**, 375–385.

Noss, R.: 1994, 'Structure and Ideology in the Mathematics Curriculum', *For the Learning of Mathematics* **14** (1), 2–10.

Poincaré, H.: 1952, *Science and Method*, Dover Publications, New York.

Polyá, G.: 1957, *How to Solve It*, Princeton University Press, Princeton, NJ.

Presmeg, N.: 1986, 'Visualization in the High-School Mathematics', *For the Learning of Mathematics* **6** (3), 42–46.

Rorty, R.: 1991, *Objectivity, Relativism, and Truth*, Cambridge University Press, Cambridge.

Rotman, B.: 1994, 'Mathematical Writing, Thinking, and Virtual Reality', in P. Ernest (ed.), *Mathematics, Education, and Philosophy: An International Perspective*, The Falmer Press, London, 76–86.

Schwartz, J. & Yerushalmy, M.: 1992, 'Getting Students to Function with Algebra', in G. Harel & E. Dubinsky (eds.), *The Concept of Function: Aspects of Epistemology and Pedagogy*, MAA Notes, v.25, Mathematical Association of America, 261–289.

Sfard, A.: 1994, 'Reification as the Birth of Metaphor', *For the Learning of Mathematics* **14** (1), 44–55.

Sierpinska, A.: 1995(a), 'Some Reflections on the Phenomenon of French Didactique', *Journal für Mathematik Didaktik* **16** (3/4), 163–192.

Sierpinska, A.: 1995(b), 'Mathematics "in Context", "Pure" or "with Applications"?', *For the Learning of Mathematics* **15** (1), 2–15.

Smith, E.: 1995, 'Where is the mind? "Knowing" and "knowledge" in Cobb's Constructivist and Sociocultural Perspectives', *Educational Researcher* **24**, 23–24.

Smith, J., diSessa, A. & Roschelle, J.: 1993, 'Misconceptions Reconceived: A Constructivist Analysis of Knowledge in Transition', *Journal of the Learning Sciences* **3** (2), 115–163.

Thurston, W.: 1994, 'On Proof and Progress in Mathematics', *Bulletin of MAA* **30** (2), 161–177.

Tsatsaroni, A. & Evans, J.: 1994, 'Mathematics: The Problematic Notion of Closure' in P. Ernest (ed.), *Mathematics, Education, and Philosophy: An International Perspective*, The Falmer Press, London, 87–108.

Tymoczko, T. (ed.): 1986, *New Directions in the Philosophy of Mathematics*, Birkhauser, Boston, MA.

Vinner, S.: 1989, 'The Avoidance of Visual Considerations in Calculus Students', *Focus on the Learning Problems in Mathematics* **11** (1/2), 149–156.

Von Glasersfeld, E. (ed.): 1991, *Radical Constructivism in Mathematics Education*, Kluwer Academic Publishers, Dordrecht.

Walkerdine, V.: 1988, *The Mastery of Reason*, Routledge, London.
Walkerdine, V.: 1994, 'Reasoning in Postmodern Age', in P. Ernest, (ed.), *Mathematics, Education, and Philosophy: An International Perspective*, The Falmer Press, London, 61–75.
Williams, S. T.: 1993, 'Mathematics as Being in the World: Toward an Interpretive Framework', *For the Learning of Mathematics* 13 (2), 2–7.
Wittgenstein, L.: 1958, *Philosophical Investigations*, Basil Blackwell, Oxford.
Yerushalmy, M.: 1993, 'Computerization in the Mathematics Classroom' (in Hebrew), *Aleh – The Israeli Journal for the Teachers of Mathematics* 12, 7–14.

Anna Sfard
The Haifa University,
School of Education,
Haifa,
Israel

HEINZ STEINBRING

EPISTEMOLOGICAL CONSTRAINTS OF MATHEMATICAL KNOWLEDGE IN SOCIAL LEARNING SETTINGS

1. THE NEED FOR AN EPISTEMOLOGICAL PERSPECTIVE ON THE OBJECT OF DIDACTIC RESEARCH

The theme of the present ICMI Study, 'Mathematics Education as a Research Domain: A Search for Identity' immediately leads to the question: 'What is the object of this research?' One direct answer that comes to mind is that it is mathematics. While this is true in some general sense, this answer is not satisfactory upon realization that the term 'mathematics' in the didactic context is full of ambiguity. In this context, at the onset, a more differentiated perspective on mathematical knowledge has to be introduced, highlighting the similarities and the differences between mathematics as a scientific discipline and mathematics as a school subject (cf. Chevallard 1991; Dörfler & McLone 1986).

University mathematics plays a decisive, yet at the same time ambiguous, role in educational research into the teaching and learning of mathematics. On the one hand, 'one notices that the structure and content of school subjects are essentially determined by the scientific disciplines, whatever the development of school subjects, of teacher education, of teacher self-concept, etc.' (Otte et al. 1979, p. 121). On the other, '[t]he science of mathematics, especially, represents today a highly specialized and dynamically developing system having its own contradictions and uncertainties which can hardly provide *immediate* clues for mathematics teaching in schools of general education' (ibid.). Overcoming this contradiction and trying to maintain meaningful links between scientific and school mathematics requires one to

consider mathematics as an active process and knowledge as action, as activity — a view which is shared generally today.... But this implies, because human cognitive activity is many sided, that an active attitude towards the social and natural reality is always given with the mathematical content, and that the content can only be learned and acquired in this context' (ibid.).

This produces a new problem for scientific investigation, because 'the pupil, the teacher, and the scientist have quite different perceptions of social and natural realities' (ibid.).

One main feature characterizing the difference between scientific mathematics and school mathematics is the different social contexts in which this subject matter is developed. These different social contexts have a crucial impact on the epistemological status of mathematical knowledge (cf. Ernest 1993). Both kinds

Sierpinska, A. and Kilpatrick, J. Mathematics Education as a Research Domain: A Search for Identity, 513–526.
© *1998 Kluwer Academic Publishers. Printed in Great Britain.*

of knowledge show a 'logical' structure, deal with mathematical signs and symbols, contain elements of mathematical argumentation and proof, and furnish, for instance, mathematical formulae that could be applied in other domains. But the two 'types' of mathematical knowledge differ in their episte-mological status: School mathematical knowledge cannot be logically deduced from scientific knowledge. It undergoes a complicated transformation process (Chevallard describes this process in his 'theory of didactic transposition', Chevallard 1991).

In order to be able to analyze differences and similarities between scientific and school mathematics, one has to expand the perspective of investigation from the 'body of knowledge' to the interrelations between mathematical knowledge and its social context (Figure. 1).

In this extended framework, links and differences between scientific and school mathematics can no longer be discussed simply on the level of the subject matter itself; it is necessary to take into account an *epistemological* approach to mathematical knowledge in different social settings. One way in which scientific mathematics might have a fundamental impact on school mathe-matics is through the epistemological constraints that have evolved during the historical development of mathematical knowledge. These constraints are struc-turally comparable to epistemological conditions arising in processes of teaching and learning mathematics in school.

The historical dimension of mathematical knowledge is regarded as an important source of improvement of our understanding of school mathematical knowledge: It puts mathematics onto a time-line of development and change and it reveals the historical links of this changing body of knowledge to different social contexts. In this respect, the historical dimension of mathematical knowl-edge seems to furnish better paradigmatic models of knowledge for school mathematics than the actual, static portrait of scientific research mathematics, because it reflects epistemological and social dependencies of the development of mathematical knowledge that are essential for teaching–learning processes of mathematical knowledge.

One main epistemological characteristic of school mathematical knowledge is its *context-specificity*. At school, children learn and understand mathematical knowledge in a context-dependent way: in 'domains of subjective experiences' (Bauersfeld 1983) or in 'micro-worlds' (Lawler 1990) and any learning of math-ematics is 'situated' (Lave 1988, 1991). According to this view, at school, math-ematical knowledge is not available on a general and abstract level, but is always embedded in specific contexts and situations in a concrete manner.

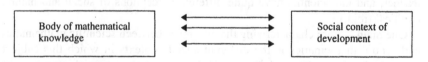

Figure 1 *Interrelations between knowledge and social context*

In contrast to this position, although scientific mathematical knowledge has been constructed in context-specific ways during earlier historical phases, since the beginning of our century at the latest, the epistemological status of mathematics has changed to viewing it as an *abstract and formal knowledge structure*. This transformation from context dependency to an abstract foundation had been achieved through the establishment of the implicit, axiomatic definition of mathematical knowledge.

The *epistemological triangle* is an appropriate tool for the analysis of mathematical knowledge in social contexts of development. Mathematical knowledge of any kind – be it in research processes, or in school learning – needs certain sign or symbol systems for its coding. Initially, these signs have no meaning on their own; the way of writing and using the signs has evolved historically and has been agreed upon through convention. For these signs to be endowed with meaning, they need, in a very general way, to have adequate *reference contexts*. This way, the mathematical meaning of concepts evolves in the interplay between *sign/symbol systems* and *reference contexts* (Steinbring 1989, 1991) (Figure 2).

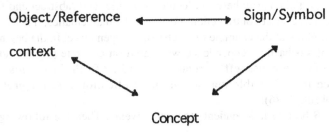

Figure 2   *The epistemological triangle*

The relations between the vertices of this triangle are neither rigid nor defined in a dependent and deductive manner. They are more of a mutually supporting and balancing system. In the ongoing development of knowledge, the explication of the sign system and the reference context changes. The early concept of probability, for instance, could be described as consisting of a sign system containing 'fraction numbers' and a reference context containing 'the ideal die'. Later, this context changed to 'statistical collectives' (von Mises) and the sign system became 'the limit of relative frequencies'. In still later developmental phases, the reference context changed to 'stochastic independent/dependent configuration' and the sign system was altered to 'implicitly defined axioms' (Figure 3).

The crucial point here is not to consider the reference context (or the object) as definitely pre-given, but to acknowledge that this context will change during the process of knowledge development in a *relational* connection. According to this fundamental change in the status of the reference context, the production of mathematical meaning in the interplay between the sign system and the reference context can be described as a process in which possible relations and

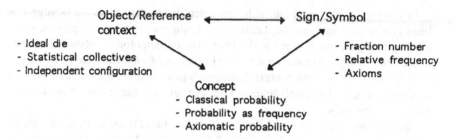

Figure 3 *The epistemological triangle applied to the concept of probability*

meanings are transferred from a relatively familiar situation (the reference context) onto a still new and unfamiliar sign system. And even the roles of 'reference context' and 'sign system' might be exchanged during the developmental process, leading to a situation in which a familiar sign system serves as a reference context for another reference context, now conceived as a sign system in some specific aspect.

Normally, the polarity between 'context-specific' and 'abstract and formal' is understood as a strict contradiction. For scientific mathematics, this can be made clear by looking at the example of stochastic independence. In the beginning, the concept of stochastic independence was based on concrete physical ideas such as 'there are no cause–effect relations', and so forth. From this one tried to deduce in a plausible manner the multiplication rule for probabilities (cf. Steinbring 1986):

Let *A, B* be two independent stochastic events. Then the following relation holds:

$$P(A \cap B) = P(A) \cdot P(B)$$

After the axiomatization of the probability concept, the definition of stochastic independence reversed this order of reasoning:

If, for two events *A, B*, the relation:

$$P(A \cap B) = P(A) \cdot P(B)$$

holds, then the two events are said to be, by definition, stochastically independent.

This reversion of definitions is often taken as evidence for the contradiction between 'context-specific' knowledge and 'abstract and formal' knowledge. First, the concept of stochastic independence was determined by context-specific physical aspects, and, later, the definition was given by a formal and abstract rule of multiplication. These two interpretations are said to be contradictory. According to this historical change, school mathematics is often said to be still in the state of concreteness, whereas scientific mathematics is said to have reached the formal and abstract level.

My intention in the following sections of this paper is to argue that this contradiction is based on a too narrow and misleading interpretation of the role of reference contexts in mathematical symbol systems, and that a perception of reference contexts from a *relational perspective* permits a comparison of school mathematics and scientific mathematics, instead of simply stating that school mathematics is context-related, whereas scientific mathematics is abstract.

2. EPISTEMOLOGICAL PROBLEMS IN EVERYDAY TEACHING: HOW CAN NEW MATHEMATICAL KNOWLEDGE BE DEVELOPED INTERACTIVELY?

In the second grade of elementary school, different basic arithmetical strategies are learned for addition and subtraction in the 'number space' up to 100. It is expected that the children will not simply use these as recipes but will understand the arithmetical relation behind the strategy (Wittmann & Müller 1990, p. 82). The aim of the arithmetical computations and exercises is to develop the elementary number concept together with a rich relational understanding in order for the students to become more flexible in arithmetical operations. One example is the following strategy that uses the 'constancy of the difference or of the sum' (Figure 4).

Arithmetical strategies using the 'constancy of the sum'

and the 'constancy of the difference':

$$23 + 58 = 21 + 60 = 81$$

$$43 - 19 = 44 - 20 = 24$$

Figure 4 *Arithmetical strategies*

The teacher discusses this strategy with a child, and she tries to clarify the underlying meaning. As the child has difficulty in grasping the concept, the teacher returns to a more elementary form of the strategy by introducing a further explanatory step (Figure 5).

Elementarizing the arithmetical strategies:

$$23 + 58 = 23 + (58 + 2) - 2 = 23 - 2 + 60 = 21 + 60$$

$$43 - 19 = 43 - (19 + 1) + 1 = 43 - 20 + 1 = 44 - 20$$

Figure 5 *Elementarizing arithmetical strategies*

In her negotiation with the child, the teacher tries to focus the latter's attention on the fact that in solving the first problem this way, one has first added too much and therefore the same amount has to be taken away again later; and in solving the second problem, one has subtracted something that has to be put back again in the end. But the student still does not understand the proper conceptual idea of this strategy. This understanding would require a generalization and an extraction of the principle of the 'constancy of the sum and of the difference' that the teacher is trying to make clear with the help of these two concrete, arithmetical examples. The given arithmetical problem is not calculated directly, but is transformed initially into an 'equivalent' arithmetical problem producing the same solution that now can be calculated much more easily (by using the properties of the decimal positional system). But the child concentrates on the step-by-step algorithm; on how to calculate the expected result.

When asked by the teacher, why, in the end, something still must be subtracted or added, the child answers: 'Well, in problems with "plus", I have to calculate "minus"; and in a task with "minus", I have to calculate "plus".' She formulates a rule for remembering the sequence of necessary operations, a mnemonic technique, without making any reference to the conceptual structure behind the procedure. The next inquiry of the teacher makes the child feel insecure about her answer and she hesitatingly asks back: 'Or rather, no, with "minus" I have to calculate "minus", and with "plus", "plus"??!.' In fact, the child lacks a conceptual basis for justifying the strategy. The more the teacher insists – and she feels obliged to ask because the student seems to have problems in understanding the strategy suggested by her – the more doubts the child starts to have. And the more the teacher splits up the arithmetical operations into smaller pieces of numbers, operations and steps, the more visible the interactive patterns of communicative routines and funneling strategies become (Bauersfeld 1978; Wood 1992).

Arithmetical strategies can be viewed from two different epistemological perspectives: on the one hand, as a *reductive strategy*, and, on the other hand, as a *developmental strategy*. A reductive strategy means that complex arithmetical problems have to be worked out by following some conventionalized set of instructions in an algorithmic manner, without any need for a conceptual justification of the consecutive steps. A developmental strategy means that complex arithmetical problems are transformed initially into equivalent (simpler) problems, and this transformation is regulated by some arithmetical relation already containing the new knowledge.

For example, the developmental strategy aimed at:

$$43 - 19 = 44 - 20$$

uses, as a new conceptual relation, the 'constancy of the difference', a general, arithmetical law, which, in this case, has to be presupposed for justifying the arithmetical strategy. If this – still new and not yet developed in detail –

arithmetical relation is not understood in a global way, every attempt to explain it locally causes the decomposition of the strategy:

$$43 - 19 = 44 - 20$$

into the small pieces of a reductive strategy:

$$43 - 19 = 43 - (19 + 1) + 1 = 43 - 20 + 1 = 23 + 1 = 24.$$

The epistemological constraints for learning and understanding new mathematical knowledge become decisive: Essentially one already has to know the conceptual idea of the new mathematical knowledge in advance before one is able to justify the strategies used for operating effectively with the new knowledge.

The child in our episode is not (yet) able to consider the two problems as being equivalent, and she cannot justify the strategy by seeing a general relation – the 'constancy of the difference (or of the sum)' – in the concrete, numerical example. In their interaction, teacher and child have entered a dilemma, changing the communication into a ritualized, funnel-like pattern: The adequate justification of the new strategy essentially requires a new conceptual idea that cannot be mediated by the teacher's simply explaining the single arithmetical steps of the strategy, because, by doing this, she degrades it to a reductive strategy that the child willingly accepts on her level of using arithmetical recipes for calculating numerical results.

The epistemological circle stating that every justification of new mathematical knowledge already requires one to have an idea in advance of the conceptual relation inherent in the new knowledge corresponds to the 'Learning Paradox' (Bereiter 1985; Steffe 1990). If students fail to use the still new and unfamiliar knowledge a priori, then, during the course of interaction with the teacher, and under the obligation confronting every lesson that the students have to learn something in any case, the great danger arises that a productive way of jointly treating mathematical knowledge degenerates into ritualized patterns of interaction.

How could one cope with the epistemological problem of the new knowledge in this case? In a very general sense, one has to clarify the relationship between the sign system and a possible reference context, instead of simply using the sign system itself as its own reference context. A productive reference context for the conceptual relation 'constancy of the difference' could be the number line together with the idea that the subtraction 43 – 19 can be interpreted as 'completing the number 19 to 43' (Figure 6).

The 'constant' arrow – giving the difference between 43 and 19 – can be moved on the number line without changing the unknown difference; moving it to a 'better' position allows one to read off directly the desired result (Figure 7).

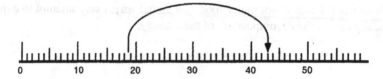

Figure 6   *Representing the task 19 + _ = 43 on the number line*

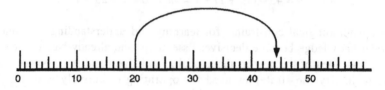

Figure 7   *Representing the task 20 + _ = 44 on the number line*

Such diagrams are examples for structural reference contexts in a corresponding epistemological triangle for describing the situation of learning the conceptually new arithmetical strategy (Figure 8).

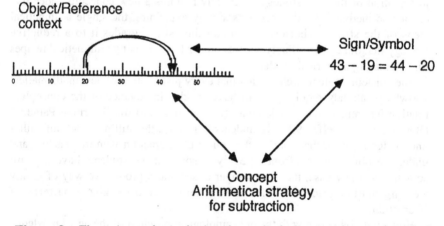

Figure 8   *The epistemological triangle applied to an arithmetical symbol system with a relational reference context*

The reference context is not given by concrete elements or empirical properties but displays a relational structure. A concrete situation could not provide an adequate basis for justifying the arithmetical strategy. The meaning for this strategy requires a relational context, given here by a geometrical diagram and the multitude of relations inherent in it.

When developing new mathematical knowledge, the construction of the new meaning will be context-dependent, but, in most cases, this development depends on concrete, empirical contexts. However it also needs reference contexts, be they specific or general, that display a relational structure. In this framework, the relation between signs and objects becomes variable in the sense that their roles as sign or as object can be exchanged. The decision about the temporal definition of what will be the sign and what will be the object is a social act that could be, for instance, negotiated in the interaction between teacher and students.

3. THE INTERACTIVE CONSTITUTION OF AN EPISTEMOLOGY OF SCHOOL MATHEMATICS: THE READING OF MATHEMATICAL SYMBOLS

Symbols, symbolic relationships and the introduction into the use and the reading of symbols are essential aspects for the formation of every culture (Wagner 1981, 1986). Mathematics deals *per se* with signs, symbols, symbolic connections, abstract diagrams and relations. The use of the symbols in the culture of mathematics teaching is constituted in a specific way, giving social and communicative meaning to letters, signs and diagrams during the course of ritualized procedures of negotiation. Social interaction constitutes a specific teaching culture based on school-mathematical symbols that are interpreted according to particular conventions and methodical rules. In this way, a very school-dependent, empirical epistemology of mathematical knowledge is created interactively, in which the reading of the symbols introduced is determined strongly by conventional rules designed to facilitate understanding, but, in this way, entering into conflict with the theoretical, epistemological structure of mathematical knowledge. The ritualized perception of mathematical symbols seems to be insufficient, according to the analysis of critical observers, because it does not bring the student to the grasp of the genuine essence of the symbol but leaves him or her on the level of pseudo-recognition.

A paradigmatic example for the specific way of reading mathematical symbols in everyday teaching is negative numbers (or directed numbers) (cf. Hefendehl-Hebeker 1993; Janvier 1983; Lytle 1992, 1994). Often, negative numbers are introduced by relating them to concrete reference contexts such as the debt model. Subsequently, negative numbers are defined by invoking analogies to the positive numbers: It is possible to operate with negative numbers in a similar way as with positive numbers. This can be made explicit in the framework of the debt model, which often is operationalized methodically by using red and black chips to represent debts and credits. In both domains, both positive and negative numbers, addition and subtraction are defined by the concrete actions of combining and removing a certain number of black or red chips.

On this basis, both sets of black and red chips are united to form one single operative set. To be able to really make this unification, the artifice of the so-called 'method of annihilating' or 'neutralizing' is introduced: An equal number of red and of black chips compensates to zero (Figure 9).

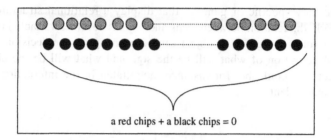

a red chips + a black chips = 0

Figure 9   *The neutralization model represented by red and black chips*

This neutralization model (Lytle 1994) seems to permit a homogeneous extension of the elementary arithmetical operations to the entire domain of positive and negative numbers while still using the familiar, empirical frame of reference.

A more detailed analysis of this representation of the zero proves the necessity for a radically different reference context. Although chips representing concrete, empirical objects are used, they constitute, in the given form, a genuine *mathematical symbol* in the same way that ciphers display a mathematical fraction or algebraic letters display variables and functions. Given by a number of chips, the zero is not ranked on the same level of meaning as, for instance, the '5' given by 5 black chips. According to this concrete, empirical interpretation, 'zero' should mean 'nothing' and must be represented by no (or zero) chips. The description of chips taken here *symbolizes* the zero, it is no longer a concrete, empirical embodiment of the zero. Already from the very beginning, this way of reading these neutralizing chips as a true and proper mathematical symbol is based on the interpretation of the 'zero' as a relation between other (mathematical) objects. 'Zero' does not represent a concrete object given by a number of chips, but a *mathematical relation*. At the same time, there are 'many' zeros, that is, infinitely many combinations of black and red chips symbolizing the zero; and this contradicts the familiar, empirical interpretation of natural numbers. In this way, the 'zero' is a kind of variable (a class of differences, as they are represented in other models for negative numbers; cf. Hefendehl-Hebeker 1993; Steinbring 1994).

Many students conceive of negative numbers as special kinds of positive numbers, but one has to be cautious with the negative sign before the number. This understanding is fundamentally reinforced by the official methodical convention in the classroom of reducing the functioning of the addition and subtraction of negative numbers to black and red chips as concrete, empirical objects. And if it is necessary to use the zero in the form of an equal number of black and red chips during the course of deducing a more general arithmetical operation, this zero is also linked exclusively to the concrete properties of empirical objects, totally ignoring the involved epistemological implications. On this basis, negative numbers are often read as mythical symbols (cf. Seeger &

Steinbring 1994). They represent something universal while simultaneously always being tied to a concrete object that cannot be generalized so easily to a functional mathematical symbol (Cassirer 1955, p. 144). The understanding of such a symbol presupposes a metamorphosis of the signs into a mathematical symbol constituting a *relation*. This shift of view only becomes effective in a process of generalization.

The innate relation encapsulated in the symbolization of the zero is extended to all numbers, which now are all interpreted as *differences* or as *relations* in the general form $a - b$. A possible shift of view on a concrete arithmetical problem, which could not be solved before, '5 – 7', now permits a solution through the generalization of the specific: { ..., 5 – 7, 4 – 6, ..., 0 – 2, ..., –3 – (–1), –4 – (–2), ... }. The problem becomes its own solution by making the operative relation (the subtraction) of the unsolvable problem the basis of the new concept in the sense of an appropriate generalization. Such a shift of view triggered off by generalization can convert a mythical and thing-connected symbol into a mathematical, relational symbol.

This second example of negotiating new mathematical knowledge in every-day teaching demonstrates the conflict between the interactively constituted *mythical epistemology* of school mathematics and the *theoretical epistemology* of mathematical knowledge. The specific requirements of teaching and learning often lead to an empirical interpretation of the reference context for a mathematical sign system or to a hypostasis or re-interpretation of relational connections to substantial properties. When being forced to mediate and to understand new mathematical knowledge, the interactive communicative patterns often try to avoid a new, relational perception of reference contexts. Processes of teaching and learning in a social setting try to maintain as long as possible the justification of the new concept on the basis of an empirical reference context, but, suddenly, there are conflicts with the theoretical epistemology, showing the necessity for a relational understanding of the reference context.

4. CONSEQUENCES FOR THE DIFFERENCE BETWEEN THE CONTEXT SPECIFICITY AND FORMAL ABSTRACTNESS OF MATHEMATICAL KNOWLEDGE

Under the obligation to obtain a definite result in learning processes, mathematics teaching tends in general to provide empirical reference contexts and to avoid relational reference contexts for sign systems. In this way, mathematics teaching in fact supports an empirical type of context and context specificity for mathematical knowledge. The first example has demonstrated how such an empirical foundation without any link to relational reference contexts accompanied by routinized interactive patterns of communication, such as the funnel pattern, changes meaningful mathematical understanding into conventionalized rules of algorithmic operations. The second example has shown that the conservation of an empirical foundation of mathematical concepts might produce a mythical interpretation of mathematical symbols that conflicts with the theoretical epistemology of mathematical knowledge because, in this way, students become accustomed to an artificially concrete understanding of mathematical concepts, and this produces epistemological obstacles to an understanding of the relational character of mathematical knowledge that is unavoidable in later confrontations with new mathematical concepts.

Against this background, it seems productive to modify the alternative between 'context specificity and formal abstractness of mathematical knowledge' into the distinction between 'empirical, explicit foundation and theoretical, implicit justification of mathematical knowledge'. This change of perspective first of all accepts that every mathematical knowledge, be it scientific knowledge or school knowledge, needs reference contexts, and, in this sense, every knowledge is context-specific. On this basis, the difference between scientific and school mathematics lies in the different types of reference contexts used in these different social contexts of development. One important difference concerns the reference contexts in school mathematics which must be adjusted to the requirements of learning and to the cognitive development of the students. But, at the same time, the didactic construction and the use of these contexts in the classroom has to enable the advancement to a relational perspective on the justification of mathematical knowledge.

The *epistemological perspective* on the body of mathematical knowledge in relation to a social context of development changes the perception from the knowledge structure as a prefabricated body to its epistemological, relational network and thus opens possible links to the specific social needs and requirements between the knowledge and the social subjects involved. In this way, an epistemological perspective makes the mathematical content accessible in a new way to research questions on educational problems of mathematics, without assuming that mathematical knowledge is definitely pre-given or that mathematical knowledge is in some way irrelevant for didactic research.

Compared to other theoretical approaches to school mathematics, we do not consider mathematics as pre-constructed or based on 'eternal' foundations from where – in principle – all school mathematical knowledge could be deduced. The process of constructing knowledge does not simply consist in adding new pieces and building them up on this given foundation. Rather, we see the construction of mathematical knowledge as requiring that the learning subject establishes relationships between signs and symbols and contexts of references (of different and changing types) and in this way develops the mathematical knowledge by him or herself. This personal interpretation of signs by reference to contexts is embedded in the learning culture of the classroom, and it is guided by the teacher and supported and questioned by other classmates. But the essential point here is the active involvement of the student in the factual construction of mathematical knowledge and meaning of knowledge.

This focus on the social and personal construction of knowledge at the same time makes it possible that sciences related to mathematics education such as psychology, sociology, cognitive sciences, etc. are not to be seen simply as additional to mathematics education but that they really enter, in a fundamental way, into the questions of mathematics education research when they are linked to the complex problems of how individual learners within the social mathematical learning culture might make sense of mathematical symbols by relating them to different contexts of reference.

ACKNOWLEDGMENT

I gratefully acknowledge the helpful comments and criticisms of Nicholas Balacheff and Anna Sierpinska on an earlier version of this paper.

REFERENCES

Bauersfeld, H.: 1978, 'Kommunikationsmuster im Mathematikunterricht – Eine Analyse am Beispiel der Handlungsverengung durch Antworterwartung', in H. Bauersfeld (ed.), *Fallstudien und Analysen zum Mathematikunterricht*, Schroedel, Hannover, 158–170.

Bauersfeld, H.: 1983, 'Subjektive Erfahrungsbereiche als Grundlage einer Interaktionstheorie des Mathematiklernens und -lehrens', in H. Bauersfeld et al. (eds.), *Lernen und Lehren von Mathematik*, Aulis, Cologne, 1–56.

Bereiter, C.: 1985, 'Towards a Solution of the Learning Paradox', *Review of Educational Research* **15**, 201–226.

Cassirer, E.: 1955, *The Philosophy of Symbolic Forms. Volume 2: Mythical Thought*, Yale University Press, New Haven, CT.

Chevallard, Y.: 1991, *La Transposition Didactique: Du Savoir Savant au Savoir Enseigné* (2nd ed.), La Penseé Sauvage, Grenoble.

Dörfler, W. & McLone, R. R.: 1986, 'Mathematics as a School Subject', in B. Christiansen, A. G. Howson & M. Otte (eds.), *Perspectives on Mathematics Education*, Reidel, Dordrecht, 49–97.

Ernest, P.: 1993, 'The Relation between Personal and Public Knowledge from an Epistemological Perspective', Manuscript for the Conference on 'The Culture of the Mathematics Classroom: Analyzing and Reflecting upon the Conditions of Change', Osnabrück, Germany, October 11–15, 1993, Exeter University, Exeter.

Hefendehl-Hebeker, L.: 1993, 'The Practice of Teaching Mathematics and Teacher Education', Manuscript for the Conference on 'The Culture of the Mathematics Classroom: Analyzing and Reflecting upon the Conditions of Change', Osnabrück, Germany, October 11–15, 1993, Augsburg University, Augsburg.

Janvier, C.: 1983, 'The Understanding of Directed Numbers' in J. C. Bergeron and N. Herscovics (eds.), *Proceedings of the Fifth Annual Meeting of the North American Chapter of the International Group for the Psychology of Mathematics Education (PME-NA). Vol. 2*, International Group for the Psychology of Mathematics Education, North American Chapter, Montréal, 295–300.

Lave, J.: 1988, *Cognition in Practice. Mind, Mathematics and Culture in Everyday Life*, Cambridge University Press, Cambridge.

Lave, J. & Wenger, E.: 1991, *Situated Learning – Legitimate Peripheral Participation*, Cambridge University Press, Cambridge.

Lawler, R. W.: 1990, 'Constructing Knowledge from Interactions', *Journal of Mathematical Behavior* **9** (2), 177–191.

Lytle, P. A.: 1992, Use of a Neutralization Model to Develop Understanding of Integers and of the Operations of Integer Addition and Subtraction, Master's thesis, Concordia University, Montréal.

Lytle, P. A.: 1994, 'Investigation of a Model Based on the Neutralization of Opposites to Teach Integer Addition and Subtraction', in J. P. da Ponte & J. F. Matos (eds.), *Proceedings of the Eighteenth International Conference for the Psychology of Mathematics Education, Vol. III*, University of Lisbon, Lisbon, 192–199.

Otte, M., Reiß, V. & Steinbring, H.: 1979, 'The Education and Professional Life of Mathematics Teachers' in UNESCO (ed.), *New Trends in Mathematics Teaching*, UNESCO, Paris, 107–133.

Seeger, F. & Steinbring, H.: 1994, 'The Myth of Mathematics', in P. Ernest (ed.), *Constructing Mathematical Knowledge: Epistemology and Mathematics Education*, Studies in Mathematics Education Vol. 4, The Falmer Press, London, New York and Philadelphia, 151–169.

Steffe, L. P.: 1990, 'The Learning Paradox: A Plausible Counterexample', in L. P. Steffe (ed.), *Epistemological Foundations of Mathematical Experiences*, Springer, New York.

Steinbring, H.: 1986, 'L'Indépendance Stochastique – un Exemple de Renversement du Contenu Intuitif d'un Concept et de sa Définition Mathématique Formelle', *Recherches en Didactique des Mathématiques* 7 (3), 5–50.

Steinbring, H.: 1989, 'Routine and Meaning in the Mathematics Classroom', *For the Learning of Mathematics* 9 (1), 24–33.

Steinbring, H.: 1991, 'The Concept of Chance in Everyday Teaching: Aspects of a Social Epistemology of Mathematical Knowledge', *Educational Studies in Mathematics* 22, 503–522.

Steinbring, H.: 1994, 'Symbole, Referenzkontexte und die Konstruktion mathematischer Bedeutung – am Beispiel der negativen Zahlen im Unterricht', *Journal für Mathematikdidaktik* 3, 277–309.

Wagner, R.: 1981, *The Invention of Culture*, University of Chicago Press, Chicago.

Wagner, R.: 1986, *Symbols that Stand for Themselves*, University of Chicago Press, Chicago.

Wittmann, E. C. & Müller, G.: 1990, *Handbuch produktiver Rechenübungen. Bd. 1: Vom Einspluseins zum Einmaleins*, Klett, Stuttgart.

Wood, T.: 1992, Funneling or Focusing: Patterns of Interaction in Mathematics Discussion. Paper presented in the Working Group 7, *Language and Communication in the Mathematics Classroom at the 7th International Congress of Mathematics Education*, Québec, August 17–23.

Heinz Steinbring
IDM University of Dortmund,
Vogelpothsweg 87,
44221 Dortmund,
Germany

ANNA SIERPINSKA & JEREMY KILPATRICK

CONTINUING THE SEARCH

In this chapter, we attempt to draw together the principal arguments made in the preceding chapters and at the ICMI Study Conference so as to provide the reader with a retrospective view of the clearly very complex issue, 'What is research in mathematics education, and what are its results?'. To that question, we are compelled to add: What does it mean to be a researcher in mathematics education? Do we have a common identity?

A theme resounding throughout the book has been that none of these questions, or the other questions arising from them, can be given a definite answer. Like research questions in mathematics education, questions about the field itself can be explored, but there is no way to resolve them once and for all in the way, for example, one might prove a theorem. Each generation must address anew what doing research in mathematics education is all about.

Even as the preceding chapters demonstrate our inability to provide definite answers, they both highlight and elaborate the progress we have made in clarifying the meanings of the questions. As the Discussion Document indicated and the Conference reaffirmed, the aim of this ICMI Study was to identify 'perspectives, goals, research problems, and ways of approaching problems' so as to allow a productive confrontation of views to occur and our mutual understanding to grow. This chapter, then, does not portray a finished and final consensus about our field. Instead, it points toward directions in which to continue the never-ending search for who we are and who we want to be.

MATHEMATICS EDUCATION AS A RESEARCH FIELD

Some of the preceding chapters address directly the fundamental question: what is the nature of research in mathematics education? The authors attempt to situate the research within categories of existing knowledge or research practice. What one generally finds in the chapters, however, are mostly statements about where research in mathematics education *should be* situated rather than where it actually *is* being situated by practitioners and their colleagues from academia. In fact, there seems to be genuine difficulty in classifying research in mathematics education within existing disciplines because of its dual theoretic-pragmatic character, which many authors stress.

Several authors remind us of the interdisciplinarity of research in mathematics education. Presmeg, for example, proposes that research in mathematics

Sierpinska, A. and Kilpatrick, J. Mathematics Education as a Research Domain: A Search for Identity, 527–548.
© *1998 Kluwer Academic Publishers. Printed in Great Britain.*

education should look at the complex phenomena of the teaching and learning of mathematics from many points of view simultaneously, *in a balanced manner*. She mentions mathematics, psychology, sociology, anthropology, history, linguistics, and philosophy. Adda argues that research in mathematics education should belong to a discipline she calls *Cognitive Science* (in the singular) that has not been established yet as such but that seems to be emerging as a result of cooperation between linguists, psychologists, computer scientists, neurobiologists, logicians, and mathematicians. She thus concedes that research in mathematics education is interdisciplinary but hopes that interdisciplinary cooperation will eventually lead to the emergence of a new entity that can provide a home for mathematics education research.

Interdisciplinarity does not usually mean that results and methods from certain disciplines should be or can be transposed wholesale into mathematics education research. The reason, according to Presmeg, is that research in mathematics education has 'not only the scientific goal of theory-building but also the pragmatic goal of the improvement of the teaching and learning of mathematics at all levels'. A similar argument is made by Wittmann, who says that 'scientific knowledge in mathematics education cannot be obtained by simply combining results from [the neighboring disciplines]; it presupposes a specific didactic approach that integrates different aspects into a new whole which takes into account its transposition into practice'.

Presmeg lists several neighboring disciplines of mathematics education without giving priority to any; she advocates a balanced view that takes a variety of perspectives into account. Other participants in the Study, however, insisted on giving a privileged status to the links of mathematics education with mathematics. For example, in his remarks at the first plenary session of the Study Conference, Guy Brousseau stated:

There exist problems in mathematics education that are irreducibly mathematical – like, for example, the choice of problems, the organization of mathematical activities for didactic purposes, the analysis of understanding in mathematics, [and] ... the structuring of the mathematical discourse. ... There is no conjunction of classical disciplines to explain the functioning of this irreducibly mathematical part of teaching.... A scientific approach to [this part] is and will be essentially the work of a mathematician.

Quoting the mathematician William Thurston, Brousseau proposed a broadening of the concept of mathematical activity to encompass not only the production of definitions, conjectures, theorems, and proofs, but also 'the communication of results, the reorganization of theories and knowledge, the formulation of questions and problems, and all that "enables *people* to understand and think more clearly and effectively about mathematics"' (Thurston 1994, p. 163, emphasis in original). This broader view of mathematical activity implies, according to Brousseau, that 'the *didactique* of mathematics would become an integral part of mathematics'. He did not mean that all of mathematics education research is mathematics or should be considered part of mathematics. Rather, he identified three components of research in mathematics education. In addition to the '*didactique* of mathematics' proper, whose task is to identify and explain the teaching and learning phenomena specific to mathematics, he distinguished between research stemming from the application of methods and concepts from other disciplines (e.g., psychology, sociology, linguistics, pedagogy, general education) not specific to mathematics, and *didactic engineering*, which is a design activity devoted to the elaboration of didactic means and teaching materials for mathematics, using the results of research from the two other components.

The design of teaching processes receives a more prominent role in the approach proposed by Wittmann, for whom the 'development and evaluation of substantial teaching units' belongs to the core of research in mathematics education. He compares the domain of mathematics education to music, engineering, and medicine, observing that in these fields, too, creative, practical activity has priority over theoretical deliberations, laboratory research, and critiques.

Other authors in this book are not concerned with attempting to situate mathematics education within or among the classical academic organization of disciplines as it exists in many universities. Ernest proposes to take what he calls a *postmodern view* that 'does not separate research and knowledge from the group of people that do this research and produce this knowledge and from the goals they attempt to reach through these'. In Ernest's view, a field of research can be defined only by the *practices considered relevant* by the people claiming to be researchers in this field. Hence, the definition of research in mathematics education should be given not in the form of a sentence that follows the Aristotelian pattern of *genus et differentiam specificam* but by 'some non-hierarchical list of practices, reflecting a multiplicity of viewpoints, theories, frameworks, methodologies and interests'. That approach would lead not to a classification in terms of academic disciplines but to a structure whose categories would be certain 'programs', 'paradigms' or 'trends'. Ernest attempts to describe and classify mathematics education research practices along those lines.

Ernest thus offers a reflection on what mathematics education research *is* rather than on what it *should be*. Another author who focuses on the status quo rather than a vision is Mura, whose chapter delineates how Canadian researchers in mathematics education see themselves with respect to other domains of research. Her survey shows that, in Canada, 'mathematics education tends to be institutionalized in a way which is not conducive to its becoming "an integral part of mathematics" as advocated by Brousseau'.

Ellerton and Clements remind us of the political dimensions of education in general and of mathematics education in particular. They argue that because the compulsory education of children around the world includes a certain internationalized version of mathematics and because many children fail to learn it, 'mathematics education researchers should pay more attention to their role of helping to create more equitable forms of mathematics education'. The political side of mathematics education and the expectations of various groups, including parents, governments, employers, and mathematicians, seem to indicate that mathematics education is not, and will never become, a purely academic discipline. Mathematics education research is a social and cultural *task* rather than a field of scientific inquiry.

As Bishop observes, that task increasingly demands strong efforts by researchers to relate their work to the practice of teaching and learning mathematics. Teachers face pressures from all sides to make their work more 'effective', and researchers have not been able to provide them with much assistance. In Bishop's view, research in mathematics education needs to be shaped not by theoretical considerations but by the questions and problems of practitioners.

THE OBJECT OF STUDY

Mura's survey also documents a variety of views among Canadian mathematics educators on the object of study in mathematics education research. For some of her respondents, that object is the teaching of mathematics, where *study* can mean either observation and analysis of the phenomena of teaching or, in addition, the search for means and approaches to help students construct mathematical concepts for themselves. Other respondents focused on the learning of mathematics. For example, research in mathematics education might mean a systematic study of how students come to know and understand mathematics. For some, mathematics education is a theoretical pursuit; for others, a practical pursuit, an art, an applied discipline aiming at improving the teaching and learning of mathematics. Still others claimed that mathematics education is not a discipline in itself but an 'integrated application of mathematics, psychology, epistemology, sociology and philosophy'. Mura concludes that the object of study in mathematics education research should be defined very broadly as, for example, 'all aspects of learning and teaching of mathematics'. She adds that the objects of study of, for example, psychology, sociology, or mathematics are similarly broadly (or even vaguely) defined, yet the identity of these domains is not in question.

In general, the other authors agree with Mura's desire not to narrow the scope of the field of study. Ellerton and Clements take an even stronger position, claiming that defining that scope can be dangerously limiting. They cite the case of the 1979 book by E. G. Begle, *Critical Variables in Mathematics Education*, whose influence, they say, 'straitjacketed thinking within the international mathematics education research community for many years'. One could, of course, take issue with this opinion. The book, although influential, was hardly a straitjacket, as witnessed by the many critiques it elicited when it was published. Nor did it prevent the development, in the 1980s, of original theories and research methodologies that had little to do with 'testing hypotheses using statistical controls and mechanisms'. It is enough to mention the developments in French *didactique* or the constructivist program endorsed by many scholars in the Anglo-American community.

It is worthwhile remembering the historical context of Begle's book. In his time, the emerging community of mathematics education researchers had to contend with a widespread 'armchair pedagogy' and the formulation of recommendations for curricula and teaching practice by people who had never taken a careful or systematic look at actual mathematics classrooms. In her chapter, Adda reminds us that such pedagogy has not completely disappeared. The quotation from Begle in Ellerton and Clements' chapter addresses precisely Adda's concern.

The chapters in the present book describe the object of research in mathematics education in broad terms, but the descriptions are quite varied. For example, Margolinas sees the object of study in mathematics education research as 'the production and spreading of mathematical knowledge'. For Presmeg, it is 'the complex inner and outer worlds of human beings as they engage in the learning of mathematics'. These worlds include classroom life, and Presmeg sees the development of mathematical concepts as 'situated within this social and mental complexity'. Rather than characterize *the* object of research in mathematics education, Ellerton and Clements prefer to list a set of problems currently important for mathematics education. The first and most important problem for them is to identify the many outdated assumptions that underlie the ways school mathematics is currently practiced.

Wittmann rejects even the concept of an object of study. He says that as a design science, mathematics education does not have an object of study; it has a field of practices that it wants to improve. This field encompasses those practices related to the teaching of mathematics, including teaching at different grade levels and teaching teachers.

RESEARCH GOALS AND DIRECTIONS

One of the questions in the Discussion Document was related to the 'aims of research' in mathematics education. But what do we mean by 'aims of research'? At least two meanings come to mind. If we focus on the *outcomes* of research, then we understand 'aims' as being the ultimate *goals* of research. If, on the other hand, we are more interested in the *process* of research or of scientific pursuit as such, then we think of 'aims' as being, for example, the directions or orientations of research activity, research problems, issues for research, or *problématiques*.

Traditionally, the goals of scientific endeavor have been divided between 'pure' and 'applied' goals. A researcher may be satisfied with understanding how things are, how they change and why, or under what conditions they change. On the other hand, he or she may want to acquire knowledge in order to take action and change the way things are. Although this distinction can be quite useful for the purposes of methodological clarification, actual research practices can, of course, seldom be classified under one rubric or the other (especially in mathematics education, where most researchers, because they are also mathematics teachers, teacher trainers, curriculum developers, etc., are part of the reality they want to study or change). Still, there is a marked difference between the two questions below:

1. What are the mechanisms of learning mathematics and the factors influencing these mechanisms? These factors may be internal (such as individual characteristics, motivation, personal goals, mathematical ability, mathematical cast of mind or lack thereof, and cultural background that matches or does not match the school culture) or external (such as components of the school setting, teaching methods, expected levels of performance, organization and presentation of mathematical topics, use of computers and calculators, the teacher's communication skills, instruction in reading mathematical symbols, and relations with other subject areas and with the home culture, ethnic culture, and everyday life).
2. What can we do to enable students to learn more mathematics that is relevant for them now and will be useful for them in their future everyday and professional lives?

The latter, seemingly more 'pragmatic', question requires considerable theory and experimentation, just as the former one does. The theory, however, must take into account the sociological and institutional dimensions of learning mathematics at school, and teaching experiments have to go outside the laboratory into the reality of everyday mathematics teaching. The pragmatic question may require anthropological studies of educational systems in different cultures; historical studies of school reform and analyses of the successes and (more often) failures of reform; studies of educational policies and their relationships to political regimes and official ideologies; critical analyses of teachers' beliefs about mathematics, students' learning, and their own role in the process of learning; studies aimed at identifying the hidden cultural and other assumptions behind certain influential survey research studies (e.g., the Second International Mathematics Study or the Third International Mathematics and Science Study); and other studies that may seem more like 'basic research' than 'applied research'.

In fact, Hiebert develops in his chapter an argument to support the claim that the aim of 'understanding' encompasses both the fundamental or theoretical aims of research and its practical aims. Thus, from his perspective, mentioning two types of goals, understanding and improving practice (as, e.g., Leder does), can be seen as redundant.

Most of the authors prefer to speak about directions of research, research programs, or paradigms rather than about the general goals of mathematics education research. For example, Lerman argues that, increasingly, researchers in mathematics education are taking a Vygotskian socio-cultural perspective on learning in which cognition is seen as constituted through discursive practice. He calls for more efforts to build coherence between theories and research methods since 'research paradigms or perspectives do not disappear, although they may perhaps become less popular'. Gravemeijer describes a research program comprising projects conducted by individual researchers to develop instructional activities and complete courses in mathematics. Within each project, a 'local' instructional theory is also developed and elaborated. Mason proposes 'inner research', a form of inquiry at the heart of all research activity in which the researcher maintains 'a conjecturing "as-if" stance towards the whole process of teaching and learning'.

Several of the chapters contain lists of relevant issues for research. These lists throw some light on what is considered to be an object of research in mathematics education. Since several items directly address meta-level issues concerning the nature of research in mathematics education, one can say that, for some authors (but not all), the domain of research in mathematics education contains research into its own philosophy, or that research in mathematics education is one of the objects of research in mathematics education.

A 1984 poll of mathematics educators' opinions on what are or should be the research problems or programs in mathematics education (summarized in Leder's chapter) revealed a focus on the processes of learning mathematics among the respondents. There was an interest in identifying the mechanisms of learning that are specific to mathematics and the factors influencing and influenced by mathematical learning. The respondents mentioned mental structures, mental as well as physical or neurological activities involved in doing mathematics, relationships among those activities, and activities involved in mathematical problem solving. Teaching processes seemed to be interesting only insofar as they contained certain factors affecting (hindering or facilitating) the learning of mathematics.

The research questions identified in the survey have been explored over the ensuing decade or so and have yielded various results. They include the distinctions between dualities such as intuitive and analytic thinking, instrumental and relational understanding, and operation and its thematization; theories of the development of mathematical operations in children, whether naturally or as a consequence of schooling; the identification of levels of understanding specific mathematical concepts and tasks; analyses of the differences between the mathematical thinking promoted at school and the situated cognition occurring at home and other out-of-school environments; and so on.

RESEARCH PROBLEMS

For many authors, the domain of research in mathematics education is defined by its research problems. In the practice of research, however, these are not ordinarily fixed questions that one poses at the outset and tries to answer by whatever means are available or acceptable in a given research paradigm. Rather, research problems are typically dynamic entities that evolve, undergoing substantial reformulation while they are being investigated. It has been said that 'the research problem can be formulated in a correct and adequate way only after it has been solved' (Cackowski 1964, p. 296).

A problem often changes most drastically when the researcher is trying to find a way to decide whether a finding can count as a solution to the problem. For example, if the problem is 'to find effective teaching methods to teach fractions to fourth graders' the meaning of *effective* will depend on the criteria used for evaluating a proposed set of methods. Another researcher may not understand *effective* in the same way, and for that researcher, the finding will not be a solution to the problem. There are other examples of this kind in Leder's chapter.

Problems can change their meaning during the inquiry process. If the general research question is 'What is the impact of computer technology on the teaching and learning of geometry?', the inquiry can lead the researcher to conclude that the question is poorly formulated because the knowledge called 'geometry' that is learned and taught using a textbook, paper, and pencil is not the same knowledge as the 'geometry' that is learned and taught through interactions with dynamic geometry software such as Cabri or Geometer's Sketchpad. Notions such as *figure, construction, theorem*, and *proof* may acquire very different meanings. Also, in the two contexts, different problems will be interesting for the teacher to propose to the students, and the students themselves will stumble upon different problems. Thus 'geometry', in the context of the research problem, becomes an ill-defined object, and the question must be reformulated. The same shift in meaning can happen to 'mathematics' in the question, cited in Leder's chapter: 'How will and should calculators and computers affect the teaching and learning of mathematics?' Perhaps, instead of asking this question, it would be safer to ask: 'What is the epistemological status and the content of the knowledge (or competence) that has emerged in such and such interactions between the teacher, the students, and such and such computer software or calculators under such and such conditions (features of the didactic design)?' Of course, this question does not read half so smoothly as the previous version.

THE ROLE OF THEORY

The Discussion Document suggested that a research domain is known by its *problématiques*. In their chapter, Ellerton and Clements propose a list of what they consider the 'main *problématiques* of mathematics education'. It may be useful to clarify the difference between a *problem* or *research question* and a *problématique*. As defined by Balacheff:

A *problématique* is a set of research questions related to a specific theoretical framework. It refers to the criteria we use to assert that these research questions are to be considered and to the way we formulate them. It is [for example] not sufficient that the subject matter being studied is mathematics for one to assert that such a study is research on mathematics teaching. A problem belongs to a *problématique* of research on mathematics teaching if it is specifically related to the mathematical meaning of pupils' behavior in the mathematics classroom. (Balacheff 1990, p. 258)

For example, if in a study of 'how people learn', the mathematical subject matter could, in principle, be replaced by some other subject matter (e.g., literature or geography), then the research problem does not belong to a *problématique* in mathematics education. It is important to note that what makes a set of research questions into a *problématique* is some common and specific theoretical framework that contains in itself the criteria for the relevance and formulation of research problems.

The insistence by funding agencies and journal editors that research projects and reports should make explicit the theoretical frameworks on which they are based – in other words, should identify the *problématique* within which their research question is situated – has met with criticism from some authors (see the chapters by Leder, Ellerton & Clements, and Hanna). Funding agencies and editors tend to favor *problématiques* that are fashionable and whose simplified versions have received wide circulation. This bias inevitably leads to an even deeper simplification and thereby trivialization of the theories underlying the *problématique*. Funding agencies, moreover, often use political or economic criteria to discriminate between 'interesting' and 'uninteresting' *problématiques*. As a consequence, the description of *problématiques* may be distorted so as to meet these 'mercenary' criteria.

The conclusion to be drawn is not that situating one's own research within a *problématique* should be abandoned as a requirement for research reports or projects. Rather, the conclusion is that the necessity, whether self-imposed or imposed by others, of adopting a currently well-known and accepted *problématique* be abandoned (however utopian that may sound).

Ellerton and Clements ask that more research be theory generating rather than theory driven. The role of explicit theory in research, therefore, is not denied. In fact, Hiebert argues that making theories explicit has several advantages: It helps researchers confront their biases, opens opportunities for debate and critique, and increases the usefulness of their data.

MATHEMATICS AND MATHEMATICS EDUCATION

Among the theoretical concerns of mathematics educators, an important place is occupied by epistemological questions related to mathematical knowledge. For Godino and Batanero, 'the epistemological and psychological analyses concerning the nature of mathematical objects play a fundamental role in addressing certain research questions in mathematics education, [such as] the question of assessment of students' knowledge, and that of the selection of didactic situations'. These authors propose a theoretical framework – indeed, an epistemology – specifically tailored for the needs of research in mathematics education.

For Brousseau and for Artigue, the questions of the meaning and genesis of mathematical notions are exactly the meeting point between mathematics education research and research in mathematics, and without this 'irreducibly mathematical part', mathematics education would not be what it is. It would become part of general education.

Steinbring even states as a matter of fact that the object of research in mathematics education is mathematics, albeit the epistemological status of this knowledge is different from that of the mathematics of research mathematicians. 'In contrast to scientific mathematics', says Steinbring, 'school mathematics is dependent on specific contexts of reference that take into account the students' everyday experiences and prior cognitive development'. Mathematics education research and practice must identify and study these contexts of reference.

Several authors stress the need to take into account the specific social and institutional contexts of growth of mathematical knowledge at school as opposed to 'scientific' mathematical knowledge (Godino & Batanero and Sfard, the latter often evoking the theory of didactic transposition elaborated by Chevallard).

The question 'What is mathematics?' was one of the most debated questions posed at the Study Conference.[1] It started with Brousseau's statement in his plenary talk that activities such as communicating mathematics and organizing mathematical knowledge are not usually considered as mathematical activities by mathematicians. Those who do not produce original mathematics are not considered to be mathematicians. Brousseau said that the challenge of mathematics education as a field of research is to find scientific means to legitimatize the above-mentioned activities, but, he added, *'cela demande une sérieuse reconsidération par les mathématiciens et les autres de ce que sont les mathématiques'* ['this requires a serious reconsideration by mathematicians and others of what mathematics is'].

'What is mathematics?' is, indeed, a serious question for mathematics education. For Ronnie Brown, one of the mathematicians invited to the Study Conference, it is even the central question of mathematics education. He writes in his chapter that without giving thought to 'mathematics itself, how it is done, what is its value, what it advances towards, and the way it advances, ... we might have Hamlet without the prince', and 'if it is unclear as to what is mathematics, what are its main achievements, and what constitutes performance in it, then what hope is there of teaching it in a clear way, or of coming up with new practical hints as to how it should be taught more effectively?'.

The question 'What is mathematics?' is implicated also in practical issues such as: What should count as mathematical knowledge for curriculum designers or educational policy makers? Is 'communication of mathematics' a mathematical activity that should be explicitly taught (and assessed)? More generally, what is 'a mathematical activity'? Many people agree that the core of research in mathematics education should be related to the identification of 'mathematical activities', their study, and the design of ways in which students can be motivated to engage in such activities. Views start to diverge at the question of what counts as mathematical activity.

In mathematics education, we often broaden the meaning of 'mathematics' (beyond what university mathematicians usually call mathematics – see the chapter by Sfard). Brousseau wanted to include communication of results and organization of mathematical knowledge, a view echoed by Ronnie Brown's argument that the process of finding appropriate forms of expression, 'of *trying to understand in order to explain to others*' (our emphasis), can play a critical role in the development of mathematics. Wittmann included in mathematics all the societal uses and modes of expression that are mathematical in nature but are not studied at a university.

THEORY AND PRACTICE

The question of the relations between theory and practice was another very hotly debated issue at the Study Conference. There was a great polarity among the standpoints. On one side were those who saw a sharp dichotomy between theory and practice (e.g., Margolinas, Brousseau, and Balacheff). On the other were those who apparently agreed with Begle and Gibb's (1980, p. 3) thesis that there is a continuum of practical and intellectual activities between the questions 'What is the case?' (theory) and 'What is to be done?' (practice) (e.g., Bishop and Hatch & Shiu).

Margolinas proposed the view that theory is separated from practice by the distinction between facts and phenomena. A fact is an isolated statement that can be verified (by statistical or other methods). For example, 'pupils have difficulties with the notion of limit' is a fact. To become a phenomenon, this statement must be embedded in a theory that will explain it. She suggested that 'research must take fact into account, and practice must take phenomenon into account'. During the discussion, Balacheff very strongly criticized the use of such 'hybrid' terms as *teacher-researcher* or *action research*. An analogy he used was the following: 'You cannot be a teacher-researcher any more than you can be your own psychoanalyst.'

In their chapter, however, Hatch and Shiu claim that, on the contrary, 'the teacher of mathematics is uniquely placed to investigate – and record – aspects of her teaching, her classroom and her students, that are hidden from others'. Citing Schön, these authors propose that, as a 'reflective practitioner', the teacher can contribute to the body of knowledge called 'mathematics education' by becoming 'a researcher in the practice context':

[The reflective practitioner], not dependent on the categories of established theory and technique... constructs a new theory of the unique case. ... He does not keep ends and means separate, but defines them interactively as he frames a problematic situation. He does not separate thinking from doing, ratiocinating his way to a decision which he must later convert to action. Because his experimenting is a kind of action, implementation is built into his inquiry. (Schön 1983, p. 68)

As teacher-researchers, teachers not only plan, teach, and create an account of their plans and teaching; they also analyze, theorize, and account for the events in the classroom, discuss their reflections with other teacher-researchers, and consult the existing literature on relevant topics.

Like Wittmann, Hatch and Shiu draw an analogy between medicine and mathematics education. Both are fields of practice, in which observer-researchers and reflective practitioners collaborate to construct a body of knowledge on the basis of theory and accumulated case studies. Unlike medicine, however, say Hatch and Shiu, 'mathematics education has few formal mechanisms for collating and disseminating practitioner-research'.

By stressing that *practitioner-research* means not only reflecting on one's practice as a teacher but also recording its aspects and explaining it on the basis of existing theories or theories in development, Hatch and Shiu dismiss the simplistic view of action research that reduces it to spontaneous innovation and reflection on practice. In her chapter, Hart points to the dangers of making teachers believe that reflection on practice is research: 'By equating reflection on practice with "research", we encourage teachers to think that this is a special activity and not one expected of *every* classroom practitioner' (emphasis in the original).

The discussion so far has been about the distinction between theory in mathematics education and the practice of teaching. But, as Vergnaud rightly points out, in mathematics education research, the practice/theory issue arises in two contexts: in mathematics (a practical activity versus a theoretical body of knowledge), and in mathematics education (teaching practice versus theory-building and research into that practice). This duality is why mathematics education needs a more general approach to the issue, a more general 'theory of practice'. Vernaud proposes possible foundations for such a theory. He seems to be saying that a notion of practice that excludes conceptualization and a notion of theory-building and comprehension that excludes certain practical competences are either naïve or do not take into account what happens in humans dealing with situations. His theory of practice is built on the two fundamental concepts of *competence* and *scheme*. Competence enables thinkers at all levels of development and experience to marshal and process information so as to generate goals, actions, and expectations. 'The concept of scheme is essential to the understanding of the cognitive structure of competences.'

For Margaret Brown, research can be of particular benefit to practitioners when it proceeds by what she terms *iterative conceptualization,* a process in which data are refined by moving back and forth between the realm of the teacher and pupil and the realm of the researcher. This approach can provide models of behavior, beliefs, and understanding that, although limited in various respects, can enable researcher and teacher alike to think in more focused ways about practice, particularly when the research is conducted by a team.

Gravemeijer draws attention to the value of developmental research in not only developing curriculum material but also producing an instructional theory specific to the domain of mathematics. Like Margaret Brown, he speaks of cyclic processes, which he terms *reflexive,* in which theory development is both input to and output from the development of courses and instructional activities. He offers some heuristics for instructional design drawn from the long experience of the Freudenthal Institute in the Netherlands in fostering mathematics education meant to be truly 'realistic' (i.e., experientially real for students).

The issue of relations between theory and practice is an important one in mathematics education. It is, of course, important from the point of view of research methodologies. But it is also important in a different sense at the level of curriculum development, where it is involved in a variety of questions, such as: How is mathematical thinking related to practical thinking? Can mathematical thinking be developed through 'learning mathematics in contexts', or 'anchored instruction', or 'realistic mathematics'? Is 'realistic mathematics' or 'mathematics in contexts' still mathematics? Once again, the implicit question is actually 'What is mathematics?', and we can see how this issue tends to surface whatever topic one addresses in mathematics education.

RESEARCH RESULTS

In discussing research questions in mathematics education, one is also implicitly examining what mathematics educators might consider an (important) result of research. The question of what kinds of results there might be in mathematics education research was explicitly addressed in the Discussion Document and in the chapter by Boero and Szendrei. The chapter by Kieran is concerned with one particular category of results, namely theories and models (or 'theoretical perspectives', to use the categorization of Boero & Szendrei).

While the categorization proposed in the Discussion Document concerns the effects that research results produce on those who are going to use them – 'energizers of practice', 'economizers of thought', 'demolishers of illusions' – Boero and Szendrei classify results according to the characteristic and contrasting features of the different products of research: 'innovative patterns' related to teaching a particular topic; 'quantitative information' about the consequences of specific didactic choices, students' difficulties, and so on; 'qualitative information' about these matters (how a certain innovation was implemented, how students learn or solve problems, etc.); and 'theoretical perspectives' or theories pertaining to the teaching and learning of mathematics.

According to Boero and Szendrei, teachers and mathematicians alike expect that results from mathematics education research will belong mainly to the categories of 'patterns of innovation' or 'quantitative information'. But, say these authors, there are certain inherent contradictions in those expectations that could be resolved with the aid of results obtained through qualitative and theoretical research, which may also be more valued by researchers. For example, innovations are expected to be effective and reproducible. One cannot understand, however, why some innovations have proved effective and others have not without identifying and analyzing the conditions for the success of an innovative project. A knowledge of these conditions must be based on long-term observations of classroom processes in interaction with the use and development of a theoretical perspective.

Boero and Szendrei do not completely deny the value of a statistical analysis of data ('quantitative information' is, most of the time, provided by such analyses) in the way, for example, that Ellerton and Clements do in their chapter. Boero and Szendrei say only that, because of teachers' interest in quantitative data, there is a risk that certain simplified versions of results obtained by (simplistic) statistical methods may become very popular and lead to flawed and even harmful educational decisions. They call for more sophisticated multidimensional statistical analyses and for a better understanding of the context and conditions in which a survey was conducted.

Some questions in mathematics education clearly call for statistical methods of data analysis. Here is one, proposed by R. Gras (1992): Several problem-solving strategies have been identified in a series of case studies done using qualitative methods. Is it possible to associate them consistently with certain mathematical conceptions that the students have? Are these conceptions evolving, and if so, how? For some years, Gras, a researcher in mathematics education working in France, has been developing an exploratory method of data analysis. He has tried to apply it in investigating students' 'proving behavior' and students' conceptions of conditional probability (Gras 1996). As Gras's work demonstrates, the issue of 'qualitative versus quantitative methods' comes down again to a question of balance. Boero and Szendrei contend that 'theoretical perspectives' and 'qualitative information' are necessary to keep the results of the statistical data analysis under control.

Mason, in his chapter, takes an entirely different perspective on research results. He sees the products of research in mathematics education as being transformations in those doing the research and in those influenced by it. Researcher and reader alike are seen as engaged in a process of self-transformation in which both research and practice provide occasions for sensitivity, for one to be aware of one's practice and to reflect on it at the same time. Effective research changes one's being; it progresses 'from the inside'. Consequently, results cannot be handed to others to 'use'; instead, they must be produced afresh by the reconstruction of one's own activity.

CRITERIA FOR EVALUATING RESEARCH

The ICMI Study has shown that mathematics educators generally feel uncomfortable with the idea of establishing once and for all a set of criteria for assessing the quality of mathematics education research. Some refuse to adopt any, even temporary (evolving) or basic, criteria. They propose instead to identify the burning problems confronting mathematics education research at a particular historical time and place and to define the specific approaches and methods for each problem separately. Such is the position of Ellerton and Clements, who vehemently dismiss the applicability to mathematics education research of such traditional criteria for evaluating educational research as relevance, validity, objectivity, originality, rigor and precision, predictability, reproducibility and relatedness.

Others, such as Hart, acknowledge the need for a set of basic criteria – stressing that the criteria have to be made public and open to scrutiny, critique, and discussion – if mathematics education is to establish itself as a profession with something to offer the users of its research. Explicit, agreed-upon criteria are also seen by Lester and Lambdin as necessary in the processes of initiating newcomers into the field.

Hart proposes and discusses five 'basic criteria' for research in mathematics education: (1) There is a problem; (2) there is evidence or data; (3) the work can be replicated; (4) the work is reported; and (5) there is a theory. This list can be connected with the traditional criteria mentioned above: Criterion 1 is related to 'relevance'; 2, to 'objectivity' and 'validity'; 3, to 'reproducibility'; and 5 to 'relatedness', at least in part. Criterion 4 – 'The work is reported' – is not listed among the traditional criteria because it is ordinarily seen as obvious: Research that has not been reported can only be evaluated by the researcher and not by an outsider. In the same sense that a proof is not a proof until it has been made public so that it can be examined and critiqued, research is not research if it is not reported.

Using the metaphor of the ship of Theseus, Lester and Lambdin suggest that what was understood as research in mathematics education 20 years ago and today can hardly be called the same object. They conclude that to keep in touch with the changing reality of this research, a continuous dialogue is necessary concerning the criteria for evaluating its quality. For now, these authors propose criteria such as 'worthwhileness', 'competence', 'openness' (i.e., making explicit the theoretical assumptions and personal biases, as well as research methodologies and techniques), 'ethics', 'credibility', and a set of 'intangibles' such as clarity, conciseness and originality.

Lester and Lambdin's criterion of ethics is echoed by Sowder, who considers some of the ethical issues that arise in research and how they can be addressed. 'Researchers have responsibilities to several groups: to participants, to funding agencies or sponsors, to the scientific community, to society at large, and last but not least, to themselves'. Conflicts in the responsibilities to different groups can raise difficult ethical dilemmas. The use of qualitative research methods that put researchers into close and lengthy contact with teachers and students has been especially powerful in making ethical issues a critical consideration for researchers and attention to ethical issues an important criterion for evaluating research.

Hanna opposes the view of Hart and of Lester and Lambdin that there should be a public and shared set of criteria. She says that in many journals (not only in mathematics education), the editors rely heavily on the personal opinions of the reviewers. She takes a realistic stance: Her chapter aims at 'examining the gap between the principles of quality of a research paper as stated in the guidelines of individual journals and the practice of the review process'. She concludes that, because a set of precise criteria would likely be impossible to attain, researchers must settle for broad guidelines so as to allow for experimentation with new paradigms and various combinations of approaches. For Hanna, the responsibility for the quality of published research will continue to rest ultimately on reviewers and even more on journal editors, and it is important that whoever takes on these roles remain aware of their responsibility.

ORGANIZATION OF THE FIELD

The preceding chapters contain information about the organization of research in a few countries represented by the authors. For example, in France, an important role in the development of research in mathematics education is played by institutes called IREMs (*Instituts de Recherche en Didactique des Mathématiques*). Institutionally, they are attached to mathematics departments at universities; a director of an IREM must be a mathematician. These institutes are also the meeting point of researchers and practicing mathematics teachers who participate in the workshops.

In Italy and Hungary, according to our authors, mathematics education is also institutionally linked to mathematics departments. Boero and Szendrei say that, in their respective countries, 'most researchers in mathematics education come from mathematics and work in mathematics departments'. Moreover, as in France, there is, at the institutional level, collaboration between university researchers in mathematics education and mathematics teachers.

The situation is a little different in the United Kingdom, where teachers are much more autonomous and their initiatives are realized through self-organization within teachers' associations. For example, the ATM (Association of Teachers of Mathematics) has supported working groups on various themes (e.g., 'the problem-solving school', 'algebraic imagery', 'teacher intervention', 'the use of LOGO in the classroom', 'implications of constructivism and psychoanalysis for the teaching and learning of mathematics'). The MA (Mathematical Association) is more interested in promoting research into the teaching of mathematics at the secondary and undergraduate levels.

In the UK, according to Hatch and Shiu, there is a tradition of action research. Teachers' own research has been promoted through national projects such as the Low Achievers in Mathematics Project (LAMP) in 1987 and the Primary Initiatives in Mathematics Education Project (PrIME) in 1991. There is also some belief that the repertoire of a beginning teacher should include a research-oriented approach to teaching practice, and Manchester Metropolitan University has developed a program incorporating that approach. At the Open University, researchers collaborate with teachers to help teachers reflect on their own practice and thus be better prepared to take control of changes in school practice.

In North America, researchers in mathematics education may be members of a department of mathematics but more often are affiliated with a school, college, or department of education. Mathematics teachers and mathematicians in the United States and Canada are increasingly involved in programs of research in mathematics education, and many teacher education programs, especially in-service programs, give teachers opportunities to engage in action research.

Romberg, in his chapter, discusses the 'invisible college' – colleagues in the same field who set priorities for research, recruit and train students, communicate, and thus monitor the structure of knowledge in the field even though they are not located at the same institution. Using his experience in the U.S. National Center for Research in Mathematical Sciences Education, he illustrates how a research program can grow out of this strategy for interaction, building a social organization that provides for collaboration not only across time and space but also across specialized interests, backgrounds, and experience.

Whether they are working with mathematicians or with other educators, mathematics education researchers complain about the difficulties they experience with their work being recognized as research and about problems in obtaining appropriate academic degrees. In their chapter, Boero and Szendrei suggest some possible reasons for these difficulties.

HISTORY OF THE FIELD

In a discussion of the present ICMI Study at the Eighth International Congress on Mathematical Education in Seville in 1996, several people suggested that this book should contain a chapter outlining the history of mathematics education as a research field. The proposal was interesting and important, but we believe that, at this point, what is needed is a book, not a chapter. There already are chapters in books outlining this issue, usually from the perspective of the community represented by the author (e.g., Kilpatrick 1992). In the present book, one can find sketches of the historical development of certain trends in three countries (France, Germany, and Italy), and one can see how complex and often non-overlapping their trajectories have been. These sketches should caution us against the inevitable simplification in any 'general world history' of research in mathematics education that could be written in 30 pages.

Vom Hofe describes the evolution of a trend in German didactics, usually referred to as 'Stoffdidaktik'. This trend has been based on the assumption that the primary task of didactics of mathematics is to identify and find ways of generating in students the 'basic ideas' (Grundvorstellungen) or intuitions of the mathematical concepts taught at school. Recent developments include research into the students' individual images of mathematical concepts. Since the 1970s, an alternative trend in German didactics of mathematics has also been developing that is treated in the chapter by Steinbring.

Arzarello and Bartolini-Bussi offer a case study of the evolution of research in mathematics education in Italy. The prevailing trend since 1988 has been research that entails innovation conducted as research and not only as classroom action. The approach is different from that characterized by Hatch and Shiu as popular in the UK because in Italy teachers work as reflective practitioners not on their own but in very close collaboration with – or one might even say, under the guidance of – mathematics education researchers from universities.

Artigue sees the roots of the present French didacticians' strong links with mathematicians and their research interests in two curriculum reforms: the overall school curriculum reform of 1902 and the New Math reform of the 1960s. Both attracted the participation of many eminent French mathematicians. In the first reform, the focus was on the mathematical subject matter, as in Germany and Italy. Artigue says, however, referring to the lectures and writings of Poincaré, 'there was a clear will to adapt the teaching contents and processes to the student's cognitive potentials'. The failure of the New Math reform movement in France demonstrated 'that mathematics expertise, complemented by some general psychological and pedagogical principles, was not enough for promoting an effective organization and management of the complex reality of teaching mathematics for all'. Other forms of knowledge needed to be developed, with greater understanding of the educational system.

TRAINING FUTURE RESEARCHERS

Gjone, drawing on his experience in the Nordic countries and elsewhere, surveys programs of study in mathematics education, relating them to studies in the basic sciences. He notes that students entering a program for mathematics education researchers come mainly from mathematics, the educational sciences, and teaching. Consequently, they may be one-sided in their preparation and in need of very different programs of further training.

The establishment of programs that coordinate the different parts of mathematics education raises a variety of issues, as Gjone points out. In Norway, for example, there are two traditions in mathematics education research – one based on extensive study of mathematics and the other on the study of general pedagogy. It has been difficult to develop programs that integrate mathematics, education, other school subjects, and teaching practice into a program having the same length as the study of a single subject. The result has been to make some lengthy programs even longer. Furthermore, because there were no academic positions in mathematics education in Norway until recently, the assessment of theses in mathematics had to be undertaken by researchers in other fields. Although he stresses the need for variety in programs, Gjone argues that the preparation of researchers in mathematics education should be like the kind of open academic study one finds in the basic sciences, in which specialization is postponed. It should not have the structure to be found in some professional programs of study in which the aim and direction are heavily prescribed at the outset.

Several chapter authors oppose the view, commonly held by mathematicians, that 'effective teaching of mathematics is essentially based on good "technical" knowledge of the topics to be taught and on the quality of the teacher as a self-made "artist"' (Boero & Szendrei). But what does 'technical knowledge' mean in this context? Very often, it means no more than formal courses in calculus, linear algebra, analysis, abstract algebra, probability and statistics, and the like. Mathematics educators claim, however, that these courses do not prepare students to teach and to do research on the teaching and learning of elementary mathematics in school. They call for courses on 'elementary mathematics' or 'school mathematics', as well as on 'street mathematics' or 'ethnomathematics' (perhaps from 'a higher standpoint'!). They also call for the establishment of experimental schools in which teachers can receive this preparation (such a school exists in France under the directorship of Brousseau).

Ronnie Brown has pointed out that when preparing to teach or communicate something, one often has to rethink and change one's understanding of it because one is faced with questions that did not arise when one was learning it as a student. These new questions are why there must be a time for this radical change of understanding to establish itself during the preparation of the teacher and researcher. Mathematics educators acknowledge the importance of rethinking one's understanding of mathematics in doing research, but they also see the importance of knowing research methodologies. Some opt for the training of researchers in methodologies used by neighboring disciplines. For example, Lester and Lambdin argue that graduate programs in mathematics education should include attention to the research traditions of such fields as anthropology, psychology, and sociology.

Others say that the specificity of research in mathematics education does not allow for simply borrowing research techniques, methodologies, or theories from neighboring disciplines: Normally an adaptation is needed, or original theories and methodologies have to be developed from within the domain of mathematics education.

MATHEMATICS EDUCATION THROUGH THE EYES OF MATHEMATICIANS

Ronnie Brown's chapter and Sfard's interview with Shimshon A. Amitsur show what mathematicians expect from mathematics education. Their view of the domain is quite different from that of insiders. Even the terms used in the title of the Study seem to have a different meaning for them. For Brown, the task of research in mathematics education is to come up with ways of teaching mathematics clearly and effectively. Teaching mathematics is an activity, and, as such, it aims at a certain output. This ideal output, or goal, is a vision of a mathematically educated person. Mathematics education research should be clear about this goal and should find ways of achieving it. This view of mathematics education research is focused on the study of mathematics and its communication and on the design of courses and texts, rather than on psychological, sociological, or philosophical and epistemological issues.

These expectations by mathematicians may strike the researcher in mathematics education as naïve, but they are real and quite well founded. The topics and issues that researchers concentrate on in their professional lives have an impact on the lectures, seminars, and courses they offer. Many researchers in mathematics education teach courses for pre-service or in-service mathematics teachers. If, because of their research interests, researchers educate teachers by discussing at length, say, the epistemological differences between constructivism and activity theory, and do not enable the teachers to analyze mathematically what they are teaching and how they are teaching it, then these researchers are probably wasting their students' time.

MATHEMATICS EDUCATION AND OTHER ACADEMIC DISCIPLINES

Mathematics educators often complain about the difficulty of relations not only with mathematicians but also with representatives of other academic disciplines, and even of general education. Echoes of these complaints were heard at the Study Conference and can be found in the preceding chapters.

Some authors have reflected on the reasons for this state of affairs. For example, Wittmann says that research in mathematics education has difficulty being awarded scientific status as a human science because it works on the borderlines of disciplines such as psychology, sociology, history, and philosophy and not at their core. Consequently, the results are considered trivial or irrelevant by the practitioners of these disciplines. Mathematics education also has difficulties as a design science because 'design sciences have traditionally received a cold shoulder in academia'.

MATHEMATICS EDUCATION AS A RESEARCH COMMUNITY

Developing international communication and cooperation is a significant problem in mathematics education, but there are no easy solutions if 'colonization' is to be avoided. Several authors raise the issue of lack of communication between communities of researchers who speak different languages or employ different theoretical perspectives. Neither of these obstacles can be easily removed. A fluent knowledge of foreign languages by university graduates, for example, requires much planning and many resources. Moreover, familiarity with a language does not necessarily entail familiarity with the culture of the researcher using that language. As Sekiguchi argues, research practice in mathematics education is itself socially and culturally situated; it is constrained by local norms, goals, and values. Arguments over epistemology or methodology that rage among the researchers in one country may be looked on with amusement, disdain, or incomprehension by their colleagues elsewhere.

There are, of course, advantages to remaining within a single research tradition. Gravemeijer attributes much of the success of the 'realistic mathematics education' program in The Netherlands to working 'within the same research tradition for such a long time'. The reports from other authors, too, reveal the value for them of the traditions in which they do their research. But virtually everyone seems interested in finding out how colleagues elsewhere are thinking about and conducting their research.

Researchers often fail to cite other work or make comparisons with other schools of research not because of a lack of goodwill but, as Boero and Szendrei note, because of true incompatibilities even in the meaning of words between different research traditions. The comparison of paradigms must be considered a research problem on its own. Boero and Szendrei call for a new kind of international meeting that would be especially devoted to this problem: discussion and comparison of different theoretical approaches to a common topic in mathematics education. As some researchers have observed, the meetings we have now either are huge marketplaces of assorted information (like the International Congresses on Mathematica Education), where discussions are bound to stay on a very superficial level, or are small family reunions, where people seem to understand each other without going into detailed explanations and justifications.

It can be argued that mathematics education may be, at base, such a locally determined field of practice that many of its research questions cannot be addressed, or even adequately posed, across the international community. It seems no accident that the easiest research questions to discuss internationally have been those dealing with curriculum topics (e.g., functions, problem solving, proof) whose interpretation is not open to much question. But students learn specific mathematical content within specific conditions of practice, and these are difficult to generalize. It may be that important research questions are ignored in the mathematics education community internationally 'because they do not relate readily to abstractions or universals, requiring instead attention to the nuances of local educational settings' (Silver & Kilpatrick 1994, p. 750).

In mathematics education, we have a field that is not as universal as mathematics has traditionally been understood to be, and we have research that often does not translate well from this classroom to a neighboring one, let alone one that might be half a world away. Nonetheless, mathematics education seems to be much like mathematics in its power to foster cooperation and collaboration across political boundaries. The search for our common identity as researchers in mathematics education is not over, but the present book attests to the existence of a field in which serious international discourse about our collective enterprise is not only possible but fruitful.

NOTE

1. The discussion of the Study Conference in this section and the next is taken, in part, from Sierpinska (1994).

REFERENCES

Balacheff, N.: 1990, 'Towards a Problématique for Research on Mathematics Teaching', *Journal for Research in Mathematics Education* **21** (4), 258–272.

Begle, E. G.: 1979, *Critical Variables in Mathematics Education*, Mathematical Association of America and the National Council of Teachers of Mathematics, Washington, DC.

Begle, E. G. & Gibb, E. G.: 1980, 'Why Do Research?', in R. J. Shumway (ed.), *Research in Mathematics Education*, National Association of Teachers of Mathematics, Reston, VA, 3–19.

Cackowski, Z.: 1964, *Problemy i pseudoproblemy*, PWN, Warsaw.

Gras, R.: 1992, 'Data Analysis: A Method for the Processing of Didactic Questions', in R. Douady & A. Mercier (eds.), *Research in Didactique of Mathematics: Selected Papers*, La Pensée Sauvage éditions, Grenoble, 93–106.

Gras, R.: 1996, *L'implication Statistique, Nouvelle Méthode Exploratoire des Données*, La Pensée Sauvage éditions, Grenoble.

Kilpatrick, J.: 1992, 'A History of Research in Mathematics Education', in D. A. Grouws (ed.), *Handbook of Research on Mathematics Teaching and Learning*, Macmillan Publishing Company, New York, 3–38.

Schon, D. A.: 1983, *The Reflective Practitioner*, Temple Smith, London.

Sierpinska, A.: 1994, 'Report on an ICMI Study: What is Research in Mathematics Education and What are its Results?', *Proceedings of the 1994 Annual Meeting of the Canadian Mathematics Education Study Group, University of Regina, Regina, Saskatchewan, June 3–7*, 177–182.

Silver, E. A. & Kilpatrick, J.: 1994, 'E Pluribus Unum: Challenges of Diversity in the Future of Mathematics Education Research', *Journal for Research in Mathematics Education*, **25** (6), 734–754.

Thurston, W. P.: 1994, 'On Proof and Progress in Mathematics', *Bulletin of the American Mathematical Society* **30** (2), 161–177.

NOTES ON AUTHORS

JOSETTE ADDA is now retired but, as emeritus professor, she continues to be a research director and to teach in the third cycle of Didactics of Mathematics at Université Paris VII-Denis Diderot. After studies in mathematics ('agrégation de mathématiques'), she began research at the University of Paris in mathematical logic but soon oriented her research towards mathematics education. She was the first in France (1976) to obtain a Ph.D. in didactics of mathematics from a department of mathematics. The topic of her thesis was: 'On difficulties inherent in mathematics and on the phenomena of incomprehension: causes and manifestations'. Her supervisors were D. Lacombe, H. Freudenthal and L. Henkin. She contributed, in 1968, to the creation of the IREMs, and later, to the creation of the third cycle in didactics of mathematics, and she was the director of research in didactics of mathematics at Université Paris 7. From 1989 to 1994 she held the position of Professor of 'Sciences de l'Éducation at the Université Lumière-Lyon 2. She has always worked at the international level (mainly with the Polish groups with Z. Krygowska and W. Zawadowski and with H. Freudenthal in The Netherlands). She has been active in the CIEAEM, she has participated in all ICMEs so far, and she contributed to the editorial board of the ESM from 1979 to 1992. She is a member of the research group on Classroom Research and is engaged in the Gender and Mathematics group as a correspondent of IOWME. She published mainly in ESM and in the proceedings of congresses and ICMI studies. Her main interest has always been the understanding of mathematics: problems of meaning and mathematical communication, the role of contents in the presentation of mathematical concepts by 'dressing' them up in pseudo-concrete contexts, and also the influence of the school context on the mathematical communication in the classroom.

MICHÈLE ARTIGUE, after obtaining a Ph.D. in mathematical logic at the University Paris 7, obtained an habilitation in didactics of mathematics at the same university. She is presently a professor at the University Institute for Teacher Training (IUFM) in Reims and director of the research team in didactics of mathematics DIDIREM, at the University Paris 7. She collaborates with various journals and is currently director of the journal *Recherches en Didactique des Mathématiques*. Her main research interest areas are the didactics of Analysis and Computer Assisted Learning.

FERDINANDO ARZARELLO is Professor of Elementary Mathematics from an Advanced Standpoint at the Faculty of Mathematical, Physical and Natural Sciences of the University of Turin, where he has been Head of the Department of Mathematics for three years. He is a member of the Scientific Commission of UMI (Unione Matematica Italiana) and of the CIIM (Italian Commission for the Teaching of Mathematics). He is a member of the PME international group on Algebra and has written a contribution for the forthcoming volume produced by that group. His research interests concern the learning and teaching of algebra at different levels of age.

MARIA G. BARTOLINI BUSSI is Associate Professor of Elementary Mathematics from an Advanced Standpoint at the Faculty of Mathematical, Physical and Natural Sciences of the University of Modena. She has contributed to the volume *Didactics of Mathematics as a Scientific Discipline*, published by Kluwer (1994). She is a member of the International Group BaCoMET and a co-director of its Project V on Professional Knowledge of Mathematics Teachers. Her research interests concern the study of social interaction and of the use of artifacts in the process of teaching and learning geometry.

ALAN BISHOP is currently Professor of Education at Monash University, Melbourne, Australia. He worked previously at Cambridge University, England, and was Editor of *Educational Studies in*

Sierpinska, A. and Kilpatrick, J. *Mathematics Education as a Research Domain: A Search for Identity*, 549–556.
© *1998 Kluwer Academic Publishers. Printed in Great Britain.*

Mathematics for many years. He is Managing Editor of the Mathematics Education Library book series for Kluwer, and was the chief editor of the *International Handbook of Mathematics Education*. His best-known research work is on visualization and spatial ability in mathematics, and on social and cultural aspects of mathematics education.

PAOLO BOERO received his degree in Mathematics in 1964. Since 1982 he has taught mathematics education at Genova University. In 1975 he started a project concerning integrated teaching of mathematics and sciences for compulsory school (grades VI–VIII, then also I–V). At present, he is the leader of the research group responsible for this project. Since 1981 he served as a member of various commissions which have been responsible for improving mathematics education in Italy and organizing Italian mathematics education research. Author of several articles published in international journals and books, his major research interests in the field of mathematics education concern problem-solving, situated teaching–learning of mathematics, mathematical proof.

MARGARET BROWN is a Professor of Mathematics Education in the University of London and is currently Head of the School of Education at King's College London. She has a first degree in mathematics from the University of Cambridge, and a Ph.D. in mathematics education from the University of London. She has a professional qualification in teaching and has taught mathematics in primary and secondary schools, before becoming involved in training teachers of mathematics. She has directed or co-directed 16 research and/or development projects, initially concentrating on learning and assessment in mathematics. More recent research has been concerned with teachers' beliefs and subject knowledge in relation to classroom practice. Professor Brown was a member of the English National Curriculum Mathematics Working Group and a director of the project that designed the first national tests in mathematics for 14 year-old pupils. She has recently been Chair of the Joint Mathematical Council of the United Kingdom and is currently the UK ICMI representative and Deputy Chair of the Royal Society Education Committee with responsibility for mathematics.

RONALD BROWN had been an undergraduate and research student at Oxford, 1953–59, where he studied under the topologist, Professor J. H. C. Whitehead. He had been Assistant Lecturer and Lecturer in Pure Mathematics at Liverpool University, 1959–64, Senior Lecturer and then Reader at Hull University, 1964–70, and since then has been Professor of Pure Mathematics at the University of Wales, Bangor. His book on *Topology* has had two editions (1968, 1988) and he has written over ninety papers, sixteen of them on teaching and popularization. He has written on general and algebraic topology, and homological algebra, but his main research area is on the extension from groups to groupoids, and on 'Higher dimensional algebra', which explores the algebraic consequences and geometric applications of expressions not confined to a linear notation. He has given lectures to the Annual Meeting of the British Association for the Advancement of Science in 1983, 1987, and 1992, the last as President of the Mathematics Section. He gave a London Mathematical Society Popular Lecture (1984) and Mermaid Molecule Discussion (1985), on 'How mathematics gets into knots', and a Royal Institution Friday Evening Discourse on 'Out of Line' (1992). He was a part of the team with Tim Porter and Nick Gilbert which prepared an exhibition 'Mathematics and knots' for the 1989 PopMaths Roadshow, which toured the UK. The booklet prepared from this is sold widely. At the same Roadshow at Leeds, he collaborated with the sculptor John Robinson in preparing a catalog and presenting an exhibition of Robinson's Symbolic Sculptures, which has since been shown at Bangor, Liverpool, Oxford, Cambridge, London, Barcelona and Zaragoza and may be seen on the internet at http://www.bangor.ac.uk/~mar007/

M. A. ('KEN') CLEMENTS, who is Professor of Mathematics Education at Universiti Brunei Darussalam, holds the MEd and PhD degrees in Education from the University of Melbourne. He has taught, supervised and carried out research, and served as a teacher education consultant in Australia, India, Malaysia, Papua New Guinea, Thailand, and Vietnam. Before taking up his appointment at the Universiti Brunei Darussalam, he worked at the University of Newcastle (Australia), Deakin University, and Monash University. Earlier in his career he was a teacher in five schools. Ken has authored or co-authored seventeen books and over 150 articles or chapters in books. His main fields of interest are mathematics education, distance education, and action research.

NERIDA F. ELLERTON was the first person to be appointed to a nominated Chair in Mathematics Education in Australia. She is currently Professor and Dean of Education at the University of Southern Queensland, in Toowoomba. She holds two PhD degrees – one in Physical Chemistry from the University of Adelaide, and the other in Mathematics Education from Victoria University of Wellington, New Zealand. Her special interests are mathematics education, distance education, teacher education, and the influences of linguistic and cultural factors on education. She has taught, supervised and carried out research in Malaysia, the Philippines, and Thailand. Before taking up her appointment at the University of Southern Queensland, she worked at Edith Cowan University, Deakin University, and Victoria University of Wellington. She is the author or co-author of ten books and over 100 articles or chapters in books. She also holds professional qualifications in photography, and her graphic art has been shown in many exhibitions.

PAUL ERNEST is Reader in Mathematics Education at the University of Exeter, UK, where he leads the master's degree program in mathematics education. He is the author of numerous publications including the book *The Philosophy of Mathematics Education* (1991). He is editor of the Falmer Press series *Studies in Mathematics Education,* and founding editor of the *Philosophy of Mathematics Education Newsletter.* His latest book is *Social Constructivism as a Philosophy of Mathematics* (1997).

GUNNAR GJONE has studied mathematics and pedagogy at universities in Norway and the United States. His doctor's thesis was on the development of the 'new math' in Norway and some other countries. He is now Professor of Mathematics Education at the University of Oslo, and has earlier worked in teacher education at several teacher training colleges in Norway. He has a background in curriculum work at a national level for all grades in the school system. His present research interests include assessment in mathematics, the use of information technology in mathematics education, and the study of curriculum development.

JUAN D. GODINO and CARMEN BATANERO are senior lecturers of Didactics of Mathematics at the University of Granada. Their Ph.D.s are in mathematics. They are interested in the theoretical foundations of Didactics of Mathematics as a scientific discipline. They are also doing research on the teaching and learning of statistics, probability, and combinatorics. Important recent publications are: 'Institutional and Personal Meaning of Mathematical Objects', *Journal für Mathematik-Didaktik* (1996); 'Intuitive Strategies and Preconceptions about Association in Contingency Tables', *Journal for Research in Mathematics Education* (1996); and 'Razonamiento Combinatorio', *Síntesis* (1994). Web sites: http://www.ugr.es/~jgodino
http://www.ugr.es/~batanero

KOENO GRAVEMEIJER has a Ph.D. in Mathematics Education. He is research coordinator at the Freudenthal Institute at the University of Utrecht. He is the main author of a Dutch primary school mathematics textbook series. Since 1986 he has been affiliated with the Freudenthal Institute where he is involved with developmental work, developmental research, and evaluation research. This includes his work for the 'Mathematics in Contexts' project of the University of Wisconsin-Madison. His research interest lies in the theory of 'realistic mathematics education' and in the methodology of 'developmental research'. This is reflected in his recent book, *Developing Realistic Mathematics Education.* One of the aspects of the realistic mathematics education theory and the role of models and symbolizations form the focus of a current research collaboration with Paul Cobb and Erna Yackel.

GILA HANNA is Professor at the Ontario Institute for Studies in Education of the University of Toronto. Her main areas of research are the nature of proof in mathematics and the role of proof in mathematics education, and gender issues in mathematics education. She is co-editor of *Educational Studies in Mathematics.*

KATHLEEN HART has had a long career in Mathematics Education. Her first degree was in mathematics and after training as a teacher she taught mathematics in secondary schools in England, Bermuda, and the United States. This phase was followed by ten years' training teachers, both primary and secondary, of mathematics. During this time, she spent summers carrying out in-service work for teachers in Uganda, West Cameroon and Jamaica and then spent a year in Thailand as a Unesco field-officer. Her research work started with an M.Phil. of the University of London, looking at 7 to 9 year-olds who were said to have a mental block against mathematics. In 1973 she went to the University of Indiana, where she obtained an Ed.D. She returned to London to lead the mathematics section of the research project 'Concepts in Secondary Mathematics and Science' (CSMS). That was followed by the research projects 'Strategies and Errors in Secondary Mathematics' and 'Children's Mathematical Frameworks', all carried out at Chelsea College (later King's College) and funded by the E.S.R.C. She obtained a Ph.D. from London University in 1980 on 'Children's Understanding of Ratio'. For two years she was an inspector of schools (HMI) and then returned to King's College to direct the curriculum development project 'Nuffield Secondary Mathematics' for six years. This project was designed to produce mathematics materials for secondary schools, based on research findings. It was stopped when the British National Curriculum was introduced. Professor Hart now directs the Shell Centre for Mathematical Education, University of Nottingham, UK.

GILLIAN HATCH is a principal lecturer in mathematics education at Manchester Metropolitan University in the UK. After graduating in mathematics, she taught first in a college of education, then in secondary schools, before moving to training primary and secondary teachers. She is also involved in in-service work with teachers, including a master's course that involves a large element of practitioner inquiry work. She has a number of specific research interests, particularly the use of games in the mathematics classroom and attitudes to proof and justification.

JAMES HIEBERT is H. Rodney Sharp Professor of Education at the University of Delaware. He received his Ph.D. from the University of Wisconsin and, since that time, has been interested in children's mathematics learning. He served as editor for *Conceptual and Procedural Knowledge: The Case of Mathematics* and co-editor for *Number Concepts and Operations in the Middle Grades.*

CAROLYN KIERAN is a Professor in the Mathematics Department of the Université du Québec à Montréal, where she teaches both mathematics and mathematics education courses, supervises the research of master's and doctoral students, and is involved in the training of future high school mathematics teachers. Having obtained a Ph.D. degree from McGill University for her research on school algebra cognition, she currently focuses her research on the use of computers in the learning of school mathematics, as well as on the development of alternate approaches to the teaching of algebra. Professor Kieran, who is past-President of the International Group for the Psychology of Mathematics Education, has also served as Chair of the Editorial Panel of the *Journal for Research in Mathematics Education* and as Vice-President of the Canadian Mathematics Education Study Group.

DIANA V. LAMBDIN is Associate Professor of Elementary and Mathematics Education in the School of Education at Indiana University-Bloomington (IU), the institution from which she earned a Ph.D. in Mathematics Education in 1988. Prior to her work at IU, she taught mathematics to middle school students in Marblehead, Massachusetts; to high school students in Ypsilanti, Michigan; and to entry level college students and prospective teachers at Iowa State University. From 1988 to 1991 she served as Administrative Coordinator of the IU Mathematics Education Development Center. She has published numerous articles and book chapters about mathematical problem-solving, assessment and evaluation, and teacher change. She served as associate editor of the Monograph series of the *Journal for Research in Mathematics Education* (*JRME*) from 1988 to 1992, as that journal's associate editor from 1993 to 1995, and since 1995 as its review editor.

GILAH C. LEDER is a professor in the Graduate School of Education at La Trobe University in Melbourne, Australia. Earlier appointments have included teaching at the secondary level, at the Secondary Teachers' College (now Melbourne University), and at Monash University. Her teaching and research interests embrace gender issues, factors that affect mathematics learning, exceptionality, and assessment in mathematics. She serves on various editorial boards and educational and scientific committees. These include President of the Mathematics Education Research Group of Australasia and membership on the Executive Committee of the International Commission on Mathematical Instruction. She is a frequent presenter at scientific and professional teaching meetings. She is currently working on large research projects on several topics: learning mathematics in a social context, exploration of the large Australian Mathematics Competition database, dimensions of gender/cultural conflict in mathematics teaching, and effects of non-traditional assessment. She edited the April 1995 Special Issue on mathematics and gender of *Educational Studies in Mathematics.*

STEPHEN LERMAN was a school teacher of mathematics for many years in England and in Israel. He is now Reader in Mathematics Education at South Bank University in London. He is President of the International Group for the Psychology of Mathematics Education and was Chair of the British Society for Research into Learning Mathematics. His research interests include: philosophy of mathematics; teachers' beliefs; teachers as researchers; equity issues; learning theories, and socio-cultural perspectives of mathematics teaching and learning. He is editor of *Cultural Perspectives of the Mathematics Classroom,* published by Kluwer in 1994 and guest editor of a Special Issue of *Educational Studies in Mathematics* entitled 'Socio-Cultural Approaches to Mathematics Learning', published in September 1996.

FRANK K. LESTER, Jr. is Professor of Mathematics Education in the School of Education at Indiana University-Bloomington. He joined the faculty at Indiana University in 1972 immediately after completing his Ph.D. in mathematics education at Ohio State University. His primary research interests lie in the areas of mathematical problem-solving and metacognition – especially problem-solving instruction. He has been a Fulbright Fellow in the Cognitive Psychology Program at the Universidade Federal de Pernambuco in Recife, Brazil. He has lectured widely in Europe, South America, and the United States on mathematical problem-solving, alternative assessment and related topics, and he has authored or co-authored several books, chapters and journal articles on these topics. From 1988 until 1992 he served as the editor of the Monograph Series of the *Journal for Research in Mathematics Education* (*JRME*) and from 1992 to 1996 as that journal's editor.

CLAIRE MARGOLINAS is 'maître de conférences' in mathematics at the Institut Universitaire de Formation des Maîtres in Clermont-Ferrand, France, where she is involved in the training of secondary school mathematics teachers. She obtained her Ph.D. in 'didactique des mathématiques' under the direction of Colette Laborde in Grenoble; in this work, she tried to synthesize the French paradigm of research in mathematic education (Margolinas 1993). She is now doing research on the situation of the teacher through ordinary classroom observations.

JOHN MASON is Professor of Mathematics Education and Director of the Centre for Mathematics Education at the Open University. He received his B.Sc. and M.Sc. at the University of Toronto, and Ph.D. in Combinatorial Geometry at the University of Wisconsin-Madison. Best known for his work on mathematical problem solving through his book *Thinking Mathematically* (Addison Wesley, 1982), written with Leone Burton and Kaye Stacey and translated into French, German, and Spanish, his main activities are directed to fostering and sustaining mathematical thinking in himself and in others. This has led him to the fundamental role of mental imagery in learning, doing, and teaching mathematics, and a concentration on ways of working with students in order that their mathematical awarenesses be educated, the locus, focus, or structure of their attention shifted, and their mathematical behavior trained, by harnessing their mathematical energies. During the last ten years, he has concentrated on developing the Discipline of Noticing as a research methodology that enables teachers in particular, and practitioners in general, to research their own practice in a systematic and epistemologically well-founded fashion. He is particularly interested in the teaching of algebra and geometry. He has recently initiated the MEME (Meaning Enquiry in Mathematics Education) Project, which aims to elucidate and promulgate technical terms, distinctions, and frameworks that people have found informs their practice in teaching mathematics.

ROBERTA MURA is a professor of Mathematics Education at Laval University. She received her Ph.D. in Mathematics from the University of Alberta in 1974. She has conducted research and published on a variety of topics including group theory, women and mathematics, the use of calculators in schools, and conceptions of mathematics.

NORMA PRESMEG is an Associate Professor of Curriculum and Instruction and Coordinator of the Mathematics Education Program at The Florida State University. Her degrees include B.Sc. in Mathematics and Physics from Rhodes University in South Africa, B.Sc. Honours in Mathematics and B.Ed. from the University of Natal. Her M.Ed. dissertation, also at the University of Natal, was an analysis of Albert Einstein's creativity with implications for mathematics education. After completing the Ph.D. degree in Mathematics Education at Cambridge University in England, she served on the faculty of the University of Durban-Westville in South Africa for five years before joining the College of Education at The Florida State University in 1990. Her research interests include the use of imagery in mathematics classrooms at high school level, and by prospective teachers; prototypes, metaphors, metonymies, and semiotic frameworks for interpretation of mathematics learning; and use of ethnomathematics and cultural backgrounds in constructing mathematical ideas.

THOMAS A. ROMBERG is the Sears Roebuck Foundation–Bascom Professor in Education at the University of Wisconsin–Madison and is the Director of the National Center for Improving Student Learning and Achievement in Mathematics and Science for the U.S. Department of Education. Dr. Romberg has a long history of involvement with mathematics curriculum reform, including work in the 1960s with the School Mathematics Study Group, in the 1970s with Developing Mathematical Processes, and in the 1980s as chair of two commissions – School Mathematics: Options for the 1990s (U.S. Department of Education) and Curriculum and Evaluation Standards for School Mathematics (National Council of Teachers of Mathematics). In the 1990s he served as Chair of the Assessment Standards for School Mathematics (NCTM). For his work on the Curriculum and Evaluation Standards Commission, he received the Interpretive Scholarship and Professional Service awards in 1991 from the American Educational Research Association. His research has focused on young children's learning of initial mathematical concepts, methods of evaluating both students and programs, and an integration of research on teaching, curriculum, and student thinking.

YASUHIRO SEKIGUCHI is an Associate Professor in the Department of Mathematics of the Faculty of Education at Yamaguchi University in Japan. He has two M.Ed. degrees, obtained in 1983 and 1985, from the University of Tsukuba, and an Ed.D. in 1991 from the University of Georgia. His dissertation research dealt with proof and refutation in secondary geometry. He currently teaches courses for preservice elementary and secondary school teachers, and for graduate students in the master's program in mathematics education. The courses include methods of teaching mathematics in secondary school, mathematics for preservice elementary school teachers, computer programming for education, and advanced studies in mathematics education.

ANNA SFARD gained her degrees in physics, mathematics, and mathematics education at the Hebrew University of Jerusalem, Israel, where she has also taught mathematics for many years. At present, she is at the School of Education at the University of Haifa. As a researcher, her main interest lies in the mathematics learning at secondary and tertiary level. Most of her recent work deals with the question of the origins of mathematical objects and with the issue of transition from operational to structural thinking (reification). She is the editor of the *Israeli Journal for Mathematics Teachers.*

CHRISTINE SHIU is a senior lecturer in mathematics education at the Open University in the UK. After graduating in mathematics, she qualified as a teacher and taught mathematics in secondary schools before moving into research and higher education. She as recently chaired the production of a suite of distance taught undergraduate mathematics courses, and, true to her belief in practitioner research, investigating personal constructions of tutoring practices in this context.

JUDITH T. SOWDER is Professor of Mathematical Sciences at San Diego State University and is currently editor of the *Journal for Research in Mathematics Education.* Her research interests include teacher change and students' development of number sense. As a co-director of the National Center for Research in Mathematical Sciences Education, she directed a national working group that focused on the learning of rational numbers and quantities. One of the resulting publications was the book *Providing the Foundation for Teaching Mathematics in the Middle Grades,* which she co-edited with Bonnie Schappelle.

HEINZ STEINBRING is Professor of Mathematics Education at the University of Dortmund. His specific field of research is the epistemological analysis of everyday classroom interaction in the elementary grades within the frame of a qualitative research paradigm. For about twenty years, he has been a researcher in mathematics education at the Institute for the Didactics of Mathematics (IDM), University of Bielefeld. His main professional interests are: the scientific foundations of mathematics education, research on the curriculum and the syllabus of primary and lower secondary school mathematics; and analysis of the specific epistemology of mathematical knowledge in classroom interaction. Much of his research work is closely linked to cooperation with mathematics teachers. Recent major publications include: 'Problems in the Development of Mathematical Knowledge in the Classroom – the Case of a Calculus Lesson', *For the Learning of Mathematics* (1993); 'Mathematical Understanding in Classroom Interaction – the Interrelation of Social and Epistemological Constraints', *Proceedings of the 3rd Bratislava International Symposium on Mathematical Education (BISME-3)* (1993); 'The Context for the Concept of Chance – Everyday Experiences in Classroom Interactions', *Acta Didactica Universitatis Comenianae* (1993); and 'Dialogue between Theory and Practice in Mathematics Education', in R. Biehler, R. W. Scholz, R. Sträßer & B. Winkelmann (eds.), *Didactics of Mathematics as a Scientific Discipline* (Kluwer Academic Publishers, 1994). And with R. Bromme, 'Interactive Development of Subject Matter in the Mathematics Classroom', *Educational Studies in Mathematics* (1994); and with F. Seeger, 'The Myth of Mathematics', in P. Ernest (ed.), *Constructing Mathematical Knowledge: Epistemology and Mathematics Education,* Studies in Mathematics Education, Vol. 4 (Falmer Press, 1994).

JULIANNA RADNAI SZENDREI received her master's degree in mathematics and physics in 1971 and a doctorate in mathematics in 1991. Since 1992 she has been head of the mathematics department at the Teacher Training College of Budapest. At the national level, she was responsible for high school mathematics education for the National Pedagogical Institute of Hungary from 1970 to 1992 and is a co-author of the current Hungarian school mathematics curriculum. Since 1989 she has served as president of the Tamas Varga Mathematics Education Foundation, which promotes cooperation among Hungarian universities and organizes nationwide courses and seminars for mathematics teachers in Hungary. She is also a leader in the Education Division of the Janos Bolyai Mathematical Society. Active in international activities, Dr Szendrei has been vice-president and conference organizer for the International Commission for the Teaching and Improvement of Mathematical Instruction (CIEAEM) and working group coordinator for ICME-VI and ICME-VII. Dr Szendrei has served as an invited lecturer on mathematics teacher education at several European and American universities. Her major research interests concern learning problems in mathematics and the role of language in mathematics teaching and learning.

GÉRARD VERGNAUD is Directeur de Recherche in the Centre National de la Recherche Scientifique (CNRS). He is Director of the Coordinated Research Group 'Didactique et Acquisition des Connaissances Scientifiques', a network of about 100 researchers in mathematics education and physics education. His main research interests are mathematics education and cognitive and developmental psychology. During the last few years he has also been interested in the study of competences in adults at work and in training. His last publications in English are: 'Multiplicative Conceptual Field; What and Why?', in G. Harel and J. Confrey (eds.), *The Development of Multiplicative Reasoning in the Learning of Mathematics* (Suny Press, 1994); 'The Nature of Mathematical Concepts', in T. Nunes and P. Bryant (eds.), *How Do Children Learn Mathematics?* (Lawrence Erlbaum, 1997); and 'The Theory of Conceptual Fields', in L. Steffe and B. Greer (eds.), *Theories of Learning Mathematics* (Lawrence Erlbaum Associates, 1996).

RUDOLF VOM HOFE has worked as a high school and college teacher for mathematics and history for about ten years. Afterwards he went to the Department of Mathematics at the University of Kassel (1989–94) where he did research on individual concept development, especially the concept of 'basic ideas' (Ph.D. in 1994). He has been teaching mathematics education at the University of Augsburg since 1994. His current research interests focus on individual concept development in relation to advanced mathematical thinking, especially the concept of function, and the aspect of cognitive reorganization within computer-based learning environments. His publications include: 'Grundvorstellungen mathematischer Inhalte als didaktisches Modell', *Journal für Mathematik-Didaktik* (1992) and 'Grundvorstellungen mathematischer Inhalte, Texte zur Didaktik der Mathematik', in N. Knoche & H. Scheid (eds.), *Spektrum* (1995).

ERICH CH. WITTMANN, gained a State Diploma in Mathematics and Physics, University of Erlangen, Germany (1964) and his Ph.D. in Mathematics, University of Erlangen, Germany (1967). He has been a professor of mathematics education at the University of Dortmund since 1970 and co-director of the project 'Mathe 2000' since 1987. His special fields of interest include foundations of mathematics education and developmental research centered on the design of substantial teaching units. He has published numerous articles and several books, and served on the editorial board of *Educational Studies in Mathematics* from 1977 to 1990. His book *Grundfragen des Mathematikunterrichts (Topics in Mathematics Teaching)* is a leading textbook in German mathematics teacher education.

INDEX

New ICMI Studies Series

KLUWER ACADEMIC PUBLISHERS – DORDRECHT / BOSTON / LONDON